MATERIALS SCIENCE AND TECHNOLOGIES

COMPREHENSIVE GUIDE FOR MESOPOROUS MATERIALS

VOLUME 2

ANALYSIS AND FUNCTIONALIZATION

MATERIALS SCIENCE AND TECHNOLOGIES

Additional books in this series can be found on Nova's website under the Series tab.

Additional e-books in this series can be found on Nova's website under the e-book tab.

MATERIALS SCIENCE AND TECHNOLOGIES

COMPREHENSIVE GUIDE FOR MESOPOROUS MATERIALS

VOLUME 2

ANALYSIS AND FUNCTIONALIZATION

MAHMOOD ALIOFKHAZRAEI
EDITOR

nova publishers
New York

Library of Congress Cataloging-in-Publication Data

ISBN: 978-1-63463-982-8

Published by Nova Science Publishers, Inc. † New York

CONTENTS

Preface

This is the second volume of the four volume set of the Comprehensive guide for Mesoporous Materials. Mesoporous materials are found abundant in biological systems and minerals and some of these materials have long been used in various industries. Recent advances at the nanoscale level has caused an increase in the use of these materials. This volume mainly discusses the analysis and functionalization of mesoporous materials.

Chapter 1 – The classification of pores with regard to their pore width is given. The standardisation and classification work concerning the surface texture of porous material is reviewed. A compilation is presented of methods which can be used to determine the total pore volume, the pore size distribution and the fractality. The survey comprises light, X-ray, electron, ion, acoustic and tunnel microscopy, mechanical sensing, classification of particle size, flow methods, volume and density measurements, intrusion, sucking and displacement porometry, gas or liquid sorption porometry, thermoporometry, desorption spectroscopy and nuclear magnetic resonance measurements. Furthermore, a method of saving time of sorption measurements is presented.

Chapter 2 – This chapter treats of the organization and functionalization of hybrid organic-inorganic mesoporous materials. These materials based on silica matrix were prepared by sol-gel process, which consists on the hydrolysis and polycondensation of silylated molecular precursors in the presence of a structure-directing agent. The authors described methods to functionalize the channel pores of ordered mesoporous silica, this of the silica framework as well as the functionalization of both of them simultaneously. Thanks to chemical modifications, they evidenced not only the accessibility but also the location and the regular distribution of the introduced organic units. This family is currently one of the best supports for exploring polyfunctional materials, which can provide a route to original nanomaterials.

Chapter 3 – A series of mesoporous titania materials modified with with tungstosilicic (TSA) were successfully synthesized using titanium isopropoxide as titania precursor and urea as low-cost template via sol-gel reactions. They were characterized by FT-IR, XRD, DTA-TGA, DRS, SEM and BET. The obtained mesoporous solids presented an average pore diameter higher than 3.1 nm. The specific surface area decreased when the TSA amount and the calcination temperature increased. The XRD patterns of the modified samples only exhibited the characteristic peaks of anatase phase. The presence of TSA retarded the crystallization of the anatase phase and its transformation into rutile phase. The FT-IR studies showed that the main heteropolyoxometallate species present in the composites is the

$[SiW_{12}O_{40}]^{4-}$ anion for the samples calcined up to 500 °C. The *band gap* energy slightly decreased as a result of the introduction of TSA into the titania matrix, though it remained almost constant with the calcination temperature increase. The 4-chlorophenol degradation behavior of the catalysts as a function of time depended on the TSA amount and the thermal treatment temperature. The amount of degraded 4-CP increased with the increment of TSA content being the sample containing 30 % TSA calcined at 600 °C the most active. The apparent reaction constant, estimated assuming a pseudo-first-order kinetics, followed the same trend. The catalysts can be reused at least three times without an important decrease in the degradation and mineralization degrees.

Chapter 4 – Pore walls in mesoporous silicates may exhibit dual, simultaneously crystalline and amorphous features. The extents of the two characteristics may be varied depending on the synthesis and further treatments. Their partitions can be followed with different methods. Examples are presented from studies performed on Fe-MCM-41 ferrisilicates synthesized with different procedures, by hydrothermal synthesis and by hydrolysis of tetraethylorthosilicate. Properties of samples are compared by applying primarily in situ Mössbauer and infrared spectroscopies complemented with XRD, BET, TEM, TPR methods. Flexibility in variations of the crystalline/amorphous parts is illustrated in dependence of the actual conditions. These structural changes can also be strongly correlated to $Fe^{3+} \Leftrightarrow Fe^{2+}$ redox processes.

Chapter 5 – The authors report the functionalization of mesoporous materials for improvement of the electrocatalytic properties. As silica and carbon based-materials are often used as template to synthesize other mesoporous, these materials are described here. The authors focus on ordered mesoporous carbons (OMC) because the functionalization and applications of mesoporous silicas have been well reviewed. They show how the ordered mesostructure of these materials is very important in the construction of (bio) sensors. Moreover, the structural capabilities of OMC at the scale of a few nanometers agree with immobilization of other electrocataytic substances. Interesting properties of these materials may open up a new approach to study the electrochemical determination of other biomolecules.

Chapter 6 – Nanoporous titanium dioxide was prepared by sol-gel technique. To control the surface area, pore size and pore volume of the prepared TiO_2, a catalyzed hydrolysis was carried out using different concentrations of silicotungstic acid (SWA) as a template. A fixed molar ratio of H_2O/Ti was used. The prepared TiO_2 was calcined at 400 or 600°C. Samples were characterized by nitrogen physisorption, X-ray powder diffraction (XRD), selected scattered electron diffraction, Raman spectroscopic analysis, Fourier Transform spectroscopy (FTIR), Thermogravimetric analysis (TGA), Differential scanning calorimetry (DSC), scanning and transmission electron microscopy (SEM and TEM). The photocatalytic activity of the prepared samples was evaluated by the degradation of alizarin yellow under UV light. The results showed that the crystallinity increases as the concentration of SWA decreases. The presence of SWA during the precipitation of TiO_2 prevents the formation of rutile phase and suppresses the crystal growth. The results showed also that the surface area increase as the concentration of SWA decreases. The samples prepared using 0.05M SWA and calcined at 600 °C showed a higher activity.

Chapter 7 – In the early nineties, quickly after the discovery of ordered mesoporous solids, some strategies for organic functionalization were developed. Concerning the organic functionalization of ordered mesoporous silicas, the conventional post-synthesis grafting from organosilane reagents on surface silanols and the direct synthesis via co-condensation of alkoxysilane and organosilane precursors have been widely studied, compared and reviewed. The recent societal tendencies such as evolution of researches in the field of materials towards the concept of eco-conception and biologically oriented applications have provoked the emergence of new organic functionalization strategies in terms of processes, reactants and controlled functionalization. The characterization of these organically-modified ordered mesoporous solids is of paramount importance for understanding the structure-properties relationships in order to produce tailor-made hybrid mesoporous materials. Some progresses have also been done in this area and especially with the powerful and well-adapted technique for the characterization of such solids consisting in solid state NMR spectroscopy. These evolutions in both organic functionalization strategies and functionalization characterization by NMR are highlighted in this book chapter.

Chapter 8 – The application of nanoscale materials and structures is an emerging area of nanoscience and nanotechnology, with broad scientific and commercial perspectives. The high surface energy of the inorganic materials, such as metal and metal oxides, mainly determined by size, shape, composition, crystallinity and morphology has attracted lots of attention in functional nano-finishing of textiles. The purpose of this work is to highlight the added-value technologies based on nanoparticles contribution, focusing on functional textiles with higher quality attributes, such as: antimicrobial properties, water repellence, electro-magnetic protection and thermoregulatory fabrics, antistatic, anti-infrared and flame retardant properties, dyeability, colour fastness, as well as durable, or permanent press finished garments, textiles with self-cleaning properties, UV-blocking, deodorizing, thermal regulator, memorizing shape, and so on. The use of the nanoscale materials in the textile manufacturing opens many opportunities towards the production of so-called “intelligent textiles”.

Chapter 9 – The authors report on the structural and optical properties of mesoporous silicon nanowires (PSiNWs) fabricated using silver ions assisted electroless etching method. The strongest photoluminescence signal has been measured from samples etched with 4.8M of HF. Reflectance were measured versus nanowire length and were inferior to 0.1%. Models based on cone shape of nanowires located in circular and rectangular bases were used to calculate the reflectance using the Transfert Matrix Formalism of *P*SiNWs layer. The modeling of the reflectance permits to explain this value by taking account into the shape of the nanowires and its porosity. Optical absorbance and transmission were also theoretically studied. The absorbance was superior to that obtained with silicon nanowires and the ultimate efficiency was about equal to 25% for normal incidence angle. These results could be applied to the potential application in low-cost and high efficiency *P*SiNWs based solar cells.

Chapter 10 – Mesoporous TiO_2 thin films had been prepared through sol-gel combined with evaporation-induced self-assembly (EISA) by using F127 and $Ti(OBu^n)_4$ as templating agent and inorganic precursor respectively. It was found that the calcinations temperature, relative humility and the mass ratio of templating agent to inorganic precursor could vary the TiO_2 mesoporous structure. The thermodynamic study for the sample had been carried out and the critical crystallite size expression of the mesoporous film based on the authors' established four-coordinate channel model was presented for the prediction of the thermal stability of the mesoporous structure at high temperature. Varying the amount of templating agent to inorganic precursor could adjust the pore size and structure of mesoporous TiO_2. When the mass ratio of F127/$Ti(OBu^n)_4$ varied from 0.1 to 0.5, the pore sizes of mesoporous TiO_2 changed from 5.4nm to 9.1nm. The pore size regulation mechanism had been discussed.

N, S co-doped TiO_2 exhibited framework with a high porosity. Both pure mesoporous TiO_2 and N,S co-doped TiO_2 exhibited higher photocatalytic activity, especially for the N,S co-doped samples.

Chapter 11 – This chapter deals with the Inversion or Inflection point(s), termed collectively *I-point(s),* which appeared in all the N_2 adsorption isotherms $V=f(P/P_o)$ of porous solids. The *I-point(s)* correspond to the equivalent point(s) of titration of the adsorbent material (analyte or titrand) by the adsorbate N_2 (titrant or titrator) at 77K. If there is just one I-point, this corresponds to the completement of physical adsorption and is useful for the determination of the specific surface area *-ssa-* of solid (S_p, m^2g^{-1}) without any knowledge or use of the traditional BET methodology. If there is a second equivalent point, this corresponds to the completement of the condensation of N_2 into the capillary pores. The *I-point(s) method* is a precise and convenient procedure for estimating the *ssa* of porous materials based on N_2 adsorption data, $V=f(P/P_o)$. Such data if plotted in the form $V(1-P/P_o)/(P/P_o) = f(V(1-P/P_o))$ may show one I-point, if mesopores are spatially mixed with micropores (case i) or two I-points if the mesopores form separate domains in the porous solid (case ii). The projection of the single I-point (case i), or the low pressure I-point (case ii), on the $(V(1-P/P_o))$ axis corresponds precisely to the volume of monolayer V_m from which the S_p values can be easily estimated ($S_p = 4.356V_m$ for N_2). Finally the variable slope of the I-plots corresponds to the *variable equilibrium constant C of adsorption,* introduced in the BET equation, which is not attainable via the classical linear treatment.

Chapter 12 – In most cases the chemisorption process is described in terms of formation of adsorbate complexes of 1 : 1 stoichiometry, *i.e.* one surface site binds one adsorbate molecule. However, there are two more possibilities: (i) one adsorbate molecule bridges more than one site and (ii) one site binds two or more adsorbate molecules. The second possibility is typical of cations in porous materials. In this chapter the authors summarize the achievements on why and when two or more identical or different molecules can be simultaneously attached to one cationic site.

Chapter 13 – Crystalline layered silicates, which possess two-dimensional nano-spaces between the layers, have been utilized for various applications, as adsorbents, catalysts/catalyst supports, ion exchangers etc. These uses are possible due to the covalent modifications on the silicate surface $SiOH/SiO^-$ groups. This chapter will examine the feasibility of synthesizing organo–inorganic composites based on layered silicates. It aims to explore the basic and advanced aspects of preparation and characterization of silylated and/or pillared interlayered clays by quaternary ammonium salts pre-swelling step, which is responsible for the controllable and designed pore structure and accessibility.

Chapter 14 – Ordered mesoporous and non-porous silica films were studied by analyzing angular and spectral dependencies of reflectivity and hemispherical elastic light scattering (HELS). The real and imaginary parts of the film index were estimated from the reflection spectra. The HELS angular distribution of the mesoporous film shows a minimum, which has been interpreted as an interference pattern coming from light scattered by continuous set of the mesopores. Another mesoporous Al_2O_2 membrane (a 2D periodic array of empty and metal-filled linear nanopores) has been studied. These mesoporous membranes are characterized by noticeable differences in reflectance. A Maxwell-Garnett's (MG) model of composite material has been used to recover the dielectric function of tin nanorods. A consistent explanation of the optical properties has been achieved using multiple scattering and metallic behaviors of the nanorods.

Chapter 15 – The interest of worldwide scientists on developing functional materials with special properties that allow their use in emerging fields as biomedicine and environmental remediation has continuously grown in the last several years. Due to their structure and surface properties mesoporous molecular sieves offer possibilities to be modified in a controlled manner to obtain important materials with specific applications in emerging fields. Silica- and carbon-based mesoporous molecular sieves modified with iron oxides (magnetite and hematite), titanium oxide (anatase), bismuth oxide (alpha-Bi_2O_3) by controlled encapsulation were developed. Their functionalities and specific applications in biomedicine and environmental remediation are discussed in separate sections.

Chapter 16 – The synthesis and characterization of Pd-poly(N-vinyl-2-pyrrolidone)/CMK-8 (Pd-PVP/CMK-8) composite and its application in Suzuki-Miyaura cross-coupling reaction as a novel heterogeneous catalyst is reported. The catalyst was characterized by FT-IR, XRD, XPS, BET, TGA, DRS UV-Vis, SEM and TEM techniques. The catalytic activity of this purely organic hybrid catalyst was compared with Pd-PVP/KIT-6 to clarify the advantages of mesoporous carbon on mesoporous silica as support. Surprisingly, the composite prepared by mesoporous carbon showed much higher activity than that of Pd-PVP/KIT-6. In addition, the results showed that the stability of Pd-PVP/CMK-8 was excellent and that it could be reused 13 times without much loss of activity in Suzuki reaction. This unique result opens new perspectives for application of mesoporous carbons as structurally defined hydrophobic catalyst support in catalytic reactions.

In: Comprehensive Guide for Mesoporous Materials, Volume 2 ISBN: 978-1-63463-982-8
Editor: Mahmood Aliofkhazraei

Chapter 1

MESOPOROUS SURFACE STRUCTURE ANALYSIS

Erich Robens[1] **and Shanath Amarasiri A. Jayaweera**#[2]*
[1]Uniwersytet Maria Curie-Skłodowska, Lublin, Poland,
[2]Stokesley, Middlesbrough, UK

ABSTRACT

The classification of pores with regard to their pore width is given. The standardisation and classification work concerning the surface texture of porous material is reviewed. A compilation is presented of methods which can be used to determine the total pore volume, the pore size distribution and the fractality. The survey comprises light, X-ray, electron, ion, acoustic and tunnel microscopy, mechanical sensing, classification of particle size, flow methods, volume and density measurements, intrusion, sucking and displacement porometry, gas or liquid sorption porometry, thermoporometry, desorption spectroscopy and nuclear magnetic resonance measurements. Furthermore, a method of saving time of sorption measurements is presented.

1. INTRODUCTION

Most solid material contains pores, in the bulk and/or on its surface. The large porous structure of a solid is important in its applications and for reaction with the environment [1-6]. Applications include storing of gases and liquids, chemical reactions with a solid surface and heterogeneous catalysis. Thus characterisation of a solid material always includes the investigation of its pore structure. The following survey brings to focus the investigation and characterisation of mesoporous structures providing a large surface area.

The majority of porous materials have pores within a wide range of size. On the other hand investigation methods are restricted into their measuring range. Therefore we can hardly

* Schlesierstr. 5, D-61381 Friedrichsdorf, Germany, hon. prof. Uniwersytet Maria Curie-Skłodowska, Lublin, Poland, *formerly*: Battelle-Institut e.V., Frankfurt am Main, Germany. erich.robens@t-online.

42 Tanton road, Stokesley, Middlesbrough TS9 5HR, UK. *formerly* Teesside University, School of Science & Engineering, Middlesbrough, United Kingdom. saajayaweera@yahoo.co.uk

define a 'true' characteristic pore. It is better to determine size distributions rather than characterising values. In each case we should indicate the measuring method used and its limits. Problems arise when measuring conditions deviate markedly from application. It is advisable to apply several different measuring methods in parallel allowing for critical validation of the results. At the end the measurements should be supplemented by practical experiments.

Measuring methods need to be validated. Validation is the process of determining the suitability of methods for providing useful analytical data with the required accuracy [7-8]. Guidelines for validation procedures are elaborated and standardized at international level [9-10]. All measurements are burdened with errors and uncertainties which should be indicated [8].

2. DEFINITIONS

Pores mostly are of irregular shape; Figure 1 shows schematically a porous grain with different types of pores.

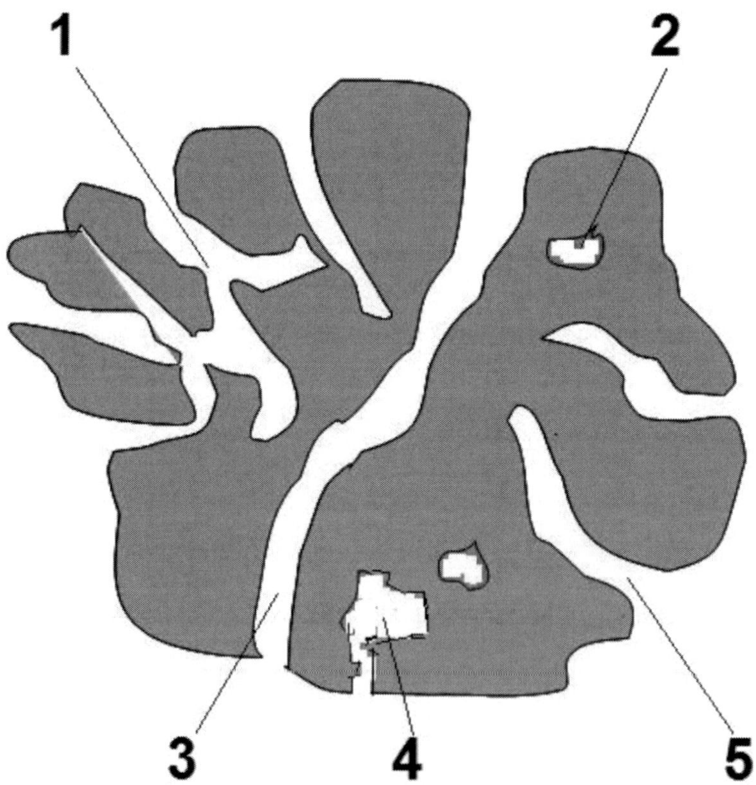

Figure 1. Diagram of a particle with different pore types: 1 inter-connecting, 2 closed, 3 passing, 4 ink bottle, 5 dead end.

Following definitions concern pore models of simple geometry. It is customary to express the average width d of a cylindrical pore as:

$$d_P = \frac{4\,V_P}{S_{smooth}} \tag{1}$$

The volume of a cylindrical pore or particle is given by:

$$V_P = \pi\, r_P^2\, h = \frac{\pi\, d_P^2\, h}{4} \tag{2}$$

The volume of spherical pore or particles is given by:

$$V_{sph} = \frac{4\,\pi\, r^3}{3} = \frac{4\,\pi\, d^3}{24} \tag{3}$$

Pores are classified by several authors: Dubinin [11], IUPAC [12], Setzer [13] (Figure 2):

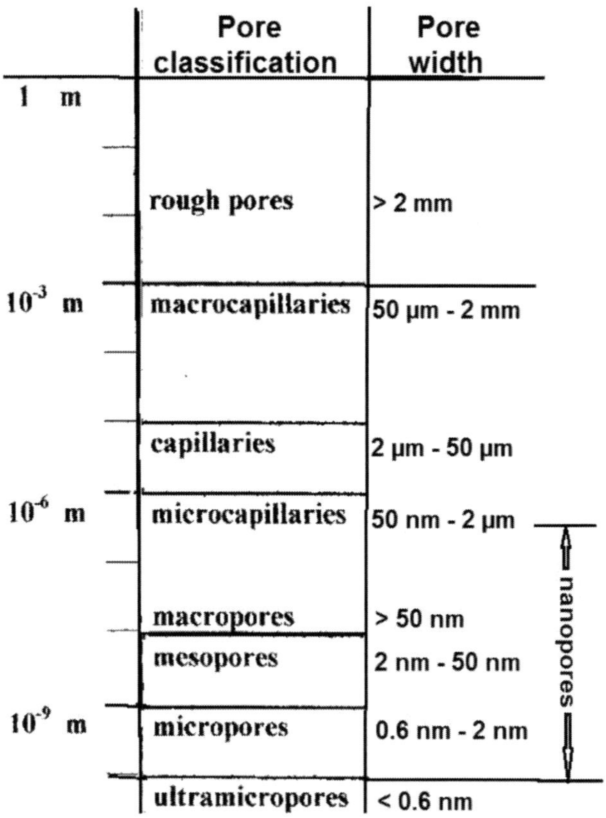

Figure 2. Pore classification according to Setzer (capillaries), IUPAC (pores), and Dubinin (ultramicropores).

3. Measuring methods

In Table 1 methods, which are used to investigate surface and internal structure of solids are presented. In Figure 3 the detection limits of the different methods are depicted. In principle the porous structure can be directly investigated or the matrix is filled with a fluid and that fluid is investigated.

Table 1. Methods characterising the surface and pore structure

Method	Instrument
Mechanical tracing	Surface roughness tester, mechanical profiler Scanning tunnel microscope (STM) Atomic force microscope (AFM)
Classification, Sedimentation	Classification (sieving, air separation)
Flow methods	Andreasen pipette Blaine apparatus Liquid flow
Optical methods	Optical profiler Microscope Electron microscope Ellipsometer Holography Small angle X-ray scattering (SAXS) Powder diffractometer
Density determination	Gas pyknometer Buoyancy in a gas, Millikan / Straubel apparatus Hydrostatic balance,
Intrusion porometry	Mercury porosimeter Water porosimeter
Gas adsorption	Gravimetric sorption apparatus Volumetric sorption apparatus Gas flow apparatus
Adsorption from solution	Quartz oscillation microbalance
Size exclusion	HPLC apparatus
Calorimetric methods	Calorimeter
Thermoporometry	Calorimeter

Tracing

By mechanical tracing a surface with a needle, we obtain the roughness factor as the relationship of the recorded length to its linear projection. From the roughness measured in two perpendicular directions we can calculate the surface area. Conversely we can calculate the roughness factor from a surface area determined e.g. by gas adsorption. The resolution limit of the tracer method is given by the tip radius. Because that pin should resist wear, only diamond can be used. A tip radius of about one micrometer can be realized.

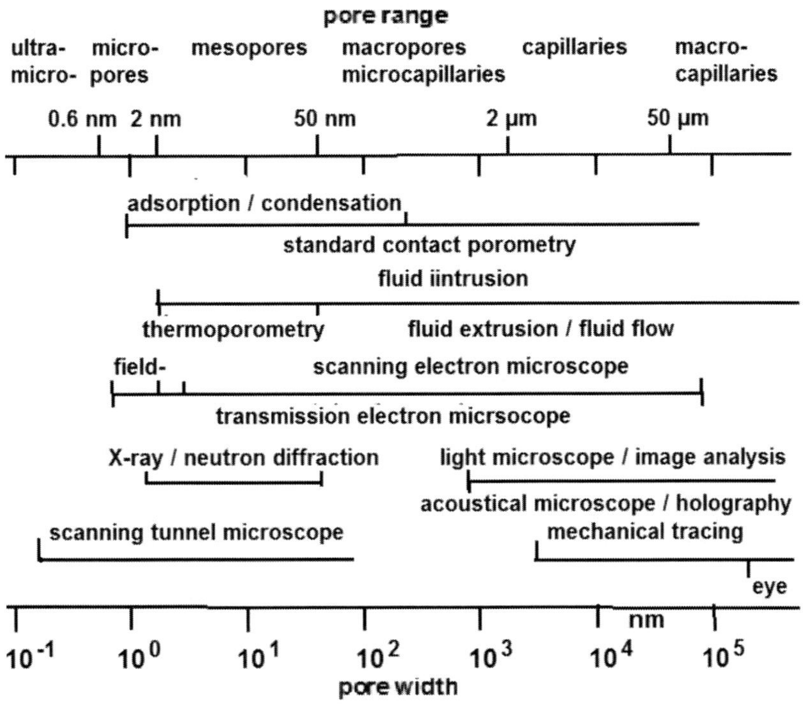

Figure 3. Detection limits of surface structure measuring methods.

Scanning probe microscopy (SPM) includes techniques in which a needle with molecular tip scans a specimen [14] (Fig 4a). The needle is glued to a flexible cantilever and is exited to vibrate (Figure 4b). An image of the surface is obtained by mechanically moving the probe in a raster scan of the surface line by line, and recording the probe-surface interaction as a function of position. On electronic command or automatically the piezoelectric actuator executes motions with a precision and accuracy at the atomic level. Using these very simple and cheap methods the molecular structure of the surface and also adsorbed layers can be revealed. With tracing methods only the surface is made perceptible whereas pores in the bulk cannot be seen.

The scanning tunnelling microscope (STM) [15] is based on the concept of quantum tunnelling. When a conducting tip is brought very near to a conducting surface, a bias (voltage difference) applied allows electrons to tunnel through the vacuum between them. The resulting tunnelling current is a function of tip position, applied voltage, and the local density of state. For an STM, good resolution is considered to be 0.1 nm lateral resolution and 0.01 nm depth resolution. The STM needs a conductive surface. If the sample is an insulator conductance can be created by a sputtered metal layer, usually gold.

The atomic force microscope (AFM) [16-17] scans the surface down to its molecular structure. Usually tungsten is used as pin and its tip may be etched down to atomic dimensions. This pin is glued to a flexible lever which is brought by means of piezo quartz to a distance below one nanometer to the sample surface. The tip of the cantilever should not contact the sample surface. Therefore the cantilever is exited either to its resonant frequency (frequency modulation) (Figure 4b) or just above (amplitude modulation) where the amplitude of oscillation is typically between 10 nanometers and a few picometers. Van der Waals forces which are strongest from 1 nm to 10 nm above the surface, act to decrease the resonance frequency of the cantilever. This decrease in resonant frequency is combined with the feedback loop system maintaining a constant oscillation amplitude or frequency by adjusting the average tip-to-sample distance.

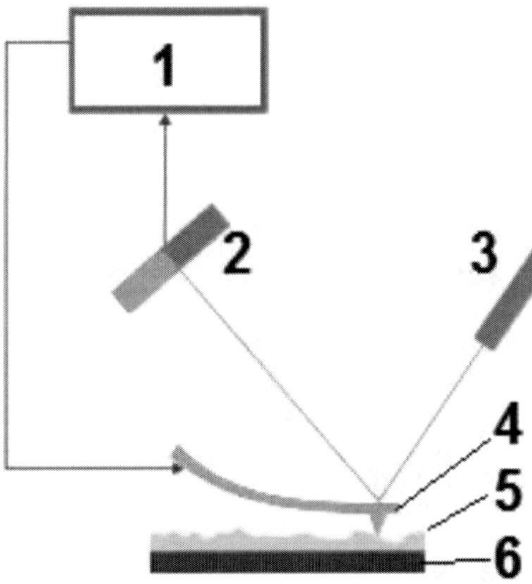

Figure 4a. Diagram of the atomic probe microscope. 1 detector and feedback electronics, 2 photodiode, 3 laser, 4 cantilever with tip, 5 sample surface, 6 piezo scanner.

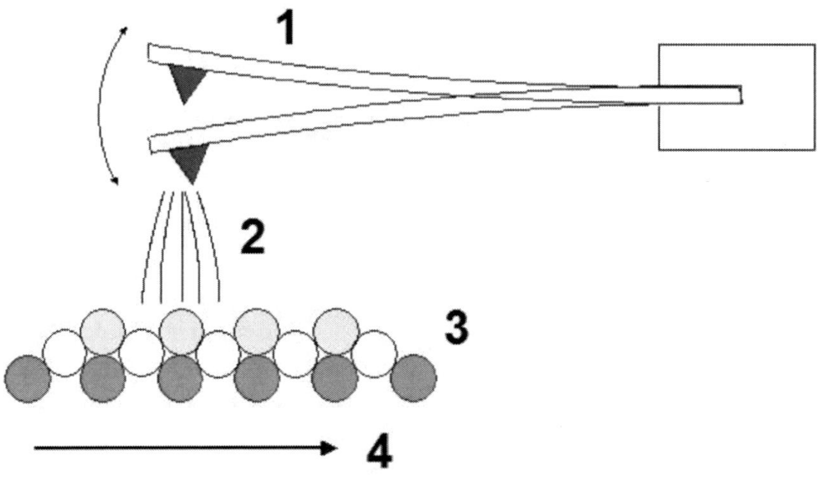

Figure 4b. Diagram of the atomic force microscope with vibrating cantilever. 1 flexible vibrating cantilever, 2 interaction force, 3 surface, 4 horizontal sample motion.

Grain Size Analysis

Porous materials are often used in the form of grains and powders. Thus, particle size analysis is an important part of an exhaustive investigation of porous matter, though pores cannot be detected in this way. Measurements usually result in a cumulative particle size curve from which the differential particle size distribution is derived. From that the specific area of the outer surface can be calculated; the inner surface of a pore system is not included.

By classification methods like sieving and air separation grains are classified with regard to their size [18-19]. Sieving is carried out in a set of sieves fitted together.

Determination of the particle size during counting is automated on various principles. Either the sample is distributed in one layer on a plane and scanned, or the particles are diluted in air or in a liquid and fed through a slit. There they are counted and size is measured optically or electrically (Coulter Counter).

Observing the sedimentation of a powder distributed in a liquid by the gravitational field or in the acceleration field of a centrifuge enables determining the grain size distribution. In sedimentation analysis particles will be uniformly suspended or sprinkled on top of a fluid of known density and viscosity. The settling is impeded by friction and this depends on size of the individual particle: smaller particles fall slower, approximately according to Stokes' law. The concentration of particles at a known distance from the top of the container is measured by stepwise sampling in the Andreasen pipette (Figure 5) or continuously monitored via extinction of X-rays or of a light beam (photo sedimentometer).

Alternatively a balance is used for continuous measurement of settling particles (Figure 6). For an irregular particle, the equivalent settling rate diameter is slightly smaller than the volume diameter. As a consequence of an adhering, immobile liquid layer, surface roughness and pores do not contribute; though due to its lower apparent density porous particles have a somewhat lower settling rate. From the cumulative curve measured the differential grain size distribution can be derived.

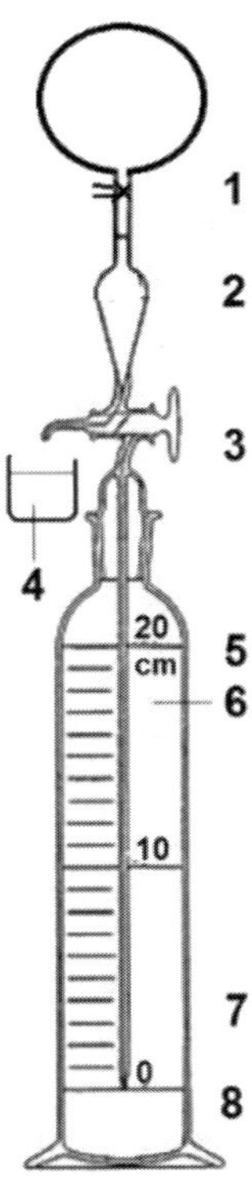

Figure 5. Andreasen pipette. 1 Peleus ball, 2 test sample volume, 3 triple cockstop, 4 test sampling bin, 5 suspension level, 6 suspension, 7 pipette, 8 glass cylinder.

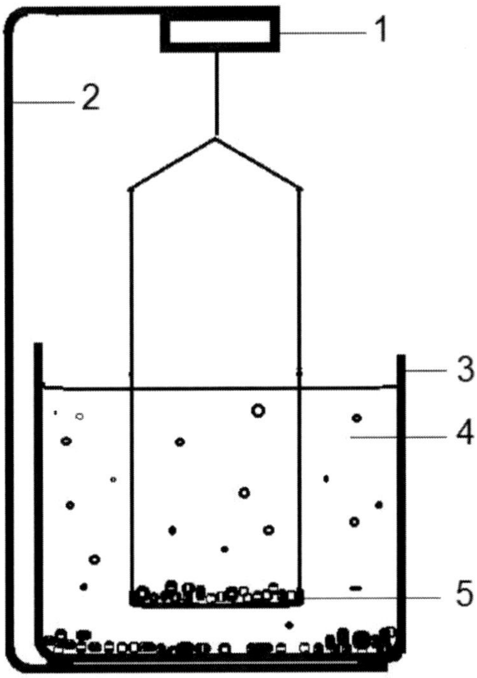

Figure 6. Principles of the gravimetric sedimentation measurement of the particle size distribution. 1 balance, 2 stand, 3 beaker, 4 liquid with suspended particles, 5 balance pan with deposited particles.

Mixtures of particles of different size are separated in a laminar flow through a capillary [20]. Sub-microscopic particles suspended in a viscous fluid within a capillary tube will sample all radial positions accessible to it with equal probability, provided the residence time of the particle in the flow field is long. Due to its finite size, however, the approach of the particle centre to the inner surface of the capillary tube will be limited to a distance equal to its radius. As a consequence, the particle is excluded from the slowest moving streamlines; this implies that the average velocity will exceed the average velocity of the eluant and also that of the smaller diameter particles. In addition to this exclusion effect, the longitudinal and radial particle displacement is affected by the presence of the capillary wall. Capillary hydrodynamic fractionating (CHDF) applies this phenomenon to the determination of particle size distributions in the sub-micron range by moving eluant and particles through a set of capillaries, determining transit times [21]. The instrument must be calibrated by monodisperse standards.

The fineness and the specific surface area of powder material can roughly determined by measuring the air permeability. The resistance to laminar flow of air through a slightly compressed sample is measured. The Carman-Kozeny equation [22-23] connects the laminar flow resistance with the wall area, not including pores. Several variants of that simple method have been developed by Lea and Nurse [24], Rigden [25], and Blaine [26].

Using the Blaine apparatus (Figure 7) the sample is compressed to a given porosity, and a decreasing water column sucks air through the powder packing [27-30]. Also Knudsen flow is considered to characterise pore systems [31]. Similar instruments are used to determine the fineness of textile fibres. On account of its simplicity and fast response the Blaine test seems to be the only surface measuring method which is used to control manufacturing process in cement production.

All that methods register only the outer surface of the particles. Using a pulsed gas flow in a variant of the Blaine test, Barrer [32] succeeded in measuring also the porous structure.

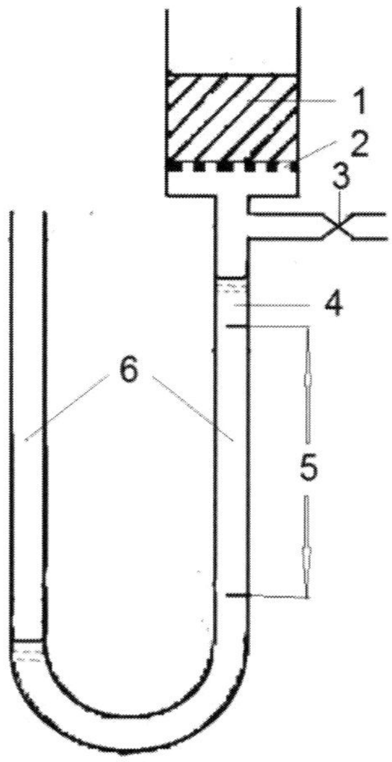

Figure 7. Blaine test apparatus to measure the specific surface area. 1 sample, 2 filter plate, 3 suction valve, 4 suction liquid, 5 measuring distance, 6 glass U-tube.

Optical Methods

The roughness of a surface can be assessed by microscopic comparison with reference tables. Particles as small as 0.2 nm could be resolved by electron microscopy, whereas the limit with ultraviolet light is 10 nm. In this way with only two-dimensional pictures, the outer surface becomes visible. Three-dimensional pictures can be reconstructed by simultaneous inspection from different sides. Projecting these results onto the inner surface often leads one astray. Using holography also the spatial distribution in a flowing suspension can be observed.

A useful method for surface area and porosity measurements is small angle X-ray scattering (SAXS) [33-35]. Observing diffraction angles up to 8°, according to the Porod method, or broadening of lines, the surface in pores with diameter between 0.2 and 50 nm can be measured [36-37]. The method can be performed using a common powder diffractometer with ancillary equipment. Because of the restricted measuring range only few applications have been reported. The method supplements the adsorption method in so far as closed pores are also included in the measurements and, thus, higher values result in most cases. Comparing results of both methods we observed, that during corrosion of reactor graphite in carbon dioxide, the increasing surface area was partially due to the opening of pores [36-37]. Similarly the diffraction of neutrons may be used [38].

Density Determination

From a density measurement of a sample the volume of included closed pores can be derived if the density of the compact material is known. If the sample is of simple geometric shape, we can also calculate the total pore volume. Density determinations may be performed using liquid or gas pycnometers.

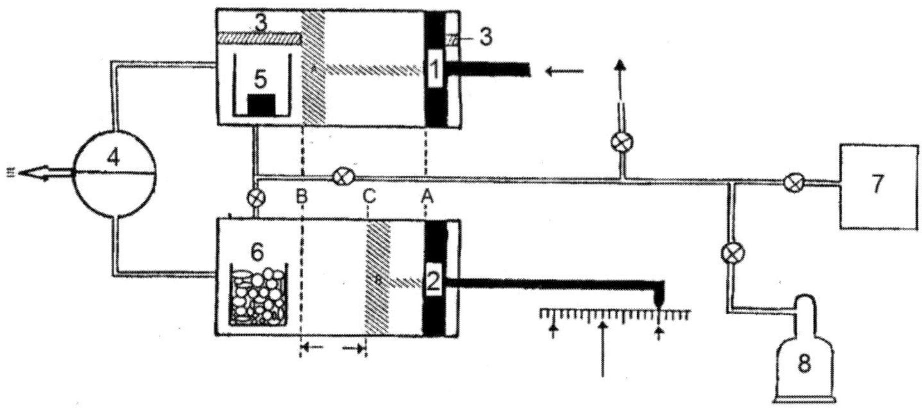

Figure 8. Diagram of a gas pyknometer for density determination of porous solids. 1 piston of the comparison chamber, 2 piston of the test chamber, 3 delimiters, 4 differential pressure gauge. 5 non-porous comparison sample, 6 porous sample, 7 gas reservoir, 8 measuring gas.

A gas pycnometer is shown schematically in Figure 8: Two cylinders, one of which containing the sample, are filled with equal volumes of helium. Subsequently the arising pressure difference is cancelled out by means of a piston. The movement of that piston is a measure for the solid sample volume including closed pores. After impregnation with a wax or in comparison to mercury volumometry also the open porosity can be determined.

Another method is based on buoyancy. In the course of thermogravimetric experiments it is customary to measure buoyancy with nitrogen gas at ambient temperature and different pressures. In liquids the hydrostatic balance is used for density determinations (Figure 9).

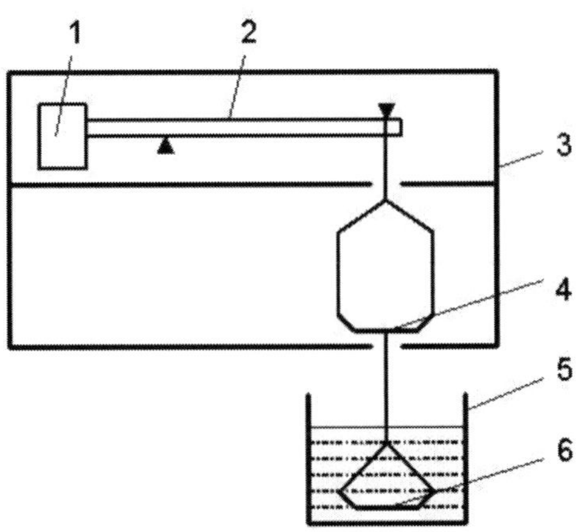

Figure 9. Diagram of a hydrostatic balance. 1 counterweight, 2 balance beam, 3 thermostated case, 4 pan for sample weighing in air, 5 beaker with liquid, 6, pan for weighing in liquid.

Intrusion Porometry

With gases or liquids simple porosity tests are made in particular for building materials. Standard method of pore volume determination in the mesopore range is mercury porosimetry. Using this method mercury is pressed against capillary forces at continuously increasing pressure into the previously degassed pores (Figure 10). The decreasing mercury column in a capillary of the sample vessel is measured and gives the respective pore volume. The Washburn equation shows the relation between pressure and pore radius. Besides simple mercury volumeters, commercial instruments make use of the hydraulic press and apply pressures up to 4 kbar so that pores with diameter of down to 5 nm can be determined [39]. An instrument of Plachenov [40] extends this to 10 kbar corresponding to a diameter of 1.7 nm. Neglecting areas of small pores, a surface area can be calculated. However, it should be considered that small pores may contribute considerably to surface area and pore volume. Similar instruments are operated also with other non-wetting liquids in particular water at hydrophobic solids.

Uncertainties are due to speculations about the contact angle. Using high pressures the pore structure of the material may be deformed. Differences in the shape of the hysteresis loop between intrusion and discharge branch using model materials could be assigned to the pore form [41]. Nevertheless, discharge curves should not be evaluated because always some mercury 'glues' to the matrix as in a clinical thermometer. The retention volume may be due to pore domains with small entrance (Figure 11).

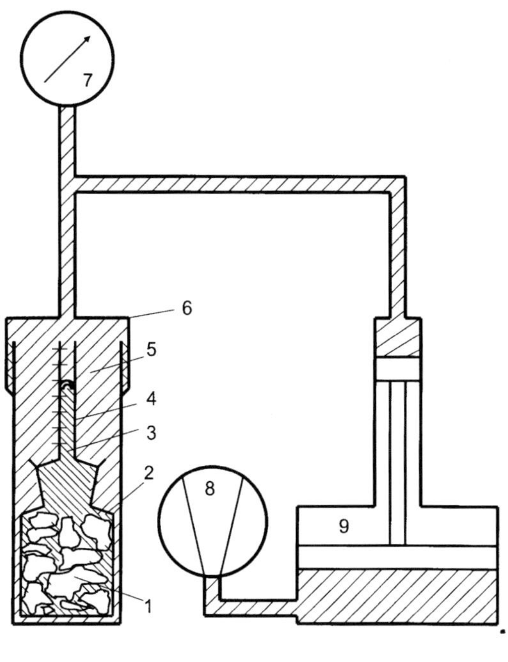

Figure 10. Diagram of a mercury intrusion porosimeter. 1 sample, 2 sample vessel, 3 mercury. 4 graduated capillary, 5 hydraulic fluid, 6 autoclave, 7 pressure gauge, 8 pump, 9 hydraulic pressure intensifier.

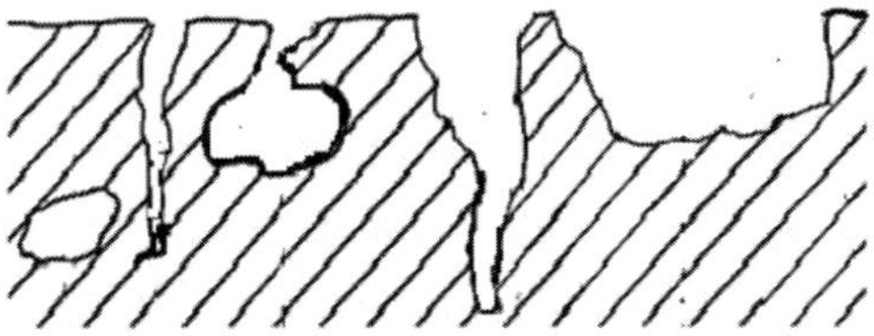

Figure 11a. Empty pore system.

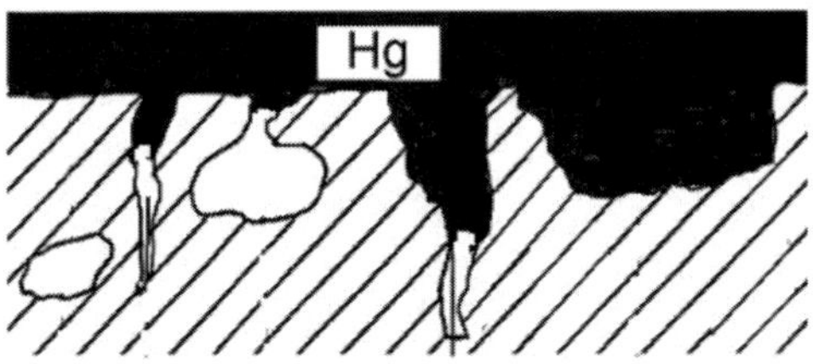

Figure 11b. Pore filling with mercury at a pressure of ~ 2 kbar.

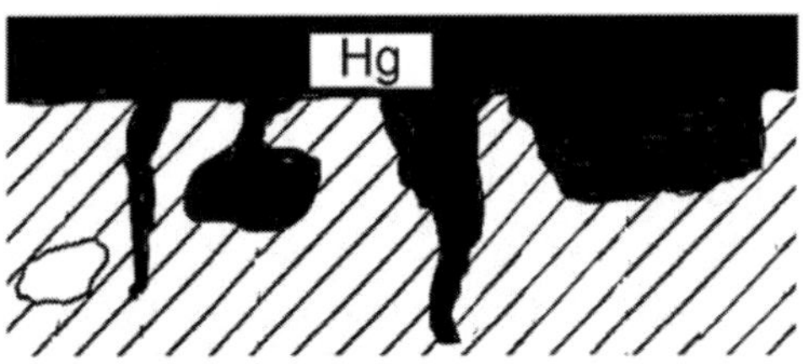

Figure 11b. Pore filling with mercury at a pressure of ~ 5 kbar.

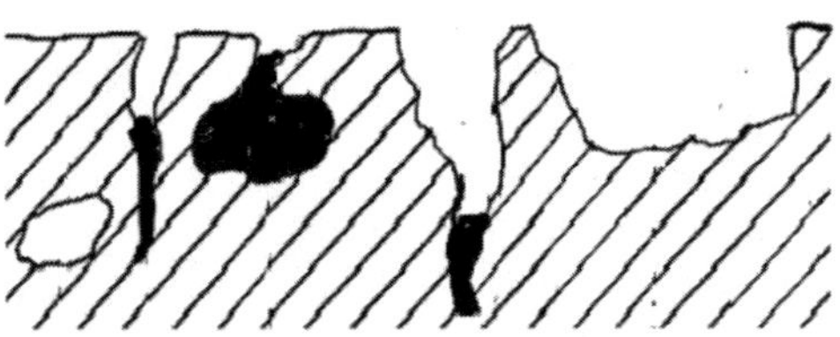

Figure 11b. Discharged pore system with some 'glued' mercury at a pressure of 0 bar.

If a porous system is filled with a wetting liquid, the pores can be depleted with a gas, applying pressure to overcome the surface energy of that liquid. According to Cantor's equation this energy is proportional to the capillary pressure and inversely proportional to the pore radius. This method is used by an apparatus of Zagar which is operated with water and air. The air flow at increasing pressure through the dry sample is compared with the flow through the sample immersed in water. Whereas the flow through the dry sample gives a straight line according to Hagen-Poiseuille's law, the other one is curved and the difference gives the pore radius distribution in the range of 1.2 to 300 µm. Non-commercial variants extend this method to a far wider range [42]. Similar methods are used to determine the pore volume of rocks, using air or water to displace oil or paraffin [43].

Gas Adsorption

Because the number of molecules is used as a measuring unit, adsorption of a gas layer can reveal the surface structure in the molecular region and can give valuable information in the investigation of chemical reactions [34-35, 44-50]. The idea of determining the surface area is based on the assumptions that the surface can be covered with a single, complete monolayer, so that its quantity can be directly measured and that the space required by one adsorbate molecule is independent of the sample material. These premises contradict each other. Due to the strong interaction, when chemisorption takes place, we obtain an isotherm with a pronounced plateau at complete coverage, but the adsorbed amount depends on the number of adsorption sites rather than on the area. In the case of weak interaction, i.e. with physisorption, the formation of multilayers begins before the monolayer is complete. The resulting isotherm is curved and the monolayer capacity must be calculated using an isotherm equation (Figure 12). Although the cross-sectional area in physisorption likewise depends on the chemical nature of the adsorbate/adsorbent, on account of the more universal applicability the weak interaction is preferred. Adsorption measurements with nitrogen or noble gases at 77 K (boiling point of liquid nitrogen) or 90 K (boiling point of liquid oxygen) have proved to be reliable. For some mesoporous materials like active carbon, water or benzene at ambient temperature may be preferred [51-52]. A general drawback of the adsorption method is requirement of degassing the sample in vacuum at elevated temperature and measurement under conditions far from practical reality.

Sorption isotherms may be measured volumetrically, gravimetrically, calorimetrically or by the gas-flow method. Cheaper equipment and easy handling leads to the volumetric technique being the most widely used method (Figs 13-14). After degassing the sample in a vacuum at elevated temperature, the volume of the preferably stepwise introduced or removed gas is measured (Figure 19) and the adsorbed amount is calculated as a function of pressure using the gas equations. High sensitivity in a restricted measuring range is achieved by difference measurements (Figure 14) [53]. Some versions of volumetric apparatus allow for simultaneous gas-pynometric density determination [54].

The gravimetric method (Figs 15-17) requires a vacuum microbalance [55]. Thermostating of the suspended sample and the sophisticated handling of the balance render some difficulties. The possibility of observing the degassing process and the direct measurement of the adsorbed amount independently from gas pressure are advantages of this method. Simultaneously the density of the sample can be measured by means of buoyancy. Of special interest is the use of the magnetic suspension balance which allows for investigations using corrosive gases [56] (Figure 16). Simple, of high sensitivity but restricted in the measuring range are (mostly helical) spring balances (Figure 17). The oscillating quartz crystal disposes on a high sensitivity in a wide measuring range, however the sample must be connected fast to the surface. It is used for special applications (Figure 20).

Because of its simple equipment the differential flow method (Figure 18) is often used despite being a secondary method, which must be calibrated using the volumetric or gravimetric method. The sample is degassed in a stream of the carrier gas helium. The measurement is made with mixtures of nitrogen with helium. The adsorption and desorption are determined by measuring the variation of the heat conductivity of the gas after having passed the sample. By using a set of different concentrations the total isotherm can be measured which requires pressures up to 10 bar.

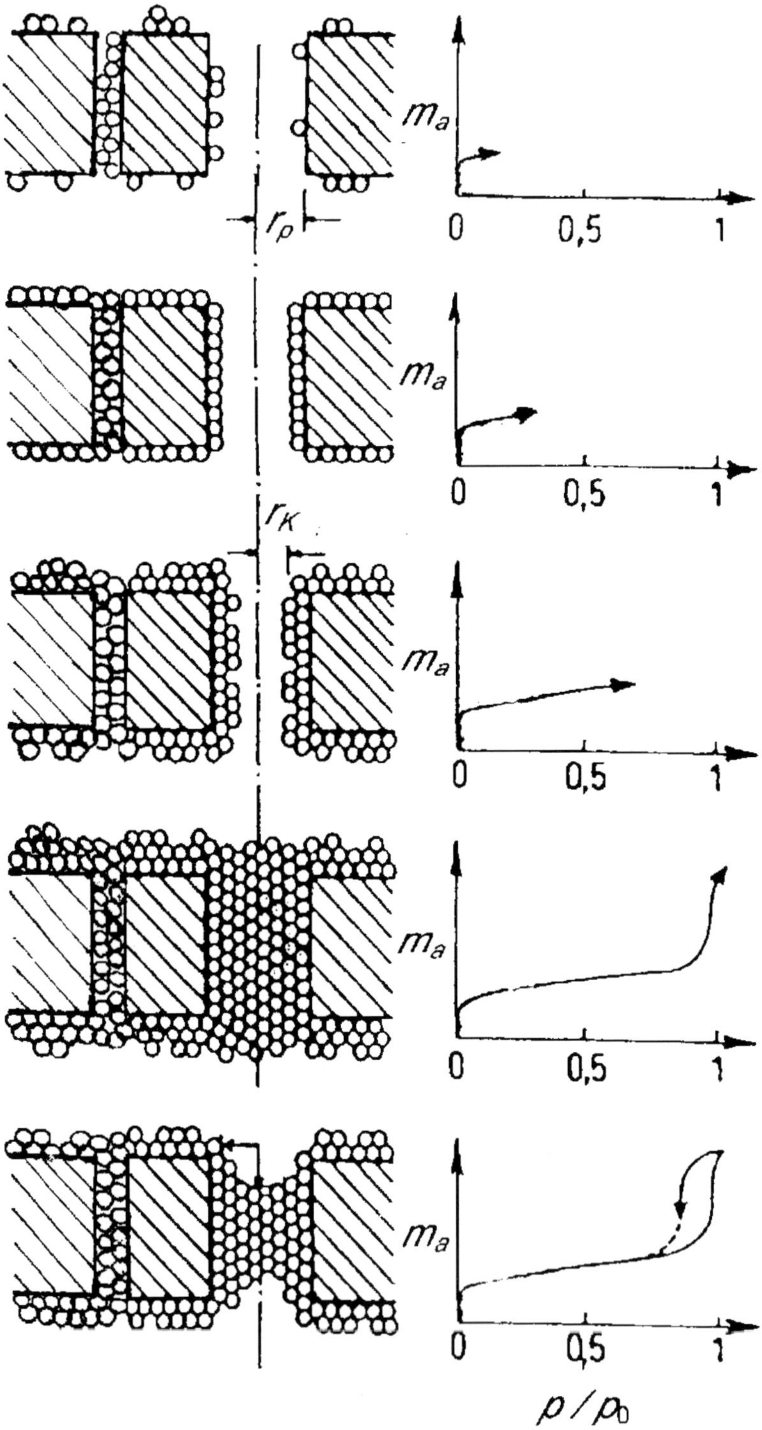

Figure 12. Gas adsorption isotherm with mesopore filling and discharge. m_a = adsorbed amount, p / p_0 = relative gas pressure.

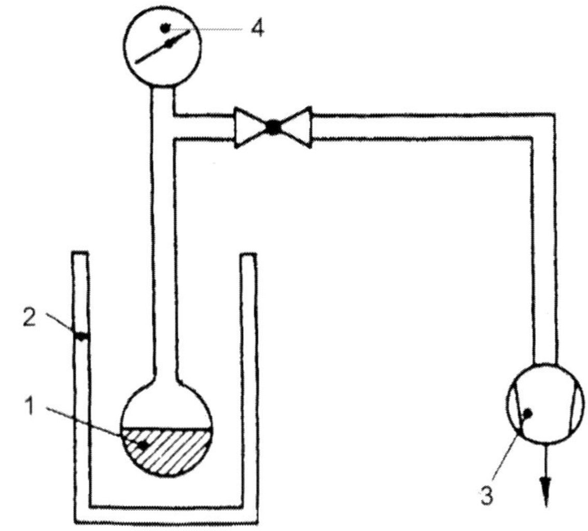

Figure 13a. Diagram of a volumetric sorption measuring apparatus. 1 sample, 2 Dewar vessel with liquefied gas 3 vacuum pump, 4 pressure gauge.

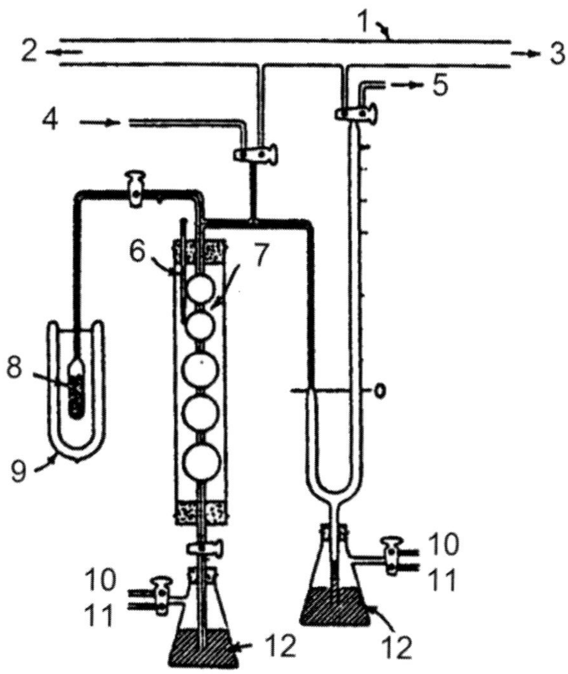

Figure 13b. Diagram of the original volumetric BET apparatus of Emmett. 1 main vacuum line, 2 to McLeod gage, 3 to mercury diffusion pump, 4 gas inlet (He or N_2), 5 to atmosphere, 6 thermometer, 7 water jacket, 8 solid sample, 9 low temperature bath (liquid N_2), 10 gas pressure, 11 vacuum, mercury inlet (confining liquid).

Adsorption can be measured also by radioactive tracer methods [57] which allows for the determination of preferred adsorption domains in heterogeneous materials. Electrode surfaces may be investigated by electrochemical measurement [58-59]. Furthermore, pore filling with water can be observed using NMR spin-lattice relaxation [60].

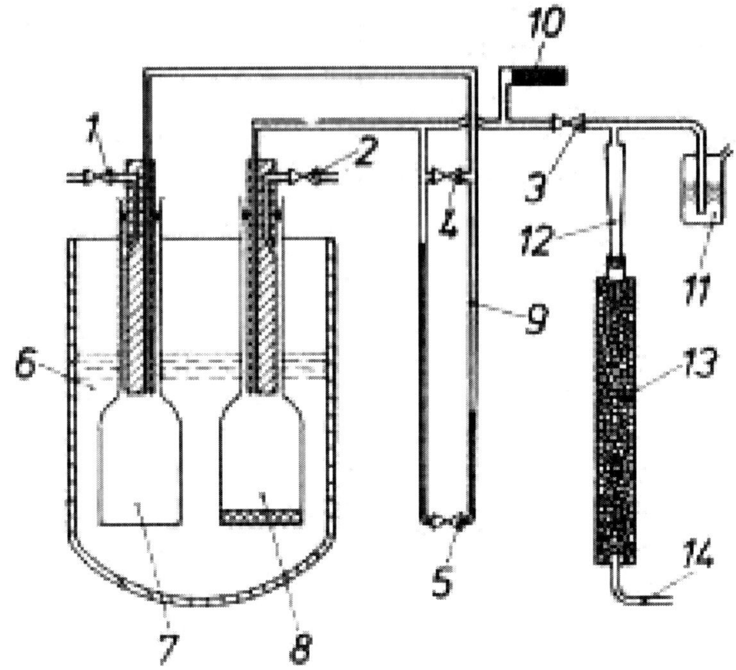

Figure 14. Diagram of the volumetric sorption difference apparatus according to Haul & Dümbgen. 1-5 valves, 6 cooling bath, 7 reference vessel, 8 sample vessel, 9 differential pressure gauge, 10 compensation volume, 11 pressure relief valve, 12 flow meter, 13 drying column, 14 nitrogen admission.

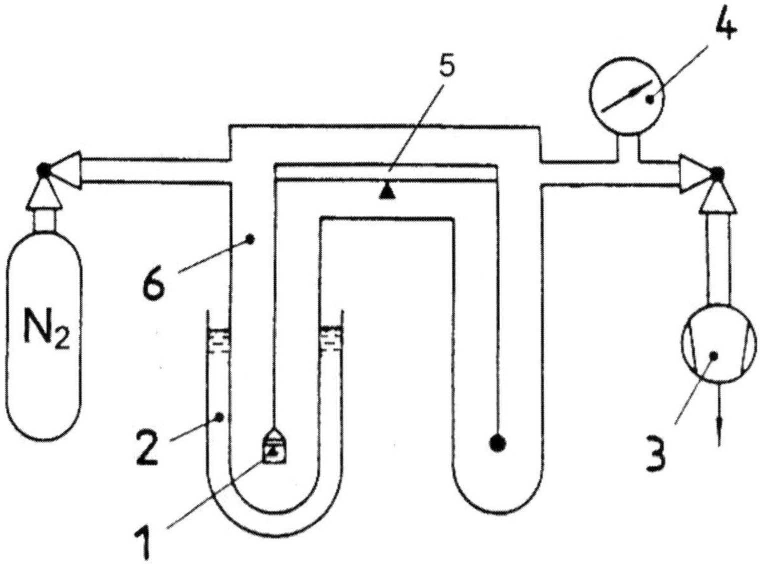

Figure 15. Diagram of a gravimetric gas sorption isotherm apparatus. 1 sample, 2 Dewar vessel with liquefied gas, 3 vacuum aggregate, 4 pressure gage, 5 microbalance, 6 vacuum vessel.

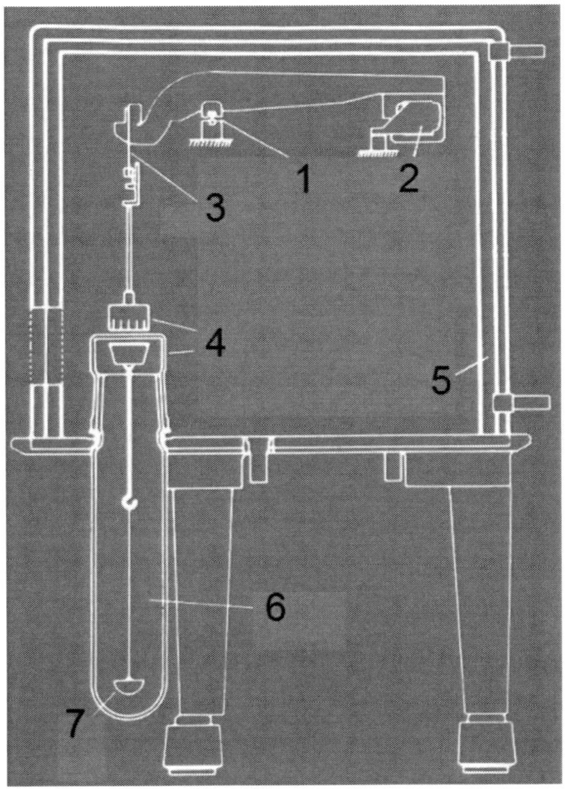

Figure 16. Diagram of a magnetic suspension balance according to Gast (Sartorius). 1 knife edge with balance beam, 2 electromagnetic measuring and restoring device, 4 controlled electromagnet and permanent suspension magnet, 5 thermostated case, 6 vacuum vessel, 7 sample.

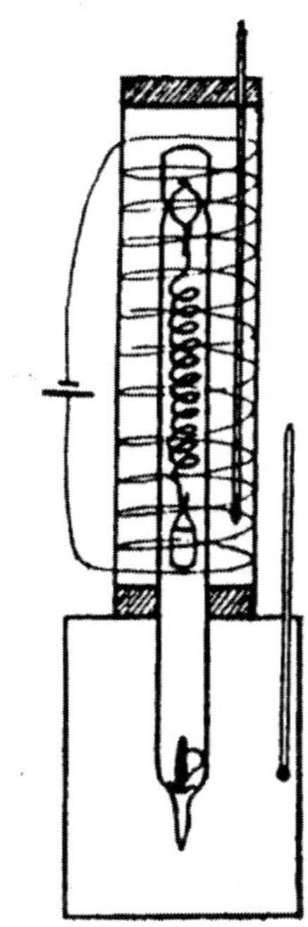

Figure 17. Diagram of a McBain / Bakr adsorption spring balance.

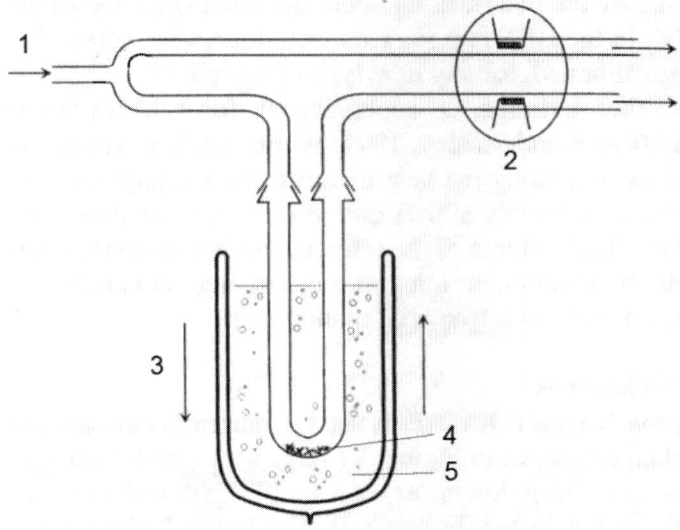

Figure 18. Diagram of a differential flow apparatus according to Nelsen and Eggertsen. 1 gas mixture, 2 differential gas flow meter, 3 adsorptive + carrier gas, 4 sample, 5 liquefied gas (N_2).

Mostly gas adsorption isotherms are measured stepwise at increasing or decreasing pressures or concentrations, respectively (Figure 19). For routine quality control of a special product also one single point of the isotherm is measured, putting up with some inaccuracy. Jäntti [61-65] recommends shortening the measurement up to 70 % by extrapolation of each measurement step using a computer. Highly accurate and slow continuous measurement [66] of the complete isotherm reveals the fine structure of the curve (high resolutions adsorption measurement = HRADS) [67] and allows for calculations of the adsorption and condensation energies [68].

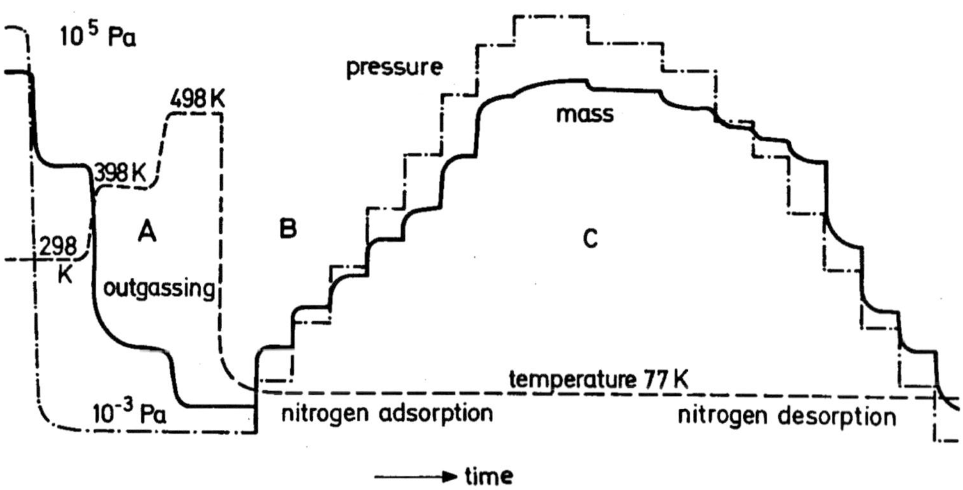

Figure 19. Record of gravimetric measurement of a nitrogen sorption isotherm at 77 K. A degassing of the, B adsorption, C desorption.

We have a variety of methods for the evaluation of adsorption isotherms [69]. The BET method calculates the monolayer capacity in the region $p/p_0 \approx 0.05$ - 0.35 of the isotherm. This method is restricted to isotherms of type II or IV according to the BDDT classification and the respective hysteresis types (Figure 20). Mostly the two-parameter BET equation is used in which it is assumed that an infinite number of layers can be formed. This, however, is not possible in porous materials and may affect the results. Using computers one can evaluate more complicated and specialised isotherm equations [70-85] taking into account that the complicated adsorption processes are always insufficiently described.

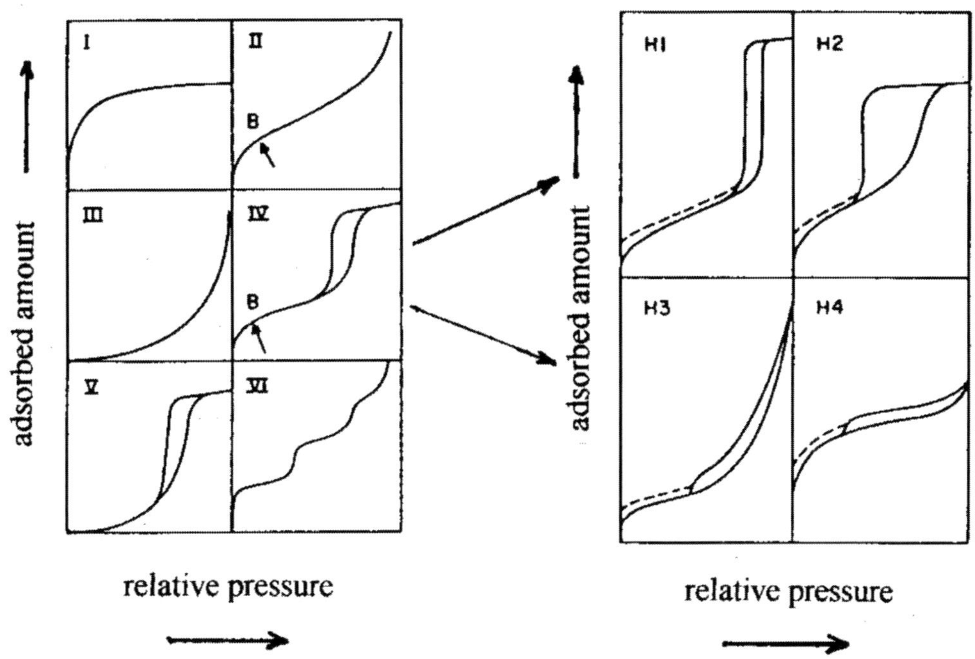

Figure 20. IUPAC classification of gas adsorption isotherms and of the hysteresis loops.

Comparison methods like t-plot [86-87] or alpha-plot [45] are based on the BET-method and yield an instructive impression of the pore content, in particular in the micropore region.

The Dubinin-Radushkevich (DR) method [11, 88-91] and the extended version by Kaganer (DKR) [90-92] use the very first beginning of the isotherm for calculation and yields results which are close to those of other methods. This is of interest because the direct adsorption at the surface is used and measuring only a very short part of the isotherm is time saving.

The different methods of estimating of the pore size distribution in the region of relative pressure above 0.4 result in similar curves and cumulative surface areas. Numerical differences may be as high as a factor of two, due to the different pore models. Therefore, also a modelless method [46, 93-94] could be applied. Many considerations were made to find relations between the shape of the hysteresis loop and the pore shape [95-97]. Suitable results are obtained only for well defined model systems [46].

Also in the micropore analysis remarkable progress can be noted [98-101].

A general problem of pore structure analysis is based in the fractal structure of most materials to be investigated. Mandelbrot's ‘fractal’ [102-103] is a special quality of symmetry:

the self-similarity of the geometric structure. Observing the surface with a microscope or a telescope at every enlargement or reduction, we see structures of fissures which are similar in its degree of roughness.

It is better to characterise porous materials by means of distributions of the volume and the surface with respect to the pore width or by the fractal dimension [104-106] as a summarising value.

Adsorption from Solution

Simple test methods are the decolourisation of solutions or the analysis of the decreasing concentration of agents like phenol as a result of adsorption [107]. The measurements are made applying titration techniques using common laboratory glass ware.

The advantage of adsorption from solutions is that the sample is not dried, which is important for gel structures, which collapse during every drying process. However, in a solution solute and solvent compete in the adsorption, whereas usually only one parameter, i.e. the concentration of the solute, can be determined [108-110]. The result is a composite isotherm which must be divided into its components [111-114]. Furthermore, the solute molecules are often large and complicated in shape.

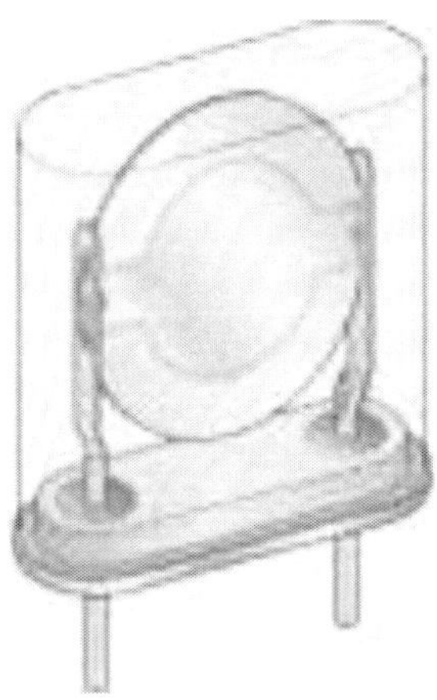

Figure 21. Quartz crystal microbalance sensor with hygroscopic covering used for moisture measurements. © Mitchell Instruments.

Size Exclusion

It seems to be a very simple idea to measure the adsorption with molecules of different size and to establish in this way a pore size distribution. The attempt has been made with gases for the micropore range but indeed this makes much effort and is time consuming [4, 115]. For the mesopore range there exists only a limited choice of molecules which are of spherical shape and indifferent in contact with the surface. Those organic molecules are dissolved in a liquid phase and therefore enable us to obtain information on the porous structures of a gel in its unaltered swollen state. The separation mechanism is governed by the relation between the molecular size of the solute and the pore size distribution of the packed gel. The evaluation of the composite isotherm and disturbances by wall effects provides difficulties. Thus, in general, size exclusion measurements are only an auxiliary means or a secondary method for texture investigations. Nevertheless, because High Performance Liquid Chromatography (HPLC) has got rapid increasing application for analysis as well as for preparation, size exclusion chromatography (SEC) is increasingly used and all the effects disturbing the results are studied intensively [116-117].

There are several possibilities to perform SEC measurements: static, dynamic and a mixed procedure [118-119]. The static method suffers from a lengthy experimental procedure for obtaining a solute exclusion curve through a series of accurate concentration measurements of polymer solutions. The practice in the dynamic chromatographic analysis of porous solids is identical to the ordinary column calibration procedure using a series of fully characterised standard polymers. The disadvantage of the SEC method is that it cannot be applied to those materials unsuitable for packing in a chromatography column. Gel materials of fibrous or bulky form, or of insufficient rigidity, cannot form a stable gel bed in a column, and only the static method can be applied to these materials. The mixed solute exclusion method was introduced by Kuga [120-121]. In this third version, a mixed solution of polymer fractions and oligomers covering a sufficiently wide molecular weight range is brought into contact with the gel sample, and the resulting differential dilution of the solutes is determined as a function of the molecular size by SEC. The method is both simple in procedure and versatile in sample form, though being less accurate than the other two methods.

Calorimetry

Whereas the micropore filling is connected in general with elevated adsorption energy, the second and subsequent layers are bound with a constant energy which is about equal to the condensation energy of the bulk liquid.

Thus, the differential adsorption heat at increasing pressure can be used to determine monolayer covering (B-point) with same reservations as with volumetric or gravimetric BET measurements. Subsequently the differential heat is proportional to the adsorbed volume and in principle the whole isotherm can be measured by means of a heat flux calorimeter (Figure 22) [122]. Furthermore, the heat of immersion of a degassed sample is proportional to the surface area and the heat per unit area for a particular solid/liquid pair, the absolute value of the specific surface area can be determined. Because of the troublesome techniques to measure differential heats of the magnitude 0.1 – 0.4 Joule this method has found only few applications [44].

Another calorimetric method, the measurement of freezing point decrease in pores according to Eyraud and Quinson has become a useful method for pore size determination [123-126]. For this 'thermoporometry' a highly sensitive calorimeter is required (Figure 22b), the handling, however, is easy. The advantage is that the sample is not dried so that also swelling materials can be investigated. With regard to the restricted size area covered the method is hardly applied [127].

In Figure 23 the freezing / thawing process is compared with the adsorption / desorption process. It should be regarded that in freezing always a liquid layer of about two molecules thickness remains at the surface.

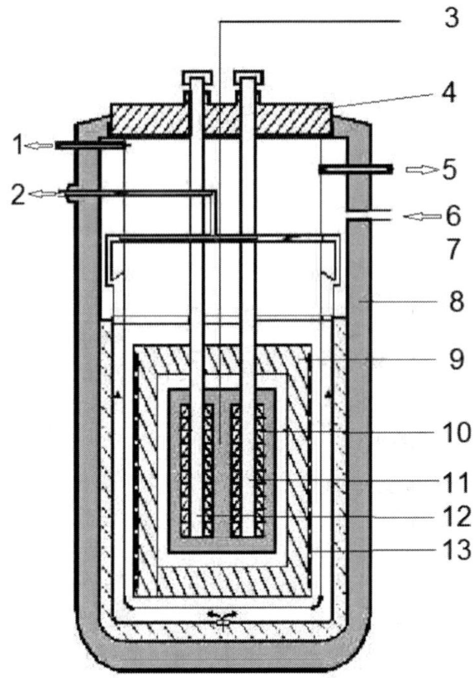

Figure 22a. Tian-Calvet Calorimeter. 1 to vacuum pump, 2 N_2 exit, 3 temperature sensor (Pt 100), 4 cover, 5 N_2 gas, 6 liquid nitrogen, 7 cover, 8vacuum tight calorimetric vessel, 9 thermostat, 10 thermocouples in series, 11 reference cell, 12 sample cell, 13 electric heating unit.

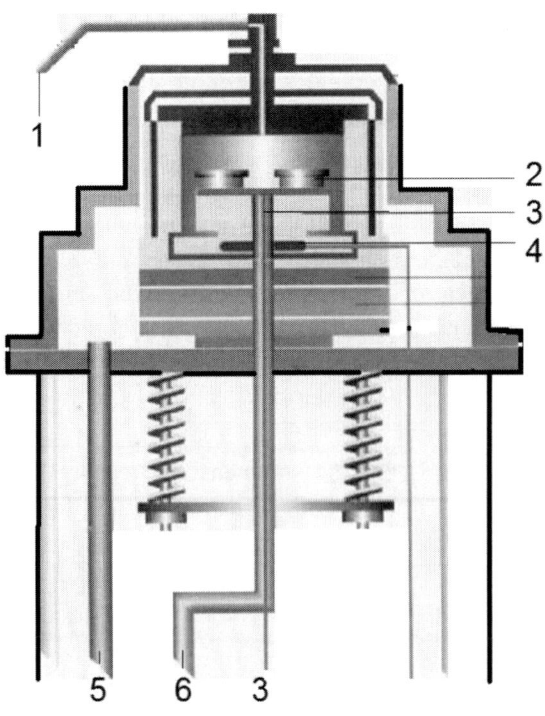

Figure 22b. Differential scanning microcalorimeter. 1 gas outlet,2 sample and reference cell, 3 thermocouple, 4 heater, 5 purge gas inlet, 6 cold nitrogen gas inlet. © Mettler-Toledo.

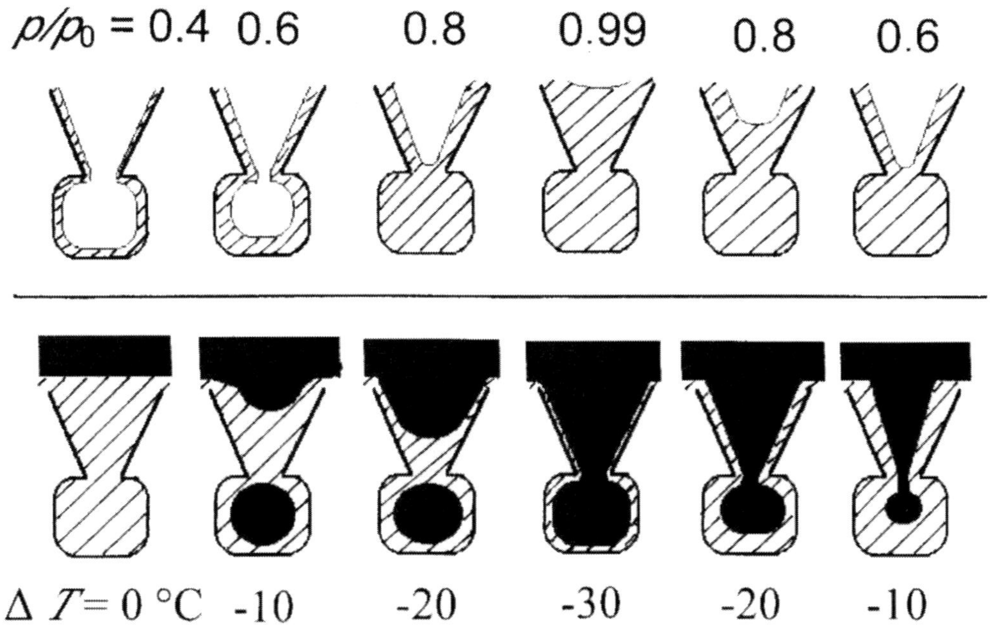

Figure 23. Comparison of gas adsorption (above) and thermoporometric water freezing process (below) in an inkbottle type mesopore. p = gas pressure, p_0 = saturation gas pressure, p/p_0 = relative gas pressure, T = temperature related to ambient temperature.

STANDARDISATION AND CLASSIFICATION

The measuring methods discussed above give empirical results only, based on the procedure of measurement and of the test medium used. The evaluation methods are based mostly on models. Therefore standardisation is inevitable. Besides recommendation of the IUPAC [12] for the sorption method, many of the measuring methods and practical test have been standardised both in the national and international frame [128-132]. In those standards usually simple equations are used for evaluation. Regarding the empirical character of almost all measuring methods discussed above such equations are sufficient for the industrial use, and more sophisticated extensions are somewhat senseless.

A variety of reference materials for the specific surface is available.

A classification of pores with regard to their width has been recommended by the IUPAC. It can be extended to smaller pores, which are not accessible to most molecules, as proposed by Dubinin [89, 133-134]. Another extension should be made in order to sub-arrange the ‘macropores’ as proposed by Setzer [13]. In this way, we obtain a scale covering the whole range of technical importance (Figure 3).

REFERENCES

[1] Knöfel, D. and F. Metz, Korrelation zwischen Porositäten und technologischen Eigenschaften bei Natursteinen, in Forschunsinitiative Morphologie und

Austauschmechanismen por"ser Stoffe (POST). Statusseminar, J.U. Keller, Editor. 1989, Universität Siegen: Siegen. p. 83-103.
[2] Gräf, H. and J. Bonzel, Über den Einfluß der Porosität des erhärteten Betons auf seine Gebrauchseigenschaften. *Zement und Beton*, 1990. 85/86: p. 29-34.
[3] Gräf, J. and J. Bonzel, Über den Einfluss der Porosität des erhärteten Betons auf seine Gebrauchseigenschaften. *Beton*, 1990. 40(7): p. 297-302.
[4] Unger, K.K., Porous Silica. 1979, Amsterdam: Elsevier.
[5] Xu, R., W. Pang, and J. Yu, Chemistry of zeolites and related porous materials: synthesis and structure. 2007, Wiley-Interscience.
[6] Beck, J.S., et al., A New Family of Mesoporous Molecular Sieves Prepared with Liquid Crystal Templates. American Chemical Society 1992. 114(114): p. 10834-10843.
[7] Riley, C.M. and T.W. Rosanke, Development of Analytical Methods. 1996, Amsterdam: Elsevier.
[8] Ellison, S.L.R., M. Rosslein, and A. Williams, eds. Quantifying Uncertainty in Analytical Measurements. 2nd ed. QUAM:2000.P1. 2000, LGC: Teddington.
[9] Green, J.M., A practical guide to analytical method validation. Anal. Chem. News & Features, 1996(5): p. 305 A-309 A.
[10] Meyer, K., P. Klobes, and B. Röhl-Kuhn, Certification of reference material with special emphasis on porous solids. *Cryst. Res. Technol.,* 1997. 32(1): p. 1175-1185.
[11] Dubinin, M.M., Characterization of adsorption properties and microporous structures of carbonaceous adsorbents, in Characterization of Porous Solids, K.K. Unger, et al., Editors. 1988, Elsevier: Amsterdam p. 127-137.
[12] Sing, K.S.W., et al., Reporting physisorption data for gas/solid systems with special reference to the determination of surface area and porosity. (IUPAC recommendations). *Pure & Appl. Chemistry, 1985*. 57(4): p. 603-619.
[13] Setzer, M.J., Zum Mikrogefüge des Zementsteins und dessen Einfluß auf das mechanische Verhalten des Betons. Zement und Beton, 1975. 85/86: p. 29-34.
[14] Wikipedia (2013) Scanning probe microscopy. http://en.wikipedia.org/wiki/Scanning_probe_microscopy.
[15] Binning, G. and H. Rohrer, Scanning tunnelling microscopy IBM Journal of Research and Development, 1986. 38: p. 4.
[16] Binnig, G., Quate, C.F. , and C. Gerber, Atomic Force Microscope. *Physical Review* Letters 1986. 9(56): p. 930-933.
[17] Besocke, K.H., Scanning tunnel microscopy - a new method in surface analysis. Fresenius Z. *Anal. Chem*., 1988. 333(4/5): p. 312.
[18] Leschonski, K., W. Alex, and B. Koglin, Teilchengrößenanalyse. Chemie Ingenieur Technik, 1974. 43: p. 23-26.
[19] Leschonsky, K., Particle size analysis and characterization of classification process., in Ullmann's Encyclopedia of Industrial Chemistry, W. Gerhartz, Editor. 1988, Urban & Schwarzenberg.
[20] Dos Ramos, J.G. and C.A. Silebi, Determination of particle size distribution of submicron particles by capillary hydrodynmaic fractionation (CHDF). *J. Colloid Interface Sci.,* 1988.
[21] Petropoulos, J.H., J.K. Petrou, and N.K. Kanellopoulos, Relative gas permeability properties of model pore networks, in Characterization of Porous Solids, K.K. Unger, et al., Editors. 1988, Elsevier: Amsterdam. p. 401-410.

[22] Carman, P.C., *J. Soc. Chem. Eng.* (London), 1938. 57: p. 225.
[23] Carman, P.C., *J.Soc.Chem.Eng.* (London), 1939. 58: p. 1.
[24] Lea, F.M. and R.W. Nurse, *J. Soc. Chem. Ind.* (London), 1939. 58: p. 277-283.
[25] Rigden, P.J., *J.Soc.Chem.Ind.*, 1943. 62: p. 1.
[26] Blaine, R.L., *Bull.Am.Soc.Test.Mater.*, 1943. 51 p. 123.
[27] ASTM, ASTM C204 - 11 Standard Test Methods for Fineness of Hydraulic Cement by Air-Permeability Apparatus. 2013.
[28] BSI, BS4359:1971 British Standard Method for the Determination of Specific Surface of Powders: Part 2: Air Permeability Methods. 2013: London.
[29] DIN 66126: Bestimmung der spezifischen Oberfläche disperser Feststoffe mit Durchströmungsverfahren - Teil 1: Grundlagen. 1969, Beuth: Berlin.
[30] DIN 66126: Bestimmung der spezifischen Oberfläche disperser Feststoffe mit Durchströmungsverfahren - Teil 2: Verfahren und Gerät nach Blaine. 1989, Beuth: Berlin.
[31] Malek, K. and M.O. Coppens, Knudsen self- and Fickian diffusion in rough nanoporous media. *Journal of Chemical Physics,* 2003. 119(5): p. 2801-2811.
[32] Barrer, R.M., N. McKenzie, and J.S.S. Reay, Capillary condensation in single pores. *J. Colloid. Sci*, 1956. 11: p. 479-495.
[33] Renouprez, A. and J. Avom, Characterization of active carbons from palm-tree fibers using nitrogen adsorption and small angle X-ray scattering, in Characterization of Porous Solids, K.K. Unger, et al., Editors. 1988, Elsevier: Amsterdam p. 49-54.
[34] Krautwasser, P., K. Koilik, and e. al., Application of SAXS in the study of microporosity of graphitic materials for high temperature nuclear reactors. Applied Crystallography onlin, 1974. 7(2): p. 223-224.
[35] Krautwasser, P., *Report J•* 1-1202. 1975, KFA Jülich.
[36] Hoinkis, E., et al., The corrosion induced change of pore size distribution and surface area of fuel element graphite. *Thermochimica Acta*, 1979. 29: p. 345-348.
[37] Hoinkis, E., et al., Corrosion- and irradiation-induced changes of a nuclear graphitic material. *J. Nuclear Materials,* 1986. 141-143: p. 87-95.
[38] Stacey, M.H., A comparison of nitrogen adsorption, small angle neutron scattering and TEM techniques for characterisation of mesoporous solids., in Characterization of Porous Solids., K.K. Unger, et al., Editors. 1988, Elsevier: Amsterdam. p. 55-65.
[39] Fischer, H., Bestimmung der Porenstruktur poröser Adsorbentien und Katalysatoren mit Hilfe der Quecksilberporosimetrie. Dissertation, . 1974, Technische Universität: Darmstadt
[40] van Brakel, J., et al., Mercury porosimetry: *State of the art. Powder Technology*, 1981. 29: p. 1-12.
[41] Connor Jr., W., et al., Measurment of morphology of high surface area solids: Inferring pore shape characteristics, in Characterization of Porous Solids, K.K. Unger, et al., Editors. 1988, Elsevier: Amsterdam. p. 273-281.
[42] Fischer, W. and D. Jahn, Eine Methode zur Ermittlung der Porengrößenverteilung. Meßtechnik 1968. 12: p. 309-313.
[43] Melrose, I.G., Characterization of petroleum reservoir rocks by capillary pressure techniques, in Characterization of Porous Solids, K.K. Unger, et al., Editors. 1988, Elsevier: Amsterdam p. 253-261.

[44] Gregg, S.J. and K.S.W. Sing, Adsorption, Surface Area and Porosity. 2. ed. 1982 / 1984, London / Moscow: Academic Press / Mir.
[45] Sing, K.S.W., Utilisation of adsorption data in the BET region., in Surface Area Determination., D.H. Everett and R.H. Ottewill, Editors. 1970, Butterworths: London. p. 25-42.
[46] Mikhail, R.S. and E. Robens, Microstructure and Thermal Analysis of Solid Surfaces. 1983, Chichester: Wiley.
[47] Rouquerol, F., J. Rouquerol, and K. Sing, Adsorption by Powders & Porous Solids. 1999, San Diego: Academic Press.
[48] Guttman, C.M. and E.A. Di Marzio, Modern size exclusion. Macromolecules 1970. 3: p. 681-691.
[49] Di Marzio, E.A. and C.M. Guttman, Sice exclusion of macromolecules. 1970. 3: p. 131-136.
[50] Kraak, J.C., et al., Hydrodynamic chromatography of macromolecules on 2 μm non-porous spherical silica gel packings. *Chromatogaphia,* 1998. 27(11/12).
[51] Juhola, A.J., Manufacture, pore structure and application of activated carbons. Part I. Kemia-Kemi, 1977. 4(11): p. 543-551.
[52] Juhola, A.J., Manufacture, pore structure and application of activated carbons. Part II. Kemia-Kemi, 1977. 4(12): p. 653-661.
[53] DIN, Partikelmeßtechnik. 4. ed. DIN Taschenbuch/CD. Vol. 133. 2004, Berlin: Beuth.
[54] Frisch, B. and W.-R. Thiele, A measuring method for the determination of density, specific surface area and porosity of powders and compacted bodies powder metallurgy international, 1986. 18(1): p. 17-21.
[55] Robens, E., S.A.A. Jayaweera, and S. Kiefer, *Balances - Instruments, Manufacturers, History.* 2013, Heidelberg: Springer.
[56] Robens, E., et al., Surface area, density and porosity measurements using the magnetic suspension balance. *Fresenius Z. Analyt. Chem.,* 1989. 333 p. 428-432.
[57] Balek, V. and J. Tölgessy, Part C:Radiometric Emanation Methods. *Thermal Analysis.* 1984, Amsterdam: Elsevier.
[58] Binder, H., et al., Elektrochemische Bestimmung der spezifischen Oberfläche und der Aktivität von porösem Platin. *Chemie-Ing.-Techn.,* 1968. 40(3): p. 586-591.
[59] Robens, E. and A. Köhling, Charakterisierung von Brennstoffzellen-Elektroden und Untersuchung der Chemisorption von Kohlenwasserstoffen in Gegenwart von Wasser. *Sprechsaal* 1984. 117 (3): p. 223-227.
[60] Smith, D.M. and D.P. Gallegos, Analysis of pore structure via spin-lattice relaxation measurements, in Characterization of Porous Solids, K.K. Unger, et al., Editors. 1988, Elsevier: Amsterdam p. 391-400.
[61] Jäntti, O., J. Junttila, and E. Yrjänheikki, Mikropunnitusajan Lyhentämisestä Ekstrapolaatiomenetelmällä. (On curtailing the microweighing time by an extrapolation method). Suomen Kemistilehti, 1970. A 43: p. 214-218.
[62] Jäntti, O., J. Junttila, and E. Yrjänheikki, On curtailing the micro-weighing time by an extrapolation method., in Progress in Vacuum Microbalance Techniques, T. Gast and E. Robens, Editors. 1972, Heyden: London. p. 345-353.
[63] Poulis, J.A., et al., A Jäntti approach for quick calculations of sorption equilibria. Z. *Phys. Chem.*, 2002. 216: p. 1123-1135.

[64] Poulis, J.A., et al., General application of Jäntti's method for the fast calculation of sorption equilibrium. *J. Therm. Anal. Cal.*, 2004. 76(2): p. 583-592.

[65] Poulis, J.A., C.H. Massen, and E. Robens, Saving time when measuring BET isotherms. *Journal of Colloid & Interface Science,* 2007. 311(2): p. 391-393.

[66] Poulis, J.A., C.H. Massen, and E. Robens, Measurement of gas adsorption with Jäntti's method using continuously increasing pressure. *J. Therm. Anal. Cal.*, 2002. 68(2): p. 719-725.

[67] Rouquerol, J., et al., A critical assessment of quasi-equilibrium gas adsorption techniques in volumetry, gravimetry or calorimetry., in Characterization of Porous Solids., K.k. Unger, et al., Editors. 1988, Elsevier: Amsterdam. p. 67-76.

[68] Müller, U., et al., High-resolution sorption studies of argon and nitrogen on large crystals of microporous zeolite ZSM-5. Fresenius' Zeitschrift für Analytische Chemie, 1989. 333: p. 433 - 436.

[69] Straube, B., Kritische Untersuchung und Standardisierung physikalisch-chemischer Meßmethoden zur Ermittlung der spezifischen Oberfläche und Porenverteilung von Adsorbentien. 1985, Johannes Gutenberg-Universität: Mainz.

[70] Brunauer, S., Adsorption of Gases and Vapors. 1943, Princeton: Princeton University Press.

[71] Brunauer, S., The Adsorption of Gases and Vapors. Vol. I: Physical Adsorption. Vol. 25. 1945, Princeton Princeton University Press. 353.

[72] Brunauer, S., L. Copeland, and D. Kantro, Theory of Langmuir and Brunauer, Emmett and Teller in The Solid-Gas Interface E.A. Flood, Editor. 1966/1967, Marcel Dekker: New York. p. 77.

[73] Brunauer, S., et al., On a Theory of the van der Waals Adsorption of Gases. *J. Amer. Chem. Soc.*, 1940. 62(7): p. 1723-1732.

[74] Brunauer, S., P.H. Emmett, and E. Teller, Adsorption of gases in multimolecular layers. *J. Am. Chem. Soc.,* 1938. 60(2): p. 309-319.

[75] Aranovich, G.L., Correction of the multimolecular adsoption isotherm. Russian *J. Phys. Chem.,* 1988. 62(11): p. 1561-1566.

[76] Aranovich, G.L., Correction of the multimolecular adsorption isotherm. *J. Phys. Chem.,* 1988. 62(11): p. 1561-1566.

[77] Aranovich, G.L., Correction of the multimolecular adsorption isotherm for limited number of layers. *Russian J. Phys. Chem.*, 1989. 63.

[78] Adolphs, J., Excess surface work - A modelless way of getting surface energies and specific surface areas directly from sorption isotherms. *Appl. Surface Sci.,* 2007. 253: p. 5645-5649.

[79] Adolphs, J. and M.J. Setzer, A model to describe adsorption isotherms. *J. Coll. Interface Sci.,* 1996. 180: p. 70-76.

[80] Adolphs, J. and M.J. Setzer, Energetic classification of adsorption isotherms. *J. Coll. Interface Sci.*, 1996. 184: p. 443-448.

[81] Kutarov, V.V., et al., Adsorption hysteresis in open slit-like supermicropores in XIII Ukrainian - Polish Symposium Theoretical and Experimental Studies of Interfacial Phenomena and their Technological Applications simultaneously with 4th COMPOSITUM conference *Hybrid Nanocomposites and their Applications*. 2012: Kyiw - Ukraine.

[82] Kutarov, V.V., et al., Adsorption measurements at lunar regolith and its evaluation using an extended Pickett equation. *J. Therm, Anal. Cal.*, 2010.
[83] Kutarov, V.V. and E. Robens, The Pickett equation analytical continuation. Adsorption, 2010.
[84] Kutarov, V.V., E. Robens, and B. Kats, Universal function for the description of multilayer adsorption isotherms. *J. Therm. Anal. Cal.*, 2006.
[85] Kutarov, V.V., Y.I. Tarasevich, and E.V. Aksenenko. Similarity theory of the condensation and vaporisation processes in open slit-like capillaries. in Actual problems of chemistry and physics of surfaces. . 2011. Kyiv, Ukraine: *Chuiko Institute of Surface Chemistry of National Academy of Sciences of Ukraine*, Kyiv.
[86] Broekhoff, J.C.P. and J.H. De Boer, The surface area in intermediate pores., in Surface Area Determination, D.H. Everett and R.H. Ottewill, Editors. 1970, Butterworths: London. p. 97-121.
[87] Broekhoff, J.C.P. and B.G. Linsen, in Physical and Chemical Aspects of Adsorbents and Catalysts, B.G. Linsen, Editor. 1970 / 1973, Academic Press / Mir: London / Moscow.
[88] Dubinin, M.M., The potential theory of adsorption of gases and vapours for adsorbents with energetically nonuniform surfaces. *Chem. Rev.*, 1969. 60: p. 235.
[89] Dubinin, M.M., Microporous structures of carbonaceous adsorbents, in Characterisation of Porous Solids, S.J. Gregg, K.S.W. Sing, and H.F. Stoeckli, Editors. 1979, Soc. of Chemical Industry: London p. 1-11.
[90] Kaganer, M.G., M.G. Kaganer, Zhur. Fiz. Khim. / *Russ J. Phys. Chem.*, 1959. 33: p. 352, 2202-2210.
[91] Kaganer, M.G., Dokl. *Akad. Nauk.*, 1961. 138: p. 405.
[92] Gottwald, B.A., Comparison of monolayer capacities from BET and Dubinin-Kaganer-Radushkevich (DKR) isotherms, in Surface Area Determination, D.H. Everett and R.H. Ottewill, Editors. 1970, Butterworths: London p. 59-61.
[93] Mikhail, R.S., S. Brunauer, and E.E. Bodor, J. Colloid Interface Sci., 1968. 26: p. 45.
[94] Mikhail, R.S. and El-Akkad, *J. Colloid Interface Sci.*, 1975. 51: p. 260.
[95] Karnaukhov, A.P., Simulation of porous materials, in Characterisation of Porous Solids, S.J. Gregg, K.S.W. Sing, and H.F. Stoeckli, Editors. 1997, *Society of Chemical* Industry: London. p. 301-311.
[96] Everett, D.H., *Capillary condensation., in Characterisation of Porous Solids*, S.J. Gregg, K.S.W. Sing, and H.F. Stoeckli, Editors. 1979, Soc. Chemical Industry: London. p. 229-251.
[97] Everett, D.H., *Pore systems and their characteristics Characterization of Porous Solids, in Characterizsation of Porous Solids*, K.K. Unger, et al., Editors. 1988, Elsevier: Amsterdam p. 1-21.
[98] Schröder, J., Nachweis und Bedeutung von Mikroporen. Teil2. GIT Fachz. Lab., 1986: p. 1095-1102.
[99] Carrot, P.J.M., et al., Characterisation of microporous carbons by n-nonane preadsoption. *Chem. & Ind.*, 1988(6): p. 371-372.
[100] Carrot, P.J.M., R.A. Roberts, and K.S.W. Sing, A new method for the determination of micropore size distribution, in *Characterization of Porous Solids*, K.K. Unger, et al., Editors. 1988, Elsevier: Amsterdam p. 89-100.

[101] Carrot, P.J.M. and K.S.W. *Sing, Assessment of microporosity, in Characterization of Porous Solids*, K.K. Unger, et al., Editors. 1988, Elsevier: Amsterdam p. 77-87.
[102] Mandelbrot, B.B., The Fractal Geometry of Nature. 1977/1983, New York: Freeman.
[103] Avnir, D., ed. *The Fractal Approach to Heterogeneous Chemistry*. 1989, Wiley: Chichester.
[104] Neimark, A.V., Determination of surface fractal dimension from adsorption experimental data. *Russian J. Phys. Chem.,* 1990. 64(10): p. 2593-2605.
[105] Farin, D.D. and D. Avnir, The fractal nature of molecule-surface interactions and reactions, in *The Fractal Approach to Heterogeneous Chemistry*, D. Avnir, Editor. 1989, Wiley: Chichester. p. 271-293.
[106] Farin, D.D. and D. Avnir, The fractal nature of molecule-surface chemical activities and physical interactions in porous materials, in *Characterization of Porous Solids*, K.K. Unger, et al., Editors. 1988, Elsevier: Amsterdam p. 421-432.
[107] Padday, J.F., *Measurement of surface area by the adsorption of dyes from solution, in Surface Area Determination*, D.H. Everett and R.H. Ottewill, Editors. 1970, Butterworths: London p. 331-340.
[108] Walter, G. and E. Robens, Characterization of porous materials, in particular Raney Platinum., in *The Physical Basis for Heterogeneous Catalysis.*, E. Drauglis and R.I. Jaffee, Editors. 1975, Plenum Press: New York. p. 363-371.
[109] Walter, G., E. Robens, and E. Zimmer, Determination of specific surface area and porosity of solids, in *Methodicum Chimicum, Analytical Methods*, F. Corte, Editor. 1974, Academic Press: New York p. 678-689.
[110] Kipling, J.J. and D.A. Tester, Adsorption from binary mixtures of non-electrolyte. *Nature* 1951. 167(4250): p. 612-612.
[111] Kipling, J.J., *Adsorption from Solutions of Non-Electrolytes*. 1965, London: Academic Press.
[112] Kipling, J.J. and D.B. Peakall, *J. Chem. Soc.*, 1956: p. 4828.
[113] Kipling, J.J. and E.H.M. Wright, *J. Chem. Soc.*, 1962: p. 855.
[114] Schay, G., On the use of adsorption measurements from solutions of non-electrolytes for surface area determination, in Surface Area Determination, D.H. Everett and R.H. Ottewill, Editors. 1970, Butterworths: London p. 273-289.
[115] Unger, K.K., G. Kreysa, and J.P. Baselt, eds. *Characterization of Porous Solids. Studies in Surface Science and Catalysis*. 2000, Elsevier Science: Amsterdam.
[116] Hagel, L., Pore size distribution, in Aqueous Size-Exclusion Chromatography, *J. of Chromatography*, P.L. Dubin, Editor. 1988, Elsevier: Amsterdam p. 149-155.
[117] Wang, Y., et al., A theoretical study of the separation principle in Size Exclusion Chromatography. Macromolecule 2010. 43: p. 1651-1659.
[118] Lathe, G.H. and C.R. Ruthven, The separation of substances on the basis of their molecular weights, using columns of starch and water. ():xxxiv. *Biochem J.* , 1955. 60(4): p. 34.
[119] Lathe, G.H. and C.R. Ruthven, The separation of substances and estimation of their relative molecular sizes by the use of columns of starch in water. *Biochem. J.*, 1956. 62(4): p. 665-674.
[120] Kuga, S., *J. Chromatography*, 1981. 206 p. 449-461.

[121] Kuga, S., Structural analysis of porous materials by measurement of size exclusion, in Aqueous Size-Exclusion Chromatography, *J. of Chromatography*, P.L. Dubin, Editor. 1988, Elsevier: Amsterdam. p. 157-190.

[122] Rouquerol, J., et al., Information brought by adsorption and immersion calorimetry to follow the porous structure of sepiolite during its thermal treatment, in Characterization of Porous Solids, K.K. Unger, et al., Editors. 1988, Elsevier: Amsterdam p. 317-322.

[123] Brun, M., et al., A new method for the simultaneous determination of the size and the shape of pores: The thermoporometry. *Thermochimica Acta,* 1977. 21: p. 59-88.

[124] Eyraud, C., J.F. Quinson, and M. Brun, The role of thermoporometry in the study of porous solids., in *Characterization of Porous Solids*., K.K. Unger, et al., Editors. 1988, Elsevier: Amsterdam. p. 295-305.

[125] Quinson, J.-F., M. Astier, and M. Brun, Determination of surface areas by thermoporometry. Applied Catalysis, 1987. 30: p. 125-130.

[126] Quinson, J.-F. and M. Brun, Progress in thermoporometry., in Characterization of Porous Solids., K.K. Unger, et al., Editors. 1988, Elsevier: Amsterdam. p. 302-315.

[127] Experimental investigation of factors affecting accuracy and resolution of pore structure of cement based materials by thermoporometry. *Journal of Zhejian University* - Science, 2013.

[128] DIN, DIN 66139 - Porengrößenanalyse - Darstellung von Porengrößenverteilungen. 2009, Berlin: Beuth.

[129] Robens, E. and K.-F. Krebs, Standardization, Reference Materials and Comparative Measurements for Surface Area and Pore Characterization in Characterization of Porous Solids II., F. Rodr¡guez-Reinoso, et al., Editors. 1991, Elsevier: Amsterdam p. 133-140.

[130] Robens, E., et al., Standardization of sorption measurements and reference materials for dispersed and porous solids, in Adsorption and its Application in Industry and Environmental Protection. Vol. 1: Application in Industry, A. Dabrowski, Editor. 1999, Elsevier: Amsterdam. p. 95-116.

[131] Dąbrowski, A., Adsorption and its Application in Industry and Environmental Protection. Surface Science and Catalysis, vol. 120 A/B. 1999, Amsterdam: Elsevier.

[132] Dąbrowski, A., et al., Standardization of methods for characterizing the surface geometry of solids. *Particle & Particle Systems Characterization*, 2003. 20(5): p. 311-322.

[133] Dubinin, M.M., *J. Colloid Interface Sci.* , 1967. 23 p. 487.

[134] Dubinin, M.M., *Advances Colloid Interface Sci.*, 1968. 2: p. 217.

In: Comprehensive Guide for Mesoporous Materials, Volume 2 ISBN: 978-1-63463-982-8
Editor: Mahmood Aliofkhazraei

Chapter 2

TWENTY YEARS OF FUNCTIONALIZED MESOPOROUS SILICA

Ahmad Mehdi*

Institut Charles Gerhardt Montpellier, Chimie Moléculaire et Organisation du Solide, Université Montpellier, Montpellier, France

ABSTRACT

This chapter treats of the organization and functionalization of hybrid organic-inorganic mesoporous materials. These materials based on silica matrix were prepared by sol-gel process, which consists on the hydrolysis and polycondensation of silylated molecular precursors in the presence of a structure-directing agent. We described methods to functionalize the channel pores of ordered mesoporous silica, this of the silica framework as well as the functionalization of both of them simultaneously. Thanks to chemical modifications, we evidenced not only the accessibility but also the location and the regular distribution of the introduced organic units. This family is currently one of the best supports for exploring polyfunctional materials, which can provide a route to original nanomaterials.

INTRODUCTION

The science of nanomaterials represents a large research domain in chemistry [1]. In this area, one of the major challenges is the finding of the ways, which permit to organize the material at nanometric scale. For this purpose, the bottom-up approach, which consists in building up of nanomaterials starting from molecular brick, is particularly appropriate for chemists. In fact, the high precision in the design of nano-objects such as molecule, clusters, oligomers, complexes nanoparticles, aggregates, etc. reached by the molecular chemistry, opens a wide way to the preparation of a large number of elemental bricks with specific property.

* Tel: 33 4 67 14 38 32; Email: ahmad.mehdi@univ-montp2.fr.

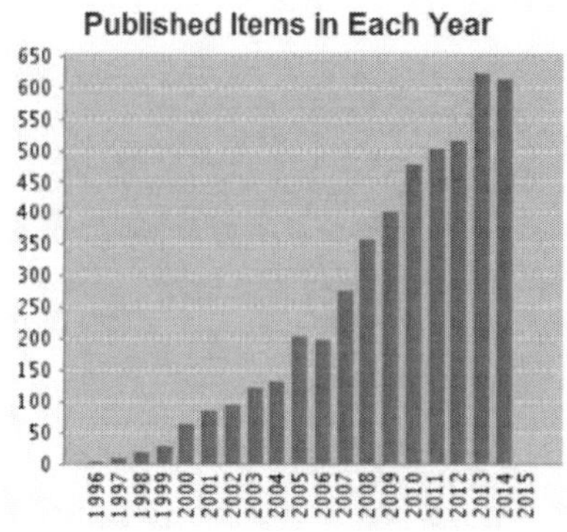

Figure 1. Evolution of publication number (November 2014) concerning functionalized mesoporous silica.

FUNCTIONALIZATION OF THE CHANNEL PORES

Two main approaches have been used to anchor organic groups onto the inner pore surface of mesoporous silicas: grafting method (post-synthesis) and the so-called co-condensation direct synthesis method. Post-synthesis grafting of an organotrialkoxysilane RSi(OR')$_3$ onto the pore surface of the mesoporous silica was the first method established for functionalization of the pore surface (Scheme 2) [15]. This method is generic and allows the incorporation of many R groups including bulky ones. However, neither the loading control nor the distributions of the functional groups are possible by this route [16], as they depend on several parameters such as the number of the residual silanol (SiOH) groups at the surface as well as their accessibility.

The direct synthesis, consists in the copolymerisation of silica precursor such as tetraethylorthosilicate (TEOS) and an organotrialkoxysilane RSi(OR')$_3$ in the presence of a structure-directing agent (Scheme 3) [17]. This one-step method, more challenging and also more attractive than the grafting one is often preferred. Indeed, it is allows to obtain a regular R group's distribution inside the channel pores as well as the loading control within the limits of the content supported by the micelle [18].

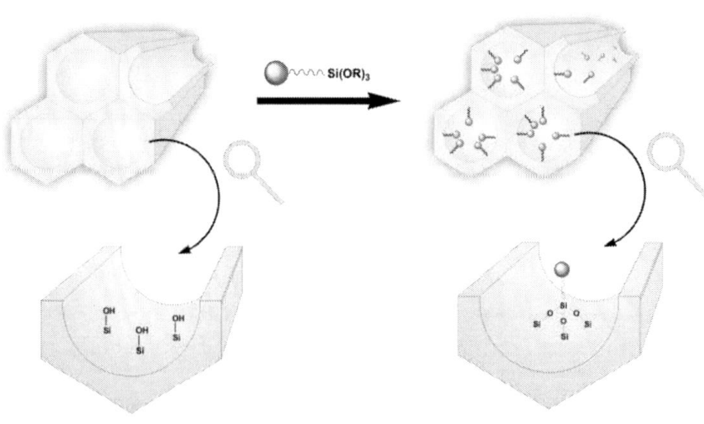

Scheme 2. Functionalization of mesoporous silica by grafting.

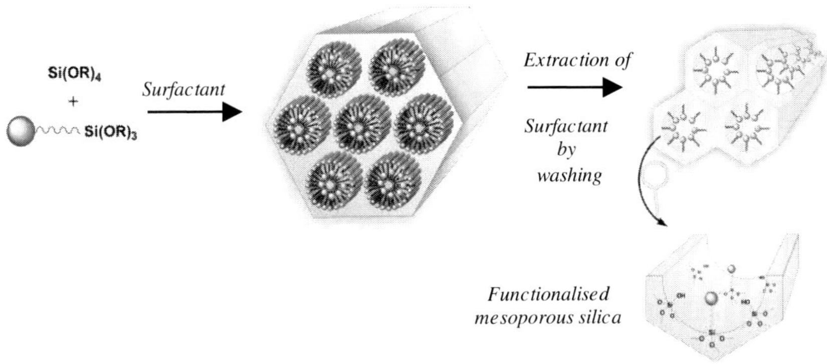

Scheme 3. Direct synthesis of functionalized mesoporous silica.

However, this method has also its own limitations. Thus, the organic R group has to be sufficiently lipophilic to enter the core micelle and not too bulky to avoid bursting. Furthermore, if the functional group is rather hydrophilic, the functionalization could occur partially within the framework.

The first used structure directing agent was a cationic surfactant such as cetyltrimethylammonium bromide (CTAB) [19]. Shortly after, the non-ionic assembly route was investigated by using primary alkylamine [20] and then the triblock copolymer as $EO_{20}PO_{70}EO_{20}$ [poly(ethylene oxide)-poly(propylene oxide)-poly(ethylene oxide)] called pluronic P123 was used very often [21]. This last surfactant was proved to be very convenient under acidic conditions for the preparation of functionalized silica with hexagonal structure. The resulting materials have generally large uniform pore sizes and high surface areas. Furthermore, they displayed also high thermal and hydrothermal stabilities, which is very important for the expected applications. Thank to the direct synthesis approach, a large variety of organic groups have been introduced with uniform distribution on the surface of the channel pores. In Table 1, some examples are given. It is worth noting that the molar % in organic group can be tailored from 2 to 10% and in some cases the organic amount can even reach 20% without destroying of the structure of the final material. Last year, A. Walcarius and L. Mercier were reported a nice review concerning the use of functionalised mesoporous materials for removal of organic and inorganic pollutants [22].

Table 1. Some examples of pore's functionalized mesoporous silica

Functionalized mesoporous material	Σ	ref
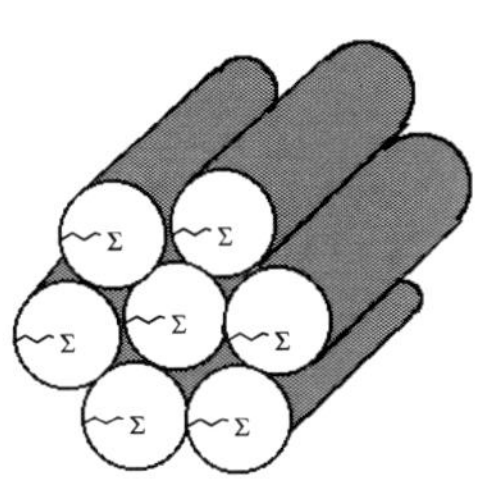	PPh_2	[18,20]
	$PO(OEt)_2$	[23]
	Cl	[24]
	SH	[21,25,26]
	NHBoc, NH_2	[27,28]
	CN, COOH	[26, 28, 29, 30]
	Acetyl	[31]
	SO_3H	[32]
	Imidazol	[33]
	Cyclam	[34]
	I	[35]

Organic groups located within the channel pores of mesoporous silica are easily accessible and can thus undergo further chemical transformations quantitatively without disrupting the ordered structure. That is an important point as it corresponds to an extension of the range of hybrid materials originating from the direct synthesis.

Thus, iodopropyl groups can undergo easily nucleophilic displacements and that much faster than the chloro groups. This property renders ordered mesoporous silica containing iodopropyl pendant groups [35] very useful for subsequent introduction of organic groups. Some post-modifications are reported in Scheme 4. Thanks to the iodo groups, it was possible to introduce within the inner pores of mesoporous material, bulky and/or hydrophilic functional groups difficult to attach by other routes. A transformation, which deserves to be noted, concerns the treatment of the material with triethylphosphite involving the introduction of phosphonate groups quantitatively within 4 h under reflux through the Arbuzov reaction (Scheme 4). In contrast, under the same experimental conditions, treatment of the corresponding material containing chloro group instead of iodo group afforded only 7% of transformation. It was the first time that an Arbuzov reaction was achieved within a mesoporous material, thanks to the iodo group, avoiding thus the tedious preparation of a trialkoxysilylalkyldiethylphosphonate.

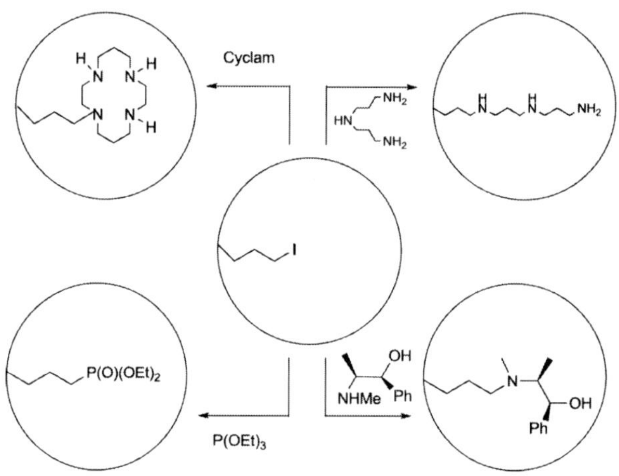

Scheme 4. Post-modification of iodopropyl functionalized mesoporous silica.

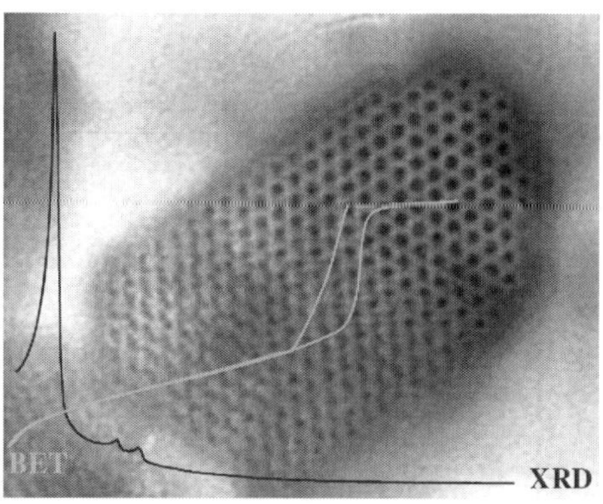

Figure 2. N_2 adsorption-desorption isotherm (BET), X-Ray Diffraction pattern (XRD) and TEM image for mesoporous material containing COOH groups.

Phosphonic and carboxylic acid groups were introduced by acidic hydrolysis of phosphonate and cyano groups respectively in highly concentrated media. After this hard treatment, the materials remained highly structured without damage of the porosity (Figure 2). As a result for example, Mn_3O_4 nanoparticles with a narrow size distribution could be prepared thanks to the mesoporous material containing carboxylic acid pendant groups (*vide infra*). Regular immobilisation of Co-fluomine onto ordered mesoporous silica has been achieved through coordination of the cobalt to imidazole groups[33] covalently attached to the silica matrix with the aim to prepare materials convenient for dioxygen uptake. For this study both the routes (*i.e.* grafting and direct synthesis) have been investigated to obtain mesoporous hybrid materials containing coordinating ligands. It was shown that, the materials prepared by the second route are the most convenient for oxygen uptake.

Direct introduction of photoluminescent ions (Eu^{III}) within mesoporous silica was successfully achieved through the chelating phosphine oxide groups resulting-themselves of the quantitative oxidation of phosphino groups anchored within the material by direct synthesis [36].

The regular distribution of organic groups within mesoporous materials, which is one of the key points concerning the materials prepared in one step, was well demonstrated in several cases [32]. For that purpose, in all cases, reactions depending on the proximity of two organic groups were used as probes. As an example, to elucidate the distribution of the sulfonic acid (SO_3H) groups on the surface of the pore channels, we used the ion-exchange reaction between the proton of SO_3H groups and NaCl. It is worth noting that SO_3H groups originated from the oxidation of SH groups within the material. The ion-exchange capacity of SO_3H groups was incomplete (57%) for the material containing 14 molar % of SO_3H units. This ion-exchange capacity increased to become 65% for the material with 10 molar % of SO_3H units while it was quantitative when only 5 molar % of SO_3H groups were introduced in the pores.

The changes in acid character of the SO_3H groups in these materials can be explained by the corresponding changes in proximity of the SO_3H moieties on the pores' surface. In the case of the material with 14 molar % of SO_3H groups, the observed partial exchange is attributed to the stabilization of nonionized groups by adjacent ionized group which prevents the ion-exchange reaction to occur. This should be due to the high number of sulfonic acid groups (2.03 SO_3H/nm^2) in close proximity to each other on the pores' surface in this material (See scheme 5 left). In contrast, in the case of 5 molar % of SO_3H groups, the SO_3H groups react as strong acids, as in a dilute solution situation. This means that the SO_3H groups must be sufficiently isolated (0.66 SO_3H/nm^2) (See scheme 5 right) to behave independently the ones from the others. Therefore, the exchange reaction becomes complete. The dramatic increase in the rate of ion-exchange as the concentration in SO_3H decreases is a good indicator of the regular distribution of SO_3H groups on the pores' surface.

Mesoporous materials functionalized with various groups regularly distributed within the pores were proved to be useful as nanoreactors for the growth, stabilization and organization of nanoparticles with controlled size [25,30,31,37,38]. This approach constitutes a promising alternative for preparing nanocomposite materials. The procedure was generally based on metallic precursor impregnation onto the mesoporous organosilica with subsequent reduction/decomposition of the metal precursor to generate the nanoparticles. In this approach, the nature of the functional groups, as well as the location and loading play a crucial role.

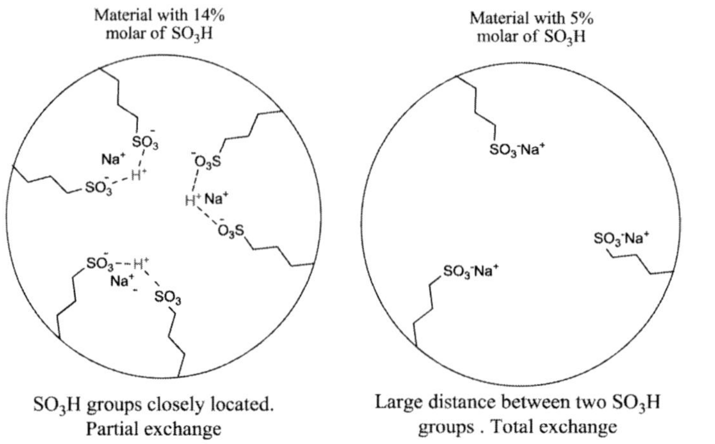

Scheme 5. Schematic representation of the distribution of SO_3H groups after NaCl treatment.

Figure 3. TEM image of material containing Fe_3O_4 nanoparticles.

Thanks to the appropriate functional groups incorporated within the channel pores of the mesostructured materials, the anchoring of the metal precursor then occurs exclusively within the internal surface of the material by strong interactions between the metal and the functional groups avoiding the growth of nanoparticles on the external surface, which would involve a broad particle size distribution. By changing the functional group/metal precursor ratio, it is possible to control the metal loading as well as the nanoparticle size. That renders this approach of particular interest as it is well known that the properties of the nanoparticles are dependent on their size. As examples, mercaptopropyl-functionalized mesoporous silicas were effective hosts for the stabilization of gold (0) nanoparticles with a narrow pore size distribution directly correlated to the pore size [25]. The formation of Mn_3O_4 nanoparticles was obtained by thermolysis of the magnetic cluster $[Mn_{12}O_{12}(C_2H_5COO)_{16}\text{-}(H_2O)_3]$ anchored to mesoporous silica functionalized with -CO_2H groups [30]. Acetylacetonatopropyl groups, introduced into the pore channels of silica by direct synthesis, allowed the growing and organisation of Fe_3O_4 nanoparticles (Figure 3) giving rise to magnetic silica [31]. By using the phosphonate groups, indium oxide nanorods as well as Ruthenium and Ruthenium oxide nanoparticles were stabilized with controllable sizes exclusively inside the pores with a homogeneous dispersion throughout the host structure.While ordered mesoporous materials were often used for applications in catalysis [39, 40]. It is worth noting that two heterogenous catalysts operating as single site were prepared thanks to functionalized mesoporous silica functionalized by direct synthesis. The first one was Ir-N-heterocyclic material which displayed a high activity in H/D-exchange reaction of acetophenone with methanol-d_4. Noteworthy, this catalytic material has catalytic performances similar to homogeneous homologues with the advantage of being easily recyclable [41]. The second is Ru-NHC material which is also an active and reusable heterogeneous catalyst for metathesis of functionalized olefins (Scheme 6) [42]. These catalysts were fully characterized by NMR solid state [43] and display high activity (TOF) and stability (TON vs. time, recycling and leaching). The versatility of such synthetic methodology and its transfer to various metals and ligands (including sensitive complexes) is a very promising approach towards a wide range of tailored made well-defined heterogeneous catalysts.

Applications of functionalized mesoporous silica are dependent not only upon the location, the amount and the accessibility of the functional groups attached in the inner pores, but also the structure of the material.

As we showed, a large variety of functionalized mesoporous 2-D hexagonal silicas have been obtained by direct synthesis method. In contrast, the number of long-range order cage-like cubic mesoporous silicas functionalized with pendant organic groups currently reported is notably inferior whatever the non ionic [44] or ionic nature of the surfactant [45]. This is probably due to stricter experimental conditions than those giving rise to the corresponding functionalized 2-D hexagonal mesoporous silicas. In addition, removing the template from functionalized cage-like mesostructures when triblock copolymers are used as structure-directing agent is still a challenge [44]. Yet, this family of ordered mesoporous materials is fascinating. Indeed, they exhibit a 3-D network of nearly spherical mesopores interconnected through small windows forming a multidirectional system, which should facilitate mass transfer and restrict pore blockage. A recent study has shown that the tridimensional cage structure of SBA-16 allows a much more rapid diffusion of heptane, cumene and mesitylene than the straight cylindrical pore shape of SBA-15 [46].

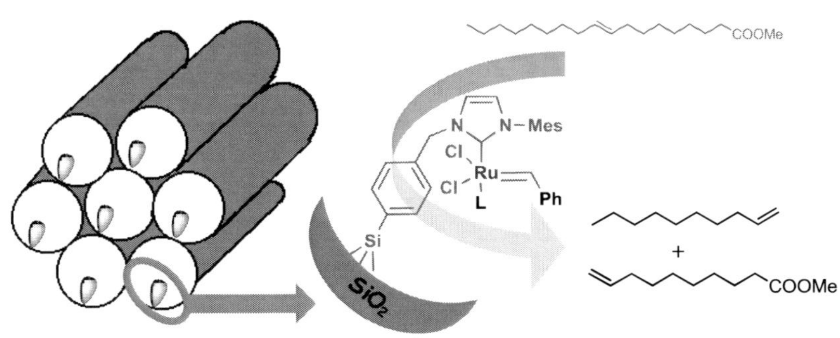

Scheme 6. Hybrid mesoporous silica as heterogeneous catalyst.

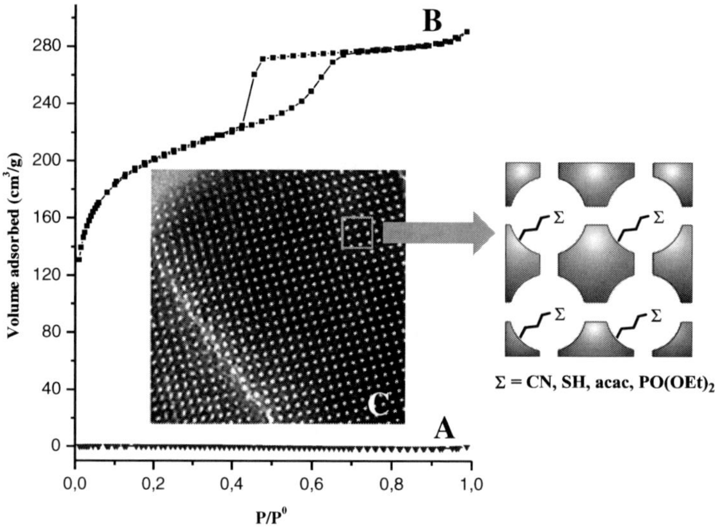

Figure 4. N_2 adsorption-desorption isotherms before (A) and after elimination of the surfactant (B) and TEM image (C) of material containing phosphonate groups.

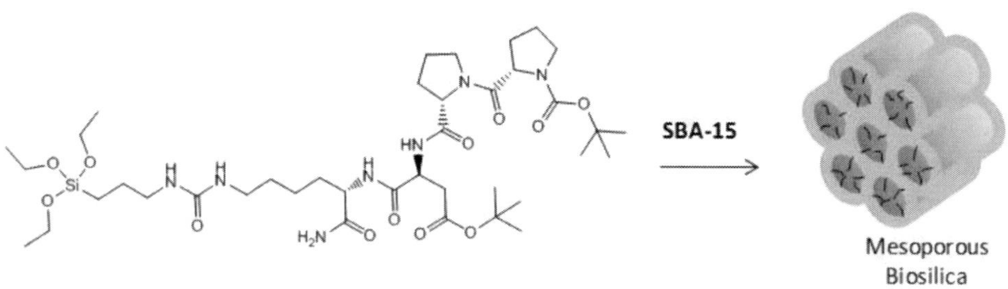

Figure 5. Preparation of mesoporous biosilica by grafting.

These features render these materials particularly attractive for many chemical or physical applications such as catalysis, separation of molecules, gas sensing or 3-D ordered nanoparticles growth.

In 2010, a generic route to functionalized, surfactant-free and well-ordered mesoporous silicas with cubic symmetry was successfully obtained by our group *via* the direct synthesis approach [47]. Several types of organic groups (CN, SH, $CH_3COCHCOCH_3$ (acac) and $PO(OEt)_2$) with different contents (3, 5 and 10% molar) have been introduced with success. Furthermore, an efficient method for complete template removal without any alteration of the neither organic moieties nor structure collapse was described (Figure 4). In addition, the accessibility and adsorption capacity of acetylacetonate groups towards lanthanide ions were also demonstrated. All combined results make these functionalized surfactant-free 3-D cubic mesoporous silicas as significant candidates in many various chemical or physical applications such as catalysis, separation of molecules, gas sensing or 3-D ordered metallic nanoparticles growth [48].

Recently, mesoporous biosilica containing peptide units have been reported. Thus, different homopeptides with controlled size were introduced in the pores channels by polymerization of various amino acid *N*-carboxyanhydrides (NCAs) thanks to amine functions [49, 50]. More interesting, the synthesis of peptide-bioorganic hybrid silica materials was achieved by grafting of hybrid peptides bearing a trialkoxysilane group. These reactive monomers are covalently attached to silica without requiring any prior surface modification. Figure 5 shows the grafting of silylated peptide sequence Pro-Pro-Asp-Lys [51].

FUNCTIONALIZATION OF THE FRAMEWORK

During the last decade, the periodic mesoporous organosilicas (PMO's) have constituted a real advance in materials science. In 1999, several groups reported independently the synthesis of a new class of mesoporous organic-inorganic hybrid materials formed by hydrolytic polycondensation of bridged organosilica precursors of general formula $[(RO)_3Si]_mR'$ ($m \geq 2$) in the presence of a structure-directing agent [52,53,54,55]. The so-called periodic mesoporous organosilicas (PMO's) were novel in that the organic fragments were located into the framework of mesoporous silica through covalent Si-C bonds (Scheme 7).

In these materials, the density of organic groups was very high, which opened an extraordinary field of investigation, as the properties of the framework (optical, electronic and magnetic) may be tailored by changing the nature of the bridging organic group R'. The first studies were focused on the formation of highly ordered PMO's by using essentially short bridging groups [56,57]. PMO materials containing bridging phenylene rings deserve a special mention as they can exhibited a hexagonal array of mesoporous and crystal-like pore walls, due to the π-π stacking of phenylene bridging groups [57].

Shortly after, efforts were made to introduce in the framework bridging groups able to induce chemical or physical (optical, electronic, magnetic) properties [57]. That was not an easy task and remains a challenge. In all cases, it supposes to be able to synthesize the appropriate molecular precursor, and then to find the right experimental conditions leading to structured materials.

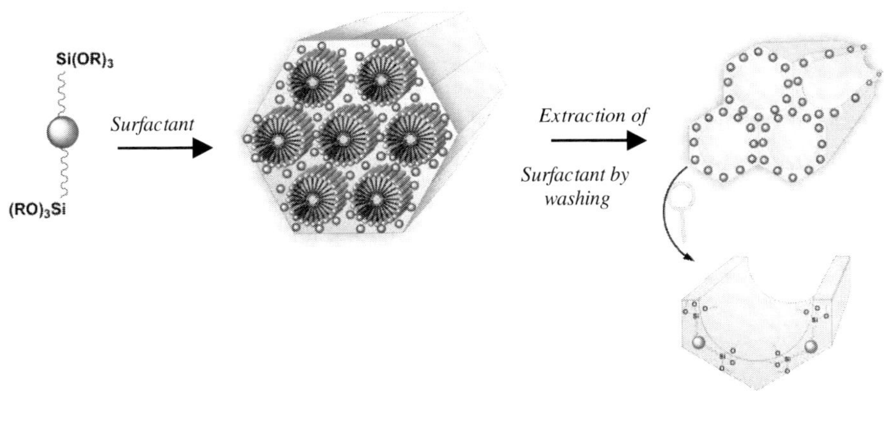

Scheme 7. Synthesis of framework functionalized mesoporous silica.

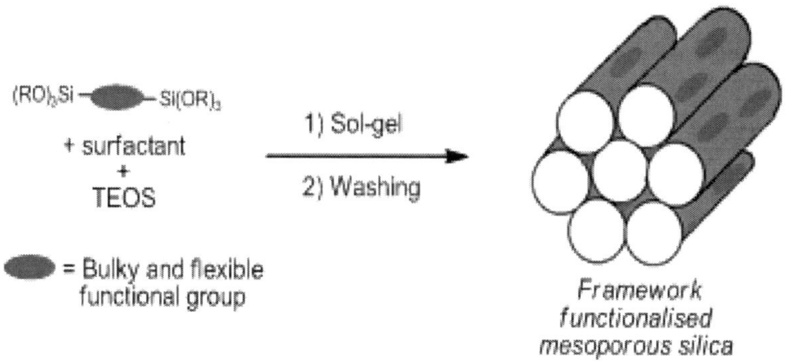

Scheme 8. Synthesis of framework functionalized mesoporous silica by co-condensation.

In addition, it worth noting that all the precursors of general formula $[(RO)_3Si]_mR'$ ($m \geq 2$) cannot give rise to periodic mesoporous materials. That could be due either to the lack of rigidity of some of them or/and to their bulky nature or/and to their too poor hydrophilic character [58,59]. All these parameters do not facilitate or even prevent the mesoscopic order, which restricts considerably the type of organic groups able to be embedded within the framework of ordered mesoporous materials. An approach allowing to overcome some of the difficulties to get PMOs consists in the preparation of materials by co-condensation of at least an organosilane of general formula $[(RO)_3Si]_mR'$ ($m \geq 2$) with mainly TEOS [60,61]. This method permits to expend not so much organosilane precursor and facilitate the mesoscopic order with keeping the location of organic groups in the framework but on condition that the organic groups are not too much hydrophobic (Scheme 8).

A large number of papers concerning materials prepared in this way were reported due to the great interest of the functionalization of the framework to tailor the properties of mesoporous silica [62,63]. Wang and co-workers were reported a review treating the multidisciplinary applications of PMO materials in the domain of chemistry, physics, materials science and engineering, biology and medicine [64].

Table 2 summarizes some examples of framework's functionalized mesoporous silica. PMO's obtained by co-condensation of TEOS as silica precursor and bridged organic precursors with heterocyclic bridging groups [65], chelating entities [66, 67], chiral groups [68,69], π-conjugated segments [70, 71], ruthenium complex such as bis(2,2'-bipyridine) {4,4'-bis {2-[3-(triethoxysilyl)propylthio]-ethyl}-2,2'-bipyridine} ruthenium (II) dichloride [72], rhodium complex [73], europium complex [74,75], ionic liquid [76], and melamine units [77], light harvesting antenna [78.79], Chromatography [63], fluorescent bridging group and their applications [80] deserve to be specially mentioned.

Hybrid materials containing chelated cations are of interest, as such systems can present physical properties (optic, electric or magnetic for example), depending on the nature of the salt. Thus, transition metal and lanthanides ions can provide such physical properties. Cyclam (1,4,8,11-tetraazacyclodecane) being well known for its remarkable binding ability towards transition metal ions [88], we established the experimental conditions leading to the incorporation of cyclam units within the silica framework by co-hydrolysis and polycondensation of TEOS and the tetrasilylated cyclam derivative in the presence of a structure-directing agent [61].

Table 2. Some examples of mesoporous silica functionalized in the framework

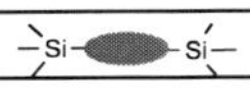	ref	Functionalized		ref
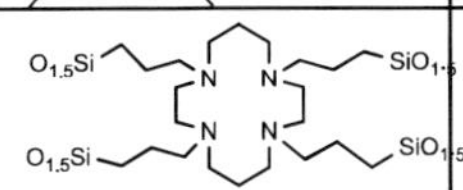	[81, 82]	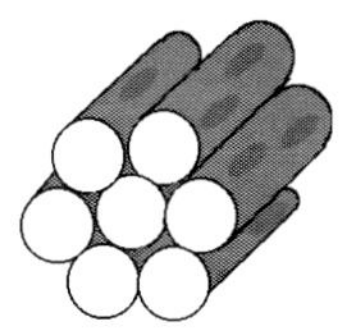	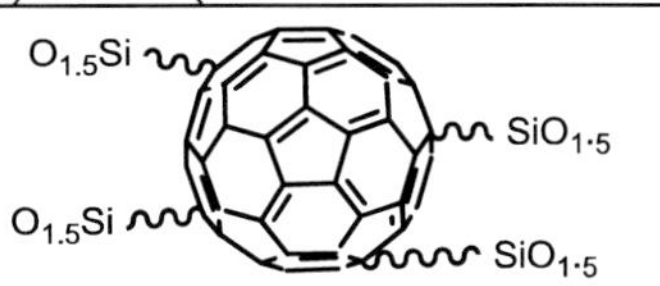	[82]
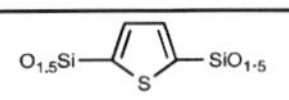	[83]		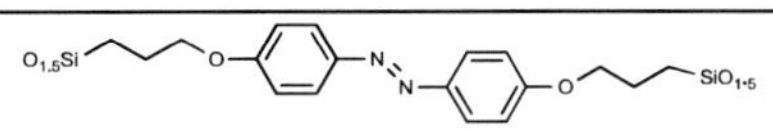	[84]
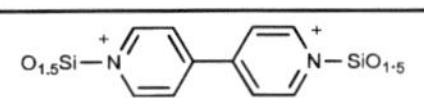	[85]		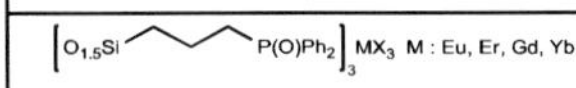	[86]
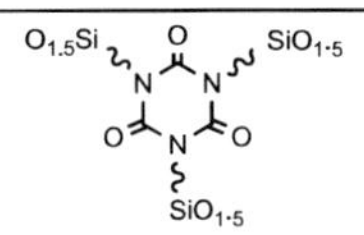	[65]		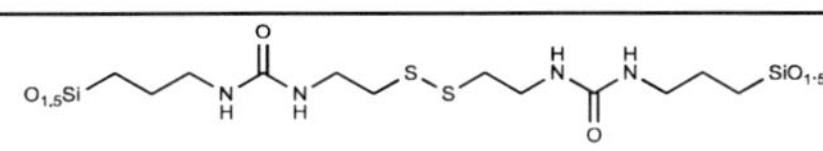	[87]

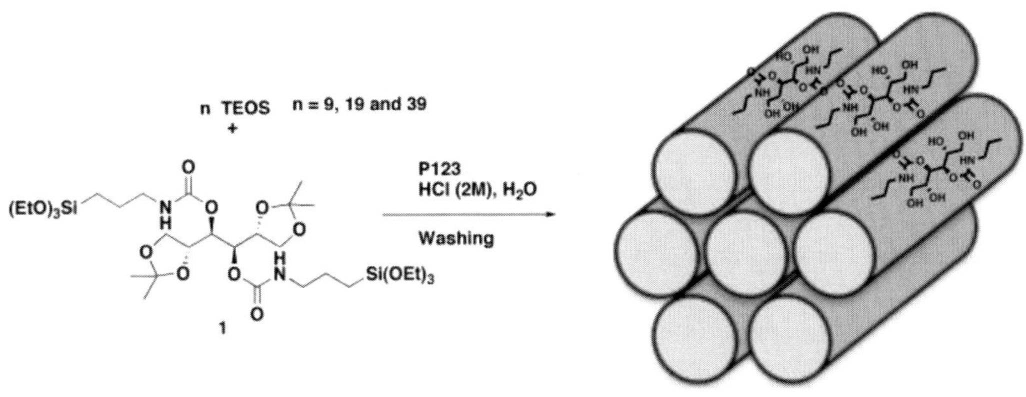

Mannitol functionalized framework

Scheme 9. One-step synthesis of mannitol functionalized framework.

This year, it was reported the preparation of mesoporous silicas functionalized with mannitol derivative in the framework [89]. These materials were obtained by co-hydrolysis and polycondensation of a bis-silylated mannitol precursor containing protected OH groups with tetraethylorthosilicate (TEOS) using the triblock co-polymer P123 as structure-directing agent. It was demonstrated that during the synthesis, the deprotection of the 1,2,5,6 OH functional groups was achieved (Scheme 9). These new functionalized mesoporous silicas with chiral framework could be very interesting for future applications in separation (ions, enantiomeric...) and as catalytic materials for asymmetric synthesis.

The determination of the location of organic groups (pores or framework) is a problem which sets by using the co-condensation approach, problem for which there is not a standard answer as the location of the organic groups depends on the nature of the molecular precursor of formula $[(RO)_3Si]_mR'$ ($m \geq 2$). Indeed, if the molecular precursor is too much hydrophobic, the R' groups can be located within the pores channels instead of being embedded in the pores walls in spite of the number of SiOR groups attached to the organic moieties [90,91]. Thus, the materials prepared from bis[3-(triethoxysilyl)propyl]tetrasulfide (TESPTS) illustrate the difficulties, which can be encountered starting from bridged silsesquioxanes: Li and co-workers [90] synthesized MCM-41-type mesoporous organosilicas with tetrasulfide ''embedded in the framework'' via the co-condensation of TESPTS and tetramethoxysilane (TMOS). They noticed that mesoporous organosilicas synthesized with TESPTS :TMOS molar ratios in the range 1: 8 to 1: 3 have extensive structural defect holes in the nanochannels.

Two years later, they reported [92] the preparation of an organosilica by co-condensation of TESPTS and TMOS in the presence of P123 as a structure-directing agent. By changing the TESPTS :TMOS molar ratio, they observed that TESPTS can act as both a swelling agent and a co-agent for the formation of a microemulsion with surfactant micelles, leading to mesophase transformations. This suggests that the location of the tetrasulfide moieties occurred in the core of the micelles. In another study reported [91] at the same time, the location of tetrasulfide moieties was depicted in the pore channels. These examples highlight the difficulty of determining the location of some kinds of functional groups within mesoporous organosilica.

This question was recently examined at the same time by Chen and co-workers [93] and by our group when we were searching for the way leading to confinement of monodisperse gold nanoparticles (AuNPs) within the framework of 2D-hexagonal mesoporous organosilica containing SH groups [87]. For that study, a bis-silylated hydrophilic molecular precursor containing disulfide unit –SS- was synthesised (see Table 2). The co-hydrolysis and polycondensation of the bridged organosilica precursor containing –SS- groups with TEOS was achieved in the presence of a non-ionic triblock co-polymer (P123) as structure directing agent. Reduction of –SS- groups of the resulting material led quantitatively to -SH functional groups.

The subsequent growth of monodisperse AuNPs was accomplished in two steps according to a methodology previously described [25] (impregnation of AuClTHT and reduction by $NaBH_4$). TEM images of the materials containing AuNPs revealed that AuNPs were mainly confined within the framework with a uniform size of about 2-3 nm, leaving the pores empty (Scheme 10). That was confirmed by BET measurements and XRD analysis. Finally, the crucial role of the –SH groups for the regular growth of AuNPs was also demonstrated, as AuNPs of small and regular size were not obtained from the material containing -SS- groups instead of -SH or from non-functionalized silicas. In other words, the growth of AuNPs of regular size within the framework of the mesoporous material was the consequence of the location of SH groups within the pore walls.

It is worth noting that, the extention of the method to other NPs is expected by adapting of the functionl groups in the walls to the NPs chosen. Thus, it will be possible to elaborate materials with two kinds of NPs perfectly located at nanometric scale, first one in the channel pores, and the second in the framework. That should involve in the future, the design of new materials involving unsuspected properties.

The problem of the distribution (regular or not) of the functional groups within the pore walls sets for materials prepared by co-condensation.

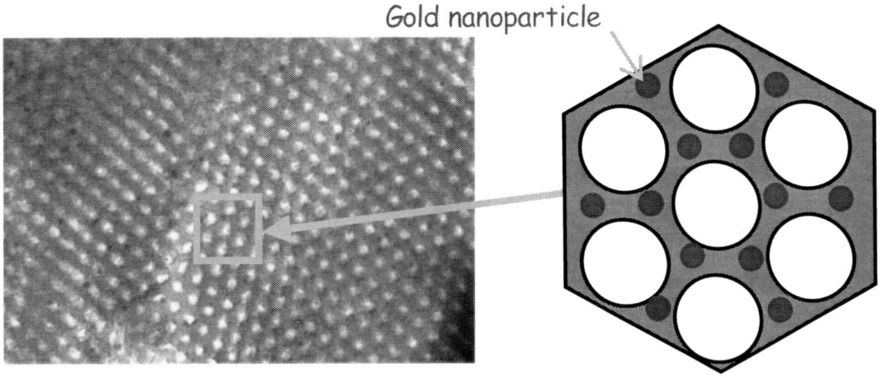

Scheme 10. Controlled growth of gold nanoparticles thanks to SH functionalized framework.

The study of the immobilization of photochromic azobenzene moieties within the pore walls of silica brought some light on this critical problem [84].

Indeed, the preparation of materials containing azobenzene moieties was of interest as they can give rise to applications as optical switches or data storage, Such materials were achieved by co-hydrolysis and polycondensation of 4,40-[(triisopropoxysilyl)propyloxy] azobenzene (see table 2) and TEOS giving rise to a worm-like structure. We showed that there is a partial and reversible trans-cis isomerization of bridged azobenzene moieties located in the framework of the mesoporous material in spite of their attachment at both ends to the silica matrix.

This process does not occur in the corresponding material prepared in the absence of surfactant. This striking result highlights the interest of the surfactant templating approach, which induces a regular dilution of the azobenzene moieties in the material.

The Multi-Functionalization

Multi-Functionalization in the Pores

As in the first part, a large variety of functional groups can be incorporated into mesoporous materials by grafting or by direct synthesis route. The introduction of two distinct organic groups together into the channel pores of mesoporous materials was also investigated but gave rise to a much more limited number of studies. Mann and co-workers [94] were the first to prove that the method of template directed co-condensation could be extended to the synthesis of MCM-41 structures containing two distinct types of covalently linked organic functionalities. Then, several authors prepared bifunctional materials with the view to improve the catalytic performance of the materials by the introduction of hydrophobic groups in addition to the functional groups [95,96,97]. Huh and co-workers [98] have synthesized by direct synthesis bifunctional mesoporous materials containing ureidopropyl (UDP) groups and 3-[2-(2-aminoethylamino)ethylamino]-propyl (AEP) groups. They observed a synergistic effect between the AEP and UDP groups in catalysis of various reactions involving carbonyl activation. Zeidan and co-workers [99] have prepared by direct synthesis a mesoporous silica containing organic acid and base groups. This material was proven to be more efficient in the aldol reaction than the materials containing only one of the two functions. Thus, a cooperative catalytic effect occurred between both of the types of functional groups. Finally, some authors prepared bifunctional materials in a two-step grafting process and tested successfully their potentiality as one pot bifunctionnal catalysts [100].

All these reports pointed out the great interest of bifunctional mesoporous materials as compared to the conventional monofunctionalized one. However, though the question of the distribution of two distinct organic groups located in the channel pores and introduced by direct synthesis has been evoked [96], it has remained a long time unsolved. There is one exception concerning the amphoteric amino acid bifunctional mesoporous silica [101]. Indeed, uniform distributions of acid and base organic groups on the mesopore surfaces were formed by interactions between the counter charged surfactant head groups and ionic parts of the structure directing agents, that is, between the cataionic and anionic surfactant head groups and the carboxyl and amino groups of the surfactant respectively (Scheme 11). A close proximity of organic NH_2 and CO_2H groups was confirmed by the formation of amide moieties. However, this interesting method cannot be extended to various organic groups.

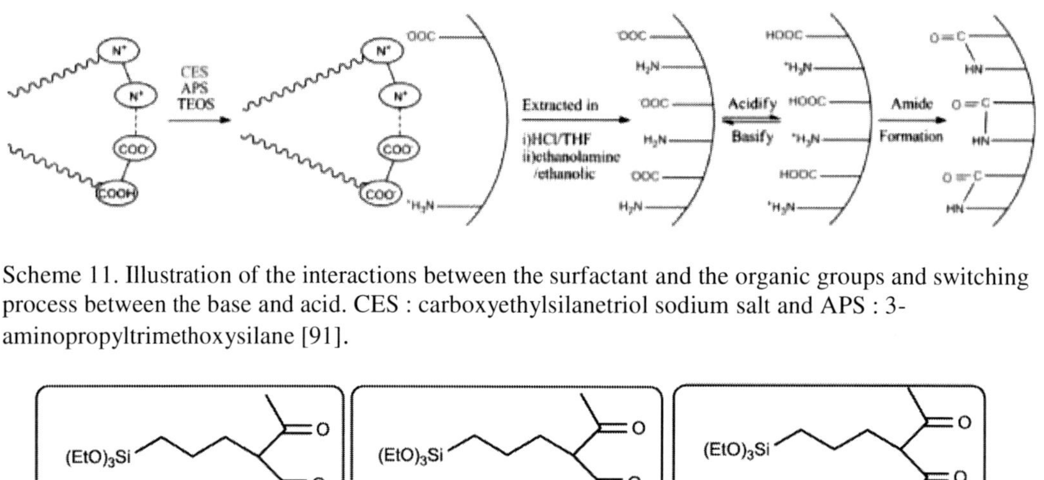

Scheme 11. Illustration of the interactions between the surfactant and the organic groups and switching process between the base and acid. CES : carboxyethylsilanetriol sodium salt and APS : 3-aminopropyltrimethoxysilane [91].

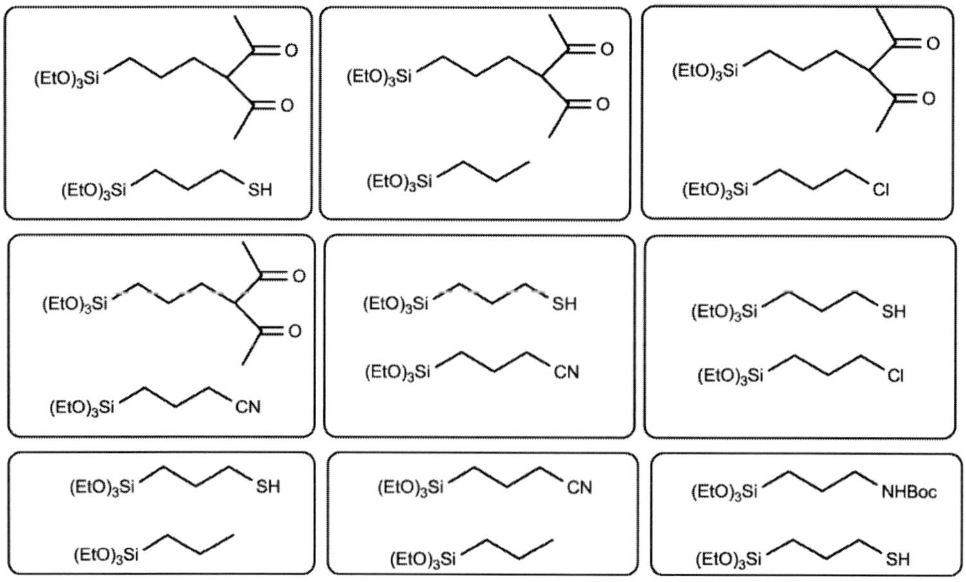

Scheme 12. Used couples for the synthesis of bifunctional materials.

Recently, we studied the distribution (regular or irregular distribution or segregation) of two distinct organic groups located within the channel pores of mesoporous hybrid materials (Scheme 12) [32]. For that purpose, we have chosen as probes organic groups, the property of which depends on the proximity of another identical group. One of them was based on the chelating property of acetyl acetonate groups (acac) towards $EuCl_3$. The other concerns the ion-exchange capability of the sulfonic acid moieties which varies with the proximity of the acid groups [102]. Thus, we first prepared monofunctional mesoporous materials by direct synthesis and showed that all the organic groups (SH and acac) are accessible and regularly distributed within the channel pores.

Then, we have prepared bifunctional mesoprorous materials containing acac and SO_3H groups (originating from SH groups). Both probes led to the same conclusion, that is to say uniform distribution of SO_3H and acac groups in the pore channels in this case. Several other bifunctional mesoporous materials have been prepared by direct synthesis with the view to determine their distribution within the channel pores by using the same probes that is to say either acac or SO_3H.

Taking into account the results obtained with the monofunctional materials, it appeared that the distribution of two distinct organic groups within the channel pores of mesoporous materials highly depends on the nature of the organic groups. We showed that the three possible types of distribution for a pair of organic groups can occur (Scheme 13): regular distribution (A), irregular distribution (B), or segregation (C). It is worth noting that the distribution of organic groups depends of their solubility within the hydrophobe core of the micelle. The main factors permitting to explain the distribution observed are probably: polarity, lipophilicity and bulkiness of organic groups. We found a regular mixture of both groups for the pair's NH_2,SH and SH,Cl in addition to the acac,SH pair. In these cases, the polarity of both organic groups is rather weak. Therefore, they should behave in the same way within the micelle. At the opposite, the segregation of both distinct organic groups was observed for three different pairs: SH,H, SH,CN and CN,H. In these last cases, the difference of polarity between the two groups could explain the segregation. Finally, an irregular distribution was observed for the pairs acac,H, aca,Cl and acac,SH.

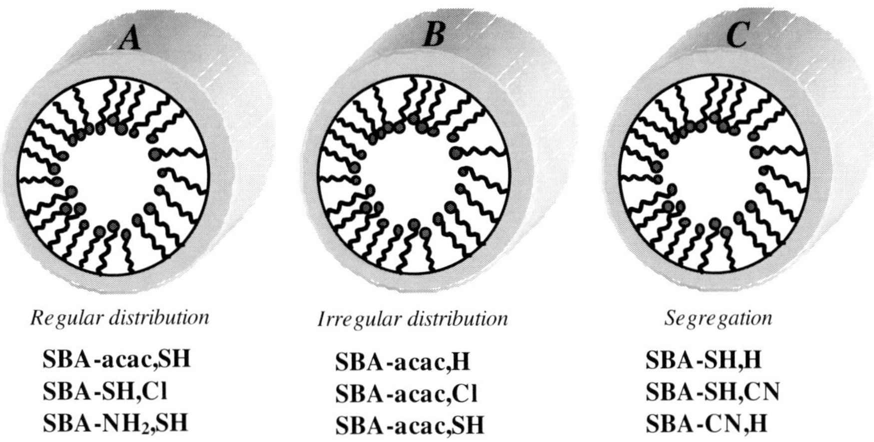

Scheme 13. Types of distribution of organic units in the bifunctional materials.

Multi-Functionalization Pores-Walls

The concept of mesoporous materials including two different organic groups, where one is terminally attached in the channel pores and the other integrated to the framework is also of great interest. Indeed, it could open the route to interactive materials that is to say, materials coupling two properties, one originating from an organic group located in the channel pores and the other in the framework at a nanometric range to each other, which involves unexpected properties. The first examples of such bifunctional materials were reported shortly after the discovery of the periodic mesoporous organosilicas (PMO's). Thus, Ozin and co-workers were the first to describe PMOs material prepared by direct synthesis and containing both bridging ethylene groups in the framework and terminal vinyl groups into the channel pores [103]. At the same time, Markowitz and co-workers depicted a variety of PMOs containing functional groups in the channel pores and bridged-ethane groups in the framework. Bifunctional organic-inorganic hybrid mesoporous materials containing bridged phenylene groups in the framework and mercaptopropyl groups in the channel pores were also reported [104]. The oxidation of SH groups into sulfonic acid groups was further achieved in order to provide catalytic active sites. More recently, a well organized bifunctional material with large heterocyclic bridging groups in the framework and mercaptopropyl ligands in the pores was described [105].

We reported the preparation of mesoporous hybrid materials containing bridged ion (Cu^{II}, Co^{II}, Eu^{III}, Gd^{III}) cyclam complexes located within the silica framework [106]. Grafting of a metal-N-triethoxysilylpropylcyclam complex inside the channel pores of these materials gave rise to hybrid materials containing two types of chelated ions (Cu^{II}-Co^{II} or Cu^{II}-Eu^{III} as examples) one located within the framework, the other in the channel pores and *vice versa* without any ion exchange reaction. We also described several examples of bifunctional organosilica containing chelating groups in the framework and reactive functional groups in the channel pores [66]. These materials were obtained by direct synthesis using co-condensation of a bridged organosilica with chelating properties, tetraethylorthosilicate (TEOS) and an organotrialkoxysilane $ZSi(OR)_3$ with Z = cyanopropyl, mercaptopropyl, chloropropyl in the presence of P123 as the structure directing agent. We investigated the influence of different parameters on the textural and structural characteristics of materials (concentration of the triblock copolymer P123, co-surfactant, and hydrothermal treatment) and the usefulness of chelating groups for complexation-decomplexation reactions (Scheme 14).

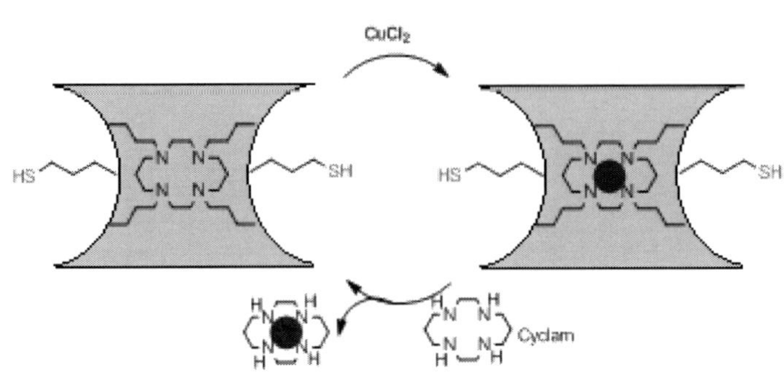

Scheme 14. Chemical modification of chelating groups within the framework.

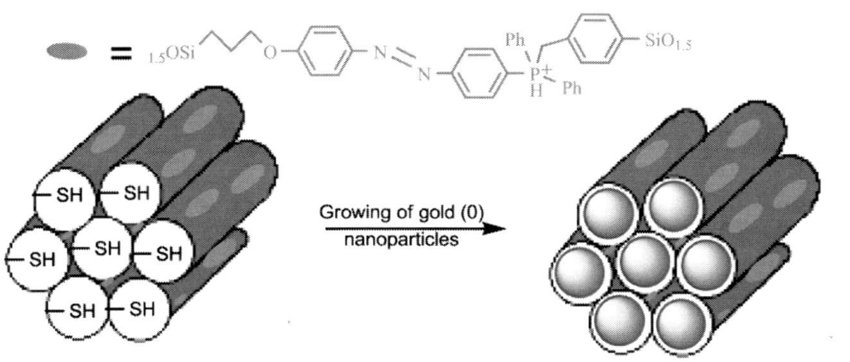

Scheme 15. Material with NLO groups in the framework and AuNPs in the pore channels.

A material containing a rather large NLO chromophore in the framework (bridged azobenzene phosphonium salt) and mercaptopropyl groups able to stabilize gold (0) nanoparticles in the channel pores of mesoporous silica (Scheme 15) [107] was also achieved in one step by using the direct lyotropic liquid crystal approach (LCA). Indeed, the presence of gold nanoparticles in a material containing NLO chromophore is expected to exalt the NLO response. Therefore, gold nanoparticles stabilized by SH groups were subsequently prepared according to a procedure described previously. The TEM image of the resulting material showed that the nanoparticles size distribution was in agreement with the pore size of the host material. The LCA methodology was in this case the appropriate method to get materials coupling two properties, which could interact, originating from two types of organic groups with distinguishable location.

In 2006, it was described the preparation of a bifunctional material containing two antagonist functional groups *i.e.* an acidic group in the framework and a basic one in the channel [108]. This novel material was synthesized by co-condensation of a bridged precursor with bi-sulfide core $(MeO)_3Si-(CH_2)_3-S-S-(CH_2)_3-Si(OMe)_3$, TEOS, and 3-*t*-butyloxycarbonyl aminopropyltriethoxysilane $(MeO)_3Si-(CH_2)_3$-NHBoc, in the presence of P123 as template under low acidic medium. The bifunctional material having an acidic framework and basic pores was obtained by appropriate chemical transformations of the resulting material *i.e.* reduction of disulfide units into SH groups followed by their oxidation into SO_3H groups and deprotection of amino groups by thermal treatment under vacuum (Scheme16). We showed that, both antagonist organic groups located at nanometric scale can cohabit independently without mutual destruction on condition to be in aprotic medium. The same type of mesoporous materials with an acidic framework and basic sites in the pores was taken again by others using a different synthetic approach [109].

Finally, we reported the synthesis of polyether-pores/framework bifunctional mesoporous materials. Triblock copolymer PEO_{20}-PPO_{70}-PEO_{20} (P123) is a good candidate to elaborate such materials because of its amphiphilic character. In fact, the hydrophilic part (Poly Ethylene Oxide) will be located in the framework while the hydrophobic part (Poly Propylene Oxide) in the channels pores. These materials were obtained by co-condensation of a mixture of silylated P123 (Si-P123), P123 and tetraethoxysilane under acidic conditions (Scheme 17). [110] The non silylated surfactant (P123) was removed by extraction with hot ethanol in a soxhlet apparatus leading to material containing only Si-P123% with tuneable molar % (0, 25, 50, 75 and 100%).

Thanks to anchored polyether groups, $Pd(OAc)_2$ salt was impregnated within the material and this was tested in heterogeneous Heck coupling reaction. The reactions proceeded in high yields. Furthermore, this catalyst may be recovered almost quantitatively by simple filtration and reused several times with limited leaching of palladium and loss of activity. Based on this recovery process, problems related to catalyst recovery and metal separation from the organic substrates can be resolved. This last feature combined with the air-stability of the catalyst may be of interest for large industrial applications, in particular for pharmaceutical products.

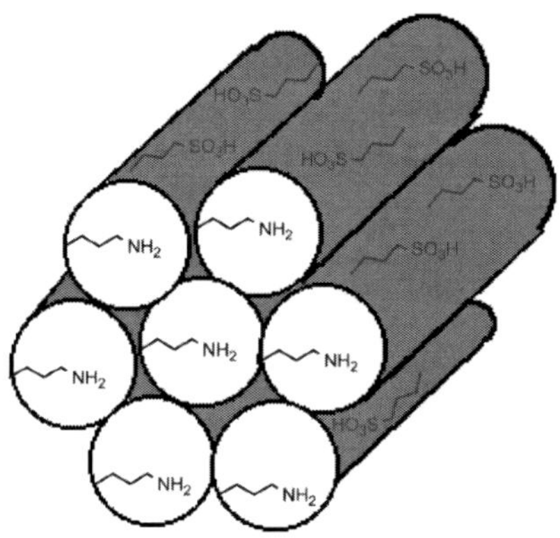

Scheme 16. Acido-basic bifunctionalized material.

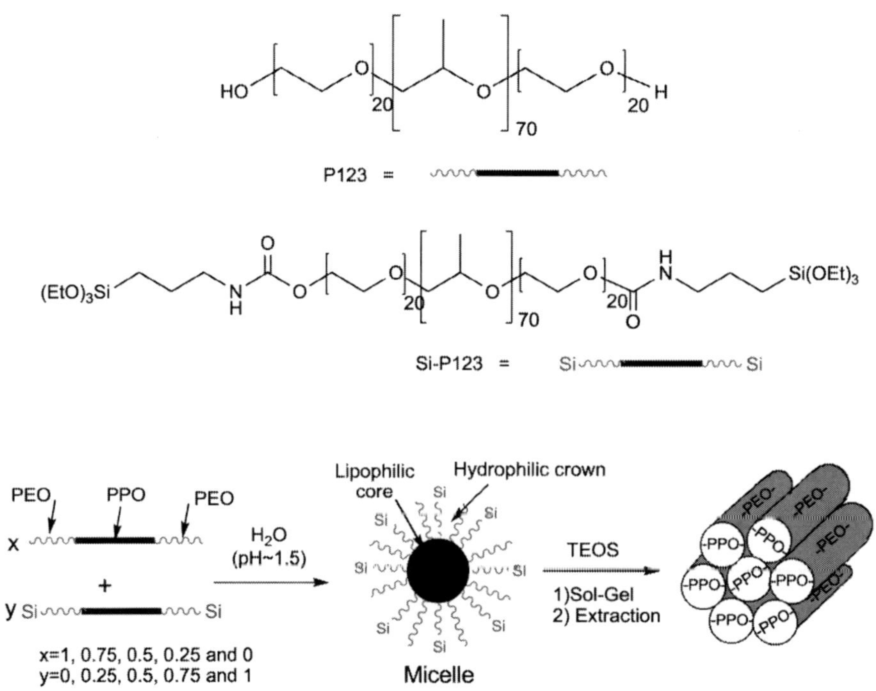

Scheme 17. Preparation of polyether functionalized materials.

These few examples show that the concept of bifunctional materials with one functional group in the framework and another in the channel pores even antagonist is realistic. That could be of great interest as it opens the route to an unlimited variety of nanomaterials with possible coupling of properties.

Since twenty years, the development of new route to synthesize and functionalize a structured oxide was largely amplified in the laboratories. However, the commercialization of these kinds of materials remains very limited in term of economic aspect. Only few examples of products can be found from several companies. We can cite, the mesostructured silica MCM-41 (hexagonal) or Al-MSU-F (cellular foam), Carbon, mesoporous nanopowder, mesoporous Niobium (V) oxide, silica nanoparticles (200 nm particle size) with 4 nm of pore size. More recently, Propylcarboxylic acid, amino or thiol functionalized silica mesoporous, nanoparticles, can be also purchased.

CONCLUSION

Twenty years after their discovery, mesoporous silicas remain an exciting research field for scientists. This chapter attempted to show the great potential offered by hybrid organic–inorganic nanomaterials prepared from molecular precursors by using the sol–gel process. It now appears that the methods permitting the introduction of specific functionalities with control of the location either in the channels pores or/and in the framework are well-known. The properties of this family of hybrid materials can be thus appropriately tailored. The possibility of multifunctionality for these materials is a particularly relevant feature. Indeed, it opens the route to interactive materials, that is, materials in which several functions susceptible to bring physical properties can interact at the nanometric scale. Finally, a large number of new hybrid nanomaterials offering various properties will be synthesized in the near future thanks to the creative force of the chemical synthesis and the imagination of the chemists.

REFERENCES

[1] G. Cao, Nanostructures and Nanomaterials. Synthesis, Properties and Applications, 2004, Imperial College Press. R. J. P. Corriu, N. Anh, Molecular Chemistry of Sol-Gel Derived Nanomaterials, 2009, Wiley-Blackwell. C. N. R. Rao, A. Müller, A. K. Cheetham, *The Chemistry of Nanomaterials*, 2010, Wiley-VCH, Weinheim.
[2] C. J. Brinker, G. W. Scherrer, Sol-Gel Science, 1990: the physics and chemistry of sol-gel processing, 1st ed.; Academic Press: San Diego. G. Kickelbick, Hybrid Materials, 2006, Wiley-VCH, Weinheim.
[3] J. C. Zhou, M. H. Chuang, E. H. Lan, B. Dunn, B. P. L. Gillman, S. M. Smith, *J. Mater. Chem.*, 2004, 14, 2311. D. Jin, K. W. Park, J. H. Lee, K. Song, J. G. Kim, M. L. Se, J. H. Jung, *J. Mater. Chem.*, 2011, 21, 3641.
[4] G. Cerveau, R. J. P. Corriu, E. Framery, *Chem. Mater.*, 2001, 13, 3373.
[5] J. Wang, W. Guo, J.-H. Bae, S.-H. Kim, J. Song, C.-S. Ha, *J. Mater. Chem.,* 2012, 22, 24681.
[6] B. A. Kamino, T. P. Bender, *Chem. Soc. Rev.,* 2013, 42, 5119

[7] N. Ehlert, P. P. Mueller, M. Stieve, T. Lenarz, P. Behrens, *Chem. Soc. Rev.,* 2013, 42, 3847.
[8] E. Magner, *Chem. Soc. Rev.,* 2013, 42, 6213.
[9] P. Yang, S. Gai, J. Lin, *Chem. Soc. Rev.,* 2012, 41, 3679.
[10] T. Yanagisawa, T. Shimizu, K. Kuroda, C. Kato, *Bull. Chem. Soc.Jpn.,* 1990, 63, 988.
[11] J. S. Beck, J. C. Vartuli, W. J. Roth, M. E. Leonowicz, C. T. Kresge, K. D. Schmitt, T. W. Chu, D. H. Olson, E. W. Sheppard, S. B. McCullen, J. B. Higgins, J. L. Schlenker, *J. Am. Chem. Soc.,* 1992, 114, 10834.
[12] C. Sanchez, B. Julian, P. Belleville and M. Popall, *J. Mater. Chem.,* 2005, 15, 3559.
[13] A. Vinu, K. Z. Hossain, K. Ariga, *J. Nanosci. Nanotech.,* 2005, 5, 347.
[14] ISI Web of Knowledge, searching by topic: functionalized mesoporous silica and create citation report.
[15] L. Mercier, T. J. Pinnavaia, *Adv. Mater.*, 1997, 9, 500; A. Cauvel, G. Renard, D. Brunel, *J. Org. Chem.,* 1997, 62, 749; J. Y. Ying, C. P. Mehnert, M. S. Wong, *Angew. Chem., Int. Ed.*1999, 38, 56; P. M. Price, J. H. Clark, D. J. Macquarrie, *Dalton*, 2000, 2, 101; F. Hoffmann, M. Cornelius, J. Morell, M. Froeba, *Angew. Chem., Int. Ed.* 2006, 45, 3216.
[16] S. L. Burkett, S. D. Sims, S. Mann, *Chem. Commun.,* 1996, 1367; A. Walcarius, C. Delacôte, *Chem. Mater.,* 2003, 15, 4181.
[17] D. J. Macquarrie, *Chem. Commun*., 1996, 196; M. H. Lim, C. F. Blanford, A. Stein, *J. Am. Chem. Soc.* 1997, 119, 4090; D. J. Macquarrie, D. B. Jackson, S. Tailland, K. A. Utting, *J. Mater. Chem*., 2001, 11, 1843; Y. Mori, T. J. Pinnavaia, *Chem. Mater*., 2001, 13, 2173; M. Kruk, T. Asefa,, M. Jaroniec, G. A. Ozin, *J. Am. Chem. Soc.,* 2002, 124, 6383; Y. Wang, B. Zibrowius, C. Yang, B. Spliethoff, F. Schüth, *Chem. Commun*., 2004, 46.
[18] R. J. P. Corriu, C. Hoarau, A. Mehdi, C. Reye, *Chem. Commun.* 2000, 71.
[19] S. L. Burkett, S.D. Sims, S. Mann, *Chem. Commun*., 1996, 1367.
[20] L. Mercier, T. J. Pinnavaia, *Chem. Mater.* 2000, 12, 188. R. J. P. Corriu, F. Embert, Y. Guari, A. Mehdi, C. Reye, *Chem. Commun*., 2001, 1116
[21] P. D. Yang, D. Y. Zhao, S. C., B. F. Chmelka, G. D. Stucky, *Chem. Mater*., 1998, 10, 2033; R. Mouawia, A. Mehdi, C. Reye, R. J. P. Corriu, *New J. Chem.,* 2006, 30, 1077.
[22] A. Walcarius, L. Mercier, *J. Mater. Chem.*, 2010, 20, 4478.
[23] R. J. P. Corriu, L. Datas, Y. Guari, A. Mehdi, C. Reye, C. Thieuleux, *Chem. Commun*., 2001, 763.
[24] R. J. P. Corriu, A. Mehdi, C. Reye, C. Thieuleux, *Chem. Mater*., 2004, 16, 159. E. A. Prasetyanto, S.-C. Lee, S.-E. Park, *Microporous Mesoporous Mater*., 2009, 118, 134.
[25] A. Bibby, L. Mercier, *Chem. Mater.* 2002, 14, 1591. Y. Guari, C. Thieuleux, A. Mehdi, C. Reye, R. J. P. Corriu, S. Gomez-Gallardo, K. Philippot, B. Chaudret, *Chem. Mater*., 2003, 15, 2017.
[26] A. S. Maria Chong, X. S. Zhao, A. T. Kustedjo, S. Z. Qiao, *Microporous Mesoporous Mater.*, 2004, 72, 33.
[27] A. Mehdi, C. Reye, S. Brandes, R. Guilard, R. J. P. Corriu, *New J. Chem*, 2005, 29, 965. T. Yokoi, H. Yoshitaki, T. Tatsumi, *Chem.Mater*., 2003, 15, 4536.
[28] S. Huh, J. W. Wiench, J.-C. Yoo, M. Pruski, V. S.-Y. Lin, *Chem. Mater*., 2003, 15, 4247.
[29] C. M. Yang, B. Zibrowius, F. Schüth, *Chem. Commun*., 2003, 1772.

[30] B. Folch, J. Larionova, Y. Guari, C. Guerin, A. Mehdi, C. Reye, *J. Mater. Chem.*, 2004, 14, 2703.
[31] V. Matsura, Y. Guari, J. Larionova, C. Guerin, A. Caneschi, C. Sangregorio E. Lancelle-Beltran, A. Mehdi, R. J. P. Corriu, *J. Mater. Chem.*, 2004, 14, 3026.
[32] R. Mouawia, A. Mehdi, C. Reye, R. J. P. Corriu, *J. Mater. Chem.*, 2008, 18, 4193.
[33] R. J. P. Corriu, E. Lancelle, A. Mehdi, C. Reye, S. Brandes, R. Guilard, *J. Mater. Chem.*, 2002, 12, 1355. T. Kang, Y. Park, K. Choi, J. S. Lee, J. Yi, *J. Mater. Chem.*, 2004, 14, 1043.
[34] R. J. P. Corriu, A. Mehdi, C. Reye, C. Thieuleux, A Frenkel, A. Gibaud, *New J. Chem.*, 2004, 28, 156.
[35] J. Alauzun, A. Mehdi, C. Reye, R. J. P. Corriu, *New J. Chem.*, 2007, 31, 911.
[36] R. J. P. Corriu, F. Embert, A. Mehdi, C. Reye, *Chem. Mater.*, 2001, 13, 4542. Q. Hu, J. E. Hampsey, N. Jiang, C. Li, Y. Lu, *Chem. Mater.*, 2005, 17, 1561.
[37] Y. Guari, K. Soulantica, K. Philippot, C. Thieuleux, A. Mehdi, C. Reye, B. Chaudret, R. J. P. Corriu, *New J. Chem.*, 2003, 7, 1029.
[38] S. Jansat, K. Pelzer, J. Garcia Anton, R. Raucoules, K. Philippot, A. Maisonnat, B. Chaudret, Y. Guari, A. Mehdi, C. Reye, R. J. P. Corriu, *Adv. Funct. Mater.*, 2007, 17, 3339.
[39] U. Diaz, D. Brunel, A. Corma, *Chem. Soc. Rev.*, 2013, 42, 4083.
[40] A. Taguchi, F. Schüth, *Microporous Mesoporous Mater.* 2005, 77, 1.
[41] T. K. Maishal, J. Alauzun, J.-M. Basset, C. Copéret, R. J. P. Corriu, E. Jeanneau, A. Mehdi, C. Reye, L. Veyre, C. Thieuleux, *Angew. Chem. Int. Ed.* 2008, 47, 8654.
[42] I. Karame, M. Boualleg, J. M. Camus, T. K. Maishal, J. Alauzun, J. M. Basset, C. Coperet, R. J. P. Corriu, E. Jeanneau, A. Mehdi, C. Reye, L. Veyre, C. Thieuleux, *Chem. Eur. J.*, 2009, 15, 11820.
[43] A. Lesage, M. Lelli, D. Gajan, M. A. Caporini, V. Vitzthum, P. Mieville, J. Alauzun, A. Roussey, C. Thieuleux, A. Mehdi, G. Bodenhausen, C. Coperet, L. Emsley, *J. Am. Chem. Soc.*, 2010, 132, 15459.
[44] R. M. Grudzien, B. E. Grabicka, D. J. Knobloch, M. Jaroniec, *Colloids and Surfaces A: Physicochem. Eng. Aspects,* 2006, 291, 139; Y. Kubota, C. Jin, T. Tatsumi, *Catalysis today*, 2008, 132, 75; H. H. Gavin, G. L. Jheng, H. M. Kao, *J. Am. Chem. Soc.*, 2008, 130, 11566.
[45] H. M. Kao, C. H. Liao, T. T. Hung, Y. C. Pan, A. S. T. Chiang, *Chem. Mater.*, 2008, 20, 2412; F. O. M. Gaslain, C. Delacote, A. Walcarius, B. Lebeau, *J. Sol-Gel. Sci. Technol.*, 2009, 49, 112; X. Wang, X. Du, H. Rao, H. Chen, *Appl. Surf. Sci.*, 2009, 255, 6073.
[46] O. C. Gobin, Q. Huang, H. Vinh-Thang, F. Kleitz, M. Eic, S. Kaliaguine, *J. Phys. Chem. C*, 2007, 111, 3059.
[47] A. Boullanger, J. Alauzun, A. Mehdi, C. Reye, R. J. P. Corriu, *New Journal of Chemistry*, 2010, 34, 738.
[48] A. Boullanger, S. Clement, V. Mendez, S. Daniele, C. Thieuleux, A. Mehdi, *RSC Advances,* 2013, 3, 725.
[49] G. Subra, A. Mehdi, C. Enjalbal, M. Amblard, L Brunel, R. J. P. Corriu, J. Martinez, *J. Mater. Chem.*, 2011, 21, 6321.
[50] S. Jebors, C. Enjalbal, M. Amblard, G. Subra, A. Mehdi, J. Martinez, *Tetrahedron*, 2013, 69, 7670.

[51] S. Jebors, C. Enjalbal, M. Amblard, G. Subra, A. Mehdi, J. Martinez, *J. Mater. Chem.B*, 2013, 23, 2921
[52] S. Inagaki, S. Guan, Y. Fukushima, T. Oshuna, O. Terasaki, *J. Am. Chem. Soc.*, 1999, 121, 9611.
[53] B. J. Melde, B. T. Holland, C. F. Blanford, A. Stein, *Chem. Mater.*, 1999, 11, 3302.
[54] T. Asefa, M. J. MacLachlan, N. Coombs, G. A. Ozin, *Nature,* 1999, 402, 867.
[55] C. Yoshina-Ishii, T. Asefa, N. Coombs, M. J. MacLachlan, G. A. Ozin, *Chem. Commun.*, 1999, 2539.
[56] F. Hoffmann, M. Cornelius, J. Morell, M. Fröba, *J. Nanosc. Nanotech.* 2006, 6, 265.
[57] S. Fujita, S. Inagaki, *Chem. Mater.*, 2008, 20, 891.
[58] M. C. Burleigh, M. A. Markowitz, M. S. Spector, B. P. Gaber, *Chem. Mater.* 2001, 13, 4760.
[59] M. Benitez, D. Das, R. Ferreira, U. Pischel, H. Garcia, *Chem. Mater.* 2006, 18, 5597.
[60] Ö. Dad, G. A. Ozin, *Adv. Mater.* 2001, 13, 1182.
[61] R. J. P. Corriu, A. Mehdi, C. Reye, C. Thieuleux, *Chem. Commun.*, 2002, 1382.
[62] P. Van Der Voorts, C. Vercaemst, D. Schaubroeck, F. Verpoort, *Phys. Chem. Chem. Phys.*, 2008, 10, 347.
[63] F. Hoffmann, M. Fröba, *Chem. Soc. Rev.*, 2011, 40, 608.
[64] W. Wang, J. E. Lofgreen, G. A. Ozin, *Small*, 2010, 6, 2634.
[65] O. Olkhovyk, M. Jaroniec, *J. Am. Chem. Soc.,* 2005, 127, 60.
[66] J. Alauzun, A. Mehdi, C. Reye, R. J. P. Corriu, *J. Mater. Chem.*, 2007, 17, 349.
[67] J. Huang, F. Zhu, W. He, F. Zhang, W. Wang, H. Li, *J. Am. Chem. Soc.*, 2010, *132*, 1492.
[68] S. MacQuarrie, M. P. Thompson, A. Blanc, N. J. Mosey, R. P. Lemieux, C. M. Crudden, *J. Am. Chem. Soc.*, 2008, 130, 14099.
[69] X. Meng, T. Yokoi, D. Lu, T. Tatsumi, *Angew. Chem.*, 2007, 119, 7492. Y. Wang, X. Liu, J. Yang, Y. Yang, L. Zhang, Q. H. Yang, C. Li, *J. Mater. Chem.*, 2009, 19, 8009.
[70] N. Mizoshita, Y. Goto, M. P. Kapoor, T. Shimada, T. Tani and S. Inagaki, *Chem.Eur. J.*, 2009, 15, 219.
[71] W. Guo, F. Kleitz, K. Cho, R. Ryoo, *J. Mater. Chem.*, 2010, 20, 8257.
[72] J. Font, P. de March, F. Busque, E. Casas, M. Benitez, L. Teruel, H. Garcia, *J. Mater. Chem.*, 2007, 17, 2336.
[73] V. Dufaud, F. Beauchesne, L. Bonneviot, *Angew. Chem., Int. Ed.*, 2005, 44, 3475.
[74] P. N. Minoofar, B. S. Dunn, J. I. Zink, *J. Am. Chem. Soc.*, 2005, 127, 2656.
[75] E. Johansson, J. I. Zink, *J. Am. Chem. Soc.*, 2007, 129, 14437.
[76] B. Karimi, D. Elhamifar, J. H. Clark, A. J. Hunt, *Chem. Eur. J.*, 2010, 16, 8047.
[77] E. A. Prasetyanto, M. B. Ansari, B. H. Min, S. E. Park, *Catalysis Today*, 2010, 158, 252.
[78] S. Inagaki, O. Ohtani, Y. Goto, K. Okamoto, M. Ikai, K.-I. Yamanaka, T. Tani, T. Okada, *Angew. Chem.*, 2009, 48, 4042.
[79] K.-I. Yamanaka, T. Okada, Y. Goto, T. Tani, S. Inagaki, *RSC Advances,* 2013, 3, 14774.
[80] N. Mizoshita, T. Tani, S. Inagaki, *Chem. Soc. Rev.*, 2011, 40, 789.
[81] J. Alauzun, A. Mehdi, C. Reye, R. J. P. Corriu, *J. Mater. Chem.*, 2007, 17, 349.
[82] W. Whitnall, L. Cademartiri, G. A. Ozin, *J. Am. Chem. Soc.*, 2007, 129, 15644.
[83] J. Morell, G. Wolter, M. Fröba, *Chem. Mater.*, 2005, 17, 804.

[84] E. Besson, A. Mehdi, D. Lerner, C. Reye, R. J. P. Corriu, *J. Mater. Chem.,* 2005, 15, 803.
[85] M. Alvaro, B. Ferrer, V. Fornés, H. Garcia, *Chem. Commun.*, 2001, 2546.
[86] E. Besson, A. Mehdi, C. Reye, R. J. P. Corriu, *J. Mater. Chem.* 2006,16, 146.
[87] E. Besson, A. Mehdi, C. Reye, R. J. P. Corriu, *J. Mater. Chem.*, 2009, 19, 4746.
[88] P. V. Bernhardt and G. A. Lawrance, *Coord. Chem. Rev.*, 1990, 104, 297.
[89] D. Herault, G. Cerveau, R. J. P. Corriu, A. Mehdi, *Dalton Trans.*, 2011, 40, 446.
[90] J. Liu, J. Yang, Q. Yang, G. Wang, Y. Li, *Adv. Funct. Mater.,* 2005, 15, 1297.
[91] O. Kwon, S. Park, G. Seo, *Chem. Commun.,* 2007, 4113.
[92] J. Liu, Q. Yang, L. Zhang, D. Jiang, X. Shi, J. Yang, H. Zhong, C. Li, *Adv. Funct. Mater.,* 2007, 17, 569.
[93] L. Chen, J. Hu, R. Richards, *J. Am. Chem. Soc.*, 2009, 131, 914.
[94] S. R. Hall, C. E. Fowler, B. Lebeau, S. Mann, *Chem. Commun.*, 1999, 201.
[95] D. J. Macquarrie, *Green Chemistry,* 1999, 195.
[96] D. Margolese, J. A. Melero, S. C. Christiansen, B. F. Chmelka, G. D. Stucky, *Chem. Mater.,* 2000, 12, 2448.
[97] E. Cano-Serrano, G. Blanco-Brieva, J. M. Campos-Martin, J. L. G. Fieorro, *Langmuir*, 2003, 19, 7621.
[98] S. Huh, H.-T. Chen, J. W. Wiench, M. Pruski, V. S.-Y. Lin, *Angew. Chem. Int. Ed.*, 2005, 44, 1826.
[99] R. K. Zeidan, S-J. Hwang M. E. Davis, *Angew. Chem. Int. Ed.*, 2006, 45, 6332.
[100] F. Goettmann, D. Grosso, F. Mercier, F. Mathey, C. Sanchez, *Chem. Commun.*, 2005, 1240.
[101] L. Han, J. Ruan, Y. Li, O. Terasaki, S. Che, *J. Mater. Chem.*, 2007, 19, 2860.
[102] J. J. Shyue, M. R. De Guire, T. Nakanishi, Y. Masuda, K. Koumoto, C. N. Sukenik, *Langmuir*, 2004, 20, 8693.
[103] T. Asefa, M. Kruk, M. J. MacLachlan, N. Coombs, H. Grondey, M. Jaroniec, G. A. Ozin, *J. Am. Chem. Soc.*, 2001, 123, 8520.
[104] Q. Yang, M. P. Kapoor, S. Inagaki, *J. Am. Chem. Soc.*, 2002, 124, 9694.
[105] O. Olkhovyk, S. Pikus, M. Jaroniec, *J. Mater. Chem.*, 2005, 15, 4285.
[106] R. J. P. Corriu, A. Mehdi, C. Reye, C. Thieuleux *Chem. Commun.*, 2003, 1564.
[107] E. Besson, A. Mehdi, V. Matsura, Y. Guari, C. Reye, R. J. P. Corriu, *Chem. Commun.*, 2005, 1775.
[108] J. Alauzun, A. Mehdi, C. Reye, R. J. P. Corriu, *J. Am. Chem. Soc.*, 2006, 128, 8719. M. Jaroniec, *Nature*, 2006, 442, 638.
[109] S. Shylesh, A. Wagener, A. Seifert, S. Ernst, W. R. Thiel, *Angew. Chem. Int. Ed.*, 2010, 49, 184.
[110] A. F. Grandsire, C. Laborde, F. Lamaty, A. Mehdi, *Applied Organometallic Chemistry*, 2010, 24, 179.

In: Comprehensive Guide for Mesoporous Materials, Volume 2 ISBN: 978-1-63463-982-8
Editor: Mahmood Aliofkhazraei

Chapter 3

MESOPOROUS NANOSPHERES OF TITANIA DIRECTLY MODIFIED WITH TUNGSTOSILICIC ACID AS CATALYSTS IN THE PHOTODEGRADATION OF 4-CHLOROPHENOL

J. A. Rengifo-Herrera, Romina A. Frenzel, Edgardo L. Soto, Mirta N. Blanco and Luis R. Pizzio*

Centro de Investigación y Desarrollo en Ciencias Aplicadas "Dr. J. J. Ronco" (CINDECA), Departamento de Química, Facultad de Ciencias Exactas, UNLP-CCT La Plata, -La Plata, Argentina

ABSTRACT

A series of mesoporous titania materials modified with with tungstosilicic (TSA) were successfully synthesized using titanium isopropoxide as titania precursor and urea as low-cost template via sol-gel reactions. They were characterized by FT-IR, XRD, DTA-TGA, DRS, SEM and BET. The obtained mesoporous solids presented an average pore diameter higher than 3.1 nm. The specific surface area decreased when the TSA amount and the calcination temperature increased. The XRD patterns of the modified samples only exhibited the characteristic peaks of anatase phase. The presence of TSA retarded the crystallization of the anatase phase and its transformation into rutile phase. The FT-IR studies showed that the main heteropolyoxometallate species present in the composites is the $[SiW_{12}O_{40}]^{4-}$ anion for the samples calcined up to 500 °C. The *band gap* energy slightly decreased as a result of the introduction of TSA into the titania matrix, though it remained almost constant with the calcination temperature increase. The 4-chlorophenol degradation behavior of the catalysts as a function of time depended on the TSA amount and the thermal treatment temperature. The amount of degraded 4-CP increased with the increment of TSA content being the sample containing 30 % TSA calcined at 600 °C the most active. The apparent reaction constant, estimated assuming a pseudo-first-order kinetics, followed the same trend. The catalysts can be reused at least three times without an important decrease in the degradation and mineralization degrees.

* Corresponding author: Fax number: 54 221 425 4477. E-mail address: lrpizzio@quimica.unlp.edu.ar.

Keywords: Mesoporous titania, Tungstosilicic acid, 4-chlorophenol, Photocatalytic degradation

1. INTRODUCTION

Titanium dioxide (titania) is a material widely used as photocatalyst because it has a relatively high chemical stability, a negligible toxicity, a low cost, and it is easily available. Titania performance in the photodegradation of contaminants contained in wastes is influenced by the crystal structure, the crystallinity, the surface area, the porosity, the surface hydroxyl density, and the *band gap* energy [1-3], among other factors.

A low surface area and the hole-electron recombination are the main effects that can lead to a low photocatalytic activity. So, an increase in the surface area and the separation of electrons and holes can improve the photocatalytic activity of titania. Sol-gel reactions using urea as a low-cost pore-forming agent were found to be a good method to synthesize mesoporous titania with suitable properties to be used as a photocatalyst [4, 5].

Many works have been reported that titania performance in the photodegradation of contaminants contained in wastes is influenced by the crystal structure, the crystallinity, the surface area, the porosity, the surface hydroxyl density, and the band gap energy [6-9], among other factors.

A low surface area and the hole-electron recombination are the main effects that can lead to a low photocatalytic activity. So, an increase in the surface area and the separation of electrons and holes can improve the photocatalytic activity of titania. Mesoporous titania with suitable properties to be used as a photocatalyst [10, 11] can be synthesized by sol–gel reactions using urea as a low-cost pore-forming agent.

To overcome the recombination of the electron–hole pairs of TiO_2-based catalysts drawback [12–15] transition metals or metal oxides were tested as electron trappers improving the photocatalytic activity. Also, it has been reported that the light absorption is extended to the visible region when large band gap semiconductors are doped with suitable transition metal ions [16], allowing the use of light from the visible region of the spectrum.

The increment of the photoactivity of titania samples modified with W has been mainly explained by considering the formation of intermediate W(V) species by means of a transfer of photogenerated electrons [17, 18]. This reduced W species could be oxidized to W(VI) by transferring electrons to oxygen [19].

Heteropolyoxometallates (POM) are employed as effective homogeneous photocatalysts in the degradation of organic pollutants in water [20 22]. They absorb strongly in the near visible and UV region of the light spectrum ($\lambda < 400$ nm). This absorption corresponds to the ligand-to-metal charge-transfer band and it can generate strongly oxidizing excited state POM* (reaction 1).

They are able to carry out the oxidation of organic substrates (S) (reaction 2) directly via charge transfer or H-atom abstraction, or indirectly through the intermediacy of solvent-derived radicals [22]. After that, the corresponding reduced POM are usually reoxidized to their original oxidation state by an electron acceptor such as dioxygen (reaction 3).

$$\mathrm{POM} + h\nu \rightarrow \mathrm{POM}^* \qquad (1)$$

$$POM^* + S \rightarrow POM^- + S^+ \quad (2)$$

$$POM^- + O_2 \rightarrow POM + O_2^- \quad (3)$$

Heteropolyoxometallates have been incorporated into TiO_2 colloids [23], added to TiO_2 suspensions [24], or anchored to TiO_2 by chemical interactions [25] with the purpose of reducing the charge recombination in UV-illuminated TiO_2. The capacity of POM as acceptors (reaction 5) of the electrons, generated in the conduction band (e_{cb}^-) together with holes in the valence band (h_{vb}^+), of UV-irradiated TiO_2 suspensions (reaction 4) was demonstrated by Park and Choi [26] using a photoelectrochemical method. The reduced POM react with O_2 to generate superoxides (reaction 3).

$$TiO_2 + h\nu \rightarrow e_{cb}^- + h_{vb}^+ \quad (4)$$

$$POM + e_{cb}^- \rightarrow POM^- \quad (5)$$

POMs have been used to modify the surface of previously prepared TiO_2 [23-26]. On the other hand, we studied the incorporation of tungstophophoric and tungstosilicic acid into the titania matrix during the TiO_2 gel synthesis, in a similar way to Li et al. [27], but using urea as low-cost template with the purpose of obtaining mesoporous materials [11, 28-30].

Event: Modification of TiO_2 surfaces by POM			
Main observed	Synthesis	Instrumental techniques characterization	Reference
Transient absorption spectra showed that TiO_2 particles containing tungstophosphoric acid (TPA) were able to transfer UV-induced Conduction band electron from the TiO_2 to TPA and catalyze the photo-reduction of methyl orange.	Sol-gel synthesis by acid hydrolysis of titanium (IV) tetraispropoxide (TTIP).	TEM, nanosecond transient spectroscopy, FT-IR, DRS	[23]
Tungstosilicic acid (TSA) and TPA were added to TiO_2 suspensions			[24]
TiO_2 was modified by several POM. Photocatalytic activity increased by incorporation of POM on TiO_2 and this was evaluated towards the oxidation of a pesticide: hexachlorobenzene and three organic dyes: congo red, methyl orange and alizarin S.	Sol-gel synthesis by acid hydrolysis of TTIP and adding several kinds of POMs and hydrothermal treatment at 200	DRS, ICP-AES, ^{31}P-MAS-NMR, XRD, TEM, specific surface area BET (SSA-BET).	[25]
Photoinduced electron transfer from UV-illuminated TiO_2 and Pt/TiO_2 suspensions to TPA was observed.	Commercial Degussa TiO_2 powder was used and TPA was added to semiconductor suspension	Electrochemical measurements were performed using a potentiostat.	[26]
TiO_2-anatase crystalline phase with micro and meso-porosity and nano-scale particles (average particle size 6 nm). Tungstophosphoric acid (TPA)	Sol-gel synthesis by acid hydrolysis of TTIP and hydrothermal treatment at 200 ºC.	XRD, FT-Raman, ^{31}P-MAS-NMR, UV-DRS, ICP-AES,	[27]

Table (Continued)

Main observed	Synthesis	Instrumental techniques characterization	Reference
Keggin structure remains intact and these materials showed visible light activity towards degradation of pesticide parathion-methyl			
TiO_2 materials modified with TPA showed mesoporous features. Herein Keggin structure was the main specie immobilized on TiO_2. Lacunar and dimeric forms of TPA were also found. Isoelectric point decreased strongly by TPA addition. Crystallization to anatase and transformation to rutile was retarded. Band gap energy decreased by addition of TPA.	Sol-gel synthesis by acid hydrolysis of TTIP and addition of urea as low-cost template. Addition of TPA at 10, 20 and 30% (w/w) and air annealing at different temperatures (100, 200, 300, 400, 500 and 600 °C).	XRD, ^{31}P- MAS-NMR, UV-DRS, DTA-TGA, BET, TEM, isoelectric point.	[28]
TiO_2 spherical particles (Sizes between 200-700 nm) modified with Tungstosilicic acid (TSA) showed only anatase crystalline structure. Raising the TSA content increased the size particle. Energy decreased by addition of TPA. The anion $[SiW_{12}O_{40}]^{-4}$ was the main specie present on TiO_2.	Sol-gel synthesis by acid hydrolysis of TTIP and addition of urea as low-cost template. Addition of TSA at 10, 20 and 30% (w/w)and air annealing at different temperatures (100, 200, 300, 400, 500 and 600 °C).	XRD, UV-DRS, DTA-TGA, SSA-BET, TEM, SEM, FT-IR, isoelectric point.	[29]
Anatase TiO_2 spherical particles modified with TPA showed average pore diameter higher than 3.1 nm. Specific surface area decreased when TPA amount and annealing temperature increased. Energy band gap measurements decreased slightly by TPA addition. Keggin anion was the main TPA specie on the samples. Photocatalytic activity was evaluated towards 4-chlorophenol degradation under UV+visible light irradiation. TiO_2 samples containing 30% (w/w) of TPA and annealed at 500 and 600 °C showed highest photocatalytic activity.	Sol-gel synthesis by acid hydrolysis of TTIP and addition of urea as low-cost template. Addition of TPA at 10, 20 and 30% (w/w) and air annealing at different temperatures (100, 200, 300, 400, 500 and 600 °C).	XRD, UV-DRS, DTA-TGA, SSA-BET, TEM, SEM, FT-IR, isoelectric point	[30]

We present here the preparation and characterization of tungstosilicic acid mesoporous titania nanospheres synthesized via sol–gel reactions and measured the catalytic activity of the materials in the 4-chlorophenol photodegradation.

Finally, we discuss the effect of the preparation variables on the structural and textural characteristics of the solids and correlated them with the pollutant degradation activity.

2. PRINCIPLES, PREPARATION AND CHARACTERIZATION

A series of mesoporous nanospheres of titania directly modified with tungstosilicic acid were successfully synthesized by using urea as low-cost template via sol–gel reactions, followed by removing the urea by extraction with water. Titanium isopropoxide (Aldrich, 26.7 g) was mixed with absolute ethanol (Merck, 186.6 g) and stirred for 10 min to obtain a homogeneous solution under N2 at room temperature, then 0.33 cm^3 of 0.28 M HCl aqueous solution was dropped slowly into the above mixture to catalyze the sol–gel reaction for 3 h. After that, 120 g of urea–alcohol–water (1:5:1 weight ratio) solution was added into the hydrolyzed solution under vigorous stirring, to act as template, together with an ethanol solution of $H_4SiW_{12}O_{40}\cdot 23H2O$ (Fluka p.a.). The amount of TSA was varied with the purpose of obtaining a TSA concentration of 0%, 10%, 20% and 30% by weight in the final solid (named TiTSA00, TiTSA10, TiTSA20, and TiTSA30, respectively). The gels were dried at room temperature, and the solids were ground into powder and extracted with distilled water for three periods of 24 h, in order to remove the urea.

Finally, the solids were thermally treated at 100, 200, 300, 400, 500, and 600 °C for 2 h, and will be hereinafter named TiTSAXXT100, TiTSAXXT200, TiTSAXXT300, TiTSAXXT400, TiTSAXXT500, and TiTSAXXT600, respectively, where XX is the TSA concentration.

The samples were characterized by the following physicochemical techniques XRD, BET, SEM, FT-IR, DRS UV-Vis.

Powder X-ray diffraction (XRD) patterns were obtained using Philips PW-1732 diffractometer with built-in recorder, using Cu Kα radiation, nickel filter, 20 mA and 40 kV in the high voltage source, scanning angle between 5 and 60 °2θ, and scanning rate of 1° per min. The crystallite size (DC) of the samples, were estimated from the XRD results using the Scherrer equation and silicon as standard for the correction of the instrumental broadening.

The specific surface area, the pore volume and the mean pore diameter of the solids were determined from the N_2 adsorption-desorption isotherms at the liquid-nitrogen temperature, obtained using Micromeritics ASAP 2020 equipment. The solids were previously degassed at 100 ºC for 2h. The specific surface area (S_{BET}) of the samples were determined using the BET method, the specific surface area of micropores (S_{micro}) were estimated by the *t*-plot method, and the average pore diameter (D_p) were obtained using the Barret-Joyner-Halenda (BJH) method.

The secondary electron micrographs of the samples were obtained by scanning electron microscopy (SEM), using Philips Model 505 equipment.

The Fourier transform infrared (FT-IR) spectra of the solids were obtained using a Bruker IFS 66 FT-IR spectrometer and pellets in KBr in the 400–4000 cm^{-1} wavenumber range.

The diffuse reflectance spectra (DRS) of the materials were recorded using a UV-visible Lambda 35, Perkin Elmer spectrophotometer, to which a diffuse reflectance chamber Labsphere RSA-PE-20 with an integrating sphere of 50 mm diameter and internal Spectralon coating is attached, in the 200–800 nm wavelength range.

The catalytic activity of the materials was evaluated in the photodegradation of 4-chlorophenol (Fluka) in water, at 25 ºC. The tests were carried out employing a 125 W high-pressure mercury lamp (with a maximum emission at about 365 nm) placed inside a Pyrex glass jacket, thermostated by water circulation, and immersed in the 4-chlorophenol (4-CP) solution contained in a 300 ml cylindrical Pyrex glass reactor. The catalyst is maintained in suspension by stirring, and air is continuously bubbled. Previously, the 4-CP solution (200 ml, 1.5 10^{-4} mol/l) containing 100 mg of catalyst was magnetically stirred in the absence of light for 60 min to ensure that the adsorption-desorption equilibrium of 4-CP on the surface of the materials is attained. The 4-CP degradation was carried on at the pH of the suspension obtained by the addition of 100 mg of catalyst to 200 ml of a 4-CP water solution. The initial and the final pH of the solutions were in the range 6.9-5.2.

During the course of the experiments, samples (3 ml) were periodically withdrawn, filtered using a Millipore syringe adapter (porosity, 0.45 μm) and then analyzed. The variation of the 4-CP concentration as a function of the reaction time was determined by a UV-visible LAMBDA 35 Perkin Elmer double-beam spectrophotometer, measuring the absorbance at 225 nm. The concentration of released chloride ions was measured by a selective Cl^- electrode (pHoenix CLO1508) with an ion meter (Consort P903). All the measurements were carried out three times.

The filtrates were extracted three times with ethyl ether, the organic layers were collected, dried with sodium sulfate, and concentrated for the determination of the intermediates using a Hewlett Packard 6890N GC/MSD. The extent of 4-CP mineralization was determined using the Total Organic Carbon, Method 10129 DR/4000 (HACH).

In order to evaluate the possibility of TSA leaching during the photocatalytic degradation of 4-CP, at the end of each experiment, the catalyst was separated by decantation, and W was determined in the liquid phase by atomic absorption spectrometry using a Varian AA Model 240 spectrophotometer. The calibration curve method was used, with standards prepared in the laboratory. The analyses were carried out at a wavelength of 254.9 nm, bandwidth 0.3 nm, lamp current 15 mA, phototube amplification 800 V, burner height 4 mm and acetylene-nitrous oxide flame (11:14).

The catalytic activity of the materials was compared with commercially available titania P25 Degussa measured in the same experimental conditions.

3. Results and Discussion

The SEM studies of the samples thermally treated at 100 °C, revealed that titania samples consist of spherical particles, whose size increases from 200 to 700 nm when the TSA content is raised. These spherical particles seem to be formed by aggregation of the nanoparticle agglomerates, as was reported by Bakarjieva et al. [31] for titania samples obtained by hydrolysis of $TiOSO_4$ aqueous solution using urea as the precipitation agent. This effect is induced by urea presence. As has already been reported [32-34], urea decomposition in aqueous solution during sol-gel synthesis might produce a smooth pH increase, leading to the formation of TiO_2 nanoparticles with a controlled morphology, in this case, spherical. The size and shape of the spherical particles remain practically unchanged during the calcination step.

According to the XRD patterns of $TiTSA00_{T100}$, $TiTSA10_{T100}$, $TiTSA20_{T100}$, and $TiTSA30_{T100}$ samples (Figure 1a) the solids are poorly crystallized and mostly amorphous. They presented weak broad peaks in the same position where the characteristic peaks of the anatase phase are placed: 25.3° (101), 37.9° (004), 47.8° (200), and 54.3° 2θ.

The crystallinity of the TiTSA00 sample increased when the temperature was raised, and the peak at 54.3° splits into two peaks at 54.0° (105) and 54.9° (211) when the temperature was above 300°C ($TiTSA00_{T400}$ and $TiTSA00_{T500}$ samples). The XRD pattern of the $TiTSA00_{T600}$ sample exhibited three new peaks at $2\theta = 27.4°$, 36.1°, and 54.2 °, as result of the partial transformation of anatase into rutile phase.

On the other hand, the XRD patterns of all the samples modified with TSA, including those calcined at 600°C (Figure 1a, b, and c, respectively), only showed the presence of titania in anatase structure, suggesting that the anatase into rutile phase transition is shifted to higher temperatures in the presence of TSA. The results showed that the presence of TSA retarded the crystallization of titania and stabilized TiO_2 in the anatase phase, in agreement with literature reports of titania modified with tungsten oxide [36, 37].

The crystallinity degree of the TSA/titania composites increased with the calcination temperature, but to a lower extent for higher TSA amount.

Neither crystalline TSA nor TSA decomposition products were observed by XRD, so TSA could be well dispersed in the titania matrix or crystallites low enough to be detected by this technique could be present.

The anatase crystallite size (D_C) of samples thermally treated at 100°C, estimated from the XRD results using the Scherrer equation and silicon as standard for the correction of instrumental broadening, showed that D_C is almost independent of the TSA content (Table 1).

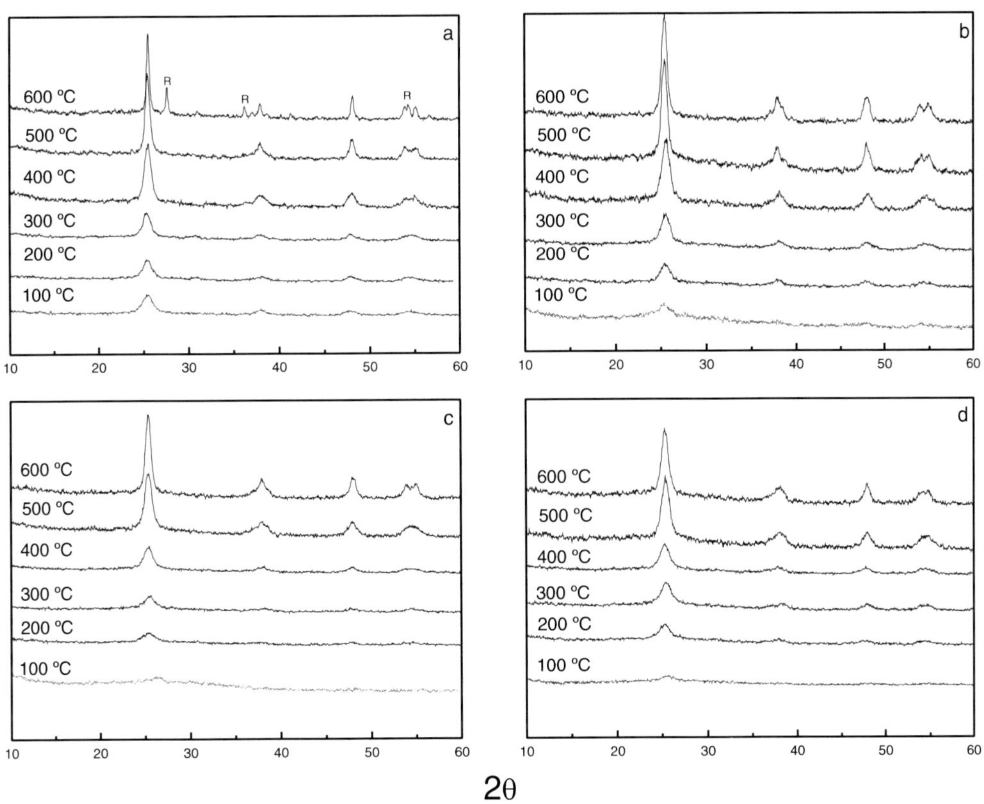

Figure 1. XRD patterns of TiTSA00 (a), TiTSA10 (b), TiTSA20 (c), and TiTSA30 (d) samples, treated at different temperatures. (R= peaks assigned to rutile phase).

Table 1. Physicochemical properties of the TSA-modified mesoporous titania treated at 100°C

Sample	S_{BET} (m^2/g)	S_{micro} [a] (m^2/g)	D_p [b] (nm)	Total pore volume [c] (cm^3/g)	D_c [d] (nm)
$TiTSA00_{T100}$	372	35	3.1	0.195	5.9
$TiTSA10_{T100}$	238	18	4.2	0.215	5.5
$TiTSA20_{T100}$	208	15	4.8	0.166	5.5
$TiTSA30_{T100}$	160	12	5.0	0.104	5.6

[a] Specific surface area of micropores was obtained from *t*-plot analysis.
[b] Average pore diameter was calculated using the BJH formula from the desorption branch.
[c] Total pore volume was calculated by the BJH method.
[d] Anatase crystallite size.

However, a D_C increase for higher calcination temperatures was observed, though the increment was lower for higher TSA content. This behavior may be attributed to crystal growth delay, which is very common in materials containing both a crystalline and an amorphous phase [38].

According to the S_{BET} and S_{micro} values (Table 1), the $TiTSAXX_{T100}$ samples are mesoporous materials with a D_p higher than 3.1 nm. The S_{BET} decreased and the D_p increased for higher TSA concentration in the solid. Taking into account the values of S_{micro}, less than 10% of the total specific surface area is due to a microporous structure.

The S_{BET} values of the TSA-modified samples treated at 100 °C decreases in the following order: $TiTSA10_{T100} > TiTSA20_{T100} > TiTSA30_{T100}$ (Table 1), and are lower than the $TiTSA00_{T100}$ value. According to the literature, the lower S_{BET} values may be due to a decrease in the cross-linking degree when the acid concentration increases [38]. These results are also in agreement with our previous reports for the specific surface area of mesoporous titania modified by the addition of TPA [11].

S_{BET} varied when the calcination temperature was increased, as is shown in Figure 2. For titania without TSA addition, S_{BET} decreased almost 60% when the calcination temperature was raised from 100 to 200°C, but the decrease was significantly lower for the TSA/titania composites. It was 28% for TiTSA10, 8% for TiTSA20, and only 1% for TiTSA30 samples.

This was attributed to the interaction of the TSA incorporated into titania, which can reduce the agglomeration by surface diffusion and/or sintering [41, 42]. Besides, similar S_{BET} values were found for TiTSA20 and TiTSA30 samples treated at higher temperatures. This result could indicate a limiting value for the above-mentioned TSA-titania interaction.

The $TiTSA00_{T100}$ FT-IR spectrum displayed bands with maxima at 3380 (very broad) and 1649 cm^{-1}, assigned to the OH stretching and bending vibrations of water that has been incorporated in the structure, and a wide band due to the Ti-O-Ti stretching vibration below 900 cm^{-1} (Figure 3a).

Similar features were observed in the range 1100-4000 cm^{-1} for the TSA/titania composites calcined at 100, 200, 300, 400, and 500 °C, though new bands appeared below 1100 cm^{-1} in these samples with TSA addition, overlapped to the characteristic wide band of titania. The new bands are placed at the same wavenumbers as those of bulk TSA, that is, at 1020, 982, 926, 884, 778, and 541 cm^{-1}, as is observed in Figure 8 for the samples treated at 100 °C. So, they can be assigned to the $[SiW_{12}O_{40}]^{4-}$ anion.

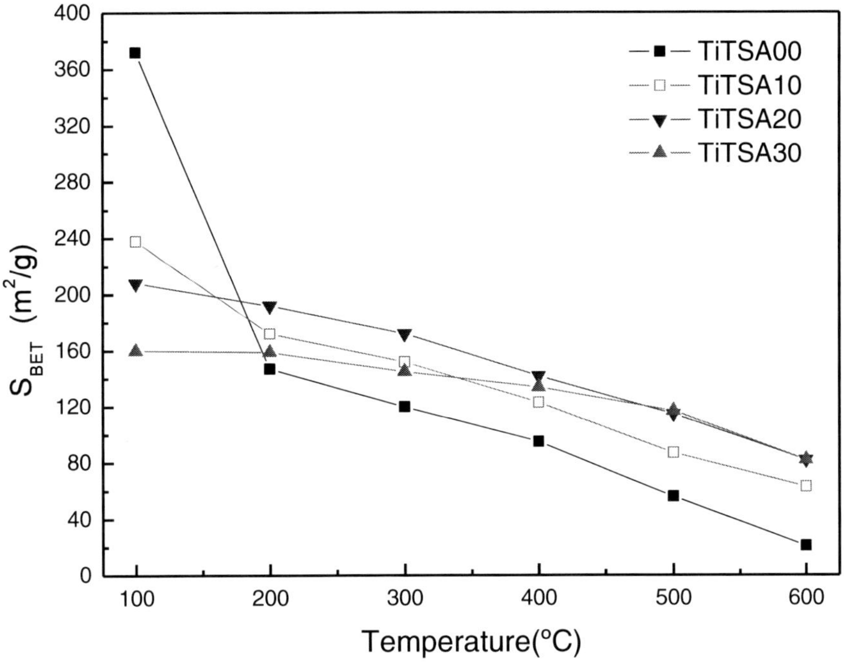

Figure 2. Variation of S_{BET} as a function of calcination temperature.

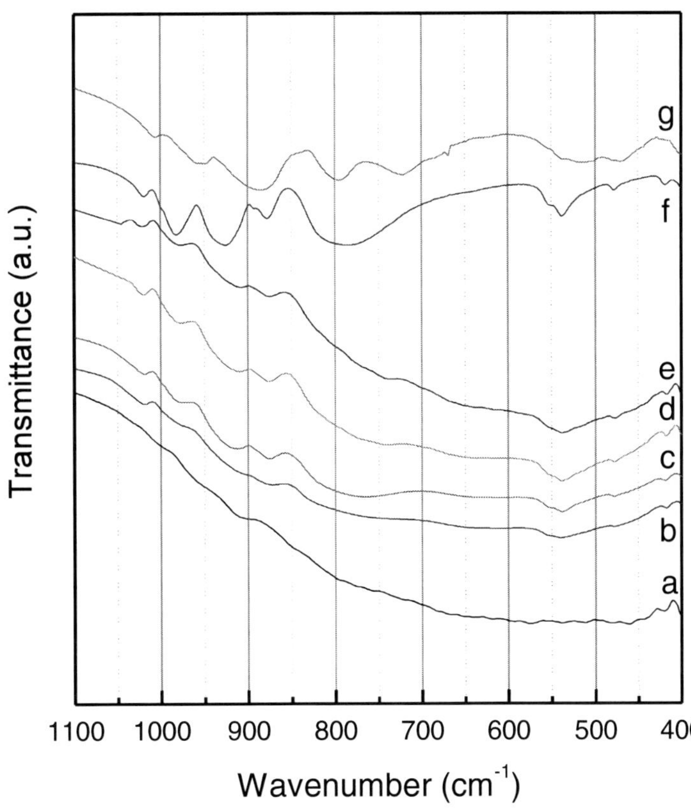

Figure 3. FT-IR spectra of TiTSA00$_{T100}$ (a), TiTSA10$_{T100}$ (b), TiTSA20$_{T100}$ (c), TiTSA30$_{T100}$ (d), and TiTSA30$_{T600}$ (e) samples; bulk TSA (f), and sodium salt of $[SiW_{11}O_{39}]^{8-}$ anion (g).

From the FT-IR studies previously reported [30], we established that the main species present in the titania modified with TSA is the $[SiW_{12}O_{40}]^{4-}$ anion that remains unchanged during the subsequent thermal treatment until 500 °C. However, the spectra of the samples treated at 600ºC and higher temperatures display a decrease of the intensity of the bands assigned to anion was observed (Figure 3), possibly as a result of a partial degradation of the Keggin-type structure. However, the main bands of the $[SiW_{11}O_{39}]^{8-}$ lacunar phase placed at 1008, 959, 946, 882, 869, 794, and 720 cm^{-1} (Figure 3g) were not observed.

The UV-Vis-DRS spectra of TiTSA00 samples presented an intense absorption in the range 200–390 nm, corresponding to charge transfer from the valence band (O 2p) to the conduction band (Ti 3d) [43]. On the other hand, bulk TSA displayed two absorption bands in the range 200–450 nm assigned to the charge transfer from bridging or terminal O 2p to W 5d (W-O-W and W-O_d, respectively) [44].

We have reported that as result of introduction of TSA into the titania matrix the UV-Vis-DRS spectra of tungstosilicic acid/titania composites samples displayed a shift of the charge transfer band to wavelength values closer to that of bulk TSA [30].

The DRS spectra of the solids thermally treated up to 400ºC showed the absorption threshold onset continuously shifted to the visible region when the TSA concentration increased, as is observed in Figure 4. However, the red shift of the absorption threshold onset is higher for the samples with TSA addition calcined at 500 and 600ºC (Figure 4).

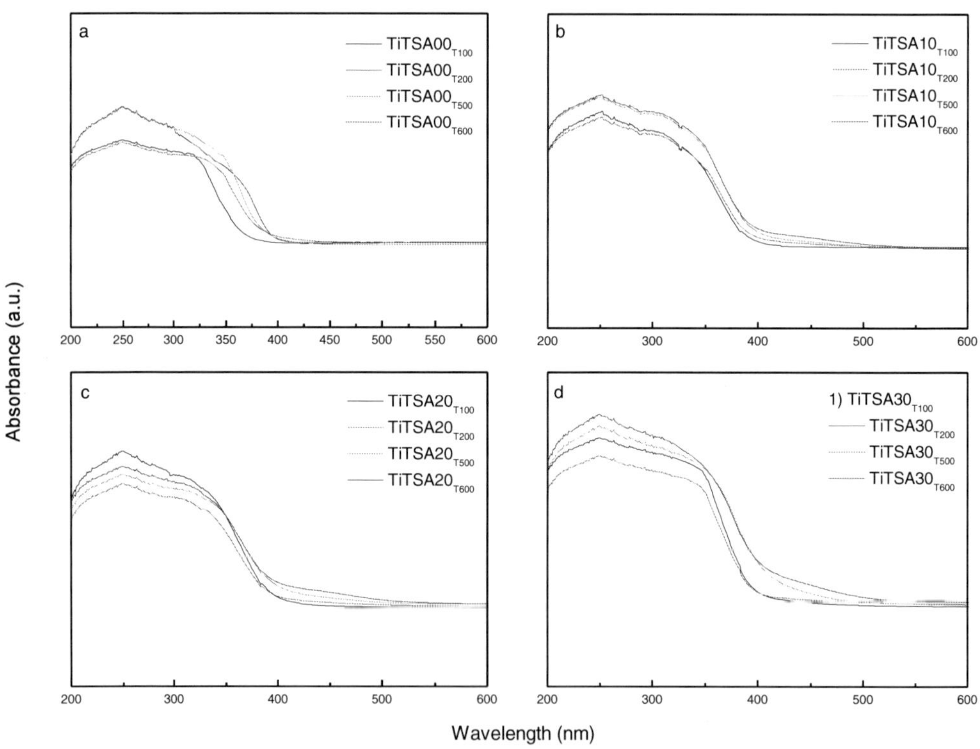

Figure 4. DRS of TiTSA00 (a), TiTSA10 (b), TiTSA20 (c), and TiTSA30 (d) samples treated at different temperatures.

The *band gap* energy (E_g) of the synthesized samples, estimated from the DRS spectra using the Kubelka-Munk remission function [45], are presented in Figure 5. For the TiTSA00 samples, E_g is in the range 3.20-2.85 eV. The values decreased with the increase in the calcination temperature. This may be caused by a higher crystallinity degree and/or the appearance of the rutile phase. The band gap energy of TiTSA00 samples calculated from the absorbance values of the DRS spectra was in the range 3.20–2.99 eV (Table 1). The values decreased when the calcination temperature increased, probably as a result of a higher crystallinity and the appearance of the rutile phase [45].

The *band gap* energy of the TSA/titania composites was lower than that of the TiTSA00 samples as a result of the introduction of TSA in the titania matrix(Figure 5); the order was the following: TiTSA00 > TiTSA10 > TiTSA20 > TiTSA30. However, the decrease was remarkable for the TiTSA10 samples, while the samples with higher TSA concentration exhibited a lower decrease. The E_g values are also lower when the calcination temperature is increased, but they change to a lesser extent than in the case of TiTSA00 sample.

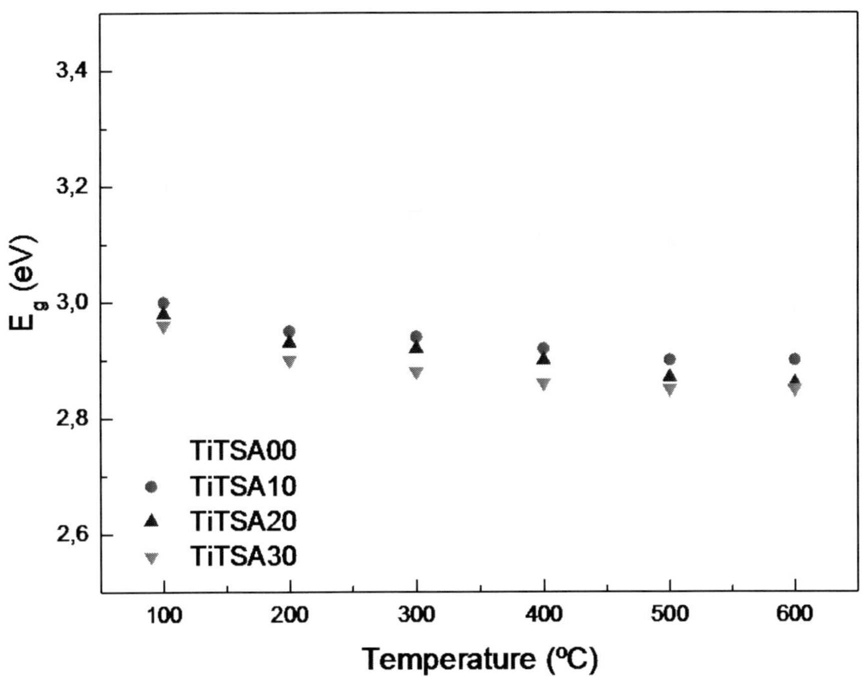

Figure 5. *Band gap* energy of TiTSA00, TiTSA10, TiTSA20, and TiTSA30 samples calcined at different temperatures.

The catalytic activity of spherical nanoparticles of tungstosilicic acid/titania composites in the photocatalytic degradation of 4-chlorophenol (4-CP) were tested at the pH of the suspension obtained by the addition of 100 mg of catalyst to 200 ml of a 4-CP water solution ($1.5\ 10^{-4}$ mol/l). The initial and the final pH of the solutions were in the range 6.9-5.2. Under this experimental conditions 4-CP (a weak acid of p*Ka* = 9.41) was mainly present in the non ionized form, and the surface of the catalysts was mostly negatively charged. As a result, 4-CP an important adsorption does not take place on the catalyst surface.

After 240 min of irradiation but without catalyst, the amount of 4-CP degraded after was only of 6 %, and it is mainly due to a photolytic process [46]. 4-CP degradation was not detected under the same experimental conditions but with the mercury lamp turned off. The reduction in the 4-CP concentration was lower than 9% at 240 min under reaction when $TiTSA00_{T100}$ was used as catalyst. However, the 4-CP degradation at the same time for the TSA-modified samples increases in parallel with the increment of tungstosilicic acid in the catalyst: $TiTSA10_{T100}$ (20 %) < $TiTSA20_{T100}$ (38 %) < $TiTSA30_{T100}$ (45 %).

Previous reports indicate that the cleavage of the Cl-aryl bond, releasing inorganic chloride to the reacting medium, is the first step during the photocatalytic degradation of 4-CP. In agreement with those reports, we found that the ratio between the amount of degraded 4-CP (C_{D4C-P}) and the amount of released chloride ions (C_{Cl^-}) as a function of the irradiation time remain practically equal to 1 during the course of the photodegradation.

Hydroquinone (HQ) and benzoquinone (BQ) have been reported as predominant aromatic intermediaries when 4-CP was photodegraded using silica-immobilized polyoxometallates and polyoxometallates, respectively [47, 20]. However, some of the intermediates detected during the 4-CP photodegradation catalyzed by $[PW_{12}O_{40}]^{3-}$ were not found or were present as traces (HQ and 4-chlorocatechol respectively) when TiO_2 was used [48]. It was argued that this fact is due to the photodegradation catalyzed by TiO_2 operates via holes and $^{\bullet}OH$ radicals, while essentially $^{\bullet}OH$ radicals take part in the case of $[PW_{12}O_{40}]^{3-}$ photocatalysis. In our experimental conditions, we found that BQ was the predominant intermediate and only traces of HQ and 4-CC were detected.

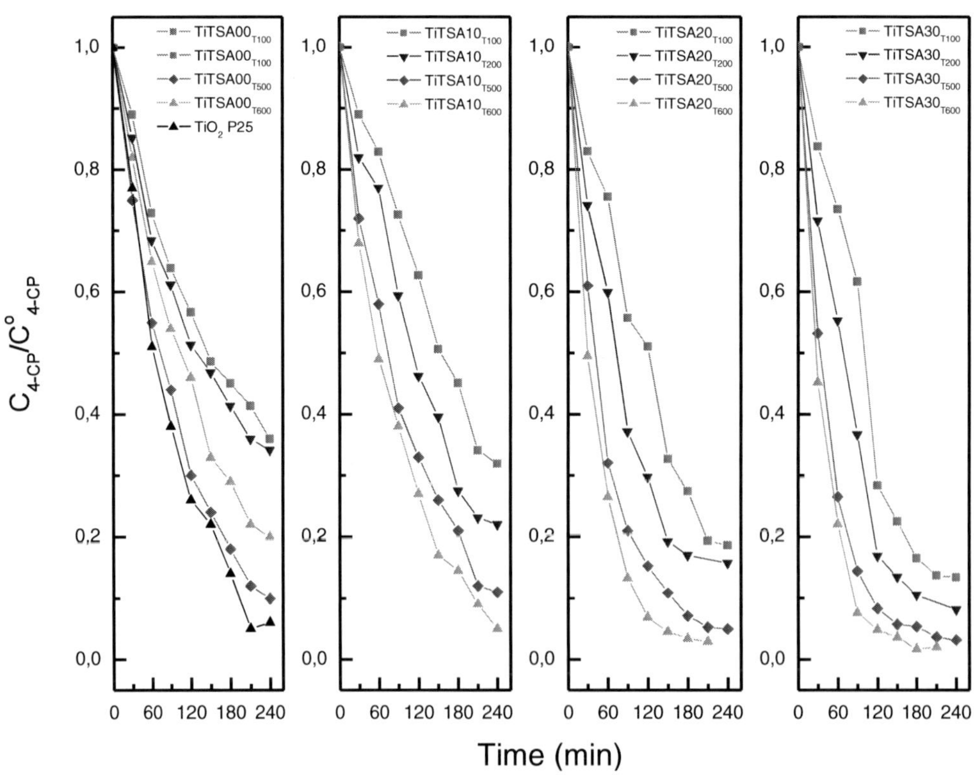

Figure 6. Photocatalytic degradation of 4-CP as a function of the irradiation time for the TiTSA samples treated at 100, 200, 500, and 600 °C, and TiO_2 Degussa P25.

The 4-CP degradation profiles using TiTSA00, TiTSA10, TiTSA20, and TiTSA30 samples calcined at selected temperatures in the range 100-600 ºC are shown in Figure 6. We can see that for TiTSA00 and TiTSAXX samples, 4-CP percentage degraded after 240 min increased when the calcination temperature was raised up to 500 and 600 ºC respectively, and decreased at higher temperature values. Furthermore, when a particular temperature is considered, the amount of degraded 4-CP increased with the increment of the TSA content. Besides, for all the samples the 4-CP percentage degraded after 240 min was higher than those obtained using samples treated at 100ºC.

The degradation profile of 4-CP using TiO_2 Degussa P25 (commonly used as reference) was very similar to those of the $TiTPA00_{500}$ and $TiTPA10_{500}$ samples.

We have used to described the photocatalytic degradation of 4-CP in aqueous systems a pseudo-first-order kinetics with respect to the 4-chlorophenol concentration [30, 49-51], which results from considering that the reaction rate follows the Langmuir-Hinshelwood model

$$r_{4-CP} = -\frac{dC_{4-CP}}{dt} = \frac{k_r\, K\, C_{4-CP}}{1 + K\, C_{4-CP}}$$

where $r_{4\text{-}CP}$ is the degradation rate, $C_{4\text{-}CP}$ is the 4-CP concentration, and k_r and K are the reaction and the adsorption constants, respectively. When $C_{4\text{-}CP}$ is low, $KC_{4\text{-}CP}$ is generally negligible and the reaction rate can be assumed as of pseudo-first-order with respect to the dye concentration. The resultant equation can be integrated considering that $C_{4\text{-}CP0}$ is the 4-CP concentration at time equal zero, so

$$\ln\left(\frac{C_{4CP0}}{C_{4-CP}}\right) = k_r\, K\, t$$

We obtained the apparent reaction constant $k_{ap}= k_r$ K values from the plots of $\ln(C_{4\text{-}CP0}/C_{4\text{-}CP})$ as a function of time. As we can see in Figure 7, the k_{ap} values of the TiTSA00 sample increased when the calcination temperature was raised up to 500ºC, us result of the higher crystallinity of the $TiTSA00_{T500}$ sample compared with that treated at 100ºC, and it is slightly lower than the value estimated for TiO_2 Degussa P25 (k_{ap} = 1.05 10^{-2} min^{-1}). This increment of k_{ap} values is attributed to a reduction of the number of defects, which act as recombination centers for the photogenerated electrons and holes [52-55]. The transformation of anatase into rutile phase and the drop of S_{BET} are responsible for the decrease of k_{ap} [56] observed for $TiTSA00_{T600}$ sample (Figure 7).

The positive effect of the modification with TSA in the catalytic activity of titania is reflected on the slight increase of k_{ap} values of the samples annealed at 100ºC ($TiTSA00_{T100} < TiTSA10_{T100} < TiTSA20_{T100} < TiTSA30_{T100}$). Keeping in mind that the S_{BET} values decreases in the reverse order and that the samples are poorly crystallized solids, the increase of k_{ap} values with the TSA content may be ascribed to the participation of TSA in the decrease of the recombination degree of photogenerated charges, as was suggested by Park and Choi [26]. Fuerte et al. [57] proposed a similar explanation to the continuous enhancement of the photocatalytic activity in the degradation of toluene when increasing the percentage of W present in the anatase structure of TiO_2 ammonium tungstate solutions.

However, the increment of the catalytic activity increment could be also due to the direct participation of TSA in the degradation of the organic substrate (reactions 1-3) and/or in the production of $^{\bullet}OH$ reactive species (reaction 6) that take part of the degradation of the organic substrate (reaction 7) [30, 58].

$$POM^{*} + H_2O \rightarrow POM^{-} + {}^{\bullet}OH + H^{+} \quad (6)$$

$${}^{\bullet}OH + S \rightarrow \text{oxidation products} \quad (7)$$

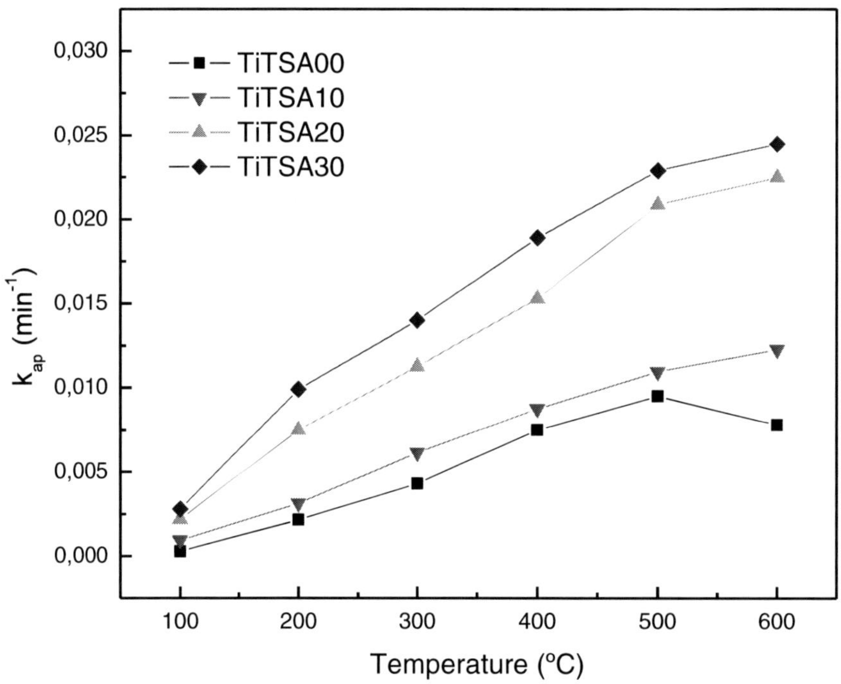

Figure 7. Apparent reaction constant kap of the 4-CP photocatalytic degradation as a function of the thermal treatment temperature.

The k_{ap} values of the TSA-modified samples increased with the rise of TSA content, though the increase is small for TiTSA10 samples, and regardless the thermal treatment temperature considered they are higher than those of TiTSA00. On the other hand, the k_{ap} values also increased with the increment of the calcination temperature from 100 up to 600°C and then decreased. The diminution in k_{ap} could be assigned to both the S_{BET} drop and the structure disruption of the $[SiW_{12}O_{40}]^{4-}$ anion during the thermal treatment, as was suggested based on the FT-IR studies.

TSA-modified samples calcined at 600°C present the highest k_{ap} values, Independently of the TSA content. These values are also higher than that of TiO_2 Degussa P25. So, it is clearly observed that the reaction rate was affected not only by the TSA amount but also by the thermal treatment, and that different factors are involved in the catalytic performance.

Taking into account the obtained results, the samples that showed the highest activity at each particular TSA content (TiTSA10$_{T600}$, TiTSA20$_{T600}$, and TiTSA30$_{T600}$) were chosen to test their reusability. Having this purpose in mind, after each photocatalytic experiment, the catalysts were easily separated from the resulting suspension by decantation, washed with distilled water, dried at 70ºC and reused. The possibility of a quick catalyst recovery by decantation is clearly an important advantage on the TiO_2 Degussa P25, that can be only separated from the suspension by centrifugation.

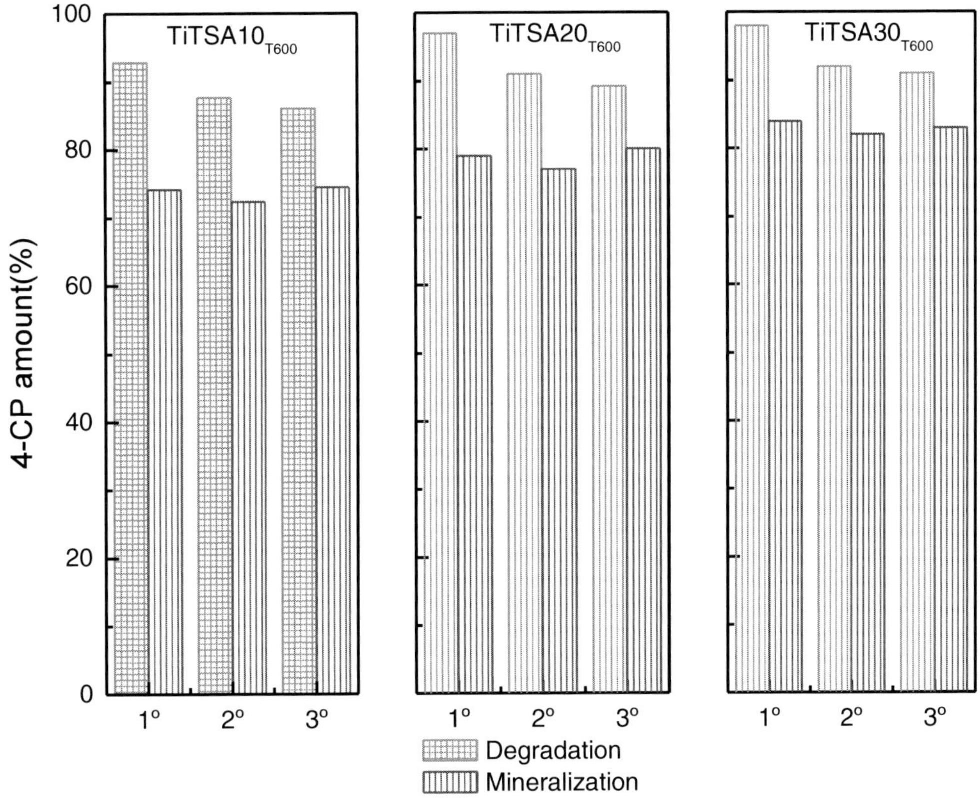

Figure 8. 4-CP degradation and mineralization degree as a function of the number cycle of.

The amount of 4-CP degraded remains practically unchanged for the TiTSA10$_{T600}$ and TiTSA20$_{T600}$ samples after each cycle of usage (Figure 8). However, in the case of the TiTSA30$_{T600}$ sample, it decreases slightly during the first reuse, and then keeps constant. For all the tested samples, the mineralization degree was lower than the amount of degraded 4-CP, due to the formation of organic intermediates previously mentioned. It also, decreases in the same way that the percentage of degraded 4-CP after each cycle of usage. We assigned the decrease to the solubilization of TSA from the fresh samples during the first catalytic test, which was established by atomic absorption spectrometry as less than 1% of the TSA content. According to these results after the third use the catalytic performance of the TiTSA20$_{T600}$ and TiTSA30$_{T600}$ samples are practically the same. In summary, the nanospheres of titania directly modified with TSA were shown to be reusable photocatalysts to degrade a toxic chlorinated phenol.

Conclusion

Mesoporous nanospheres of titania directly modified with tungstosilicic acid as were synthesized by sol-gel type reactions using urea as a pore-forming agent. The specific surface area of the materials decreased when both the TSA amount and the calcination temperature increased. The TSA-modified samples only presented titania in an anatase-type structure due to the shift of the phase transition of anatase into rutile to higher temperatures. The crystallite size increased with the calcination temperature, but lower values were obtained for higher TSA content. The Keggin structure of the $[SiW_{12}O_{40}]^{4-}$ anion remains intact only in the samples thermally treated until 500°C. A partial degradation of the Keggin-type structure takes place at higher temperatures.

The nanospheres of titania modified with TSA display higher photocatalytic activity in the 4-chlorophenol degradation than the neat TiO_2. According with the apparent reaction constant, estimated assuming a pseudo-first-order kinetics, the catalyst behavior depended on the TSA amount and the thermal treatment temperature. The sample containing 30% TSA calcined at 600°C was the most effective. On the other hand, the reused catalysts showed only a slight decrease in the degradation and mineralization degree, thus being promissory materials to aid in the photocatalytic treatment of wastewater that contain chlorinated phenols.

Acknowledgments

The authors thank L. Osiglio, and G. Valle, for their experimental help, and the financial support of CONICET and UNLP.

References

[1] A. Sclafani, L. Palmisano, M. Schiavello, *J. Phys. Chem.* 94 (1990) 829.
[2] K. B. Kyoko, S. Kazuhiro, K. Hitoshi, O. Kiyomi, A. Hironori, *Appl. Catal. A: General* 165 (1997) 394.
[3] S. Sakulkhaemaruethai, S. Pavasupree, Y. Suzuki, S. Yoshikawa, *Mater. Lett.* 59 (2005) 2965.
[4] J.-Y. Zheng, J.-B. Pang, K.-Y. Qiu, Y. Wei, *Micropor. Mesopor. Mater.* 49 (2001) 189.
[5] L. R. Pizzio, *Mater. Lett.* 59 (2005) 994.
[6] X. Li, J.W. Cubbage, T.A. Tetzlaff, W.S. Jenks, J. *Org. Chem.* 64 (1999) 8509-8524.
[7] A. Sclafani, L. Palmisano, M. Schiavello, *J. Phys. Chem.* 94 (1990) 829.
[8] K.B. Kyoko, S. Kazuhiro, K. Hitoshi, O. Kiyomi, A. Hironori, *Appl. Catal. A: Gen.* 165 (1997) 394.
[9] S. Sakulkhaemaruethai, S. Pavasupree, Y. Suzuki, S. Yoshikawa, *Mater. Lett.* 59 (2005) 2965.
[10] J.-Y. Zheng, J.-B. Pang, K.-Y. Qiu, Y. Wei, *Micropor. Mesopor. Mater.* 49 (2001) 189.
[11] V. M. Fuchs, L. Méndez, M.N. Blanco, L.R. Pizzio, *Appl. Catal. A: General* 358 (2009) 73.

[12] W. Lee, Y. M. Gao, K. Dwight, A. Wold, *Mater. Res. Bull.* 27 (1992) 685.
[13] A. Sclafani, M. N. Mozzanega, P. J. Pichat, *Photochem. Photobiol. A: Chem.* 59 (1991) 181.
[14] W. Lee, Y. R. Do, K. Dwight, A. Wold, *Mater. Res. Bull.* 28 (1993) 1127.
[15] S. Ikeda, N. Sugiyama, B. Pal, G. Marci, L. Palmisano, H. Noguchi, K. Uosaki, B. Ohtani, *Phys. Chem. Chem. Phys.* 3 (2001) 267.
[16] M. Anpo, *Catal. Surv. Jpn.* 1 (1997) 169.
[17] Y. R. Do, W. Lee, K. Dwght, A. World, *J. Solid State Chem.* 108 (1994) 198.
[18] K. Tennakone, O. A.Haperuna, J. M. Bandara, W. C. B. Kiridena, *Semicond. Sci. Technol.* 7 (1992) 423.
[19] G. Marci, L. Palmisano, A. Sclafani, A.M. Venezia, R. Campostrini, G. Carturan, C. Martín, V. Rives, G. Solana, *J. Chem. Soc. Faraday Trans.* 92 (1996) 819.
[20] A. Mylonas, E. Papaconstantinou, *J. Photochem. Photobiol.* 94 (1996) 77.
[21] C. Hu, B. Yue, T. Yamase, *Appl. Catal. A: General* 194-195 (2000) 99.
[22] R. R. Ozer, J.L. Ferry, *J. Phys. Chem.* B 104 (2000) 9444.
[23] M. Yoon, J.A. Chang, Y. Kim, J.R. Choi, K. Kim, S.J. Lee, *J. Phys. Chem.* B 105 (2001) 2539.
[24] R. R. Ozer, J.L. Ferry, *Environ. Sci. Technol.* 35 (2001) 3242.
[25] Y. Yang, Y. Guo, C. Hu, Y. Wang, E. Wang, *Appl. Catal. A: General* 273 (2004) 201.
[26] H. Park, W. Choi, *J. Phys. Chem.* B 107 (2003) 3885.
[27] L. Li, Q.-Y. Wu, Y.-H. Guo, Ch.-W. Hu, *Micropor. Mesopor. Mater.* 87 (2005) 1.
[28] V. M. Fuchs, E. L. Soto, M. N. Blanco, L. R. Pizzio, *J. Colloid Interface Sci.* 327 (2008) 403.
[29] M. N. Blanco, L.R. Pizzio, *Appl. Surf. Sci.* 256 (2010) 3546.
[30] M. N. Blanco, L.R. Pizzio, *Appl. Catal. A: General* 405 (2011) 69.
[31] S. Bakardjieva, J. Šubrt, V. Štengl, V. Balek, M.J. Dianez, M.J. Sayagues, *Appl. Catal. B Environ.* 58 (2005) 193.
[32] G. J. Soler,-Illia, M. Jobaggy, R.J. Candal, A.E. Regazzoni, M.A. Blesa, *J. Disper. Sci. Technol.* 19 (1998) 207-228.
[33] G. J. Soler-Illia, R.J. Candal, A.E. Regazzoni, M.A. Blesa, *Chem. Mater.* 9 (1997) 184-191.
[34] M. Zhou, J. Yu, H. Yu, J. Mol. *Catal. A Chem.* 313 (2009) 107-113.
[35] N. Phonthammachai, T. Chairassameewong, E. Gulari, A.M. Jamieson, S. Wongkasemjit, *Micropor. Mesopor. Mater.* 66 (2003) 261.
[36] X.-F. Yu, N.-Z. Wu, H.-Z. Huang, Y.-C. Xie, Y.-Q. Tang, *J. Mater. Chem.* 11 (2001) 3337.
[37] H. Yang, D. Zhang, L. Wang, *Mater. Lett.* 57 (2002) 674.
[38] E. Ortiz-Islas, T. López, R. Gómez, M. Picquart, D.H. Aguilar, P. Quintana, *Appl. Surf. Sci.* 252 (2005) 853.
[39] K.M.S. Khalil, T. Baird, M.I. Zaki, A.A. El-Samahy, A.M. Awad, *Colloid Surf.* A 132 (1998) 31.
[40] S. Eibil, B. C. Gates, H. Knozinger, *Langmuir* 17 (2001) 107.
[41] S. M. Kumbar, G.V. Shanbhag, F. Lefebvre, S.B. Halligudi, *J. Mol. Catal. A: Chem.* 256 (2006) 125.
[42] R.D. Shannon, *Acta Crystallogr.* A 32 (1976) 751.

[43] P. I. Villabrille, L. R. Pizzio, P. G. Vázquez, C. V. Cáceres, M. N. Blanco, *Current Topics in Colloid Int. Sci.* 6 (2003) 43.
[44] S. P. Tandom, J. P. Gupta, *Phys. Stat. Sol.* 38 (1970) 363.
[45] M.A. Barakat, H. Schaeffer, G. Hayes, S. Ismat-Shah, *Appl. Catal. B: Environ.* 57 (2005) 23.
[46] B. Yue, Y. Zhou, J. Yuxu, Z. Wu, X. Zhang, Y. Zou, S. Jin, *Environ. Sci. Technol.* 36 (2002) 1325.
[47] P. Kormali, A. Troupis, T. Triantis, A. Hiskia, E. Papaconstatinou, *Catal. Today* 124 (2007) 149.
[48] C. Lettmann, K. Hindenbrand, H. Kisch, W. Macyk, W.F. Maier, *Appl. Catal.* B: Environ. 32 (2001) 215.
[49] G. Sivalingam, M.H. Priya, G.Madras, *Appl. Catal. B: Environ* 51 (2004) 67.
[50] Y. Cheng, H. Sun, W. Jin, N. Xu, *Chem. Eng. J.* 128 (2007) 127.
[51] K. J. Jung, S.B. Park, *Appl. Catal. B: Environ.* 25 (2000) 249.
[52] H. Kominami, J. Kato, S. Murakami, Y. Ishii, M. Kohno, K. Yabutani, T. Yamamoto, Y. Kera, M. Inoue, T. Inui, B. Ohtani, *Catal. Today* 84 (2003) 181.
[53] J. G. Yu, H.G. Yu, B. Cheng, X.J. Zhao, J.C. Yu and W.K. Ho, *J. Phys. Chem.* B 107 (2003) 13871.
[54] C. Guillard, J. Disdier, J. M. Herrmann, C. Lehaut, T. Chopin, S. Malato, J. Blanco, *Catal. Today* 54 (1999) 217.
[55] C. Su, B.-Y. Hong, C.-M. Tseng, *Catal. Today* 96 (2004) 119.
[56] A. Fuerte, M.D. Hernández-Alonso, A.J. Maira, A. Martínez-Arias, M. Fernández-García, J.C. Conesa, J. Soria, G. Munuera, *J. Catal.* 212 (2002) 1.
[57] S. Antonaraki, T. M. Triantis, E. Papaconstantinou, A. Hiskia, *Catal. Today*, 151 (2010) 119.

In: Comprehensive Guide for Mesoporous Materials, Volume 2 ISBN: 978-1-63463-982-8
Editor: Mahmood Aliofkhazraei

Chapter 4

DUAL, AMORPHOUS AND CRYSTALLINE CHARACTERISTICS IN PORE WALLS OF MESOPOROUS FERRISILICATES

Károly Lázár[1] and Ágnes Szegedi[2]
[1]Centre for Energy Research, Hungarian Academy of Sciences, Budapest, Hungary
[2]Research Centre for Natural Sciences, Hungarian Academy of Sciences, Budapest, Hungary

ABSTRACT

Pore walls in mesoporous silicates may exhibit dual, simultaneously crystalline and amorphous features. The extents of the two characteristics may be varied depending on the synthesis and further treatments. Their partitions can be followed with different methods. Examples are presented from studies performed on Fe-MCM-41 ferrisilicates synthesized with different procedures, by hydrothermal synthesis and by hydrolysis of tetraethylorthosilicate. Properties of samples are compared by applying primarily in situ Mössbauer and infrared spectroscopies complemented with XRD, BET, TEM, TPR methods. Flexibility in variations of the crystalline/amorphous parts is illustrated in dependence of the actual conditions. These structural changes can also be strongly correlated to $Fe^{3+} \Leftrightarrow Fe^{2+}$ redox processes.

INTRODUCTION

Stability of synthetic mesoporous silicates depends stongly on the structure of their pore walls. Synthesis of these materials usually takes place in aqueous media, using some silicon source and long chain surfactants, to provide a mean to form micelles in the first stage of synthesis. The surfactant and most of the water is removed in further steps. The final strucure of pore walls, in particular the thickness and the extent of crystallinity may vary in them depending on the conditions of the synthesis.

One of the substances of this type, MCM-41, was prepared successfully in the early 1990-s. Its synthesis and characterisation ha been studied in detail and are reported in numerous publications since then (e.g. [1, 2]). Besides the main constituent, silicon, other elements can also be incorporated into the pore walls in small amounts. Iron plays also role among these additives, since these microporous ferrisilicates may have applications in various fields (sorption, catalysis, etc.). Iron can be incorporated into ferrisilicates only in limited amounts. Attempts to introduce iron in higher proportions result usually in agglomeration and formation of separate iron oxide entities (magnetite, hematite).

Two types of coordinations were principally identified for iron embedded into pore walls, tetrahedral and trigonal ones (the latter can occasionally be completed with adsorbed water to octahedral symmetry). The procedure of the synthesis and the method of removal of water have strong influence on the structure of pore walls [3, 4]. Tetrahedral siting is analogous to that characteristic for FW sites in zeolites, where tetrahedral symmetry is imposed by the local crystalline structure, generating simultaneously a negative charge, or forming a Bronsted acidic site by addition of a proton. On the other part, the trigonal siting corresponds better to the expected original symmetry of oxygen around iron. In correspondence, this siting is characterised by smaller extent of crystallinity and exhibits Lewis acidity. Silanol groups may also be present in significant amounts in different regions of the pore walls due to the different degree of crystallinity. The amount of silanol groups can be changed with various treatments (calcination, evacuation). Then, e.g. by heating and evacuation water is removed from the structure, the number of direct Si-O-Si bonds increases, in correspondence the extent of local order and crystallinity is also enhanced.

Presence of silanol groups in the neighbourhood of iron opens a possibility for propagation of redox processes as well. Silanol groups may give a path for fast proton transfer via jumping of protons from one -OH group to another one by forming temporary $–O-(H_2)^+$ species [5]. The mobile proton may easily mediate charge transfer processes with participation of iron, as well [6]. Thus, facile redox processes occuring on iron may indicate iron locations in in flexible, less-crystalline stucture with containing silanol groups in significant amounts.

The emphasis in the recent chapter is laid on presentation and comparison of various structures in which the extent of the local order is varied in course of various treatments. The primary tool of analysis is Mössbauer spectroscopy, complemented with other techniques (IR spectroscopy, XRD, TEM, specific surface measurements, temperature programmed reduction). In particular, Fe-MCM-41 substances prepared by two different syntheses procedures are compared and changes in coordination and oxidation states of iron are followed. Their redox behaviour is also analysed, and comparisons are made with microporous ferrisilicates in certain instances. Bond strength of iron species is also referred from comparison of recoilless factors. Principally, flexibility and interconversions of various local coordination states of iron are demonstrated and described.

SYNTHESES

First steps in syntheses of ferrisilicates usually proceed in a biphasic system, which is composed from micelles of long chain amphiphilic surfactans and an aqueous phase in-between the micelles. The structure of micropores formed in later stages is imposed by the arrangement of micelles, and, further on, the crystallisation of the solid silicate structure takes place in the aqueous phase. The stages of crystallisation depend stongly on the experimental conditions.

In most cases the used amphiphilic surfactants contain tetraalkyl ammonium ions, and usually at least one of the alkyl groups is a long chain (containing at least 10 carbon atoms). The source of the silicon framework in the aqueous phase is either simple sodium silicate or tetraethyl (or –methyl) orthosilicate. The stages of crystallization depend on the charge balances among the ammonium and silicate ions. Addition of other simple anions (e.g. fluoride) may also influence the processes [4]. pH in the aqueous phase is also important in this aspect. Iron should also be introduced into the aqueous phase for preparing ferrisilicates, which influences the charge balances further since speciation of iron hydroxocomponents is significantly different from that of silicon [7]. Further on, the ionic radii of Si^{4+} and Fe^{3+} ions are also different, thus only a limited amount of iron can be inorporated into the siliceous framework by isomorphous substitution [8]. Hydrothermal syntheses usually proceed in two stages. First the silicon source (TEOS) and the surfactant are mixed in acidic aqueous media, where hydrolysis of the orthosilicate proceeds. Iron is also usually added to the mixture in this stage. After a certain equilibration (stirring) the pH is increased to alkaline region (~ 8) where gel is formed. In the last stage crystallisation is usually performed in teflon lined autoclaves in autogeneous pressures at ~ 120 - 140 °C. Iron containing MCM-41 structures are usually synthesized by this way with maximum iron contents of ~ 10 or even 16 wt % [9, 10]. Another variant is the template exchange method. In this instance iron is introduced separately, subsequent to the completion of the formation of the siliceous structure. Then, in a second stage the ionic surfactant is replaced in alcoholic (ethanolic) media to ionic species containing iron [11, 12].

In a more recent variant of syntheses neutral surfactants (triblock copolymers, e.g. poly(ethylene glycol)20-poly(propylene glycol)70-poly(ehylene glycol)20) are used to form micelles. In this instance more acidic pH can be used, resulting in the stable SBA-15 mesopores [13, 14].

Additional possibility is the modification of boundary of micelles. Crystallisation of monodisperse spherical ferrisilicate particles commences in the very beginning, if ethanol is added to the primary synthesis mixture, there is no further need for subsequent long-term hydrothermal crystallisation [15, 16].

Short summary of some typical synthesis procedures of mesoporous ferrisilicates with a few typical references typical are given in Table 1.

Syntheses of Samples Used in the Present Study

Fe-MCM-41 samples prepared with distinctly different two techniques were selected for our detailed recent studies. Namely, one type was synthesized with conventional hydrothermal method, and preparation of the second type was carried out in ethanol/water media. In the first instance three different samples were prepared by using sodium silicate solution from synthesis mixtures of nominal compositions („SS method" - see Table 2.) First the template, cetyltrimethyl ammonum bromide (C_{16}TMABr), was dissolved in water, then acidic solution of ferric sulphate was added under stirring.

Table 1. Short overview of some typical synthesis procedures of mesoporous ferrisilicates

Method	Basic components	Characteristics	References
Three stage hydrothermal	- ionic surfactant, - TEOS *, - iron salt	- hydrolysis in acidic pH - condensation in basic Media - hydrothermal synthesis	4, 9, 10
Combined with template exchange for iron species	(as above)	iron is introduced subsequently by replacing the primary ionic surfactant	11, 12
Neutral surfactant in acidic media	Pluronic 123 **	condensation at low pH	13, 14
Ethanol/water media	- ionic surfactant, - TEOS *, - iron salt	hydrothermal crystallisation is not necessary	3, 15,16

* TEOS: tetraethyl orthosilicate.
** Pluronic 123 : triblock copolymer.

Table 2. Nominal and actual compositions of SS samples

x	y	z	Si/Fe atomic ratios, actual
0.04	0.128	0.160	12.3
0.02	0.188	0.100	20.1
0.005	0.204	0.083	139

The composition of nominal synthesis mixture is:
1 SiO_2: x Fe_2O_3: y Na_2O: z Na_2SO_4: 0.29 C_{16}TMABr: 50 H_2O.

Finally the solution of sodium silicate was added slowly. The formed light yellow gel was placed to a teflon lined small autoclave. The hydrothermal synthesis was carried out for keeping the mixture at 100 °C in the autoclave for a week. Then, the formed product was washed and separated by centrifugation, and dried at ambient temperature.

The last column above displays the actual Si/Fe ratios as determined by atomic absorption spectroscopy after dissolving the product in mixture of HCl and HF.

Tetraethylorthosilicate (TEOS) was used for silica source in ethanol-water mixtures for the synthesis of samples of second type („OS samples"). The synthesis is fast with the particular feature that the product is composed from monodisperse spherical particles [16]. Three samples were prepared with nominal compositions

1 TEOS: x $Fe(NO_3)_3$: 0.3 C_{16}TMABr: 11 NH_3: 255 H_2O: 58 EtOH

with x= 0.1, 0.05 and 0.02. First, C_{16}TMABr was dissolved in water, then ethanol was added. TEOS was given as the third component. $Fe(NO_3)_3$ was added in the following stage under stirring, then pH was set to ~11 by adding NH_4OH solution. The mixture was stored overnight. The product was separated and washed and dried next day. Further details are described in [16]. Actual Si/Fe ratios were also determined in the OS samples, they were 15, 24, 52 in the order of decrease of x, respectively.

Both types of syntheses were finished by removal of template with careful calcination in nitrogen stream. The temperature was raised slowly to 480 – 540 °C with 2°C/min ramp, samples were kept at this temperature first in nitrogen for half an hour, and in air for two hours afterwards. Further on, samples will be denoted by their synthesis method and Si/Fe ratios (e.g. Fe-MCM-41-SS(20)).

METHODS FOR STRUCTURAL CHARACTERISATION OF MESOPOROUS FERRISILICATES

For structural characterisation of ferrisilicates both types of methods used for general characterisation of mesoporous structures, and methods related primarily to studying the particular environment of iron can be used.

The former ones may collect information either from large-scale regions, e.g. X-ray diffraction, XRD to investigate the extent of crystallinity, specific surface area (sorption, BET) measurements, or information from observing atomic environments, like UV-VIS, infrared, EXAFS spectroscopies, or high resolution TEM. There are methods which are particularly linked to studying states of iron in ferrisilicates. E.g. temperature programmed reduction (TPR) measurements may provide information on reducibility of ferric/ferrous species. The acidity generated by substitution of ferric ions into tetrahedral sites can also be investigated by sorption of basic molecules (ammonia, pyridine).

For instance, the mentioned framework (FW) and extra-framework (EFW) sitings can be distinguished by analysing vibration bands of adsorbed basic molecules, (e.g. pyridine), since the Bronsted and Lewis acidic adsorption sites (indicating the presence of FW and EFW iron, respectively) can be separated [17]. Strenghts of acidities of sites can be characterised by the desorption temperatures of the sorbed basic molecules. More sophisticated methods (EXAFS, EPR etc.) are also extensively used for studying porous ferrisilicates [3, 4, 18, 19].

Among other techniques Mössbauer spectroscopy can also be a promising tool to study the environment of iron ions in homogeneous ferrisilicates since the method is particularly sensitive to analyse and identify oxidation and coordination states of iron. (The corresponding characteristic spectral parameters are the isomer shift (IS), and quadrupole splitting, (QS), respectively.) Advantages of this technique have already been utilized successfully for analysis of iron containing layered silicate minerals [20]. Framework (FW) and extra-framework (EFW) sitings of iron can also easily be distinguished by the method in microporous zeolites [21-24]. The technique was also applied in studies of mesoporous ferrisilicates [3, 11, 25]. Beside the widely applied two mentioned parameters, IS and QS, a further information related to the bonding strength of iron in the porous ferrisilicates can also be extracted from the spectra by evaluating the temperature dependence of spectral area. Namely, the probability of the Mössbauer effect can explicitly be described in dependence of the measuring temperature in the frame of the Debye model of vibrations of the solid lattice

[26]. This approach has already been used for evaluating the bonding of iron in microporous Fe-FER and Fe-ZSM-5 [27], and can be used for mesoporous Fe-MCM-41, as well [28].

Principal features of techniques used for investigation of ferrisilicates are summarized in Table 3. The literature with the applied methods is abundant, only a few typical references are included for illustration or further reading.

Methods Applied for Characterisation of the Presented OS and SS Samples

X-Ray Diffraction

Powder diffractograms of samples were recorded with a Philips 1810/1870 diffractometer by using monochromatised CuK_α radiation in $2\Theta = 0.02$ steps. d_{100} plane distances as well as cell constants (a_0) were determined by using the $a_0 = 2\ d_{100} / \sqrt{3}$ relation in correspondence with the hexagonal symmetry.

Table 3. Principal features of techniques used for characterisation of mesoporous ferrisilicates

Method *	Principal information	References
General techniques for studying mesopores		
XRD	crystallinity (> 10 nm)	4, 9, 11, 16
Sorption measurements (BET isotherms)	specific surface area, pore size distribution	9, 14, 16, 25
UV-VIS	vibration of functional groups, or parts of crystal skeleton	3, 9, 12, 25
FT-IR	as above	9, 11, 18, 25
EXAFS, XANES	electron density	3, 9, 12, 18
TEM	morphology,	9, 16, 29
Techniques using iron as a local probe		
TPR	reducibility of iron species	10, 16, 18
NH_3 or pyridine sorption	strenght of acidity (Lewis/Bronsted distinction)	16, 30
EPR	symmetry around ferric ions	3, 4, 11, 12
Mössbauer	oxidation and coordination states of iron	3, 11, 25, 29

* Abbreviations.
XRD: X-ray diffrraction.
BET: Brunauer-Emmet-Teller sorption isotherms.
UV-VIS: Ultra-violet and visible spectroscopy.
FT-IR: Fourier transformad infra red spectroscopy.
EXAFS: X-ray fine structure spectroscopy.
XANES: X-ray near K-edge spectroscopy.
TEM: Transmission electron microscopy.
TPR: Temperature programmed reduction.
EPR: Electron paramagnetic resonance.

Specific Surface Area, Pore Size Distribution and Wall Thickness

Specific surface areas and pore size distributions were determined from nitrogen adsorption isotherms measured with an automatised Quantachrome NOVA apparatus at 77 K. Adsorption data were used for calculation of specific surface areas by using the BET procedure, and the pore size distributions were determined from the desorption curve with application of the Barett-Joyner-Halenda method [31].

Pore wall thicknesses (WTH) can simply be estimated by combining the XRD (cell constant, a_0) and sorption data (mean of the pore diameter, Φ_{BJH}), namely WTH = a_0 - Φ_{BJH}. Another approach was also proposed for determination of channel diameter (W_d) for mesoporous substances with cylindrical pores:

$$W_d = c\, d_{100}\, (\rho\, V_p / (1 + \rho\, V_p))^{1/2}, \qquad \text{Eq (1)}$$

where c is a geometrical factor (c = 1.213), ρ is the density of the amorphous silica (2.2 g/cm^3), and V_p is the pore volume determined in the sorption measurements [32]. Another approach is also proposed for calculation, which can be more suitable to determine the specific surface area of mesoporous subtances [33]. Namely, the $S_d = 4\ V_p\ /\ W_d$ approximation was also considered.

TEM Measurements

Philips CM20 Twin transmission electron microscope was used with 200 kV acceleration voltage to obtain images on the samples.

Infrared Spectroscopy Combined with Pyridine Sorption

Pyridine sorption measurements were performed with IR detection to determine the acid strength of the adsorption sites, in order to distinguish between the the Lewis and Bronsted centres [17]. Thin self-supporting platelets (2-3 mg/cm^2) were pressed and placed into the measuring cell described in [34]. Samples were evacuated first in vacuum (10^{-3} Pa) at 520 K, then various in situ treatments were performed in various atmospheres (in CO, H_2, O_2 at 670 – 770 K) to modify the oxidation and coordination states of iron embedded into the pore walls. Pyridine was sorbed in the next stage at 470 K and at 6 mbar, then sample wafer was cooled to 370 K. Desorption of pyridine at various temperatures was followed by recording Fourier Transformed Infra Red (FTIR) spectra of samples afterwards.

Temperature Programmed Reduction (TPR)

TPR measurements were also carried out. Samples were pretreated in O_2/He (20:80) mixtures at 570 K. The TPR curves were obtained by measuring the consumption of hydrogen with a thermal conductivity detector in a stream of H_2/Ar (10:90, 20 ml/min) mixture. The detector signal was calibrated against the stoichiometric CuO → Cu reduction for quantitative determination of the amount of consumed hydrogen.

Mössbauer Spectroscopy

Mössbauer spectra were recorded with a KFKI spectrometer by using a $^{57}Co/Rh$ source with 1 GBq activity. Samples were placed into an in situ cell providing means to perform treatments in various atmospheres up to 660 K, and recording spectra at as low as 80 K (liquid nitrogen), without exposing the sample to air during sequential series of treatments and measurements [35]. Spectra were decomposed to Lorentzian shape lines, the accuracy of positional data is 0.03 mm/s.

Results of Characterisation of Applied Samples

X-Ray Diffraction

Mesoporosity of samples was confirmed by XRD measurements. Samples display reflections in the low 2Θ region, indicating the primary mesoporous order (Figure 1). No reflections were observed in the 2Θ > 10 deg region. d_{100} plain distances and a_0 unit cell constants were also determined from the difractograms (Table 4).

Surface Area and Pore Size-Distribution

Nitrogen adsorption isotherms and pore diameter distributions are shown for SS and OS samples in Figures 2a and 2b respectively.

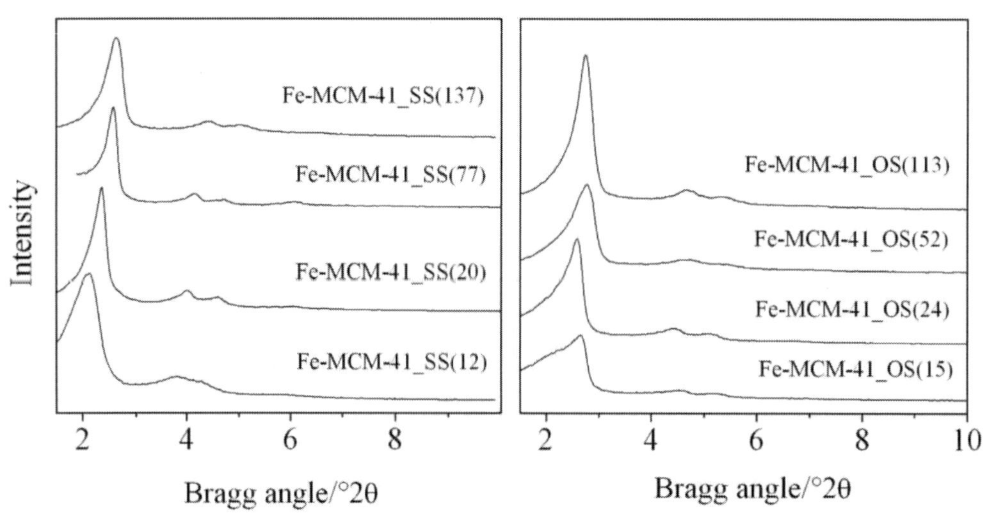

Figure 1. XRD diffractograms of Fe-MCM-41 samples.

Table 4. Characteristic data of SS and OS samples

Values of plane distances from (100) reflections, d_{100} , unit cell parameters, a_0 pore volumes, V_p, specific surface areas from N_2 adsorption measurements by BET method, S_{BET}, specific surface areas estimated from, $S_d = 4\ V_p/W_d$, centre of pore diameter distribution, Φ_{BJH}, diameter of primary pores esimated by Eq (1), W_d, wall-thicknesses, $WTH(I) = a_0 - \Phi_{BJH}$, and $WTH(II) = a_0 - W_d$

Sample	d_{100} (nm)	a_0 (nm)	V_p (cm^3/g)	S_{BET} (m^2/g)	S_d (m^2/g)	Φ_{BJH} (nm)	W_d^3 (nm)	WTH(I) (nm)	WTH(II) (nm)
SS-137	3.38	3.90	0.794	1265	971	2.43	3.27	1.47	0.63
SS-20	3.80	4.38	0.891	1397	950	2.66	3.75	1.72	0.63
SS-12	4.16	4.80	0.642	894	665	2.83	3.86	1.97	0.94
OS-52	3.37	3.89	0.690	1121	871	2.15	3.17	1.74	0.72
OS-24	3.47	4.01	0.622	1004	778	2.31	3.20	1.70	0.81

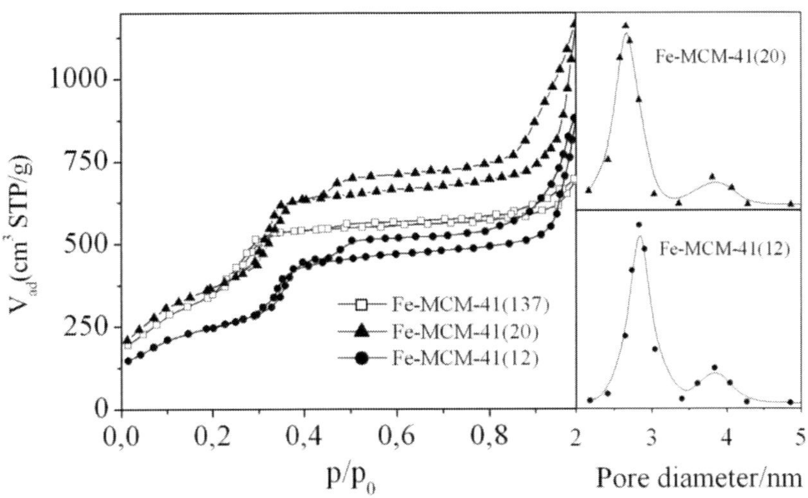

Figure 2a. Sorption isotherms and pore diameters of SS samples.

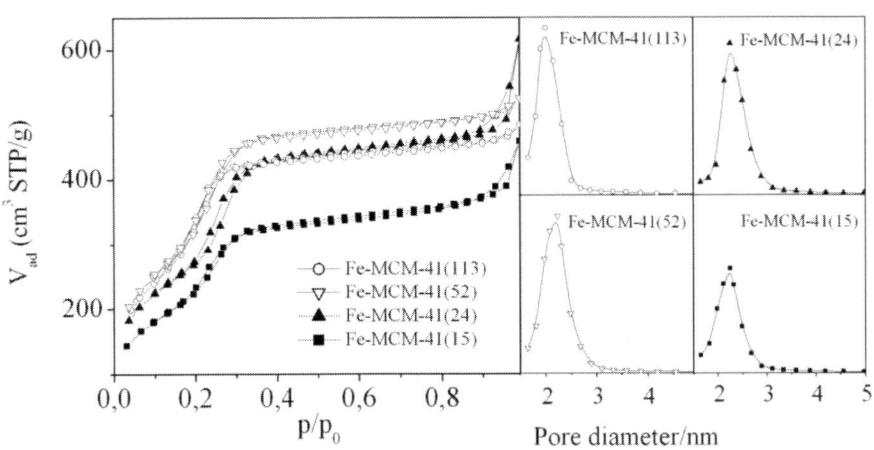

Figure 2b. Sorption isotherms of OS samples

Pore diameters for SS samples are larger than for OS ones (2.9 vs. 2.2 nm, respectively). Furthermore, secondary pores also appear in SS samples. Specific surface area values and pore diameter values were also calculated from the adsorption isotherms (Table 4).

Transmission Electron Microscopy – Morphology

High resolution TEM images are shown in Figure 3. SS sample exhibits irregular facets, whereas the OS sample is more oredered with spherical morphology.

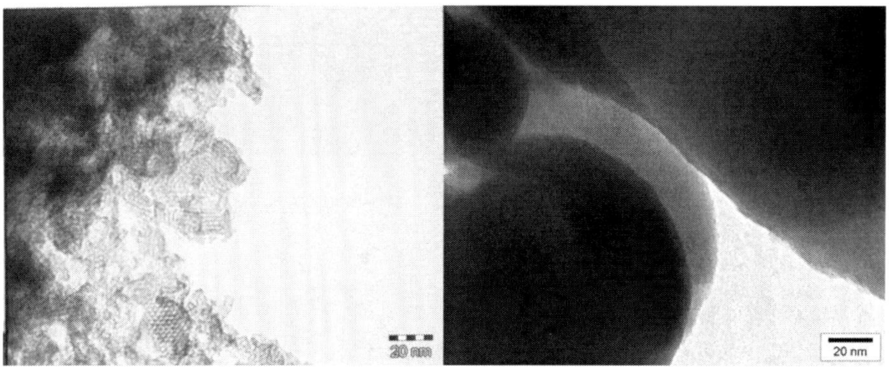

Figure 3. TEM images of Fe-MCM-4-SS(20), and OS-(113) particles (left, and right, respectively).

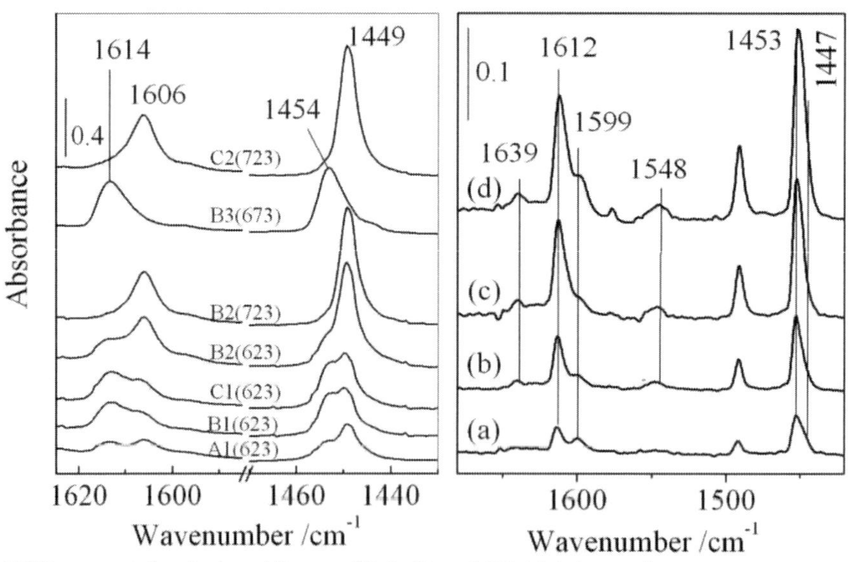

Figure 4. FTIR specra of sorbed pyridine on SS (left) and OS (right) samples.

SS samples: adsorption at 473 K and subsequent degassing at 373 K on Fe-MCM-41-SS(139), -SS(20) and –SS(12) samples, marked with A, B and C, respectively, following treatments in high vacuum, reduction with CO and subsequent oxidation in O_2/He, marked with 1,2,3 respectively. Temperatures of treatments are indicated in parentheses.

OS samples: spectra collected after sorption at 473 K and subsequent degassing at 423 K.

(a) -OS(113), (b) –OS(52) –, (c) –OS(24), (d) – OS(15).

FTIR - Pyridine Sorption

FTIR spectra of sorbed pyridine were recorded on various samples in order to identify various coordination states of iron, to distinguish between Lewis and Bronsted acidic centres. 1595-1615 and 1440-1450 cm^{-1} pyridine bands can be used for assignments of pyridine attached to Lewis acidic sites, whereas pyridinium ions formed on Bronsted sites are characterised by bands appearing at ~ 1550 cm^{-1} region [30]. Collected spectra are shown in Figures 4a and 4b.

No bands were detected in the ~1550 cm^{-1} region in spectra of SS samples. (Because of the lack of signal this range was even omitted from the figure in order to display the Lewis region in better resolution). This band appears with intensities increasing monotously with iron content in OS samples. Primary components in the spectra belong to the 1595-1615 and 1440-1450 cm^{-1} regions demonstrating the prevalence of Lewis sites in both types of samples.

Effects of reduction/oxidation are also reflected in positions of bands in spectra of SS sample, i.e. the strengths of Lewis acidity are different for Fe^{2+} and Fe^{3+}. For instance, reduction with carbon monoxide results in predominance of 1606 and 1449 cm^{-1} bands for pyridine sorbed on Fe^{2+} in Fe-MCM-41-SS(20) sample (B2(723) spectrum in Figure 4a). After oxidation to Fe^{3+} with O_2/He treatment these bands are shifted to higher frequencies (1614 and 1454 cm^{-1}, B3 spectrum).

As for the OS samples, a semi-quantitative comparison can be made by comparing the spectral areas characterising Lewis and Bronsted sites (Table 5.) The intensities between the Lewis and Bronsted sites can not be directly compared since the absorption coefficients my be different. However, the „Lewis" and „Bronsted" intensities can be compared among different samples separately. These intensties chage monotously with the iron content. The signal intensities can be normalised to the iron content – the values show a certain scatter, but do not exhibit any particular tendency (Table 5.)

Temperature Programmed Reduction (TPR)

Reducibilities of samples were studied by TPR measurements. The detector signals obtained for hydrogen consumption are shown in dependence of temperature for SS and OS samples in Figure 5. Reduction of Fe_2O_3 (hematite) is also shown for comparison in the left panel of Figure 5. In the case of hematite the first peak below 640 K is clearly separated, it corresponds to the $Fe_2O_3 \rightarrow Fe_3O_4$ (magnetite) process. The further two-stages of reduction to FeO (wustite) and later to metallic iron are not well resolved.

Table 5. Amounts of pyridine, sorbed on Bronsted and Lewis sites of OS samples

Sample	$I_{Bronsted}$ (a.u.)	I_{Lewis} (a.u)	Fe content mmol/g	$I_{Bronsted}$ (relative to	I_{Lewis} iron content)
OS-52	0.048	0.662	0.314	0.153	2.11
OS-24	0.135	1.233	0.667	0.202	1.85
OS-15	0.203	1.979	0.991	0.203	2.00

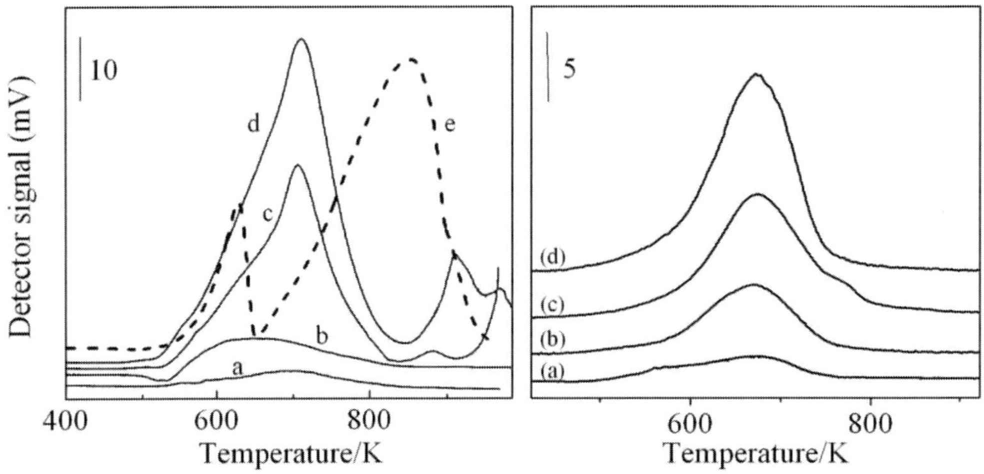

Figure 5. Temperature programmed reduction curves of SS (left) and OS (right) samples.
Si/Fe ratios in SS samples: 139 (a), 77 (b), 20 (c) 12 (d). TPR curve for Fe_2O_3 is also shown for comparison in dashed line (e). Si/Fe ratios in OS samples: 113 (a), 52 (b), 24 (c), 15 (d).

Table 6. Stoichiometry of TPR measurements for $Fe^{3+} \rightarrow Fe^{2+}$ reduction

Si/Fe ratio	Fe content (mmol/g)	H_2 consumed (mmol/g)	e/Fe
SS samples			
137	0.119	0.039	0.66
77	0.215	0.082	0.76
20	0.788	0.394	1.00
12	1.254	0.677	1.08
OS samples			
113	0.145	0.071	0.98
52	0.314	0.151	0.96
24	0.667	0.383	1.15
15	0.991	0.545	1.10

Signals recorded for the Fe-MCM-41 samples are single peaks, corresponding to one step $Fe^{3+} \rightarrow Fe^{2+}$ reduction. It is worth to note that reduction peaks for SS samples are less symmetrical than those for OS samples. Reduction of SS samples covers a broader temperaute range at the start of reduction, indicating that the environments of iron ions may have broader variety.

Quantitative data can also be extracted from the TPR records. With assuming one step $Fe^{3+} \rightarrow Fe^{2+}$ reduction from the amount of consumed hydrogen below 870 K the stoichiometrical equivalens are shown in Table 6.

Mössbauer Spectroscopy

Spectra were recorded on SS samples after calcination, evacuation and treatments in carbon monoxide (as for mild reduction) and after treatment in hydrogen (as for stronger reduction, allowing formation of –OH groups as well). Spectra were collected at ambient temperature following the treatments, except the last treatment in hydrogen, when the samples were immediately cooled from 670 to 80 K to preserve better the reduced state of samples (Figure 6). Data extracted from decompositions of spectra are shown in Table 7.

Spectra of SS samples recorded after the first stage (calcination, and subsequent storage in open air) exhibit doublets characteristic for octahedrally coordinated ferric ions. These doublets display medium QS values, indicating rather symmetrical environment, i.e. probably the presence of coordinated water (which can be adsorbed partly from the moisture of open air).

The evacuation at 670 K results in partial distortion of the symmetry, in correspondence with the increase of the QS value for one species (QS > 1.65 mm/s component). This process most probably can be attributed to the removal of sorbed water from the vicinity of tetrahedrally coordinated ferric ions. More precisely, in presence of sorbed water the charge compensation may be provided via hydroxonium ions (H_3O^+), and the removal of water may leave back bare protons, which will be attached to only one oxygen in the vicinity of iron, resulting in the increase of the charge asymmetry, and in formation of Bronsted acidic centres.

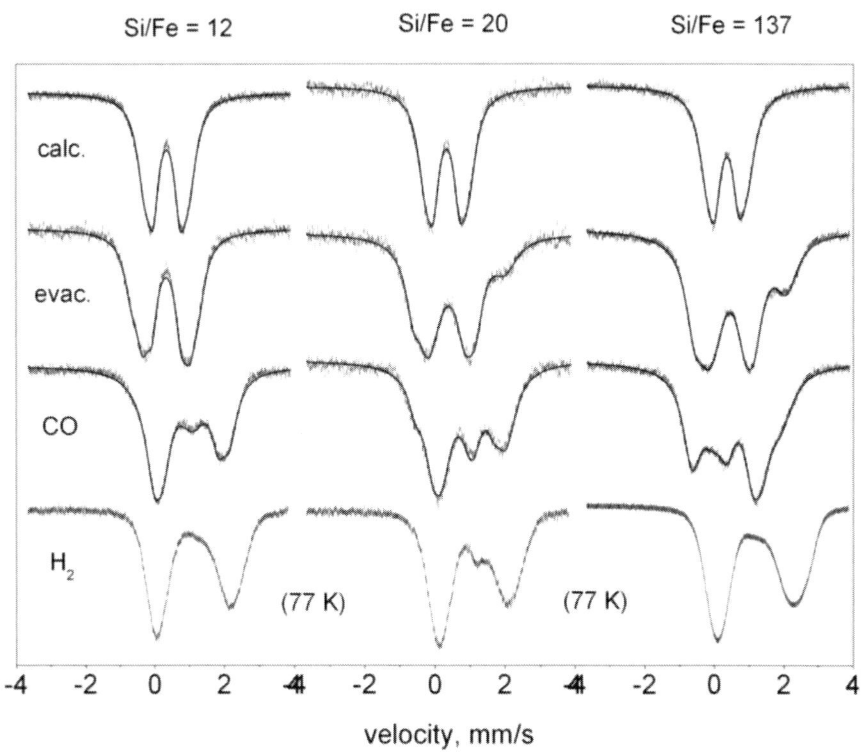

Figure 6. Mössbauer spectra of Fe-MCM-41-SS samples collected after the 670 K treatments indicated. Spectra were collected at room temperature – except the 77 K measurements following the 670 K treatments in hydrogen (bottom spectra).

Table 7. Data extracted from Mössbauer spectra Fe-MCM-41-SS samples shown in Figure 6. (IS: isomer shift, related to metallic α-iron, mm/s; QS: quadrupole splitting, mm/s; RI: relative spectral contribution, %)

Si/Fe		12			20			137		
Treatm	Comp	IS	QS	RI	IS	QS	RI	IS	QS	RI
As rec	Fe^{3+}	0.33	0.82	47	0.33	0.76	46	0.35	0.68	40
	Fe^{3+}	0.31	1.25	45	0.34	1.23	54	0.37	1.13	60
	Fe^{2+}/Fe^{3+}	0.61	1.09	8						
Evac	Fe^{3+}	0.29	1.94	21	0.27	1.78	26	0.28	1.68	44
670 K	Fe^{3+}	0.32	1.36	53	0.33	1.08	53	0.34	1.04	27
	Fe^{3+}	0.34	0.84	26						
	Fe^{2+}				1.03	1.95	21	1.07	2.03	30
CO	Fe^{3+}							0.23	1.76	45
670 K	Fe^{2+}/Fe^{3+}	0.50	1.16	41	0.55	2.25	10			
	Fe^{2+}/Fe^{3+}				0.63	0.85	46			
	Fe^{2+}	0.92	1.83	31	0.97	2.09	44	0.87	1.03	24
	Fe^{2+}	1.20	1.90	28				0.92	1.97	32
H_2	Fe^{2+}				0.83	0.95	20	0.84	1.14	13
670 K	Fe^{2+}	1.09	1.43	29				0.93	2.12	27
	Fe^{2+}	1.09	2.18	39	1.01	1.94	38	1.20	2.25	31
(m77K)	Fe^{2+}	1.33	2.35	32	1.28	2.14	41	1.50	2.34	29

Thus, appearance of the iron species exhibiting high QS value can be related directly to presence of Bronsted acidic centres, similarly as described for microporous Fe-MFI zeolite [36]. Other effect of the evacuation is the $Fe^{3+} \rightarrow Fe^{2+}$ reduction observed on another, loosely bonded species detected in 20 – 30 % of iron in Fe-MCM-41(SS-20) and (SS-137) samples. This feature may probably be connected with removal of water combined with simultaneous charge transfer processes, as proposed in [6].

Modest reduction was performed with a 670 K treatment in carbon monoxide in the next stage. CO is supposed to extract oxygen from breakable bonds. The extent of reduction depends on the iron content as data of Table 6 reveal. Almost all iron is reduced at least to an intermediate mixed $Fe^{2+/3+}$ state (IS ~ 0.5 mm/s) beside the regular ferrous state (IS ~ 0.9-1.2 mm/s) in the high iron-content SS-12 sample. In contrast, 45 % of iron retains the original ferric state in tetrahedral Bronsted acidic locations in the low iron-content sample (SS-137).

The further stage of reduction was performed in hydrogen at 670 K. In this case reduction may also proceed by forming –OH groups from the oxidic species (beside the direct removal of oxygen). All ferric iron is reduced to ferrous state in hydrogen (in correspondence, the earlier Bronsted acidic centres are also converted to another species). At least three or four different coordination states exists for ferrous iron as the decomposition of spectra indicates. One to three Fe^{2+} components also appear with low QS values ($0.8 < QS < 1.4$ mm/s) in each sample, indicating the distorted environment, which can be interpreted most probably with stabilisation of ferrous ions with low symmetry in the pore walls.

Pairwise recording of spectra at 300 and 80 K was also accomplished after stages in a series of treatments performed on SS-137 sample. From the comparison of the intensities of spectra recorded at different temperatures the value of Θ_D connected to bond strengths can also be estimated. Namely, the intensities of spectra are correlated with the probability of the Mössbauer effect, f_A, and this f_A is a function of the Debye-temperature, Θ_D. as mentioned also in the Introduction. In a first approximation Θ_D values can also be estimated from the comparison of spectral areas recorded at 80 and 300 K (this approach was already used for microporous Fe-FER and Fe-ZSM-5 samples [27]). The same comparison of spectral areas can be performed for Fe-MCM-41(SS-137) spectra. The corresponding I/BL (Intensity/Base line) values are collected in Table 8. Corresponding spectra and more details are reported in [28].

From the $(I/BL_{80})/(I/BL_{300})$ ratios the slope of $f_A \sim f(\Theta_D)$ can be approximated and Θ_D estimated. Using graphical evaluation $\Theta_D \sim 200$ K can be estimated from the data shown in Table 8. This value is significanlty smaller than the value obtained for microporous Fe-FER in a similar procedure ($\Theta_D \sim 260$ K [27]).

The $(I/BL_{80})/(I/BL_{300})$ values shown in Table 8 display a certain scatter. In fact, the listed four pairs of spectra are composed from different mixtures of ferric and ferrous components. Each species may have different Θ_D. In a more accurate procedure Θ_D could be determined for each different species comparing their intensities obtained from the respective separate spectral areas in the 300 and 80 K spectra. (For our following considerations in the Discussion paragraph the average value of Θ_D will be sufficient.)

Recyclation and/or reactivation of mesoporous ferrisilicates can be advantageous during the usage of these materials (e.g in sorption processes or in use as catalysts). In this regard it is worth to investigate whether structural changes may proceed during cycling of treatments. A repetition in two series of treatments (calcination, evacuation, reduction and repeated evacuation) was performed on the SS-137 sample. The corresponding spectra are shown in Figure 7 and data extracted from the decompositions are collected in Table 9. (Spectra were recorded at various temperatures to maintain the original coordination states of iron developed during the particular treatment.)

Table 8. I/BL values extacted from comparisons of 80 and 300 K spectra of SS-137 sample

	I/BL	(%)	$(I/BL_{80})/(I/BL_{300})$
Temp. of meas.	300 K	80 K	
Treatment			
As. Rec (calc.)	1.91	4.10	2.15
Evac. 670 K	1.21	4.22	3.49
CO 670 K	2.49	4.80	1.93
Hydrogen (2nd)	2.59	5.34	2.06

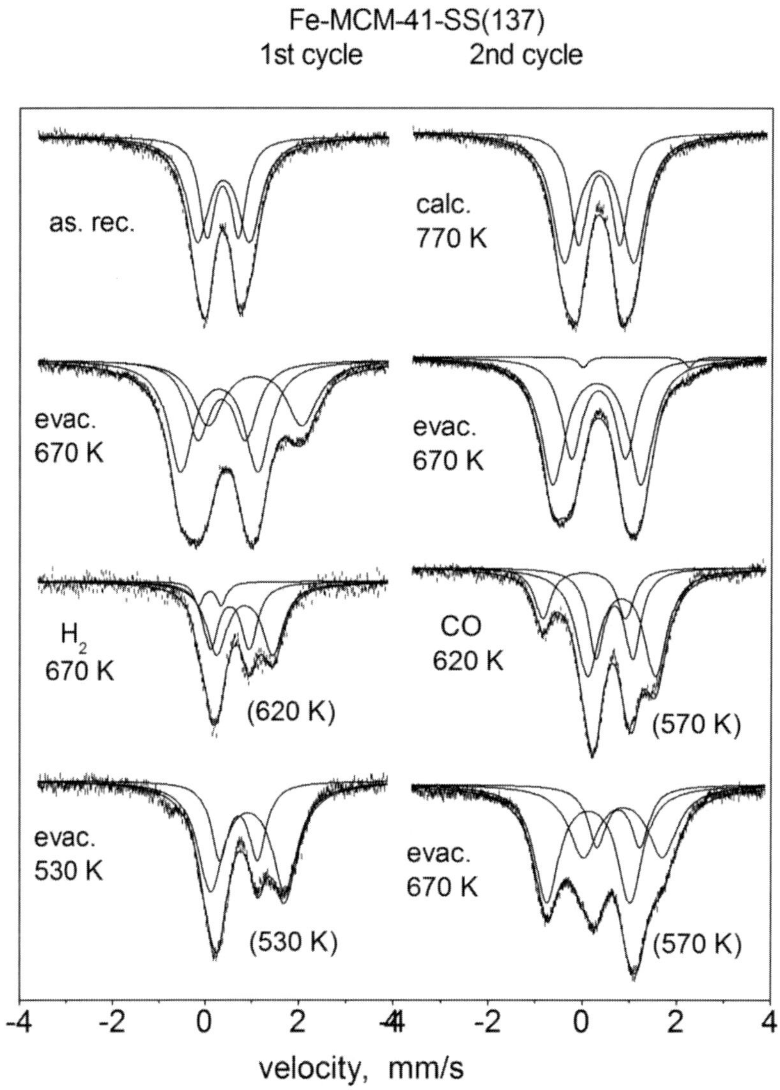

Figure 7. In situ Mössbauer spectra of SS(137) sample displaying the effects of repetiton in series of treatments. Spectra in the two upper rows were collected at room temperature, otherwise at temperatures marked in parentheses.

Behaviour of iron in the first series (left side in Figure 7) is similar to that shown in the previous series in Figure 6. Certain difference can be found in the data for the components detected after the reduction in hydrogen, which can be attributed to the difference of the temperatures of collection of spectra in hydrogen (77 K in Figure 6, vs. 620 K in Figure 7).

The effects of cyling became apparent in the second series, the partition of iron among various sites changes. The second calcination in air at 770 K following the reduction and the evacuation in the first cycle results in certain rearrangement of iron ions. The changes are more observable upon the following (3rd) evacuation.

Table 9. Mössbauer parameters extracted from the in situ spectra recorded after repetition of series of treatments on Fe-MCM-41-SS(137) sample. IS: isomer shift, related to metallic α-iron, mm/s; QS: quadrupole splitting, mm/s; RI: relative spectral contribution, %. Spectra were collected at temperatures indicated in parentheses

	1 st cycle				2 nd cycle			
Comp.	Treatment	IS	QS	RI	Treatment	IS	QS	RI
Fe^{3+}	as rec.	0.35	0.68	40	Calc.(2nd)	0.33	0.88	62
Fe^{3}	(m:300 K)	0.37	1.13	60	(m:300 K)	0.32	1.48	38
Fe^{3+}	Evac. (1st)	0.28	1.68	44	Evac 3rd)	0.29	1.88	60
Fe^{3+}	670 K	0.34	1.04	27	670 K	0.33	1.15	38
Fe^{2+}	(m:300 K)	1.07	2.03	30	(m:300 K)	1.12	2.27	2
Fe^{3+}	H_2	0.09	0.50	8	CO	0.02	1.77	18
Fe^{2+}	670 K	0.53	0.84	37	620 K	0.68	0.78	30
Fe^{2+}	(m:620 K)	0.85	1.21	54	(m:570 K)	0.83	1.45	52
Fe^{3+}	H_2+Evac.				Evac(4th)	0.13	1.78	48
Fe^{2+}	(2nd)530K	0.73	0.81	32	670 K	0.78	0.91	18
Fe^{2+}	(m:530 K)	0.92	1.61	68	(m:570 K)	0.87	1.70	34

The portion of the Bronsted acidic high QS component ($1.7 < QS < 1.9$ mm/s) component increased noticeably. Furthermore, in contrast to the first evacuation, practically no $Fe^{3+} \rightarrow Fe^{2+}$ reduction can be observed in the third one, probably the structure is more compact and abundance of –OH groups is more limited in the start of the second series. The treatment in carbon monoxide results in a limited ferric to ferrous reduction similarly as found in the previous case (shown in Figure 5). The spectrum recorded after the final, fourth evacuation displays a rearranged state – large portion of iron (ca. 50 %) is located in Bronsted acidic Fe^{3+} tetrahedral sites and only the other half is reduced to ferrous species present in two types of coordination. These spectra collected in this cycling clearly illustrate the flexibility of the Fe-MCM-41 structure – coordinations easily transform from one to another, in correspondence with the actual reaction conditions.

In the further stage of analysis the behaviour of the OS samples obtained in a fast synthesis based on hydrolysis of TEOS in water/methanol mixtures was studied. Similar series of sequential treatments were applied as previously for the SS samples, with different iron contents as well (Si/Fe ratios with 15, 24 and 52). (The temperature of measurements for collection of spectra after the reducing treatment in carbon monoxide and in hydrogen was again a lifted one, 490 K, in order to preserve the reduced state by avoiding the adsorption of any water formed eventually in the reduction process.)

Distinct differences can be revealed for OS samples when comparing the spectra with those of SS ones.

First, it is apparent that $Fe^{3+} \rightarrow Fe^{2+}$ reduction practically does not take place during the first evacuation. This feature can probably be attributed to the more compact structure of pore walls in OS samples, containing –OH groups in smaller amounts. Another significant difference is that the reduction in hydrogen does not completely reduces the total amount of the tetrahedral ferric component (large QS), it is present in smaller or larger amounts (5 – 21 %) after this treatment.

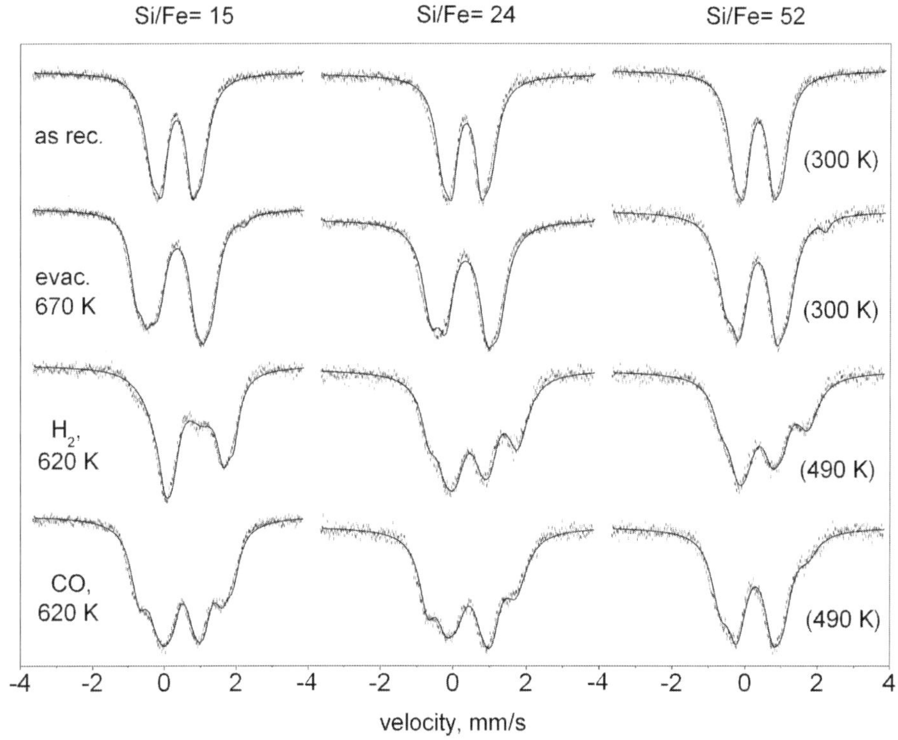

Figure 8. In situ Mössbauer spectra collected on Fe-MCM-41-OS samples after treatments indicated. Spectra were recorded at temperatures indicated in parentheses.

This feature is also in accordance with the previous interpretation, namely, in a more compact ordered structure the reduction of tetrahedrally coordinated iron embedded into the crystalline parts of the pore walls can be less enhanced than in the case of the more flexible, less crystalline SS samples.

Discussion

Mesoporous Structure – General Scope

The mesoporous structures for all the studied SS and OS samples are confirmed. XRD showed only the low angle diffraction peak, from BET measurements high specific surface area and 2.2 – 2.9 nm pore diameters were evidenced. The ordered array of channels clearly appears in the TEM images. Certain differences among SS and OS samples can be revealed (cf. Table 3, BET measurements in Figure 2, TEM images in Figure 3, etc.).

Local Structure around Embedded Iron

Local structure observed in the vicinity of iron ions may probably be characteristic for larger regions, too.

Table 10. Mössbauer parameters extracted from decompositions of in situ spectra of Fe-MCM-41(SS) sample shown in Figure 8. IS: isomer shift, relative to metallic α-iron, mm/s; QS: quadrupole splitting, mm/s; RI: relative spectral contribution, %. Spectra were collected at temperatures indicated in parentheses

Si/Fe =		15			24			52		
Treatment	Comp	IS	QS	RI	IS	QS	RI	IS	QS	RI
As rec	Fe^{3+}	0.33	1.34	60	0.34	1.28	63	0.34	1.26	63
calc.	Fe^{3+}	0.34	0.82	40	0.34	0.78	37	0.34	0.79	37
Evac.	Fe^{3+}	0.14	1.83	27				0.30	1.70	48
660 K.	Fe^{3+}	0.17	1.34	28	0.29	1.84	56	0.33	1.02	49
	$Fe^{2+/3+}$	0.54	1.50	42	0.34	1.11	44			
(m:300 K)	Fe^{2+}	1.09	2.20	3				1.12	2.17	4
H_2	Fe^{3+}				0.17	1.68	21	0.18	1.72	15
620 K	Fe^{3+}	0.34	2.10	5	0.27	1.01	37	0.27	0.98	50
	$Fe^{2+/3+}$	0.43	1.09	34						
	Fe^{2+}	0.96	1.89	27						
(m:490 K)	Fe^{2+}	0.94	1.40	34	0.91	1.69	42	0.88	1.70	35
CO	Fe^{3+}	0.19	1.77	29	0.18	1.78	32	0.19	1.73	38
620 K	Fe^{3+}	0.31	1.04	32	0.30	1.02	35	0.26	1.00	49
	Fe^{2+}	0.82	1.56	24	0.91	1.62	34	0.83	1.85	12
(m:490 K)	Fe^{2+}	1.08	1.65	15						

Namely, iron is probably distributed evenly in the pore walls, presence of associated oxidic particles was not detected either with any suitable method (XRD, TEM or Mössbauer spectroscopy). Further on, the TPR patterns exhibit a single reduction step attesting for one general type of embedded iron.

However certain distinctions with regard to the local structure around iron can be revealed, based on the analysis of pyridine sorption and Mössbauer data. Namely, different coordinations can be identified – as reflected in correspondence with the the distinction of Lewis and Bronsted bands appearing in the FTIR spectra. Lewis band can be attributed to the trigonal iron, located in less ordered environment, whereas the band at ca. ~ 1548 cm^{-1} wave number characterises the tetrahedrally coordinated iron forming Bronsted acidic centres. This tetrahedral coordination can be imposed to iron ions in a more ordered, rigid crystalline environment.

These distinctions are also in correspondence with data extracted from the corresponding in situ Mössbauer spectra. Namely, different types of Fe^{3+} components appear, depending on their coordinations. Iron species located in more symmetric local environment (e.g. in octahedral coordination with O^{2-} or –OH ligands) are provided with enough space, i.e., the structure around them is less ordered within few more atomic layers. These centres can be characteristic for the Lewis acidic sites and may exhibit the smaller quadrupole splitting values. In contrast, if the local order is more expressed and the pore wall is crystalline in a few atomic layer thick region, the microenvironment close to iron may impose even strict tetrahedral coordination in correspondence with the optimal array of [$SiO_{4/2}$] tetrahedra. This component exhibits the large QS value in the Mössbauer spectra (1.7 < QS < 2.0 mm/s) and the 1548 cm^{-1} band in the FTIR spectra. The distinction of these two components demonstrates the simultaneous existence of the partly amorphous and the partly crystalline regions in pore walls of mesoporous ferisilicates. Namely the local coordinations with more available space and higher symmetry (low QS) are characteristic for the locally amorphous environment, whereas the strict local smaller (tetrahedral) symmetry is imposed by the crystalline microenvironment.

Flexibility of Environments Reflected in $Fe^{3+} \leftrightarrow Fe^{2+}$ Redox Processes

Both ferric and ferrous ions can be embedded into the pore walls. Various figures illustrate the variations in the Fe^{3+}/Fe^{2+} ratios, depending on the treatments and experimental conditions. The change in the Fe^{3+}/Fe^{2+} ratio can be traced from pyridine FTIR measurements from change of the intensities of characteristic pairs of bands indicating Lewis sites at 1606 and 1449 cm^{-1} or 1614 and 1454 cm^{-1} generated by Fe^{2+} or Fe^{3+} centres respectively, in Figure 4. The appearance of various Fe^{2+} and Fe^{3+} species is also demonstrated in Mössbauer spectra presented from Figure 6 to Figure 8.

Variety of possible available environments during $Fe^{3+} \leftrightarrow Fe^{2+}$ reversible processes is also apparent. Fe^{3+} ions located in high symmetry sites embedded into partly amorphous regions can be reduced to ferrous state with treatments in carbon-monoxide or hydrogen at 630 – 650 K. Various coordinations may be occupied by Fe^{2+} ions in the pore walls as can be derived from the broad variety of IS and QS pairs obtained in the Mössbauer spectra. In contrast, $Fe^{3+} \rightarrow Fe^{2+}$ reduction cannot be performed in iron sites embedded to the crystalline regions, as the presence of the large QS Fe^{3+} component in the spectra obtained after the 630 – 650 K reducing treatments reveal. Thus the distinction of the partly amorphous and partly crystalline local environments can also be illustrated by the differences reflected in the extents of the ferric to ferrous reduction.

The opposite process, the $Fe^{2+} \rightarrow Fe^{3+}$ oxidation was also studied in the repetition of series of treatments. For instance, all the iron was reduced in the SS sample in a previous treatment in hydrogen. A subsequent repeated calcination resulted in partial formation of crystalline microenvironments, resulting in formation of iron sites with low, tetrahedral symmetry, where the ferric state was retained during a treatment in a carbon monoxide in 650 K (Figure 8). Thus the apparent variation of the local environments clearly demonstrates that the local structure is flexible, depending on the experimental conditions either the amorphous characters or under different conditions - the more crystalline features may prevail.

Comparison of Behaviour of Iron Embedded into Mesoporous and into Microporous Ferrisilicates

Characteristics of iron located in microporous ferrisilicates have been briefly mentioned in the Introduction already. Principal feature of the microporous structures is their complete crystallinity. This condition clearly determines the possible sitings of iron ions. Namely they can be located either in framework substituted (FW) sites in tetrahedral coordination (T), or in extra-framework (EFW) sites providing charge compensation for the previous Al^{3+}, Fe^{3+}, etc. centered in uncompensated $[TO_{4/2}]^-$ tetrahedra. Fe^{3+} => Fe^{2+} reduction of ferric iron in EFW sites may proceed, whereas the ferric to ferrous reduction in the FW sites is not allowed since the diameter of the Fe^{2+} ions exceeds the size of ions which can be incorporated without severe distortion of the siliceous microporous lattice made of $[SiO_{4/2}]$ tetrahedra (Si^{4+} : 40, Fe^{3+} : 63, Fe^{2+} : 77 pm [8]). In correspondence, FW to EFW relocation takes place once FW ferric ion is reduced to ferrous state.

These features of microporous ferrisilicates are reflected both in pyridine sorption FTIR and Mössbauer spectra as well. For instance, expressed absorbance can be detected in Fe-MCM-22 samples at the 1545 cm^{-1} band in pyridine sorption measurement performed under similar experimental conditions as used in the present study. This attests for the apparent presence of the Bronsted acidic tetrahedral sites in Fe-MCM-22 [37]. Mössbauer spectra recorded on various crystalline microporus ferrisilicates are also in accordance with all the observations mentioned above [38]. Further on, 260 K was found for the the value of the average Debye temperature (Θ_D) characterizing the bonding strength in Fe-FER [27].

On the other side, a primary feature of the mesoporous ferrisilicates is their shared, partially amorphous, partially crystalline structure as illustrated in detail in the preceeding paragraphs. In the amorphous regions of the pore walls Fe^{3+} => Fe^{2+} reduction may proceed without relocation of iron, Fe^{2+} can be stabilised in various coordinations. The opposite Fe^{2+} => Fe^{3+} process may also take place in the pore walls, as is demonstrated on the Fe-MCM-41-SS(137) sample in Figure 7. (This process cannot be observed in microporous crystalline ferrisilicates.) Further on, features of crystalline structures can also be detected simultaneously in microenvironments of iron. In certain instances pyridine bands were observed at 1548 cm^{-1} attesting for the tetrahedral coordination of iron forming Bronsted acidic sites in OS samples (Figure 4b). Similarly, this tetrahedral coordination caracterised with the large QS value (> 1.7 mm/s) was also found in several Mössbauer spectra. The partition between the crystalline and amorphous regions can easily be modified, it strongly depends on the actual conditions. For instance, the reduction/oxidation treatments and evacuation have also a strong influence on this partition and on the structure of the developing microenvironments. The significantly lower average value of the Debye temperature (Θ_D ~ 200 K) attests also for the less ordered dual, partially crystalline and partially amorphous structure of the mesoporous MCM-41ferrisilicates in comparison to the microporous ones (Θ_D ~ 260 K for Fe-FER).

Variations in the Extent of Crystallinity – Comparison of SS and OS Samples

The dual character, the simultaneous presence of amorphous and crystalline microenvironments is manifested for both types of samples. More closely, the amorphous character is more expressed for the samples of SS series. Combination of the XRD and sorption data shows that the pore walls are thicker for the OS samples than for SS ones (Table 3). The difference in the extent of ordering is reflected also in the pore diameter distribution graphs shown in Figure 2. SS samples exhibit pores with c.a. 2.5 – 2.8 nm diameter with a secondary smaller peak at c.a. 3.8 nm. The pores in OS samples are more narrow, and more ordered, with only one peak at c.a. 2.2 nm. Further on the OS sample displays also more ordered structure than the SS counterpart, as presented in TEM images of Figure 3. The hydrothermal synthesis of samples in aqueous media may allow incorporation of silanolic groups in larger proportion into the thinner pore walls. In contrast, the hydrolysis of TEOS in a media containing methanol may result in more ordered structure and thicker of the pore walls and smaller share of silanolic groups in the Fe-MCM-41(OS) samples.

This general difference is inherited by the local environment of iron ions embedded into the pore walls. For example the difference of the extent of ordering is also reflected in the absence of Bronsted acidic sites in pyridine sorption measurements on SS samples compared to the modest presence in OS samples (Figure 4). Differences can also be revealed in Mössbauer spectra. Characteristic feature of the SS samples is that almost all the ferric iron can be reduced to ferrous state in 650 K reductions, whereas noticeable part of iron retains its ferric state in similar treatments of OS samples.

Structures of samples of both SS and OS series are flexible, they respond to the reduction/oxidation treatments not only with the ferric ⇔ ferrous transitions but with changing the proportions between the amorphous and crystalline regions as well.

CONCLUSION

Simultaneous existence of dual, amorphous and crystalline features in pore walls of mesoporous ferrisilicates is demonstrated in Fe-MCM-41 samples. In particular, samples obtained with two different preparation methods are compared, namely samples from a hydrothermal synthesis (SS method) and another series of samples prepared by hydrolysis of tetraethylorthosilicate (OS method). Their mesoporous structures have been proven by XRD, sorption and TEM measurements. Particular attention is devoted to the characterisation of the local environments of the embedded iron. Iron species coordinated in tetrahedral local symmetry is supposed to characterise a dominantly crystalline microenvironment. On the other hand, iron located in trigonal and/or octahedral coordinations is considered to indicate partly amorphous microenvironments. The local environments in the pore walls are rather flexible, they can easily be interconverted. These changes are also strongly correlated to reversible redox processes as in situ FTIR and Mössbauer spectroscopies reveal.

A short comparison with ordered crystalline microporous ferrisilicates is also presented, the characteristic differences between the crystalline microporous and the dual amorphous/crystalline mesoporous ferrisilicates investigated under same experimental conditions are discussed. Finally, features of SS and OS samples are also compared and extents of the partitions among the amorphous and crystalline parts in the two types of samples are also considered.

ACKNOWLEDGMENT

The authors are thankful to Hermann K. Beyer, Gabriella Pál-Borbély and Magdolna R.-Mihályi for their continuous support and for valuable discussions.

REFERENCES

[1] Kruk, M.; Jaroniec, M.; Sakamoto, Y.; Terasaki, O.; Ryoo, R.; Ko, C.H. *J. Phys. Chem. B,* 2000, 104, 292-301.

[2] Schacht, S.; Janicke, M.; Schüth, F. *Micropor. Mesopor. Mater.,* 1998, 22, 485-493.

[3] Tuel, A.; Arcon, I.; Millet, J.M.M. Millet *J. Chem Soc., Faraday Trans.*, 1998, 94, 3501-3510.

[4] Alves, M.; Pastore, H. O. *Micropor. Mesopor. Mater.*, 2001, 47, 397-406.

[5] Lebedeva, O. E.; Sachtler, W. M. H. *J. Catal.*, 2000, 191, 364-371.

[6] Lázár, K.; Pál-Borbély, G.; Szegedi, Á.; Beyer, H. K. *Stud. Surf. Sci. Catal.*, 2002, 142, 1347-1354.

[7] Brookins, D. G.; Eh-pH Diagrams for Geochemistry, Springer-Verlag, Berlin, 1988.

[8] Mihályi, R. M.; Szegedi, Á.; Lónyi, F. In Silica and Silicates in Modern Catalysis Halasz, I. Ed; Transworld Research Network, 2010, pp. 213-255.

[9] Li, B.; Xu, J.; Liu, J.; Zuo, S.; Pan, Z.; Wu, Z. *J. Colloid Interface Sci.*, 2012, 366, 114-119.

[10] Mokhonoana, M. P., Coville, N. J.; *Materials,* 2009, 2, 2337-2359.

[11] Bourlinos, A. B; Karakassides, M. A.; Petridis, D. *J. Phys. Chem. B,* 2000, 104, 4375-4380.

[12] Wang, Y., Zhang, Q., Shishido, T., Takehira, K.; *J. Catalysis,* 2002, 209, 186-196.

[13] Lázár, K., Calleja, G., Melero, J. A., Martinez, F., Molina, R.; *Stud. Surf. Sci. Catal.*, 2004, 154, 805-812.

[14] Han, Y., Meng, X., Guan, H., Yu, Y., Zhao, L., Xu, X., Yang, X., Wu, S., Li, N., Xiao, F.-S.; *Micropor. Mesopor. Mater.,* 2003, 57, 191-198.

[15] Grün, M.; Lauer, I.; Unger, K. K. *Adv. Mater.*, 1997, 9, 254-257.

[16] Szegedi, Á.; Kónya, Z.; Méhn, D.; Solymár, E.; Pál-Borbély, G.; Horváth, E. Zs.; Bíró, P. L.; Kiricsi, I. *Appl. Cat. A:General*, 2004, 272, 257-266.

[17] Kosslick, H.; Lischke, G.; Walther, G.; Storek, W.; Martin, A.; Fricke, R. *Micropor. Mater.*, 1997, 9, 13-33.

[18] Stockenhuber, M.; Hudson, M. J.; Joyner, R. W. *J. Phys. Chem. B,* 2000, 104, 3370-3374.

[19] Béland, F.; Echchahed, B.; Badiei, A.-R.; Bonneviot, L., *Stud. Surf. Sci. Catal.*, 1998, 117, 567-574.

[20] Burns, R. G. *Hyperfine Interact.,* 1994, 91, 739-745.

[21] Taboada, J. B.; Overweg, A.; Crajé, M. W. J.; Arends, I. W. C. E.; Mul, G.; van der Kraan, A. M. *Micropor. Mesopor. Mater.,* 2004, 75, 237-246.

[22] Dubkov, K. A.; Ovanesyan, N. S.; Shteinman, A. A.; Starokon, E. V.; Panov, G. I. *J.Catal.*, 2002, 207, 341-352.

[23] Sobalik, Z.; Tabor, E.; Nováková, J.; Sathu, N. K.; Závéta, K. *J. Catal.,* 2012, 289, 164-170.

[24] Lázár, K. In Zeolites: Synthesis, Chemistry and Applications; Andreyev, M. K.; Zubkov, O. L. Ed.; Nova Science Publishers, 2012, 251-260.

[25] Samanta, S.; Giri, S.; Sastry, P. U.; Mal, N. K; Manna, A.; Bhaumik, A. *Ind. Eng. Chem. Res.*, 2003, 42, 3012-3018.

[26] Vértes, A.; Korecz, L.; Burger, K. *Mössbauer Spectroscopy*; Akadémiai Kiadó: Budapest, 1979.

[27] Lázár, K.; Kotasthane, A.N.; Fejes, P. *Catal. Lett.*, 1999, 57, 171-177.

[28] Lázár, K.; *Hyperfine Interact.* 2014, 226, 99-109.

[29] Selvam, P., Mohapatra, S.K.; *J. Catalysis*, 2006, 238, 88-99.

[30] He, N-Y.; Yang, C.; Lu, Z-H. *Stud. Surf. Sci. Catal.*, 2002, 141, 459-466.

[31] Barrett, E.P.; Joyner, L.G.; Halenda, P.P. *J. Am. Chem. Soc.* 1951, 73, 373-380.

[32] Kruk, M.; Jaroniec, M.; Sayari, A. *Langmuir* 1997, 13, 6267-6273.

[33] Franke, O.; Schulz-Ekloff, G.; Rathousky, J.; Starek, J.; Zukal, A. *J. Chem. Soc. Chem. Commun.* 1993, 724-726.

[34] Karge, H.G.; Niessen, W. *Catal. Today*, 1991, 8, 451-465.

[35] Lázár, K.; Matusek, K.; Mink, J.; Dobos, S.; Guczi, L.; Vizi.Orosz, A.; Markó, L.; Reiff, W.M. *J. Catal.* 1984, 87, 163-178.

[36] Lázár, K.; Borbély, G., Beyer, H. *Zeolites*, 1991, 11, 214-222

[37] Mihályi, R.M.; Lázár, K.; Kollár, M.; Lónyi, F.; Pál-Borbély, G., Szegedi, Á. *Micropor. Mesopor. Mater.* 2008, 110, 51-63.

[38] Lázár, K.; Szegedi, Á.; Pál-Borbély, G.; Kotasthane, A.N.; Fejes, P. *Catal. Today* 2005, 110, 239-246.

In: Comprehensive Guide for Mesoporous Materials, Volume 2 ISBN: 978-1-63463-982-8
Editor: Mahmood Aliofkhazraei

Chapter 5

FUNCTIONALIZATION OF MESOPOROUS MATERIALS

***Jean Chrysostome Ndamanisha*[1,2] *and Liping Guo*[2]**
[1]University of Burundi, Bujumbura, Burundi
[2]Northeast Normal University, Faculty of chemistry, China

ABSTRACT

We report the functionalization of mesoporous materials for improvement of the electrocatalytic properties. As silica and carbon based-materials are often used as template to synthesize other mesoporous, these materials are described here. We focus on ordered mesoporous carbons (OMC) because the functionalization and applications of mesoporous silicas have been well reviewed. We show how the ordered mesostructure of these materials is very important in the construction of (bio) sensors. Moreover, the structural capabilities of OMC at the scale of a few nanometers agree with immobilization of other electrocataytic substances. Interesting properties of these materials may open up a new approach to study the electrochemical determination of other biomolecules.

1. INTRODUCTION

The first synthesis of an ordered mesoporous material was described in 1969. However, due to a lack of analysis, the remarkable features of this product were not recognized [1, 2]. Table 1 shows how mesoporous materials were discovered. In 1992, a similar material was obtained by a scientist in Mobil Oil Corporation and the remarkable features of this novel type of silica were discovered [3]. This opened up a whole field of research. MCM-41, which stands for Mobil Composition of Matter No. 41, shows a highly ordered hexagonal array of unidimensional pores with a very narrow pore size distribution [4, 5] and the synthesis mechanism used a silica source and a surfactant.

Table 1. Different works done in the discovery of mesoporous materials

Works	References
Synthesis of an ordered mesoporous material without description	1,2
Discovery of MCM-41	3
Description of the MCM-41 pores and pore size distribution	4,5
Synthesis of other mesoporous materials following the co-operative assembly pathway	6- 8
Synthesis of other mesoporous materials following the true liquid crystal templating pathway	9
Synthesis of other mesoporous materials following the nanocasting	10,11
Application of mesoporous materials	12-18
Synthesis of mesoporous matrials using others as templates	17-19

The discovery of ordered mesoporous solids (like MCM-41) and related materials in the early 1990s has been a breakthrough in materials engineering, and since then there has been impressive progress in the development of many new mesoporous solids. The mechanisms were based on templating and were related to the one used for the original MCM-41 synthesis. Depending on the synthesis conditions, the silica source or the type of surfactant used, many other mesoporous materials can be synthesized following the co-operative assembly pathway [6- 8]. In addition to the co-operative pathways, the true liquid crystal templating pathway [9] and nanocasting using already formed ordered mesoporous materials as hard templates [10, 11] have been developed. Many reviews have been published covering various aspects of ordered mesoporous materials, such as their synthesis, surface modification, application as host materials, and in catalysis [12-18]. Therefore, in this chapter, it is not necessary to come back to these well described subjects.

Here, we will describe the general pathways used to introduce other functions in or on mesoporous materials with a focus on silica and carbon materials because they are often used as template to synthesize others. Ordered mesoporous carbon (OMC) is usually synthesized with ordered mesoporous silica as a template [19] as it is shown in Figure 1. The ordered mesoporous silica is prepared from a surfactant. From an ionic surfactant such as a quaternary ammonium an ordered mesoporous material is formed based on the interaction between the positively charges surfactant and the negatively charged silicate species in the solution [5]. It is interesting to remember that it is possible to obtain different types of ordered structures, owing to the diversity in the achievable 3-dimensional (3D) structures of the ordered mesoporous silica (OMS) templates [20]. The different ordered mesostructures of the silica-based materials mainly depend on the ordered aggregates of amphiphilic surfactants.

In particular, CMK-1 carbon with *I41/a* (or lower) symmetry and CMK-4 carbon with cubic *Ia3d* symmetry are synthesized using MCM-48 and FDU-5 silica templates respectively because these templates are of *Ia3d* symmetry. CMK-2 carbon with cubic *Pm3n* symmetry is obtained using an SBA-1 template of *Pm3n* space group. CMK-3 and CMK-5 carbons with 2D hexagonal (*p6mm*) symmetry are obtained using an SBA-15 template (p6mm symmetry). Moreover, it is possible to synthesize OMC with cubic Im3m symmetry using for instance the SBA-16 silica template [20]. The different symmetries of mesoporous silicas are well described in the review of Wan and his coworkers [20]. From Figure 1, it is clear that one mesoporous material can be a template for another but the synthesis starts with mesoporous silica (SBA-15) or mesoporous carbon (CMK-3). Also, these materials have been more used and functionalized than others for many applications. Therefore, we will take mesoporous silcas and mesoporous carbons as representatives of other mesoporous materials.

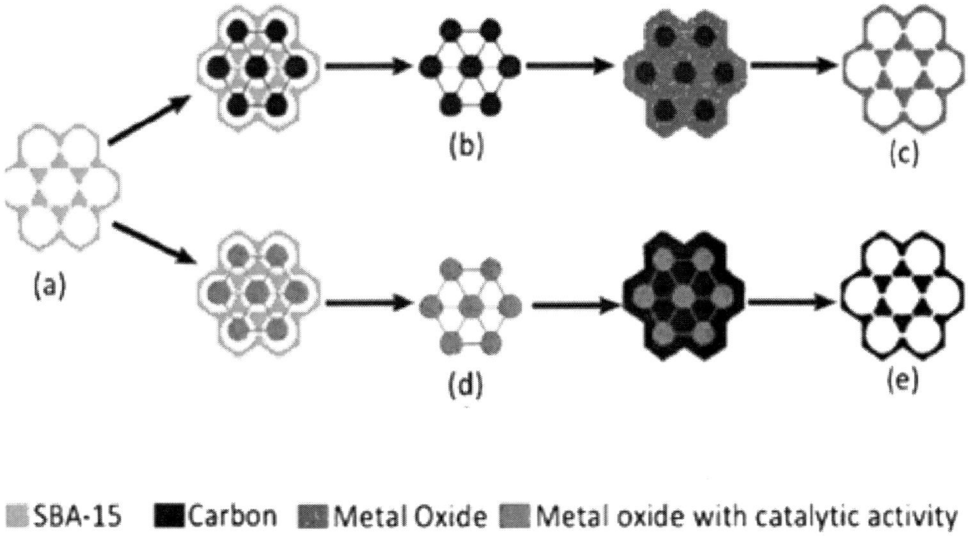

Figure 1. Synthesis of different mesoporous materials. SBA-15 silica (a), CMK-3 mesoporous carbon (b), mesoporous metal oxide (c); mesoporous metal oxide (d) obtained from SBA-15 silica preparing Graphitic-OMC (e) [19].

2. FUNCTIONALIZATION OF MESOPOROUS MATERIALS

In order to functionalize mesoporous materials, two pathways are generally used. The first one is the generation of more existing functional groups. Some mesoporous materials, especially OMC, have functional groups that play an important role in the electrocatalytic properties of these materials. Therefore, if the groups are increased, the electrocatalytic properties of the materials are improved. The second pathway is the incorporation of other functional groups in the mesoporous materials. Many active substances can be immobilized on the mesoporous materials with different means as it is shown in Table 2.

2.1. Generation of More Existing Groups

Ordered mesoporous silicas are not often used as catalysts as such. That is why it is useless to generate more existing groups. Contrary to most porous silica- based materials, the mesoporous carbons are conductors which makes them potential novel materials for investigating the electrochemical behavior of substances. The investigation on the electrochemical activity of OMC showed that the biomolecules could be detected at this material due to the presence of the oxygen-containing functional groups and edge plane-like defective sites available on the OMC materials. It is then possible to affirm that those groups are very important for the electrocatalytic activity of OMC [21].

Table 2. Different species incorporated for the functionalization of mesoporous materials

Functionalization /Species incorporated	References
Generation of oxygen-containing functional groups	21
Generation of edge plane-like defective sites	21, 23
Generation of large BET surface area	24
Generation of large pores' diameter	24
Generation of large pores' volume	25
Al, Ga, Fe, Cs, Zn,	28- 31
Pt	32-38, 44, 47
Iron oxide	39
Rh	42, 43
Polynuclear molecular transition metal complexes	45,46
Co	48
Au	49
MnO_2	50
Iron oxide	51, 53
Cu_2S	52
NiO	54
Polyoxometalate (P_2Mo_{18})	55
Cerium (III) 12-tungstophosphoric acid (CePW)	56
Ferrocenecarboxylic acid (Fc)	57
Binuclear cobalt phthalocyaninehexasulfonate sodium salt (bi-CoPc)	59
Polymer-metal complex of poly-o-toluidine (POT)	60
Fullerene (C_{60})	61
Tetrathiafulvalene (TTF)	62
Nile Blue (NB)	63
Polyaniline (PANI)	64, 65
Ionic liquid	66, 67, 68

As those functional groups are available on OMC (the Raman spectrum has proved their presence [22], it became important to study their influence on the electrocatalytic properties. For this goal, two kinds of experiments were carried out [21]. First, OMC was oxidized with HNO_3 to generate a high concentration of oxygen-containing groups and the bands attributed to OH groups became stronger and shifted to higher wavenumbers. The material obtained (OMC-Ox) was used to modify a glassy carbon (GC) electrode to observe the influence of oxygen-containing groups. In the second experiment, OMC was treated electrochemically to see the influence of edge plane defect sites. The results demonstrated that the ordered structure of OMC was very important for its electrochemical and electrocatalytic properties. The destruction of this structure resulted in the decrease of its electrochemical and electrocatalytic properties. The influence of the oxygen-containing functional groups and edge plane-like defective sites available on the OM was then confirmed and their generation improves the electrocatalytic properties of OMC. An acceptable sensor based on the structure of OMC has been developed for the detection of glutathione (GSH) and cysteine (CySH). In order to detect CySH in the presence of GSH, it has been proposed to use the amperometric method at 0.4 V, where the interference by GSH is well reduced. Song and coworkers [23] found the same conclusions with dopamine (DA). An OMC with carboxylic groups was prepared with concentrated nitric acid treatment at 80°C for 24 h. The obtained OMC was used to detect DA in the presence of ascorbic acid.

The discussion shown above may come to a conclusion that the intrinsic electrocatalytic properties of OMC attributed to oxygen containing functional groups and edge plane defect sites take advantage of the large surface area and open porous structure. The direct detection of the analytes above results in well-defined voltammetric signals occurring at lower overvoltage than other carbon materials.

The large surface area and the widely open and ordered mesostructure of OMC are expected to offer high preconcentration efficiency and fast diffusion processes which should lead to a very sensitive voltammetric or amperometric detections. Moreover, the pore size and pore volume play an important role in the electrocatalytic properties. Therefore, if it is possible to improve those characteristics, the OMC properties are improved.

Usually, SBA-15, template in the synthesis of OMC is prepared at a temperature in the range of 348–373 K. If we raise the temperature at 398 K, pores' diameter and Brunauer–Emmett–Teller (BET) surface area of the obtained OMC become larger. The influence of the BET surface area on the electrocatalytic activity of OMC [24] has been studied. The results showed that OMC with high BET surface area exhibits improved electrocatalytic activity. Additional active sites are available on OMC-modified electrode with a higher BET surface area for electrocatalysis of morphine [24].

Moreover, along with the decrease of the sucrose to silica weight ratio in the synthesis, the increase in the pore volume, pore size and surface heterogeneity of the different OMC samples might make more surface area electrochemically accessible and more exposure of edge-plane-like defective sites [25].

2.2. Introduction of Other Functional Groups in the Mesoporous Materials

Much more frequently, additional catalytic functions are introduced, by incorporation of active sites in the walls or by deposition of active species on the inner surface of the material. The advantage of using ordered mesoporous solids in catalysis are the relatively large pores which facilitate mass transfer and the very high surface area which allows a high concentration of active sites per mass of material. There are many possible pathways to modify silica-based materials when one wants to give them a new catalytic function as schematically shown in Figure 2 and details can be found in the review of Tagushi et al. [26] and that of Hartmann [27]. Those pathways can also be used for OMC.

2.2.1. Introduction of Heteroatoms

Frequently, heteroatoms are introduced in mesoporous silicas. This is maybe due to the rigid silanol group absent in mesoporous carbons. In order to introduce heteroatoms in mesoporous silicas, many of the modification steps make use of the silanol groups present on the silica surface of the ordered mesoporous silica, because they are the anchor sites for metal species or silane coupling agents.

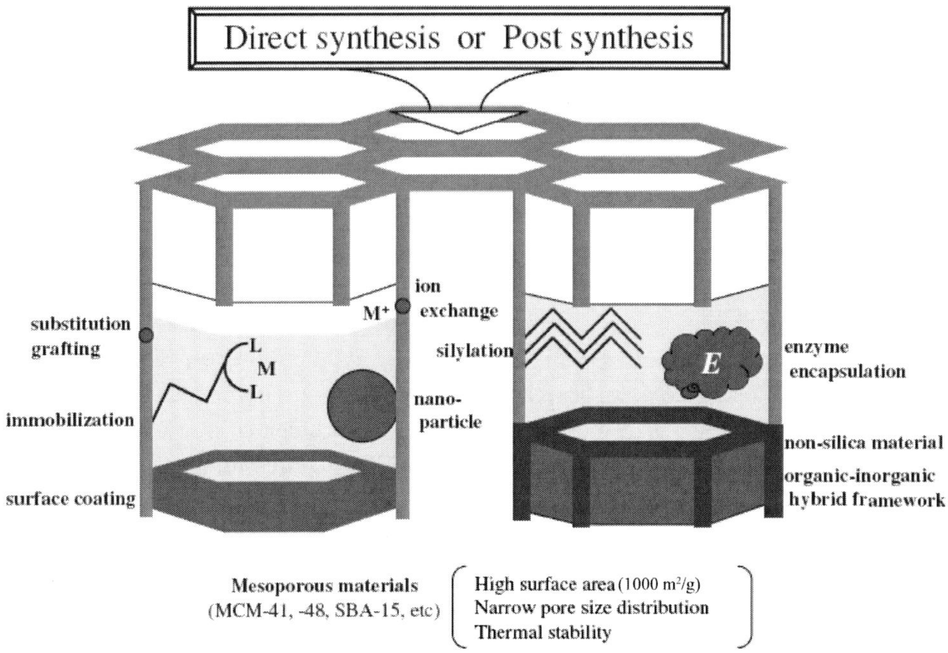

Figure 2. Schematic sketch of the various methods for the functionalization of mesoporous silicas [26].

It has been shown that the introduction of heteroatoms such as B, Fe, Ga, Ti and Al in the siliceous framework is very important because substitution of silicon allows fine tuning of the acidity or creation of redox properties. Modification of the framework composition is possible by the direct synthesis or by post treatment of an initially prepared silica mesoprous material. Tagushi et al. [26] have well reviewed the different species incorporated in silica mesoporous materials. However, let us show how the acidity is generated and how the redox properties are generated.

Incorporation of aluminum is of special interest if, as in zeolites, this results in the formation of Brønsted acidity and ion exchange sites. Since the O–Al–O angle is less flexible than the O–Si–O angle, Al-MCM-41 materials are commonly less well ordered on the mesoscale and show a broader pore size distribution than their pure silica analogues [28, 29]. Adsorption of bases such as ammonia and pyridine on Al-MCM-41 allows to determine the strength of acid sites via temperature programmed desorption and FTIR [28]. It is interesting to note that different heteroelements provide different strength of the resulting acid sites. Therefore, the Brønsted acid strength of Al-, Ga-, and Fe-substituted MCM-48 investigated were in the order of Al > Ga > Fe [30].

The substitution of trivalent elements in a silica framework requires the presence of extra framework charge compensation cations. Such counter ions are usually exchangeable and can be used to introduce novel functionality into the mesoporous silicate. For example, the cations in aluminosilicates were successfully exchanged with Cs^+, Zn^{2+}or Fe^{3+}, and the resulting materials show catalytic activity in the Diels–Alder reaction of cyclopentadiene and methyl acrylate [31].

2.2.2. Introduction of Nanoparticles

Nanoparticles can be incorporated in mesoporous silica and mesoporous carbons. The mechanism is almost similar for the two groups of mesoporous as it is shown in Figure 2. In the following, some pathways of such introduction are briefly described.

Metal, metal oxide and sulfide nanoparticules can be deposited on ordered mesoporous materials because, in general, they cannot grow to sizes larger than the pore size. The other advantage is that the metal nanoparticles have noble metal particles in the nanometer ranger. Many pathways for depositing those particles can be used. Directed inclusion of metal particle in the synthetic gel of mesoprous materials has been used. In the case of Pt, Junges et al. prepared Pt/MCM-41 by adding different platinum precursors to the synthetic gel [32]. It has been found that the neutral Pt precursor, $[Pt(NH_3)_2Cl_2]$, was more suitable for the preparation of small particles compared to the cationic $[Pt(NH_3)_4](NO_3)_2$ or anionic $K_2[PtCl_6]$ species. Under neutral synthesis conditions, in the direct synthesis of Pt/MSU-1, anionic $(NH_4)_2PtCl_4$ was found to be preferable over cationic or neutral precursors [33].

Using this direct method, there are well known methods for producing high Pt dispersions on carbon materials [34- 36]. However, the loading amount of Pt is limited by the surface area of the carbon support and surface of the carbon matrix often must be functionalized to increase the hydrophilicity.

These materials are promising matrix for high dispersion of Pt nanomaterials [34, 35, 37] and accessible ordered pores can provide sufficient room for mass transport. But in those reports, the preparations of Pt nanoparticles were achieved by complicated synthetic routes. Kim and co-workers reported the preparation of Pt/OMC catalyst by the formic acid reduction method in organic solvent tetrahydrofuran under reflux condition [38]. The problem is that organic solvent is toxic and flammable, and evaporates easily to create noxious smog.

The direct method has also been used to incorporate iron oxide in the OMC using ferrocene derivative as a precursor. The obtained material was without any decrease in the surface area [39]. The decomposition of ferrocene derivative yields iron species that facilitate the formation of the carbon framework.

The conventional methods, such as impregnation techniques, often yield nanoparticles of various sizes in the mesopores and on the external surface of the support materials. Yao et al. introduced a new technique designated as vacuum evaporation impregnation. The MCM-41 was suspended in an aqueous solution of $[Pt(NH_3)_4](NO_3)_2$ at pH 5, heated at 70°C and the water was evaporated in vacuum. The obtained material shows a very high Pt dispersion (1.03) at a loading level of 1.3 wt%, measured by CO chemisorption [40]. The use of supercritical fluids, which have excellent properties such as low viscosity, high diffusivity and controllable solubility of dissolved species, are a novel method to overcome the limitations of conventional solvents [41]. Recently, the synthesis of Pt and Rh–Pt alloy particles supported on FSM-16 via supercritical CO_2 has been reported [42, 43].

Bo et al. [44] reported a facile and economical route to synthesize Pt/OMC nanocomposite via reduction of H_2PtCl_6 by formic acid in water at room temperature without polymeric stabilizers and surface functionalization. The chemically and mechanically stable 2D mesopores of OMC could act as an ideal nanosized reactor for in situ growth of Pt nanoparticles.

Immobilization of polynuclear molecular transition metal complexes in porous materials is also a promising method for the preparation of supported metal catalysts, because fine tuning of the metal complexes in terms of electronic states and steric environment is substantially easier than with the metal salts used in the conventional catalyst preparation. Metal carbonyl complexes react with the surface silanol groups at elevated temperature. Continuous adsorption of $Co_2(CO)_8$ on MCM-41 in a flow reactor resulted in a maximum Co deposition of 20 wt%. The metal loading increased up to 41wt% using the pulse deposition method [45, 46].

2.2.3. Incorporation of Electroactive Substances in OMC

Many other compounds have been supported on ordered mesoporous silicas and on OMC. For mesoporous silicas, details can be found in the review of Tagushi et al. [26]. However, it is better to come back to the functionalization of OMC with other substances because this subject has not been reported. There are many substances incorporated to functionalize OMC and some of them are briefly described:

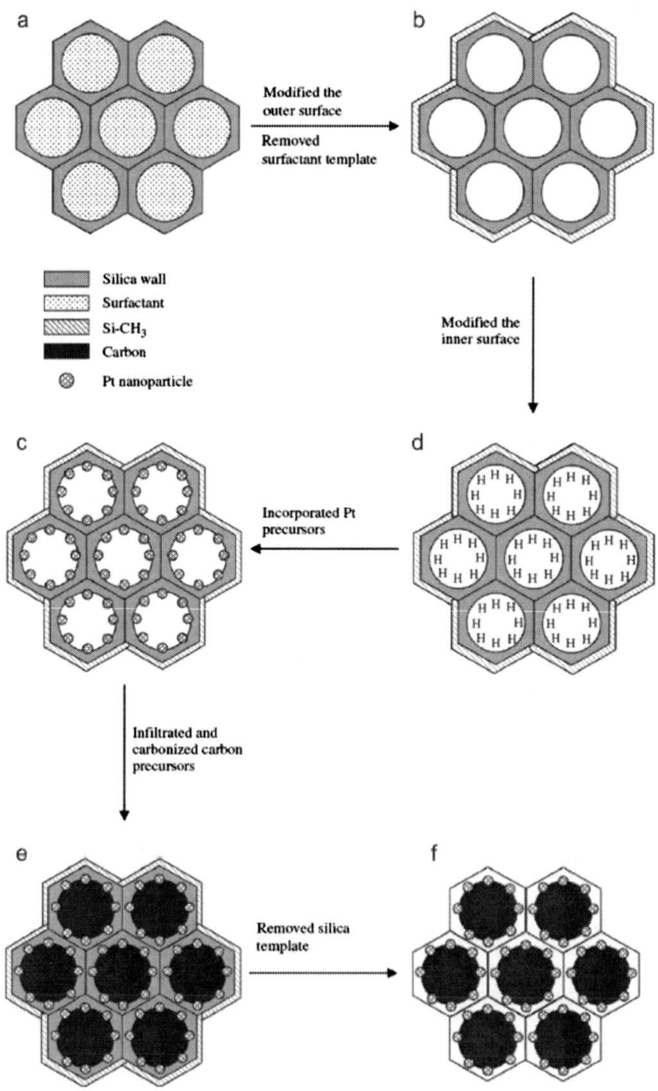

Figure 3. Schematic illustrations of synthesis procedure for Pt–OMC samples. Copyright 2011 Elsevier. Reproduced with permission from Ref. [47].

(1) Noble metal and metal oxide sulfide nanoparticles can be incorporated in the mesopore network. In general, cconventional techniques for embedding platinum or mix-metal nanoparticles in the porous carbon supports typically invoke post-synthesis treatments, such as impregnation. Nevertheless, owing to their chemically inert nature, most nanostructured carbon materials normally have a lack of desirable functional groups on their surfaces and hence the uncontrolled growth of Pt nanoparticles supported on these supports are vulnerable to aggregation during the synthetic process and electrocatalytic reactions, due to weak interaction between metal particles and supports. To overcome this problem, Liu and Chen [47] reported a novel procedure, based on the pyrolysis of carbon and Pt precursors in functionalized SBA-15 mesoporous silica. Figure 3 shows how the method is used.

First, SBA-15 (Figure 3a) is modified with Si–CH_3 groups on the outer surface, followed by removal of surfactant template with solvent extraction. Second, the inner surface of SBA–15 CH_3 (Figure 3b) is modified with a homogeneous layer of Si–H groups on the silica walls. Third, Pt precursors ($PtCl_6^-$) are introduced into the channels of SBA–15H (Figure 3c) to obtain Pt–SBA-15(Figure 3d) in which oxidized Pt is reduced by the Si–H groups homogeneously dispersed on the pore surface. Finally, the Pt–SBA-15 is infiltrated with carbon precursors, then subjected to carbonization at 1073K under vacuum (Figure 3e), followed by removal of the silica template with hydrofluoric acid solution, The obtained material is washed and dried to obtain the final Pt–OMC product (Figure 3f). Compared with the original OMC modified electrode, the Pt/OMC-modified electrode displayed improved current response towards hydrogen peroxide.

The direct technique has also been used by Hou et al. [48] for the immobilization of cobalt oxide nanoparticles (Co) in the wall of the ordered mesoporous carbon. The mesostructure of the nanocomposite material remained highly ordered and intact. SBA-15 silica, which contained partially polymerized and carbonized sucrose, was impregnated with 5 mL ethanol containing 0.5 g of cobalt nitrate hexahydrate. The silica template was then removed. A sensitive GSH sensor was developed based on the OMC–Co/GC electrode and it showed a high sensitivity and a remarkably low detection limit. Moreover, OMC–Co/GC modified electrode could be used for selective amperometric determination of GSH in the presence of glucose, dopamine and uric acide.

Electrochemical method displays an easy and rapid alternative for the immobilization of noble metal nanoparticles to OMC in a short time. In this way, Wang et al. [49] used an electrodeposition method to attach gold nanoparticles (GNPs) on OMC. The OMC fixed on a substrate was immersed into solutions of 1mM $HAuCl_4$ in 0.5 M H_2SO_4 and a potential step from 1.10 V (vs. Ag/AgCl) to - 0.10 V (vs. Ag/AgCl) was applied for 15 s. The average diameter of GNPs was 20–30 nm. It has been found that gold nanoparticles were well fixed and uniformly dispersed on OMC. No aggregation of them was observed. A nonenzymatic hydrogen peroxide (H_2O_2) sensor based on GNPs–OMC/GC electrode has been fabricated using the method [49]. It exhibited good reproducibility and long-term stability.

It is also possible to synthesize nanoparticles inside the pore channels of OMC. In this way, a sonochemical method has been successfully used in order to incorporate MnO_2 nanoparticles inside the pore channels of OMC. Modification of the intrachannel surfaces of OMC to make them hydrophilic enables $KMnO_4$ to penetrate the pore channels. At the same time, the modification changes the surface reactivity, enabling the formation of MnO_2 nanoparticles inside the pore channels by the sonochemical reduction of metal ions [50].

Other oxide nanoparticles have been incorporated in OMC [51]. The performance of ordered mesoporous carbon containing iron oxide (OMC–Fe) has been compared to OMC and the properties of the new material have been found to be improved. It has been demonstrated that OMC–Fe combines electrochemical properties of OMC and iron oxide species. Electrocatalytic properties towards H_2O_2 are enhanced because of the synergetic effect of OMC and iron oxide species. The results are based on the combination of OMC and Fe_3O_4 because it has been shown that the reduction of H_2O_2 occurs at OMC but its detection is less sensitive than at OMC–Fe. This high sensitive determination can be explained by the OMC–Fe structure. The presence of iron oxide species results in the increase of the surface area. Having a catalytically active surface, OMC (in OMC–Fe) can promote the H_2O_2 reduction and provide a base for the mediation for H_2O_2 at OMC–Fe modified electrode with Fe_3O_4 as a mediator. In that case, the oxidative valence state (Fe^{3+}) and the reductive valence-state (Fe^{2+}) on Fe_3O_4 are reduced and oxidized respectively. The electrochemical determination of H_2O_2 at the OMC–Fe modified electrode was more sensitive than that at OMC electrode because the sensitivity for OMC–Fe was 8.4 μA mM^{-1} and not only OMC substantially offered smaller signals, but also the response is affected by regenerating the surface of the electrode. A sensitive and stable H_2O_2 sensor was developed based on OMC–Fe electrode with a large determination range (7– 4000 μM) and a good reproducibility.

A simple and facile synthetic method to incorporate copper sulfide (Cu_2S) nanoparticles inside the mesopores of OMC has also been reported [52]. A hybrid of iron oxide nanoparticle modified ordered OMC was successfully prepared via a controllable two-step route [53]. This excellent electrochemical performance of the FeOx/OMC hybrid makes it a candidate anode material for commercial lithium-ion batteries. The prominent improvement of electrochemical performance can be attributed to the synergistic effects of the FeOx/OMC composites.

A nonenzymatic amperometric sensor of hydrogen peroxide based on the Cu_2S/OMC nanocomposite modified GC electrode was then developed. The combination of the unique properties of Cu_2S nanoparticles and the ordered mesostructure of OMC matrix resulted in the excellent electrocatalysis for hydrogen peroxide. The peak potential for reduction of hydrogen peroxide at Cu_2S/OMC/Nafion/GC electrode was more positive than that at OMC/Nafion/GC electrode, shifting positively by about 70 mV with higher cathodic response. Luo et al. [54] developed a highly sensitive and selective nonenzymatic glucose sensor based on electrodepositing NiO nanoparticles on ordered mesoporous carbon (named as NiO/OMC) modified GC electrode. The NiO/OMC/GC electrode electrode showed excellent electrochemical activity toward the oxidation of glucose. The calibration curve for glucose determination was linear in the range of 2 - 1000 μM with a fast response time (<2 s), and a high sensitivity of 834.8 μA mM^{-1} cm^{-2} and a low detection limit (S/N = 3) of 0.65 μM were obtained.

(2) Inorganic mediators have been immobilized. Zhou et al. [55] developed a convenient and efficient method for the functionalization of OMC using polyoxometalate $H_6P_2Mo_{18}O_{62}\cdot xH_2O$ (P_2Mo_{18}). OMC was heated at 250 °C for 0.5 h, and added to cool P_2Mo_{18} aqueous solution immediately. The mixture was centrifuged to get P_2Mo_{18}/OMC. It has been found that the P_2Mo_{18} was finely dispersed on the OMC supporter with a loading level of 18%. The electrochemical behavior of the modified electrode was studied in detail, including pH-dependence, stability and so on. The cyclic voltammtry (CV) and amperometry studies

demonstrated that P_2Mo_{18}/OMC/GC electrode has high stability, fast response and good electrocatalytic activity for the reduction of nitrite, bromate, iodonate, and hydrogen peroxide

In the same way, Cerium (III) 12-tungstophosphoric acid (CePW) has been immobilized on OMC [56]. After laoding of CePW, the global particle shape and the delicate structure of the OMC host were maintained. The CePW/OMC modified GC electrode showed an enhanced electrocatalytic activity [56]. This property was applied in the determination of some biomolecules. Especially, the detection and determination of the guanine (G) in the presence of adenine (A) was achieved with good stability, low detection limit and reproducibility.

(3) Organometallic catalysts can also play this role. OMC was functionalized with ferrocenecarboxylic acid (Fc) and the OMC–Fc [57]. Here, the procedure is a little different from the usual incorporation of substances on the OMC channel. In step 1, HNO_3 oxidizes the OMC whereas in the second it acts as a catalyst of the formation of OMC–Fc ester bond. The OMC–Fc modified electrode improved electrocatalytic properties compared to OMC alone [57]. Ascorbic acid was detected in the presence of dopamine with a separation peak-to-peak of 170 mV. The same modified electrode has been used in the electrochemical determination of uric acid (UA). A similar surface modification has been developed with silica mesoporous [58].

Liu et al. [59] developed an interesting method to adsorb Binuclear cobalt phthalocyaninehexasulfonate sodium salt (bi-CoPc) onto didodecyldimethylammonium bromide shown in Figure 4. It has been found that the smooth surface of DDAB/OMC becomes rough after (DDAB)/OMC film by ion exchange and bi-CoPc is then immobilized on OMC. The procedure is the ion exchange in bi-CoPc buffer solution. Oxygen, hemoglobin and 2-mercaptoethanol were detected with this bi-CoPc/DDAB/OMC modified electrode with higher sensitivity and low detection limits.

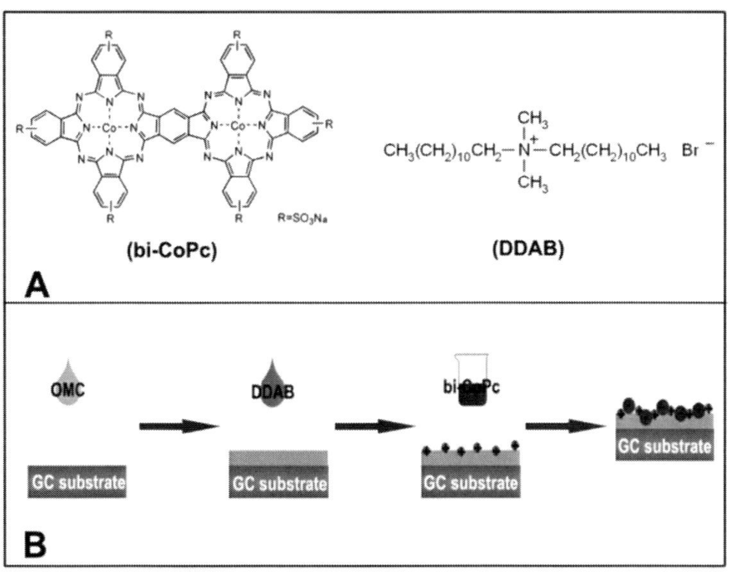

Figure 4. Chemical structure of the binuclear cobalt phthalocyaninehexasulfonate sodium and didodecyldimethylammonium bromide molecule and (B) schematic representation of bi-CoPc/DDAB/OMC film modified on GC electrode [59].

Polymer-metal complex of poly-o-toluidine (POT) with cobalt supported on OMC was prepared by the reaction of cobalt nitrite and POT in the presence of OMC. This new POT-Co-OMC composite was used as a cathode material for oxygen reduction reaction. The POT-Co-OMC catalyst also exhibits excellent anti-crossover effect to methanol oxidation in the oxygen reduction reaction (ORR), superior to the commercial Pt/C catalyst. [60]

(4) Organic redox mediators, molecular and polymeric ones have been immobilized by impregnation or electropolymerization. It has been demonstrated that the integration of OMC and fullerene (C_{60}) can provide a remarkable synergistic augmentation of the current [61]. After the loading of C_{60}, the global particle shape and the delicate structure of the OMC host were maintained. Nevertheless, for OMC-C_{60}, the C_{60} nanoparticles on the channel surface of OMC can be hardly observed by scanning electron microscopy (SEM), which could be due to the nanosize and the fine dispersion of C_{60} in OMC. It has been demonstrated that the integration of OMC and C_{60} can provide a remarkable synergistic augmentation of the current [61]. To illuminate the concept, eight kinds of inorganic and organic electroactive compounds were employed to study the electrochemical response at an OMC–C_{60} modified glassy carbon (OMC–C_{60}/GC) electrode. The modified electrode showed more favorable electron transfer kinetics than OMC/GC, carbon nanotube modified GC, C_{60}/GC, and GC electrodes. Such electrocatalytic behaviour at OMC–C_{60}/GC electrode could be attributed to the unique physicochemical properties of OMC and C_{60}, especially the unusual host–guest synergy of OMC–C_{60}.

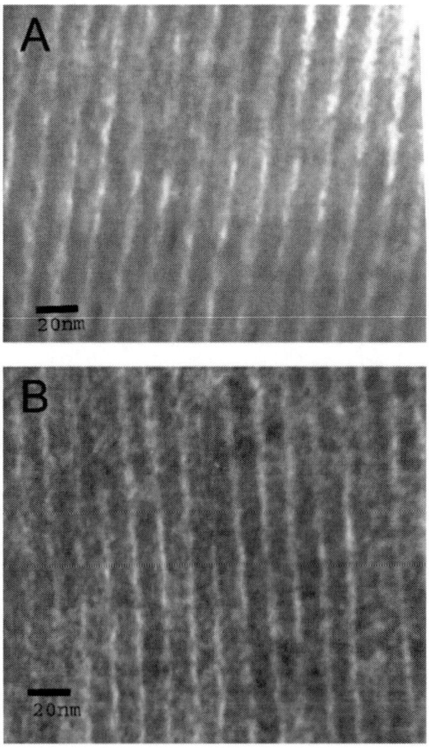

Figure 5. TEM images of OMC(A) and OMC-TFF (B) [62].

Impregnation has also been used to immobilize an excellent electron donor, tetrathiafulvalene (TTF) on OMC [62]. Though the incorporation of TTF results in the reduced surface area in OMC-TTF, the ordered structure is kept. The preservation of this ordered structure is confirmed in Figure 5A and Figure 5B where transmission electron microscopy (TEM) images of OMC and OMC-TTF are shown. One can see that the two materials have an ordered structure. The presence of TTF particles is indicated by the darker areas in the region of the OMC-TTF composite (Figure 5B). It has been demonstrated that the electrocatalytic behavior of OMC–TTF is attributed to the unique physicochemical properties of OMC and TTF [62]. An amperometric oxygen biosensor has been constructed based on OMC–TTF and it exhibited good response to dissolved oxygen with a large linear range and a very low detection limit. The interferences of AA and UA were suppressed.

Nile Blue (NB) modified electrode can be prepared by immersing the OMC/GC electrode into NB solution (0.05 M, pH 6.8 PBS) for 5 min. Then, the NB molecules are adsorbed on the OMC/GC electrode. NB/CNT and NB/graphite powder modified GC electrodes were prepared in the same way in the control tests [63]. A novel amperometric NADH sensor based on a Nile blue (NB)/ordered mesoporous carbon (OMC) composite (NB/OMC) electrode can then be prepared. Under a lower operation potential of −0.1 V, NADH can be linearly detected with an extremely lower detection limit of 1.2 μM (S/N = 3).

Conducting polymers such as polyaniline (PANI) has been used to functionalize OMC [64] and the structure of OMC-PANI is very similar to that of OMC. Moreover, the properties of PANI are also preserved as it has been confirmed by the electrochemical behaviour. The OMC–PANI electrode was used to detect Cu^{2+} and Pb^{2+} [64]. The detection limit was 6.0×10^{-9} M for Cu^{2+} and 4.0×10^{-9} M for Pb^{2+}. Moreover, Yan et al. [65] prepared hierarchical polyaniline nanowires/ordered bimodal mesoporous carbon (PANI/OBMC) composites via chemical oxidative polymerization. Electrochemical performance of the hierarchical PANI/OBMC composite as supercapacitor electrode materials has been evaluated by cyclic voltammetry and galvanostatic charge-discharge techniques. The hierarchical composite with 60 wt% PANI possesses the highest specific capacitance of 517 F g^{-1} and outstanding cycling stability with a capacitance retention of 91.5% after 1000 cycles.

(5) Ionic liquids with their high chemical stabilities, relatively high ionic conductivity and wide electrochemical windows can be used to functionalize OMC. They form gel with OMC by grinding. In this way, a quick and sensitive biosensor based on functionalized OMC and an ionic liquid has been developed for the first time for the detection of DA and UA in the presence of a large amount of ascorbic acid [66]. Similarly, Zhu et al. [67] prepared OMC modified carbon ionic liquid electrode (CILE). CILE was prepared by mixing graphite powder with 1-ethyl-3-ethylimidazolium ethylsulphate ([EMIM]$EtOSO_3$) and liquid paraffin. Using Nafion as a binder, a double-stranded (dsDNA) biosensor based on Nafion–OMC/CILE was constructed. Under the optimal conditions, the oxidation peak current increased with dsDNA concentration in the range of 10.0–600.0 μg mL^{-1} by differential pulse voltammetry (DPV) with the detection limit of 1.2 μg mL^{-1}. Zhai et al. [68] developed a novel gel based on the polymerization of a composite containing toluidine blue O (TBO) functionalized OMC by a π-π stacking interaction and ionic liquid (IL). The poly (toluidine blue O) (PTBO)-OMC-IL composite film distributes almost homogeneously with unique structure on the surface of substrate. The PTBO-OMC-IL gel modified GC electrode provides a positively synergistic effect among PTBO, OMC and IL on the electrochemical oxidation of

NADH. The NADH could be linearly detected from 1.0 μmol L^{-1} up to 6.0 mmol L^{-1} with the LOD of 0.4 μmol $L^{-1}$and fast response time of 2 s.

2.3. Functionalization of Mesoporous Materials for the Construction of Electrochemical Biosensors

Here we will use the IUPAC definition of a biosensor. "Electrochemical biosensor is a self-contained integrated device, which is capable of providing specific quantitative or semi-quantitative analytical information using a biological recognition element (biochemical receptor) which is retained in direct spatial contact with an electrochemical transduction element" [69]. Therefore, we call electrochemical biosensor, a sensor that integrates a biorecognition element into the sensor [70].

Since heterogenized molecular catalysts are considerably easier to handle, retrieve, and re-cycle than their homogeneous counterparts, immobilization of homogeneous catalysts on solid supports is of high commercial interest. Ordered mesoporous silica has often been used for immobilization of molecular catalysts due to the high surface area. Often, immobilized catalysts suffer from reduced activity upon immobilization, but occasionally, the catalytic performance might improve as compared to the homogeneous case [71, 72]. Details can be found in the review of Tagushi et al [26]. Here electrochemical biosensors based on OMC are detailed because those materials are recently reported. Table 3 shows different biomolecules incorporated in OMC for the construction of electrochemical biosensor. The mesoporous materials can be used with usual pore size or it can be enlarged.

Table 3. Biomolecules incorporated in OMC for the construction of electrochemical biosensor

Biomolecules incorporated	Type of OMC	References
Hemoglobin (Hb)	OMC with usual pore size	73
Glucose oxidase (GOD)	OMC with usual pore size	74, 75
Alcohol dehydrogenase (ADH)	OMC with usual pore size	75
myoglobin	bicontinuous gyroidal mesoporous carbon (BGMC) with large pore size	76
Glucose oxidase (GOD)	highly ordered 3D-mesoporous carbon material with large pore size	77
Glucose oxidase (GOD)	FDU-15 with large pore size	78
Hemoglobin (Hb)	FDU-15 with large pore size	79

2.3.1. Immobilization of Biomolecules on Mesoporous Materials with Usual Pore Size

With its large pore, OMC is also able to load simple active molecules and several examples are available. Using chitosan as an effective linker between CMK-3 and GC electrode surface, Chen and coworkers [73] constructed a novel biosensor for the determination of H_2O_2. The biosensor was based on a {Hb/CMK-3}6 film where hemoglobin

(Hb) is adsorbed on OMC through layer-by-layer assembly. When OMC is used as a carbon paste, immobilization of the active molecules becomes easier. This procedure has been used by Zhu et al. [74]. For the construction of this biosensor using an enzyme, glucose oxidase (GOD) was incorporated in OMC by mechanical mixing OMC, glucose oxidase and mineral oil to fabricate a more sensitive and selective glucose biosensor. The obtained electrode (OMC–GOD) was compared to that constructed with carbon nanotube (CNT) and GOD (CNT–GOD). The OMC–GOD electrode provided an evidently larger current signals with a sensitivity augmentation of 6.3-fold compared to the CNT–GOD electrode (18.42 $\mu A\ M^{-1}$ and 2.92 $\mu A\ M^{-1}$ for OMC–GOD and CNT–GOD electrodes, respectively) so as to decrease the detection limit from 0.13 mM (CNT–GOD electrode) to 0.072 mM (OMC–GOD electrode). This was due to the substantial decrease in the over voltage of the hydrogen peroxide oxidation along with the facile incorporation of GOD into the composite matrix.

The incorporation of GOD can also be achieved with Nafion as a linker in the construction of an enzyme-based device. For example, Zhou et al. [75] described a simple method for the construction of electrochemical alcohol dehydrogenase (ADH) and GOD-based biosensors for the determination of nicotinamide adenine dinucleotide (NADH) and H_2O_2. The high density of edge-plane-like defective sites and high specific surface area of OMC could be responsible for the electrocatalytic behaviour at OMC modified GC electrode, which induced a substantial decrease in the overpotential of NADH and H_2O_2 oxidation reaction compared to CNT modified GC electrode. The oxidation potential appeared at about 0.6 V, 0.4 V and 0.2 V for GC electrode, CNT/GC electrode and OMC/GC electrode, respectively. Especially, as an amperometric glucose biosensor, Nafion/GOD–OMC/GC electrode showed large determination range (500 - 15,000 μM), high sensitivity (0.053 nA μM^{-1}), fast (9 s) and stable response.

2.3.2. Incorporation of Active Molecules in OMC with Large Pore

One can figure out that OMC with large pore and/or large surface area can load easily other active molecules. Therefore, it is possible to enlarge the pore, it will become easier to immobilize active biomolecules. In this way, a three-dimensional (3D) ordered mesoporous carbon, a bicontinuous gyroidal mesoporous carbon (BGMC) has been used to immobilize myoglobin [76]. In contrast with carbon nanotubes and general carbon mesoporous materials, BGMC is of a relatively isotropic graphited structure and thus can more effectively enhance the heterogeneous electron transfer. The electrocatalytic activity of immobilized protein on BGMC was used to detect H_2O_2.

Similarly, You et al. [77] demonstrated a strategy of a protein entrapment within mesoporous carbon matrices to probe the electrochemistry of glucose oxidase. The large surface area and remarkable electro-catalytic properties of mesoporous carbon materials make them suitable candidates for high loading of protein molecules and the promotion of heterogeneous electron transfer. Moreover, it has been noted that highly ordered 3D-mesoporous carbon material exhibited larger adsorption capacity for glucose oxidase and the immobilized enzymes retained a higher bioactivity as it is shown in Figure 6. The assembled glucose biosensor showed a broader linear response scale and higher detection sensitivity.

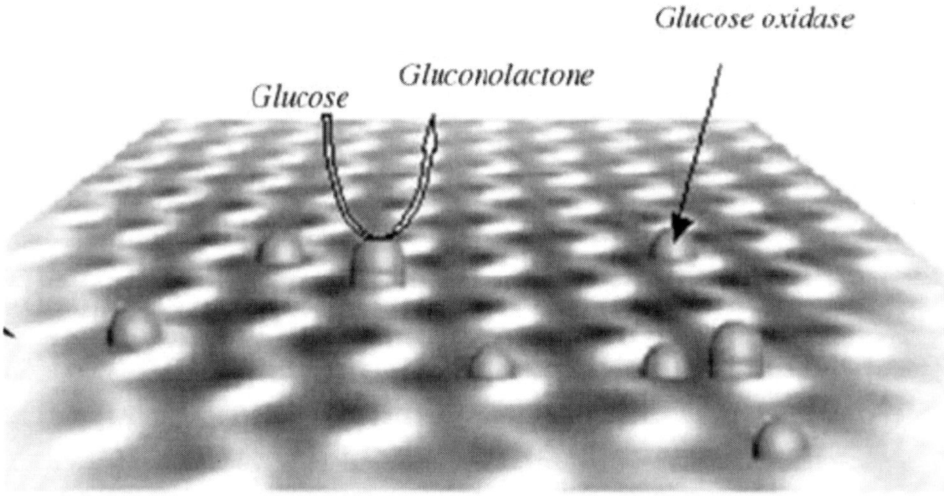

Figure 6. Adsorption of glucose oxidase on ordered 3D-mesoporous carbon material for the detection of glucose [77].

The pore volume can be enlarged again to incorporate enzymes easily. In this way, an ordered mesoporous carbon (called FDU-15) was used by Wang et al. [78] to immobilize GOD. FDU-15 is a mesoporous material with uniform tubular channels varying from 2.8 to 7.4 nm. Furthermore, it was found that, in the presence of O_2, GOD immobilized on Nafion and FDU-15 matrices could produce a linear response to glucose.

Pei et al. [79] claimed that Hb immobilized on FDU-15 film had performed direct electrochemistry and retained high electrocatalytic efficiency toward H_2O_2. The obtained Hb/FDU-15 electrode exhibited good analytical performance features, such as a wide determination range and low detection limit for H_2O_2 determination.

Conclusion

In this chapter we have shown how mesoporous materials are functionalized. Mesoporous silcas and mesoporous carbons are usually templates in the synthesis of other mesoporous materials. Therefore, the ordered mesostructure is almost similar and the consequence is that the mechanism of functionalization has the same principles. That is why this chapter has described the mesoporous silicas and mesoporous carbons with a focus on ordered mesoporous carbon (OMC). In fact, mesoporus silicas have been well described in many reviews and articles. Actually, the research on OMC is opening up a new challenge and approach to explore the properties of these materials.

As shown in the previous section, OMC has many important characteristics that, when improved, play an important role in the electrocatalytic properties of these materials. Especially, the high density of edge plane-like defective sites (EDSs) and oxygen-containing groups are the main functional groups that play an important role in the electrocatalytic properties for the design of those sensors. However, the large surface area and the widely open and ordered mesostructure of OMC can be applied in the construction of many sensors and biosensors.It has been shown that, the generation of those more existing characteristics results in the improved electrocatalytic properties of OMC and the construction of more sensitive sensors.

Introduction of electroactive substances in OMC is possible due to its large pore and/or large surface area. It has been noted that this immobilized substances maintain their activity and the electrocatalytic properties of the constructed composite are improved. More sensitive and stable (bio) sensors can be designed.

This point, however, has many points to be addressed for future developments. First, we need to understand exactly the electrochemical behavior of this material. In fact we have described the influence of pore size, surface area, functional groups on OMC. However, other works are still needed in this way. For example, there are many factors that affect the performance of the functionalized material and each contribution must be known. In particular, rationalization of the observed electrocatalytic effects, apparently different from one material to another one, would help at selecting to most appropriate OMC for target applications. Second, more than 75% of the applications used the CMK-3 material and more than 95% of the studies were made on the basis of a single material type. Only few works investigated, in the same time, several materials exhibiting distinct structural and/or porosity parameters. The future will certainly require a study of the other ordered mesoporous carbons for the analytical performance of the resulting devices.

REFERENCES

[1] V. Chiola, J.E. Ritsko, C.D. Vanderpool, US Patent No. 3 556 725, 1971.

[2] F. Di Renzo, H. Cambon, R. Dutartre, Micropor. *Mater.* 10 (1997) 283.

[3] J.S. Beck, C.T.-W. Chu, I.D. Johnson, C.T. Kresge, M.E. Leonowicz, W.J. Roth, J.W. Vartuli, WO Patent 91/11390, 1991.

[4] C.T. Kresge, M.E. Leonowicz, W.J. Roth, J.C. Vartuli, J.S. Beck, *Nature* 359 (1992) 710.

[5] J.S. Beck, J.C. Vartuli, W.J. Roth, M.E. Leonowicz, C.T. Kresge, K.D. Schmitt, C.T.W. Chu, D.H. Olson, E.W. Sheppard, S.B. McCullen, J.B. Higgins, J.L. Schlenker, *J. Am. Chem. Soc.* 114 (1992) 10834.

[6] A. Monnier, F. Schu¨ th, Q. Huo, D. Kumar, D. Margolese, R.S. Maxwell, G. Stucky, M. Krishnamurty, P. Petroff, A. Firouzi, M. Janicke, B. Chmelka, *Science* 261 (1993) 1299.

[7] Q. Huo, D. Margolese, U. Ciesla, P. Feng, T. Gier, P. Sieger, R. Leon, P.M. Petroff, U. Ciesla, F. Schuth, G. Stucky, *Nature* 368 (1994) 317.

[8] Q. Huo, D. Margolese, U. Ciesla, D. Demuth, P. Feng, T. Gier, P. Sieger, A. Firouzi, B. Chmelka, F. Schuth, G.D. Stucky, *Chem. Mater.* 6 (1994) 1176.

[9] G.S. Attard, J.C. Glyde, C.G. Goltner, *Nature* 378 (1995) 366.

[10] R. Ryoo, S.H. Joo, S. Jun, *J. Phys. Chem.* B 103 (1999) 7743.

[11] A.H. Lu, W. Schmidt, A. Taguchi, B. Spliethoff, B. Tesche, F. Schuth, Angew. *Chem. Int. Ed.* 41 (2002) 3489.

[12] X. He, D. Antonelli, Angew. *Chem. Int. Ed.* 41 (2002) 214.

[13] A.P. Wight, M.E. Davis, *Chem. Rev.* 102 (2002) 3589.

[14] A.Walcarius, *Chem. Soc. Rev.* 42 (2013) 4098-4140

[15] G.J. de A.A. Soler-Illia, C. Sanchez, B. Leveau, J. Patarin, *Chem. Rev.* 102 (2002) 4093.
[16] T.Y. Ma, L. Liu, Z.Y. Yuan, *Chem. Soc. Rev.* 42 (2013) 3977.
[17] A. Wingen, F. Kleitz, F. Schu¨ th, in: M. Baerns (Ed.), Basic Principles in Applied Catalysis, *Springer,* Berlin, 2003, p. 281.
[18] F. Schuth, *Chem. Mater.* 13 (2001) 3184.
[19] D. Yuan, F. Zeng, J. Yan, X. Yuan, X. Huang, W. Zou, RSC Adv. 3 (2013) 5570.
[20] Y. Wan, Y. Shi, D. Zhao, Chem. *Mater.* 20 (2008) 932.
[21] J.C. Ndamanisha, J. Bai, B. Qi, L.P. Guo, *Anal. Biochem.* 386 (2009) 79.
[22] M. Zhou, J. Ding, L.P. Guo, Q.K. Shang, *Anal. Chem.* 79 (2007) 5328.
[23] S. Song, Q. Gao, K. Xia, L. Gao, *Electroanalysis* 20 (2008) 1159.
[24] X. Bo, W. Xie, J.C. Ndamanisha, J. Bai, L.P. Guo, *Electroanalysis* 21 (2009) 2549.
[25] Y. Hou, L.P. Guo, G. Wang, *J. Electroanl. Chem.* 617 (2008) 211.
[26] A. Tagushi, F. Schuth, Micropor. Mesopor. *Mater.* 77 (2005) 1.
[27] Hartmann, *Chem. Mater.* 17 (2005) 4577.
[28] C.Y. Chen, H.X. Li, M.E. Davis, Micropor. *Mater.* 2 (1993) 17.
[29] C.T. Kresge, M.E. Leonowicz, W.J. Roth, J.C. Vartuli, J.S. Beck, *Nature* 359 (1992) 710.
[30] [30] H. Landmesser, H. Kosslick, U. Kurschner, R. Fricke, *J. Chem. Soc. Faraday Trans.* 94 (1998) 971.
[31] M. Onaka, R. Yamazaki, Chem. Lett. (1998) 259.
[32] U. Junges, W. Jacobs, I. Voigt-Martin, B. Krutzsch, F. Schuth, *J. Chem. Soc. Chem. Commun.* (1995) 2283.
[33] M.A´. Aramendia, V. Borau, C. Jime´nez, J.M. Marinas, F.J. Romero, *Chem. Commun.* (1999) 873.
[34] S.H. Joo, S.J. Choi, I. Oh, J. Kwak, Z. Liu, O. Terasaki, R. Ryoo, *Nature* 412 (2001) 169.
[35] W.C. Choi, S.I. Woo, M.K. Jeon, J.M. Sohn, M.R. Kim, H.J. Jeon, *Adv. Mater.* (2005) 446.
[36] B. Escobar, R. Barbosa, M.M. Yoshida, Y. V. Gomez, *J. Power Sources 243* (2013) 88.
[37] Y. Sun, L. Zhuang, J. Lu, X. Hong, P. Liu, *J. Am. Chem. Soc.* 129 (2007) 15465.
[38] Y. Jung, S. Kim, S.-J. Park, J.M. Kim, Colloids Surf. A: Physicochem. *Eng. Aspects* 313–314 (2008) 167.
[39] J.C. Ndamanisha, Y. Hou, J. Bai, L.P. Guo, Electrochim. *Acta* 54 (2009) 3935–3942.
[40] N. Yao, C. Pinckney, S. Lim, C. Pak, G.L. Haller, Micropor. *Mesopor. Mater.* 44. (2001) 377.
[41] J.A. Darr, M. Poliakoff, *Chem. Rev.* 99 (1999) 495.
[42] H. Wakayama, N. Setoyama, Y. Fukushima, *Adv. Mater.* 15 (2003) 742.
[43] P.L. Dhepe, A. Fukushima, M. Ichikawa, *Chem. Commun.* (2003) 590.
[44] X. Bo, J.C. Ndamanisha, J. Bai, L.P Guo, Talanta 81 (2010) 339.
[45] S. Suvanto, T.A. Pakkanen, L. Backman, *Appl. Catal.* A: Gen. 177 (1999) 25.
[46] S. Suvanto, J. Hukkama¨ki, T.T. Pakkanen, T.A. Pakkanen, *Langmuir* 16 (2000) 4109.
[47] S.H. Liu, S.-C. Chen, J. *Solid. State Chem.* 184 (2011) 2420–2427.

[48] Y. Hou, J.C. Ndamanisha, L.P. Guo, S.J. Peng, J. Bai, Electrochim. *Acta* 54 (2009) 6166.
[49] L. Wang, X. Bo, J. Bai, L. Zhu, L.P. Guo, *Electroanalysis* 22 (2010) 1.
[50] S. Zhu, H. Zhou, M. Hibino, I. Honma and M. Ichihara, Adv. Funct. *Mater* 15 (2005)381.
[51] J.C. Ndamanisha, Y. Hou, J. Bai, L.P. Guo, Electrochim. *Acta* 54 (2009) 3935.
[52] X. Bo, J. Bai, L. Wang, L.P. Guo, Talanta 81 (2010) 339.
[53] Z. Li, X. Wang, C. Wang, L. Yin, *RCS Adv*. 3 (2013) 17097-17104.
[54] L. Luo, F. Li, Li. Zhu, Y.P. Ding, Z. Zhang, D.M. Deng, B. Lu, Colloids Surf B *Biointerfaces* 102 (2013) 307.
[55] M. Zhou, L.P. Guo, F.Y. Lin, H.X. Liu, Anal. *Chim. Acta* 587 (2007) 124–13.
[56] L. Liu, J.C. Ndamanisha, J. Bai, L.P. Guo, *Electrochim. Acta* 55 (2010) 3035–3040.
[57] J.C. Ndamanisha, L.P. Guo, Biosens. *Bioelectron.* 23 (2008) 1680–1685.
[58] R.L. Brutchey, D.A. Ruddy, R.K. Anderson, T.D. Tilley, *Langmuir* 21 (2005) 9576.
[59] L. Liu, L.P. Guo, X. Bo, J. Bai, X. Cui, Anal. Chim. *Acta* 673 (2010) 88.
[60] J. Ju, X. Bo, H. Wang, Y. Zhang, C. Luhana, L. Guo, *Electrochem. Commun.* 25 (2012) 35-38.
[61] M. Zhou, J. Guo, L.P. Guo, J. Bai, *Anal. Chem.* 80 (2008) 4642.
[62] J.C. Ndamanisha, X. Bo, L.P. Guo, *Analyst* 135 (2010) 621.
[63] L. Zhu, R. Yang, X. Jiang, D. Yang, *Electrochem. Commun.* 11 (2009) 530.
[64] Z. Guo, S. Li, X.M. Liu, Y.P. Gao, W.W. Zhang, X.P. Ding, *Mater. Chem. Phys.* 128 (2011) 238.
[65] Y. Yan, Q.Cheng, Z. Zhu, V.Pavlinek, P. Sah, C. Li, J. *Power Sources* 240 (2013) 544-550.
[66] J. Dong, Y. Hu, S. Zhu, J. Xu, Y. Xu, *Anal. Bioanal. Chem.* 396 (2010) 1755.
[67] Z. Zhu, X. Li, Y. Zeng, W. Sun, *Biosens. Bioelectron.* 25 (2010) 2313.
[68] X. Zhai, Y. Li, G.Liu, Y. Cao, H. Gao, C. Yue, N. Sheng, *Sens. Actuators*, B 178 (2013) 169.
[69] D.R. Theavenot, K. Toth, R.A. Durst, G.S. Wilson, *Pure Appl. Chem.* 71 (1999) 2333.
[70] M. Pumera, A. Ambrosi, A. Bonanni, E.K. Chng, H.L. Poh, *Trends Anal. Chem.* 29 (2010) 954.
[71] P.P. Knops-Gerrits, D. De Vos, F. Thibault-Starzyk, P.A. Jacobs, *Nature* 369 (1994) 543.
[72] B. Pugin, J. Mol. Catal. A: *Chem*. 107 (1996) 273.
[73] J.J. Feng, J.J. Xu, H.Y. Chen, Biosens. *Bioelectron*. 22 (2007) 1618.
[74] L. Zhu, C. Tian, D. Zhu, R. Yang, *Electroanalysis* 20 (2008) 1128.
[75] M. Zhou, L. Shang, B.L. Li, L.J. Huang, S.J. Dong, *Biosens. Bioelectron.* 24 (2008) 442.
[76] C.P. You, X. Yan, J.L. Kong, D.Y. Zhao, B.H. Liu, *Electrochem. Commun.* 10 (2008) 1864.
[77] C.P. You, X. Xu, B.Z. Tian, J.L. Kong, D.Y. Zhao, B.H. Liu, Talanta 78 (2009) 705.
[78] K. Wang, H. Yang, L. Zhu, Z. Ma, S. Xing, Q. Lv, J. Liao, C. Liu, W. Xing, Electrochim. *Acta* 54 (2009) 4626.
[79] S. Pei, S. Qu, Y. Zhang, *Sensors* 10 (2010) 1279.

In: Comprehensive Guide for Mesoporous Materials, Volume 2 ISBN: 978-1-63463-982-8
Editor: Mahmood Aliofkhazraei

Chapter 6

CONTROLLING STRUCTURE OF MESOPOROUS TiO_2 AND PHOTOCATALYTIC ACTIVITY FOR ALIZARINE YELLOW DEGRADATION

***Sawsan A. Mahmoud*[1,*], *Ágnes Szegedi*[2] *and Ahmed K. Aboul Gheit*[1]**
[1]Processes Development Department, Egyptian Petroleum Research Institute, Cairo, Egypt
[2]Research Centre for Natural Sciences, Hungarian Academy of Sciences, Budapest, Hungary

ABSTRACT

Nanoporous titanium dioxide was prepared by sol-gel technique. To control the surface area, pore size and pore volume of the prepared TiO_2, a catalyzed hydrolysis was carried out using different concentrations of silicotungstic acid (SWA) as a template. A fixed molar ratio of H_2O/Ti was used. The prepared TiO_2 was calcined at 400 or 600°C. Samples were characterized by nitrogen physisorption, X-ray powder diffraction (XRD), selected scattered electron diffraction, Raman spectroscopic analysis, Fourier Transform spectroscopy (FTIR), Thermogravimetric analysis (TGA), Differential scanning calorimetry (DSC), scanning and transmission electron microscopy (SEM and TEM). The photocatalytic activity of the prepared samples was evaluated by the degradation of alizarin yellow under UV light. The results showed that the crystallinity increases as the concentration of SWA decreases. The presence of SWA during the precipitation of TiO_2 prevents the formation of rutile phase and suppresses the crystal growth. The results showed also that the surface area increase as the concentration of SWA decreases. The samples prepared using 0.05M SWA and calcined at 600 °C showed a higher activity.

OVERVIEW

The dyeing wastewaters discharge a large number of hazardous pollutants, in which, the percentage of azo dyes accounts for about 70% of all dyestuffs by weight [1]. Azo dyes and

* Corresponding author: Sawsan A. Mahmoud, e-mail: sawsanhassan2003@yahoo.com.

their intermediates, e.g., aromatic amines, are toxic, carcinogenic and mutagenic, which may pose a potential health hazard tohumans [2,3]. Now adays, the dyeing wastewaters are mostly treated by biological processes. However, the biological degradation of azo dye is difficult. Furthermore, azo dye could inhibit the microbial activity and cause cell death at high concentration. Moreover, the effluent of biological treatment of dyeing wastewater was proved to be carcinogenic and mutagenic [4]. To improve the biodegradability of azo dyeing wastewaters, combining advanced oxidation processes (AOPs) with biological process was pursued in recent years [5]. AOPs, defined as those technologies that utilize the hydroxyl radical ($OH^{\cdot}$) for oxidation, including ozonation [6], Fenton oxidation [7,8], electrochemical oxidation [9], or UV/H2O2 [10], followed by aerobic process are efficient for azo dyes degradation. Ionizing radiation (e.g., gamma radiationan delectron beam) is a special kind of AOP [10], which can produce approximately equal amounts of oxidizing species (dOH and H2O2) and reducing species (ea_q and Hd) through water radiolysis. This technology has been receiving more and more attention for removal and/or pretreatment of azo dyes and other toxic pollutants [11-17] but this technology is more expensive than chlorine or ultraviolet disinfection. Because of this gamma irradiation is unlikely to be applied.

Photocatalytic degradation and mineralization of organic and inorganic pollutants by means of semiconductor TiO_2 have been extensively studied in order to solve environmental problems relating to waste waters and polluted air [18-25]. Among various metal oxide semiconductors, TiO_2 has been in the focus of photocatalysis under UV irradiation because of its physical and chemical stability, low cost, ease of availability, non-toxicity, and electronic and optical properties.

Azodyes are abundant class of synthetic, colored, organic compounds, characterized by the presence of one or more azo bonds (-N=N-). Synthetic textile dyes of the azo family represent an important part of the world production of synthetic dyes. They are degraded into potentially carcinogenic amines [26, 27]. Moreover, their color causes an aesthetic problem in receiving waters. The degradation mechanism of dye using TiO_2 can be described by the following steps:

$$TiO_2 + h\upsilon \rightarrow h^+ + e^- \quad (1)$$

$$h^+ + OH^- \rightarrow \cdot OH \quad (2)$$

$$e^- + O_2 \rightarrow O_2^- \quad (3)$$

$$dye + h\upsilon(UV) \rightarrow dye^* \quad (4)$$

$$dye\cdot^- + O_2 \text{ (or } \cdot OH \text{ or } O_2\cdot^-) \rightarrow \text{smaller organic molecules, } CO_2\text{, etc.} \quad (5)$$

Nowadays, progress is focusing on to prepare anatase TiO_2 phase of high surface area and controlled pore size, aiming to enhance the charge transfer efficiency. Generation of these highly controlled materials can be achieved by combining sol–gel techniques with the evaporation process taking place in the presence of a surfactant templating agent [28a,b]. After elimination of the organic template and controlled crystallization of the network [29], one ends up with an organized arrangement of anatase nanoparticles shaped in mesoporous framework with narrow pore size distribution adjusted by the conditions and the surfactant molecular weight [30]. Table 1 summarizes some representative highly organized mesoporous TiO_2 developed by research groups with emphasis on the precursors, different templates used, calcination temperatures, the obtained surface area and pore volume.

Table 1. Examples of some previous studies using different templates for powdered mesoporous TiO_2 preparation

Precursor	Template	Symbol	Calcination temperature, (°C)	S_{BET} m^2/g	Pore volme cm^3/g	Ref
$Ti(O^nBU^{)}{}_{4\text{-}}T$	non	-	25	411	0.35	[31]
			300	189	0.03	
		-	500	16	0.09	
		-	700	-	-	
$Ti(O^n Pr)_{4\text{-}}T$	non	-	25	473	0.37	
		-	300	235	0.21	
		-	500	13	0.09	
		-	700	-	-	
$Ti(O^{\cdot} Pr)_{4\text{-}}T$	non	-	25	470	0.37	
		-	300	174	0.22	
		-	500	10	0.07	
		-	700	-	-	
$Ti(OEt)_{4\text{-}}T$	non	-	25	299	0.23	
		-	3 00	210	0.21	
		-	500	10	0.05	
		-	700	-	-	
Ti oxysulfate Hydrate	sulfonated poly (ether-ether ketone)	-	25	125	-	[32,33]
$(Ti(OiPr)_4$	Polyvinyl-pyrrolidone	-	900	air 4.4 H_2, 132.9 (Sp)*	-	[43]
$TiCl^4$	(1-hexadecyl) trimethyl- Ammonium bromide (CETAB)	(R),	450	2.17	-	[35]
		TM200,		141.65	0.186	
		TM400,		84.13	0.159	
		TM550,		87.03	0.161	
		TM700,		93.9	0.169	
		TM800,		101.75	0.174	
$TiCl^4$	non	as-prepared	500	87.69	0.16	[36]
		Pure TiO_2	500	39.27	0.14	
		Commercial TiO_2	500	72.12	-	
Ti(Iv)-isopropoxide	hexadecylamine HAD	beads P-25	500	86.6	-	[37]
				47.0		
Ti(IV)-Isopropoxide	non	PT64	350	156 (Sp)*	-	
		P-25		53	-	[38]
$Ti(OBU)_4$	ethylene glycol	Meso-0	300	43	-	[39]
		Meso-10		52	-	
		Meso-20		55	-	
		Meso-30		62	-	
		Meso-40		65	-	
		Meso-50		69	-	
		Meso-60		71	-	
		Meso-65		75	-	

Table 1. (Continued)

Precursor	Template	Symbol	Calcination temperature, (°C)	S_{BET} m²/g	Pore volme cm³/g	Ref
		Meso-70		76	-	
		Meso-80		85	-	
		Meso-90		88	-	
		Meso-100		90	-	
Ti(IV)-Isopropoxide	P123, F127	S1	500	223.76	-	
		S2		192.06	-	[40]
Ti-isopropoxide	P123	-	350	166	0.43	[41]
$Ti(OBU)_4$	CTAB	pure mesoporous Urea-doped pure	-	-	0.0057	
		mesoporous	-	-	0.0013	[42]
$Ti(OC(CH_3)_3)_4$	F127	-	450	154	0.29	[43]
Ti-isopropoxide	F127	M-TiO2	550	120	0.34	[44]
		M-TiO2		85	0.32	
		M-TiO2-F127		31	0.18	
		Nonporous TiO2		7	0.04	
		P25		45	0.17	
Ti-isopropoxide	diethylenetriamine	-	400	153	-	[45]
$Ti(OBU)_4$	PVA-I	IT-200	200	200.26	0.14	[46]
		IT-300	300	144.34	0.15	
		IT-400	400	61.14	0.13	
		PIT-200	200	190.22	0.19	
		PIT-300	300	180.05	0.13	
		PIT-400	400	44.62	0.13	
Ti-isopropoxide	Tween 20+ PEG1000	-	-	242 (sp)*	-	[47]
Ti-isopropoxide	1-tetra- decylamine	TiO2-01	450	84.26	0.279	[48]
		TiO2-21		125.11	0.181	
		TiO2-11		184.70	0.181	
		TiO2-12		121.79	0.174	
		TiO2-14		110.29	0.160	
		Anatase-TiO_2		8.66	0.03	
		P-25		50.0	0.262	
Ti-Sulfate	CETAB	-	-	161.2	0.33	[59]
Tetraisopropyl Orthotitanate $Ti(OCH(CH_3)_2)_4$ Laurylamine	1-butanol + Hydrochloride (LAHC)	TiO_2	500	58.9	0.119	[50-52]
			600	18.4	0.04	
			700	3.2	-a	
			750	2.8	-a	
		P-25	-	65.0	-a	
$Ti(OCH(CH_3)_2)_4$	Urea + Acetylaceton	TiO_2	450	104.7(Sp)	-	[53]
			500	71.9	-	
			550	26.9	-	
$Ti(OBU)_4$	PEI	S1	400	118	-	[54]
		S2	500	69	-	
		P25	-	50	-	
	OA	S3	160[h]	43	-	
	PEG	S4	-	57	-	

TM according to the size a- N_2 adsorption- desorption isotherms correspond to IUPAC type II pattern
h-hydrothermal temperature sp: specific surface area

Generally, a sol-gel method based on the hydrolysis of titanium alkoxide is widely used to synthesize TiO_2 nanoparticles [55]. However, this method encounters some problems, such as weak anatase crystallinity and poor monodispersity. In addition, nanocrystalline TiO_2 prepared by the sol-gel method undergoes both phase transformation and crystallite size growth even at relative low temperature [56]. To apply TiO_2 particles synthesized from the sol-gel process as photocatalytic catalysts, it is important to maintain high anatase crystallinity [57]. A previous study reported that high degree of anatase crystallinity can be achieved without high temperature calcination when TiO_2 particles were synthesized at low temperature due to fast hydrolysis and slow condensation [58]. In addition, many studies regarding the evaluation of the photocatalytic activity of TiO_2 nanoparticles have been conducted (e.g., [59–61]); however, few focused on the correlation between parameters used in the synthesis process of TiO_2 nanoparticles and their photocatalytic activity [60- 62].

Therefore, in this study we pay attention to the effect of the synthesis condition using silicotungstic acid in different concentration as a template on the physical properties of mesoporous TiO_2, in particular on its crystal composition, thermal stability, pore size distribution, surface area, and photocatalytic activity. The prepared materials were calcined at various temperatures and their crystalline composition, thermal stability, pore size distribution, surface area and photocatalytic activity were compared.

METHODOLOGY

Materials

Titanium (IV) isopropoxide, silicotungstic acid $H_4[W_{12}SiO_{40}]$, and the absolute ethanol were provided from Sigma-Aldrich CO.USA. The analytical grade alizarin yellow $C_{13}H_9N_3O_5$ was used without further purification.

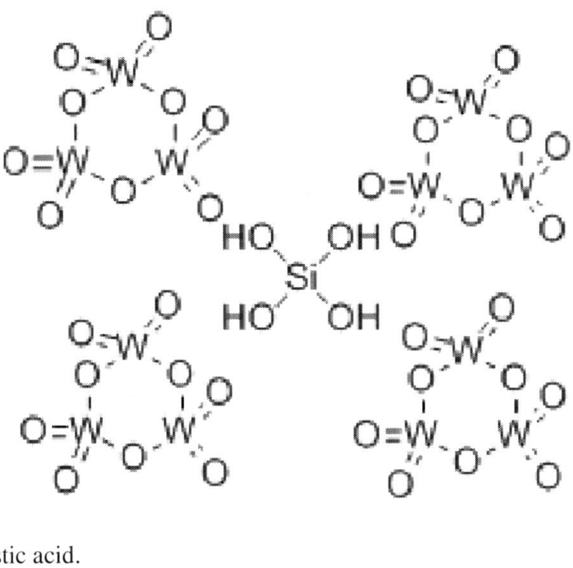

Structure of silicotungstic acid.

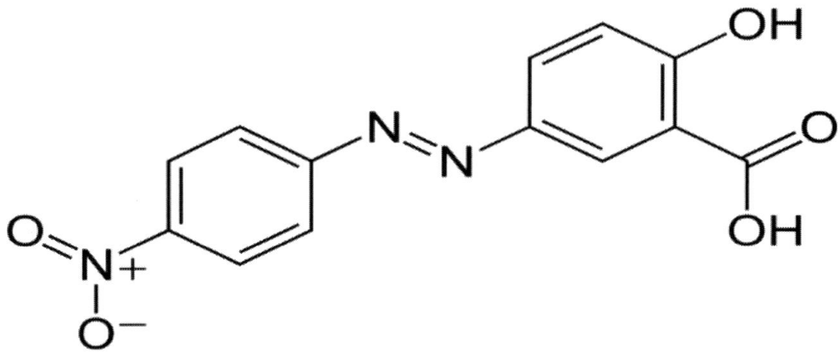

Structure of alizarin yellow.

Preparation of Titanium Dioxide

Titanium dioxide was prepared using sol-gel method as follows: 3.6 g of titanium tetraisopropoxide was diluted by absolute ethanol with constant stirring in a weight ratio of 1/7. After 15 min, 0.48 ml of silicotungstic acid (SWA) was added. The resulting suspension was stirred for three hours at room temperature, followed by the addition of 26 moles of waters per mole of Ti with vigorous stirring for two hours followed by the evaporation of water and ethanol at 78°C. The resulting precipitate was washed with deionized water and ethanol, drying under vacuum at 80°C overnight, and then calcined at 400 or 600°C for three hours.

Characterization of TiO_2

Nitrogen physisorption isotherms were measured at -196°C using a Quantachrome Nova Win 2 apparatus. The average pore size and pore size distribution were calculated using the Barrett-Joyner-Halenda (BJH) method from the desorption branch of the isotherms.

The crystalline phase of the TiO_2 formed was determined applying a Philips RB diffractometer in reflection mode using MoKα radiation at 40 kV, 40 mA.

The crystalline phases of TiO_2 were also investigated by scattered electron diffraction using a MORGAGNI 268D type (100 kV; W filament; point-resolution = 0.5 nm) transmission electron microscope

Fourier transform infrared spectroscopy was performed using (Nicolet Is-10 FTIR, Nicolet Is-10; Thermo Fisher Scientific). The KBr technique was carried out approximately in a quantitative manner since the weight of sample and that of KBr, were always kept constant.

Fourier transform infrared spectroscopy (FTIR) of the prepared samples were measured (Nicolet Is-10 FTIR, Nicolet Is-10; Thermo Fisher Scientific).

The surface morphology of the prepared TiO_2 was performed on JEOL JEM 3500 electron microscope.

Thermal analysis DSC/TGA was carried out using simultaneous 600 Q DSC-TGA USA under air atmosphere with a heating range of 10 °C/min.

Transmission electron microscopy (TEM) studies were carried out using a JEOL JEM-1230 electron microscope operating at 120 kV scaning electron microscopy.

The pore structure of TiO_2 was investigated by a MORGAGNI 268D type (100 kV; W filament; point-resolution = 0.5 nm) transmission electron microscope.

Photodegradation of Alizarin Yellow

500 ccm of an aqueous solution containing 10 ppm of high purity Alizarine yellow was subjected to UV irradiation using a 6 watt lamp at a wavelength of 254 nm. All photodegradation experiments were conducted in a batch reactor. The UV lamp was placed in a cooling silica jacket and placed in a jar containing the polluted water. The catalyst (0.05 g) was stirred in the solution using a magnetic stirrer at a controlled reaction temperature of 25 °C during the experimental period. At different irradiation time intervals, samples of the irradiated water were withdrawn for analysis and the dye concentration was determined by absorption mode (using calibration curve of absorbance & the concentration of the dye) 8500 series UV-Visible spectrophotometer.

Results and Discussion

Effect of SWA Concentration and Calcination Temperature on the Physicochemical Structure of TiO_2

Figure 1 (a&b) shows the nitrogen adsorption-desorption isotherms, and the pore size distribution of a TiO_2 sample calcined at 400°C. The isotherms are of type IV with H2 type hysteresis loop classified by IUPAC [63] with closure point ranging between p/p_0=0.3-0.4 and a quite steep condensation step indicating the mesoporous nature of the solid. The H2 type hysteresis loop is generally attributed to pores of narrow entrance and wider body, i.e. ink bottle like shape. The amount of adsorbed nitrogen increases simultaneously with decreasing in SWA content indicating increasing adsorption capacity of the samples. The decrease in surface area with increasing SWA content can be attributed to a lower cross-linking degree during the sol-gel synthesis process [64, 65]. These features are reflected in surface parameters, pore volume and average pore radius especially for the samples embedded low SWA content. Figure 3/a also shows the widening of the hysteresis loop which is more pronounced after introduction of SWA, and increases with decreasing SWA content implying the successful role of SWA in creating mesoporous system. It should be emphasized that the surface parameters are related to SWA content and the crystalline features of the samples (Table 2). Figure 1/b shows the pore size distribution for the calcined samples at 400°C. It is obvious that as the SWA increases bimodal pores are formed. This indicates highly incorporated SWA in the TiO_2 structure.

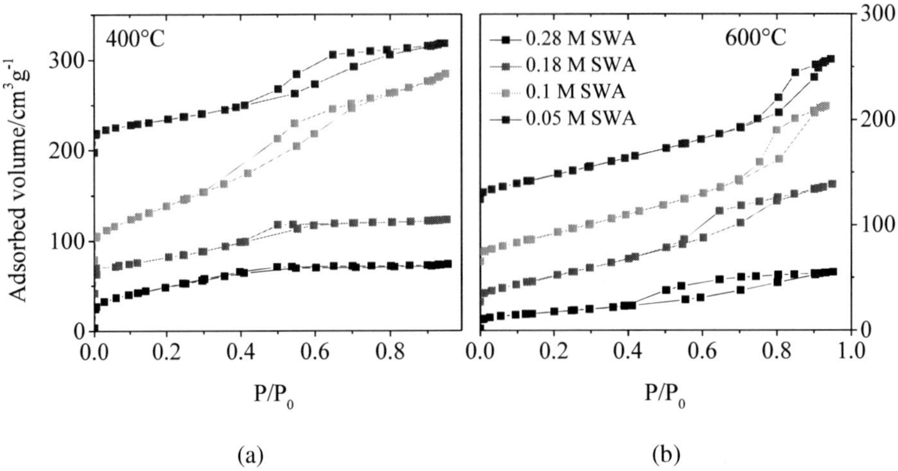

Figure 1. Nitrogen physisorption isotherms for TiO_2 prepared by different concentrations of SWA, calcined at 400°C, and at 600°C for 3 h.

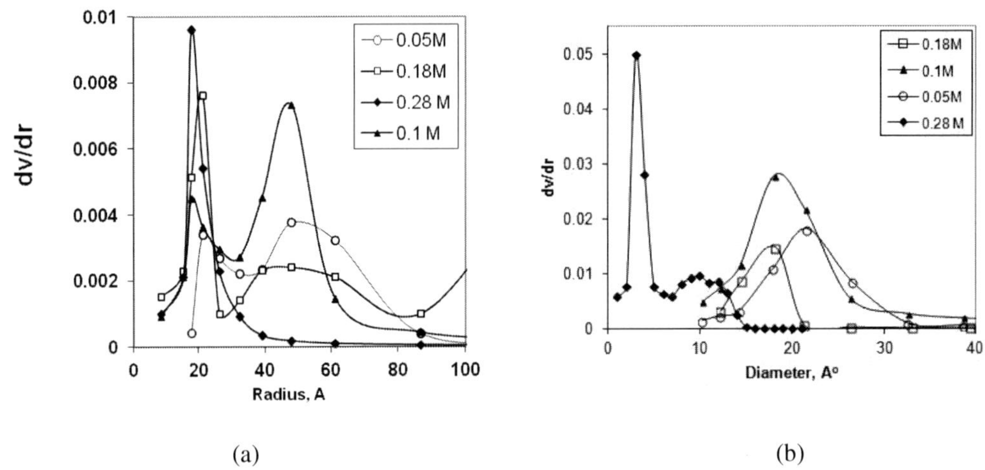

Figure 2. Pore size distribution for TiO_2 prepared by different amounts of SWA, calcined at (a): 400°C and (b) 600°C.

Table 2. Textural properties of TiO_2 calcined at 400°C using different concentrations of SWA

SWA conc. (Mol/L)	BET surface area (m^2/g)	Totalpore volume (cm^3/g)	Average pore diameter (nm)	Crystallite size (nm)
0.28	140.0	0.11	10.9	18.2
0.18	153.0	0.12	18.2	15.0
0.1	173.7	0.18	18.1	12.1
0.05	256.6	0.31	21.6	9.1
Ref. a	93	0.10	5.8	-

[a] Mesoporous TiO2 was prepared using phosphotungstic acid assisted hydrolysis and polycondensation reactions of tetrabutyl titanate, followed by removal of the template via washing with water and ethanol. G.S. Luo et al. Microporous and Mesoporous Materials, 96 (2006) 301–306.

Table 3. Textural properties of TiO_2 calcined at 600°C using different concentrations of SWA

SWA conc. (Mol/L)	BET surface area (m^2/g)	Total pore volume (cm^3/g)	Average pore diameter (nm)	Crystallite size (nm)
0.28	61.9	0.08	3.6	24.5
0.18	107.0	0.17	4.3	18.3
0.1	102.0	0.21	9.7	14.3
0.05	132.9	0.23	9.7	11.5

The experimental results show that the surface area is influenced by the annealing temperature, whereas the mesopore size and the total pore volume are influenced by the amount of SWA and the annealing temperature (Table 3). The average mesopore diameter and the total pore volume decrease with increasing the concentration of SWA. Increasing the calcination temperature the surface area and the total pore volume are decreasing rapidly due to the growth of the crystallite size (Figure 2 a&b).

The X-ray diffraction (XRD) patterns of the calcined TiO_2 material prepared under various concentrations of SWA calcined at 400°C and 600°C are shown in Figure 3 a&b. All the reflections can be assigned to anatase phase without any indication the presence of any crystalline SWA. Hence, SWA was successfully incorporated into TiO_2 structure, rather than existing as a free solid acid. The peak found at 11.5 °2θ is the most intensive peak for the anatase phase. This peak is relatively broad due to the nano-sized dimension of the crystals. The particle sizes estimated by the Sherrer equation are varying between 9-18 nm, depending on the SWA concentration for samples calcined at 400°C and between 12-36 nm for samples calcined at 600°C. It should be emphasized that SWA plays a vital role in controlling the particle growth and the nature of crystalline phases. The addition of small amount of SWA is considered as a prime factor in stabilizing the anatase phase and hinders its transformation to the more stable rutile phase. Anatase appears to be the most photocatalytically and electrochemically active crystallographic form of TiO_2, opposed to the rutile and brookite forms; herein, the prepared mesoporous TiO_2 materials may have far-reaching implications for its use as a photocatalytic material.

Selected area scattered electron diffraction (Figure 4) also supported that with increasing temperature the crystallinity of the anatase phase also increases. It is also obvious that the anatase crystallinity increases as the SWA content increases.

Figure 5 shows the Raman spectra of TiO_2 (0.05 M SWA) sample calcined at 600 °C. The Figure shows a peak at about 146 cm^{-1} is the strongest of all the observed bands, which is attributed to the main Eg anatase vibration mode. Moreover, vibration peaks at 147 cm^{-1} (Eg, weak), 398 cm^{-1} (B1g), 518 cm^{-1} (A1g) and 639 cm^{-1} (Eg) are presented in the spectra for the sample, which indicates the presence of crystalline anatase TiO_2 [66]. This result in agreement with the data obtained from XRD.

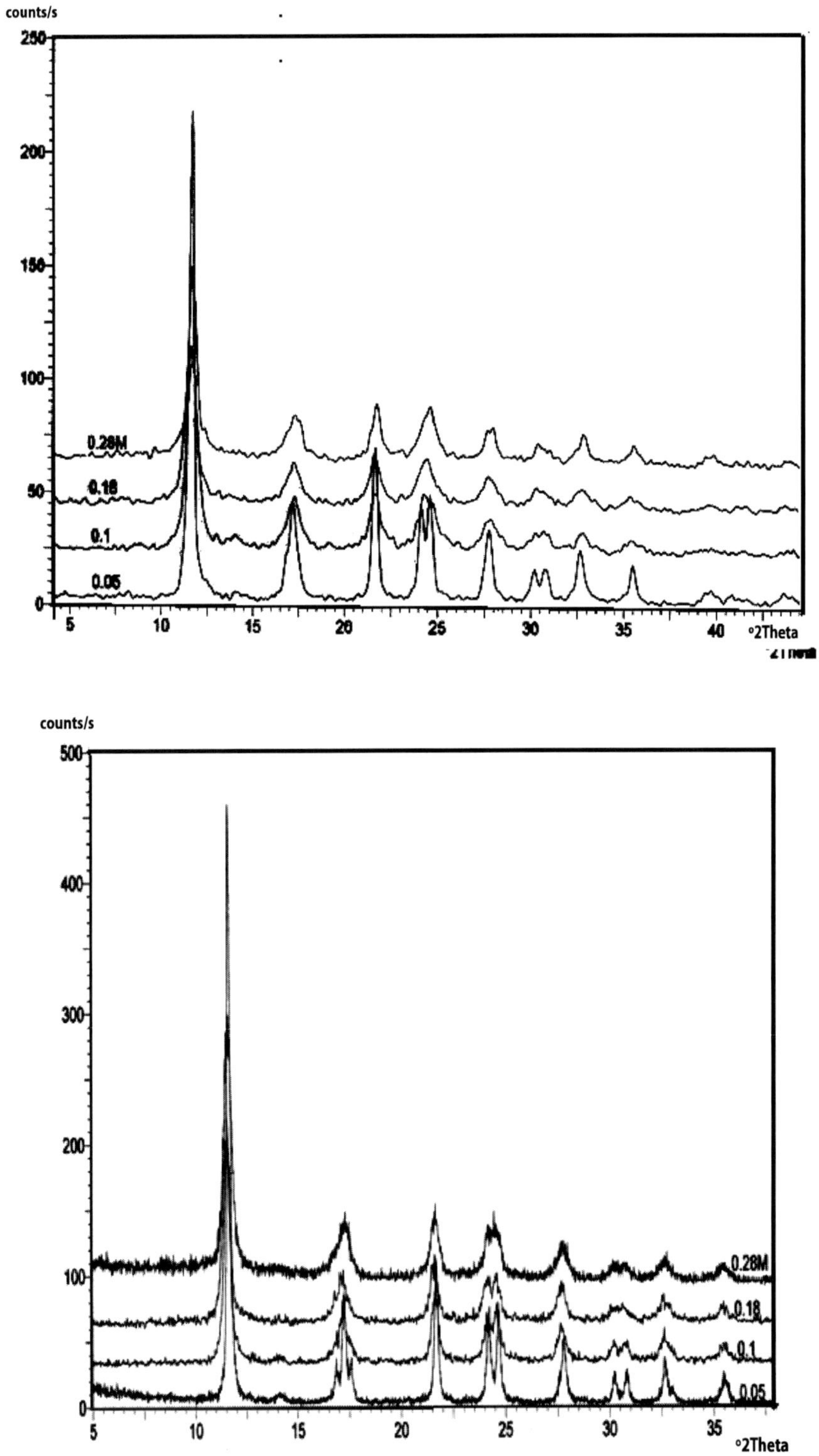

Figure 3. XRD patterns of TiO_2 prepared using different SWA concentrations, calcined at (a): 400°C, and (b) 600°C for 3 h.

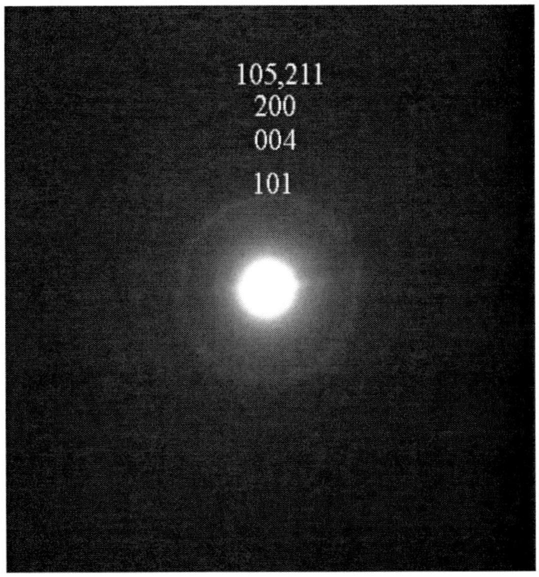

Figure 4. Electron diffraction pattern for TiO_2 prepared using 0.1 M SWA, heat treated at 600°C.

FT-IR spectra of the TiO_2 (0.05M) sample calcined at 600 °C for 3 h is shown in Figure 6. In this spectrum, the peaks around 3374 and 1626 cm^{-1} can be attributed to the stretching vibration of hydroxyl in surface-adsorbed water Ti–OH of the mesoporous materials [67]. Ti–O stretching at 608 cm^{-1} indicate the presence of TiO_2 anatase phase. On the other hand, no band characteristic of SWA is found in TiO_2 materials, suggesting the absence of SWA crystalline form in TiO_2 materials. This is consistent with the XRD results discussed above.

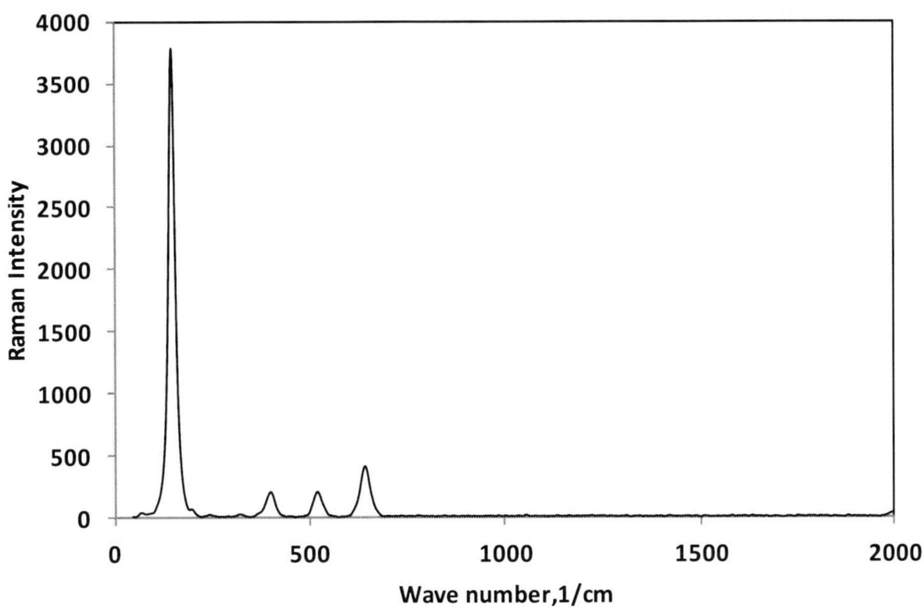

Figure 5. Raman spectra for TiO_2 prepared using 0.05M SAW and calcined at 600 °C for 3 hours.

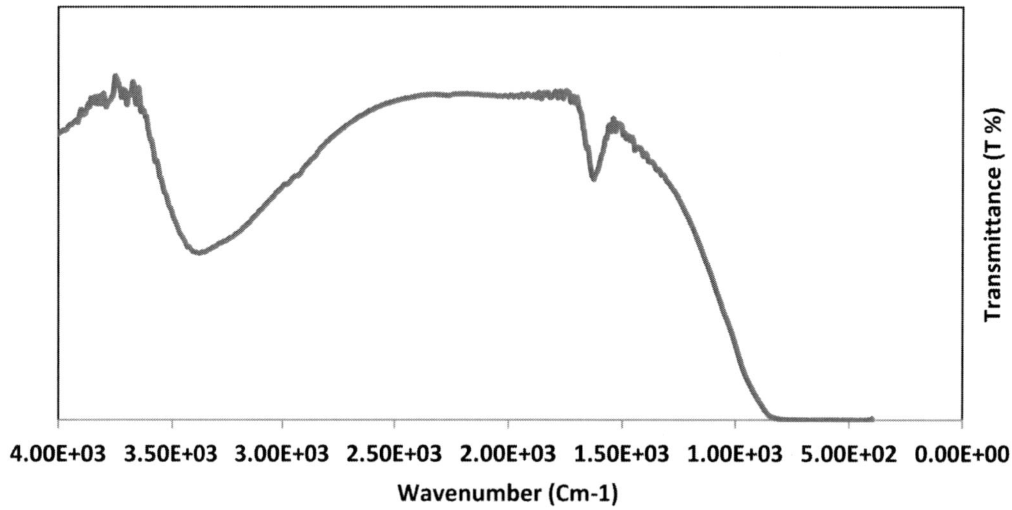

Figure 6. FTIR Spectroscopic analysis for TiO_2 prepared using 0.05 M SWA and calcined at 600 C for 3 hours.

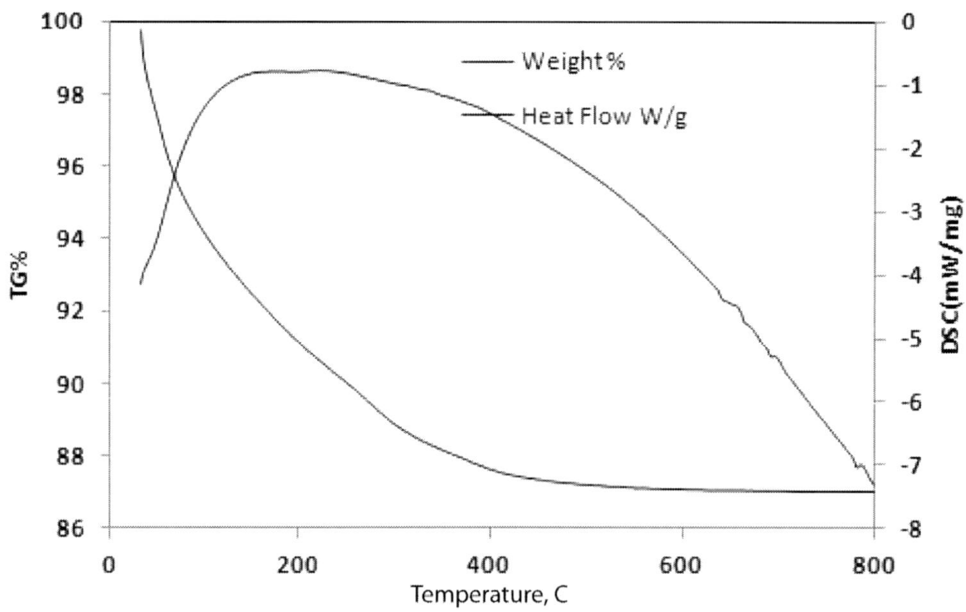

Figure 7. TG-DSC curve for TiO_2 prepared using 0.05 M SWA.

The thermal behavior of TiO_2 was investigated with a TG/ DSC technique at temperature range from room temperature to 800 °C, and the TG–DSC patterns of the is shown in Figure. 7. It is well known that the thermal behavior of TiO_2 usually depends on the chemical composition, preparation condition and existing phases [68–70]. The DSC curve showed a very broad exothermic band at 120-323 °C due to loss of water, and the decomposition of organic substance. Nevertheless, very small exothermic bands centered at 659, 700 and 777 °C for 0.05 M SWA can be attributed to crystalline phase transformation from anatase to rutile.

The TG curve indicates that the total weight loss was 12.2%. The TG curve can be divided into two stages. The first stage is range from room temperature to 400 °C with a mass loss of 12.55%, which represents the dehydration, loss of residual solvent and organic substance. The second mass loss of 0.44% in the temperature range of 400–800 °C is owning to the gradual removal of the organic residues. It can be concluded from TG–DSC results that the phase of sample TiO_2 is mainly anatase, and using the SWA can retard the transformation from anatase to rutile.

Surface morphology of the prepared TiO_2 samples calcined at 400 and 600°C are presented in Figure 8. The SEM images show that TiO_2 particles have well-defined spherical morphology, with a diameter range of 7–8 μm.

TEM images in Figure 9 show the aggregation of nanoparticles with particle size of 11-24 nm. A disordered wormhole-like pore structure can be observed, which is formed by the agglomeration of TiO_2 nanoparticles. The accessible pores are randomly connected, lacking discernible long-range order in the pore arrangement among the small TiO_2 particles. This indicates that the mesoporosity is mainly due to the interparticle porosity rather than intraparticle porosity.

Figure 8. SEM images for TiO_2 prepared by different SWA concentrations: a) 0.28 M, b) 0.18 M, c) 0.1 M and d) 0.05 M and calcined at 400°C for 3 h.

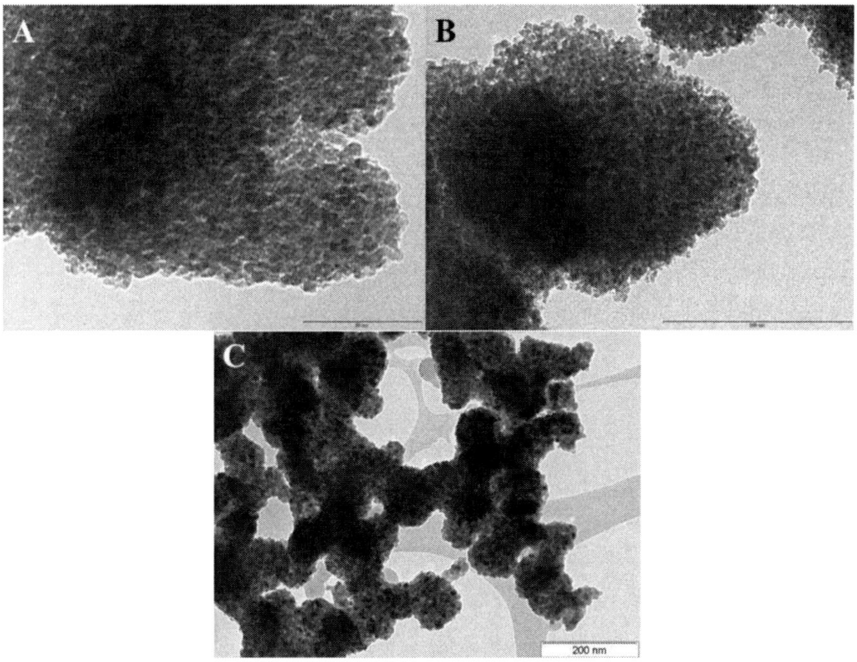

Figure 9. TEM images for TiO_2 prepared by different SWA concentrations: a) 0.1 M, b) 0.05 M and c) 0.28 M and calcined at 600°C for 3 h.

Mechanism of Formation Mesoporous TiO_2

A possible mechanism for the formation of well-dispersed mesoporous TiO_2 particles might be effective aggregation of the TiO_2 particles in the presence of SWA. The SWA plays an important role in the formation of a stable sol–gel network with ordered array of Ti and O, without leading to the shrinkage and collapse of the network. The basic unit of SWA is the heteropolyanion of Keggin structure $H_4[W_{12}SiO_{40}]$, which is the most commonly studied heteropolyanions because of its great acid strength, stability, and availability compared to those of other polyoxometalates. A Keggin unit of SWA [71] is composed of a central SiO_4 tetrahedron surrounded by twelve WO_6 octahedra. There are four types of oxygen atoms in the Keggin unit: four oxygen atoms of the central tetrahedron, 12 oxygen atoms that bridge addenda atoms not sharing a central oxygen atom (corner-sharing, 12 oxygen atoms that bridge two addenda atoms sharing the same central oxygen atom (edge-sharing), and twelve terminal oxygen atoms associated with a single addendum atom [72]. A hydrogen bond is easily formed between the proton on the Keggin unit and the oxygen atom of the water molecule. Bidentate water adsorption includes formation of a second hydrogen bond between a hydrogen atom of the water molecule and a second oxygen atom of the Keggin unit. Hydrogen bonds between SWA and hydroxylated tetrabutyl titanate are the essential factor to form this mesoporous structure. It is reported that mesoporous silica could be prepared by the hydrogen-bonding interaction of an alkylamine (S^0) head group and hydroxylated TEOS (I^0) [73]. Similarly, surfactants with head groups that could interact with the titanium alkoxide will lead to TiO_2 mesostructures. This indicates that, without the formation of ordered micelles by hydrophilic and hydrophobic surfactants, SWA can be used successfully as a self-assembling template via formation of hydrogen bonds. Consequently, the titanium precursor is hydrolyzed and condensed around the self-assembled PWA structure combined with water at a controlled rate, resulting in the formation of highly porous crystalline TiO_2. The materials lack long-range ordering of pores and have higher amounts of interparticle mesoporosity, because the long-range effects of the electrostatic interaction that could control the packing of micellar rods are absent.

Photocatalytic Degradation of Alizarin Yellow

Titanium dioxide is considered as a promising photocatalyst extensively involved in partial or complete mineralization of various hazardous organic dyes through the active species on its surface. Once titanium dioxide is illuminated with UV light, electrons are excited from the lower energy valance band to the conducting band, which leads to the formation of positive holes and negative electrons on the catalyst surface. The holes react with water or hydroxyl ions producing hydroxyl radicals. Electrons in the conduction band (e_{CB}^-) on the catalyst surface can reduce molecular oxygen to a superoxide anion. Hydroxyl (HO•), hydrogen peroxide (HO_2•), and superoxide (•O_2^-) radicals are considered the reactive species that oxidize the organic compounds adsorbed on the oxide surface [18].

These generated radicals are usually react with the adsorbed dye molecules followed by the formation of several intermediates including radicals, and radical cations finally mineralized into carbon dioxide, water and inorganic nitrogen from nitrate ion.

The response mechanism of the electrons and holes of the Ti catalyst is expressed as follows:

$$\text{Ti (metal oxide)} + h\nu \rightarrow \text{Ti} (e^- + h^+) \quad (2)$$

$$\text{Ti} (h^+) + \rightarrow \text{Ti} + RX^{\cdot+}_{ad} \quad (3)$$

$$\text{Ti} (h^+) + H_2O \rightarrow \text{Ti} + {}^{\cdot}OH + H^+ \quad (4)$$

$$\text{Ti} (h^+) + OH^- \rightarrow \text{Ti} + {}^{\cdot}OH \quad (5)$$

The electrons will, therefore, be transmitted via substances absorbed on the surface of the Ti catalyst such as H_2O, dye, and ^-OH. In addition,

$$\text{Ti(metal oxides)} (e^-) + O_2 \rightarrow \text{Ti} + O_2^{\cdot-} \quad (6)$$

$$O_2^{\cdot-} + H^+ \rightarrow HO_2^{\cdot}, \quad (7)$$

$$2\,O_2^{\cdot-} + 2\,H^+ \rightarrow 2\,{}^{\cdot}OH + O_2, \quad (8)$$

$$2\,{}^{\cdot}OH \rightarrow O_2 + H_2O_2 \quad (9)$$

$$\text{Ti (metal oxides) (e-)} + H_2O_2 \rightarrow \text{Ti} + OH- + {}^{\cdot}OH \quad (10)$$

Organic compounds absorbed on the surface of Ti catalyst then oxidized by $^{\cdot}OH$.

The photocatalytic activity of the prepared samples was determined by degradation of x M alizarin yellow solution under UV-light irradiation (Figure 10 a & b). For 0.28 M SWA sample no degradation of alizarin yellow was observed over a long period (3 h). For 0.18M SWA sample already a low degradation activity can be found. As the concentration of SWA decreases the photocatalytic activity increases to reach 100% after 180 min using 0.05M SWA. The photoactivity of TiO_2 in degradation process is influenced by several parameters such as the phase composition, band gap energy, surface area, particle size, crystallinity, electron hole recombination rate [74-76]. Table 4 Summarizes some results of different Azo Dyes photodegradation by several research groups with emphasis on the dyes, degradation %, irradiation time, the irradiation source and the catalyst used [77-85]. High surface area mesoporous titania is considered a reactive photocatalyst in degradation process owing to the facility of alizarine yellow dye molecules to adsorb on the various active surface centers and its ability to diffuse through the large pore structure [86-88]. Figure 10b represents the photocatalytic degradation of the prepared catalyst calcined at 600°C. The same trend can be observed as for samples annealed at 400°C but with higher activity. 100% degradation was obtained in 120 min using 0.05 M SWA. Han et al., 2012 [89] reported that, if specific surface area and pore volume of TiO_2 nanoparticles were dominant factors controlling the photocatalytic activity of TiO_2, the activity would have decreased with calcination temperature; however, the activity showed the maximum value at 700°C in his study. The trend is likely attributed to the combined effect of anatase crystallinity which showed the highest value at 600°C, and to crystallite size which did not excessively increased up to 600°C.

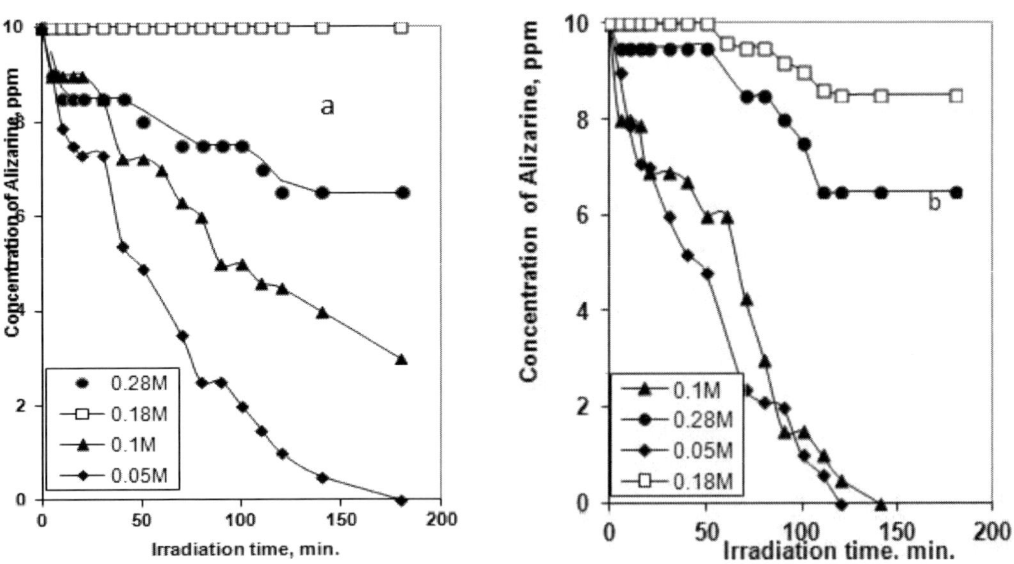

(a) (b)

Figure 10 a & b. Photocatalytic degradation of alizarin yellow using mesoporous TiO_2 a) at 400°C and b) at 600°C.

Table 4. Summary of photocatalytic degradation studies of azo dyes

Dye	Degradation %	Time, min	Irradiation source	Catalyst	Ref
*PR 180	90	120	UVC	ZnO	[77]
	85		UVC	ZnO/TiO_2	
	80		UVA	ZnO	
	75		UVA	ZnO/TiO_2	
*1Mo	90	140	UV	TiO_2-SiO_2	[78]
	90		Natural sunlight		
*2MB	1.8	90	UV-254nm	photolysis	[79]
	39	360		*3LOM, *5HAP	
	54	1440		*4ORM, HAP	
*6RB5	86	150	254	Mesoporous TiO_2	[48]
	81		306		
	69		352		
*73R	77	150	254		
	67		306		
	62		352		
mix.*8AY+AB	92.5	-	UV	Mesoporous TiO_2	[80]
*9RB	95	90	UV	0.02V-2La/TiO_2	[81]
C.I.Acid Red,18	100	480	Solar light Simulating source	TiO_2/hydrogel nanocomposite	[82]
C.I. Acid blue IIs	100	480			
C.I. Reactive black 5+ C.I. direct blue 78	100	480			
Reactive yellow 17	55	480			
Reactive yellow 125	85.91 91.64	250 250	Vis UV	Z-TiO_2-N	[83]
Alizarine yellow R	100	60	UV	Homogeneous H_2O_2	[84]
Alizarine yellow R	-	-	-	Effect of ·OH	[85]

* PR 180: PRO 180
*1: MO Methyl orange
*2: MB:Methylene blue
*3: LOM: Low oxygen media
*4: HOM: High oxygen media
*5; HAP: hydroxyl apatite
*6; RB5 : reactive black dye
*7:3R; Ramazol Brilliant Orange
*8: AY+AB; Acid yellow+ Acid blue
*9: RB : Clemson

CONCLUSION

In the present study the photocatalytic degradation of alizarin yellow by TiO_2 nanoparticles, synthesized under different concentrations of silicotungstic acid, and calcination temperatures by sol-gel process, was investigated. The correlation between physical properties and photocatalytic activity of TiO_2 nanoparticles according to the different synthesis conditions is as follows:

- As the concentrations of silicotungstic acid increases the crystallinity of the anatase phase is slightly increases and the surface area decreases.
- Alizarin yellow degradation experiments revealed that the photocatalytic activity of TiO_2 nanoparticles was maximized at the calcination temperature of 600°C.

However, surface area and pore volume of TiO_2 nanoparticles gradually decreases with increasing calcination temperature, indicating that textural properties are not critical for controlling the photocatalytic activity of TiO_2 nanoparticles. The trend is likely due to the combined effect of the anatase crystallinity, showed the highest value at 600°C and the crystallite size did not excessively increase up to 600°C.

REFERENCES

[1] Singh, K; Arora, S. Removal of synthetic textile dyes from wastewaters: a critical review on present treatment technologies, *Crit. Rev. Environ. Sci. Technol.*, 41, 807–878, 2011.

[2] Turesky, RJ. Interspecies metabolism of heterocyclic aromatic amines and the uncertainties in extrapolation of animal toxicity data for human risk assessment, *Mol. Nutr. Food Res.*, 49, 101–117, 2005.

[3] Dumont, J; Jossé, R; Lambert, C; Anthérieu, S; Le Hegarat, L; Aninat, C; Robin, M; Guguen-Guillouzo, CA. Guillouzo, Differential toxicity of hetero- cyclic aromatic amines and their mixture in metabolically competent Hepa RG cells, *Toxicol. Appl. Pharmacol*, 245, 256–263, 2010.

[4] Lima, ROA; Bazo, AP; Salvadori, DMF; Rech, CM; Oliveira, DP; Umbuzeiro, GA. Mutagenic and carcinogenic potential of a textile azo dye processing plant effluent that impacts a drinking water source, Mutat. Res. Genet. Toxicol. Environ. *Mutag.*, 626, 53–60, 2007.

[5] Wang, J; Xu, LJ. Advanced oxidation processes for wastewater treatment: formation of hydroxyl radical and application, *Crit. Rev. Environ. Sci. Technol*, 42, 251–325, 2012.

[6] Fu, Zhimin; Zhang, Yugao; Wang, Xiaojun. Textiles wastewater treatment using anoxic filter bed and biological wriggle bed-ozone biological aerated filter, *Bioresour. Technol.*, 102, 3748–3753, 2011.

[7] Tantak, N; Chaudhari, PS. Degradation of azo dyes by sequential Fenton's oxidation and aerobic biological treatment, *J. Hazard. Mater.*, 136, 698–705, 2006.

[8] Lodha, B; Chaudhari, S. Optimization of Fenton-biological treatment scheme for the treatment of aqueous dye solutions, *J. Hazard. Mater.*, 148, 459–466, 2007.

[9] Guo, J; Zhou, J; Wang, D; Xiang, X; Yu, H; Tian, C; Song, Z. Correlation of anaerobic biodegradability and the electrochemical characteristic of azo dyes, *Biodegradation*, 17, 341–346, 2006.

[10] Sudarjanto, G; Keller-Lehmann, B; Keller, J. Optimization of integrated chemical–biological degradation of a reactive azo dye using response surface methodology, *J. Hazard. Mater.*, 138, 160–168. 2006.

[11] Jianlong, W; Jiazhuo, W. Application of radiation technology to sewage sludge processing: a review, *J. Hazard. Mater.*, 143, 2–7, 2007.

[12] Sharma, KK; O'Neill, P; Oakes, J; Batchelor, SN; Rao, BSM. One-electron oxidation and reduction of different tautomeric forms of azo dyes: a pulse radiolysis study, *J. Phys. Chem. A*, 107, 7619–7628, 2003

[13] Hu, J; Wang, JL. Degradation of chlorophenols in aqueous solution by g- radiation, *Radiati., Phys.Chem.*,76, 1489–1492. 2007

[14] Ting, TM; Jamaludin, N. Decolorization and decomposition of organic pollutants for reactive and disperse dyes using electron beam technology: effect of the concentrations of pollutants and irradiation dose, *Chemosphere*, 73, 76–80. 2008.

[15] Wojnarovits, L; Takacs, E. Irradiation treatment of azo dye containing wastewater: an overview, *Radiat. Phys. Chem.*, 77, 225–244. 2008.

[16] Vahdat, A; Bahrami, SH; Arami, M; Motahari, A. Decomposition and decoloration of a direct dye by electron beam radiation, *Radiat. Phys. Chem.*, 79, 33–35. 2010.

[17] Yu, S; Hu, J; Jianlong, W. Gamma radiation-induced catalytic degradation of p-nitrophenol (PNP) in the presence of TiO_2 nanoparticles. *Radiat.Phys.Chem*, 79, 1039–1046, 2010.

[18] Hoffmann, MR; Martin, SC; Choi, W; Bahnemann, DW. Environmental Applications of Semiconductor Photocatalysis, *Chem. Rev.M* 95 (1995) 69–96.

[19] Hogfelt, A; Gratzel, M. Light-Induced Redox Reactions in Nanocrystalline Systems. *Chem. Rev.* 95 (1995) 49–68.

[20] Fox, MA; Dulay, MTH. eterogeneous photocatalysis, *Chem. Rev.*, 93, 341–357. 1993.

[21] Pelizzetti, E; Serpone, N. Photocatalysis–Fundamentals and Applications, Wiley, New York, 1989.

[22] Heller, A. Acc., Chemistry and Applications of Photocatalytic Oxidation of Thin Organic Films. *Chem. Res.*, 28, 503–508, 1995.

[23] Anpo, M; Yamashita, H. in: M. Anpo (Ed.), Surface Photochemistry, Wiley, Chichester, 117–164, 1996

[24] Kamat, PV. Photochemistry of Nonreactive and Reactive Semiconductor, Surfaces, *Chem. Rev.*, 93, 267–300, 1993

[25] Ollis, DF; Al-Ekabi, H. Photocatalytic Purification and Treatment of Water and Air, Elsevier science publisher, *Amsterdam*, 49-65, 1993.

[26] Chung, KT; Stevens, JR. Degradation of azo dyes by environmental microorganisms and helminthes, *Environ. Toxicol. Chem.*, 12, 2121–2132, 1993.

[27] Nerud, F; Baldrian, P; Gabriel, J; Ogbeifun, D. Decolorization of synthetic dyes by the Fenton reagent and the Cu/pyridine/H_2O_2 system, *Chemosphere*, 44, 957–961. 2001.

[28] (a) Soler-Illia, GJAA; Sanchez, C; Lebeau, B; Patarin, J. Chemical strategies to design textured materials: From microporous and mesoporous oxides to nanonetworks and hierarchical structures, *Chem. Rev.*, 102, 4093, (2002).
(b) Soler-Illia, G; Crepaldi, EL; Grosso, D; Sanchez, C. Block Copolymer –Templated

Mesoporous Oxides, *Curr. Opin. Colloid& Interf. Sci.*, 8, 109, 2003.

[29] Grosso, D; Soler-Illia, GJAA; Crepaldi, EL; Cagnol, F; Sinturel, C; Bourgeois, A; Brunet-Bruneau, A; Amenitsch, H; Albouy, P; Sanchez, C. Highly Porous TiO_2 anatase Optical Thin Films with Cubic Mesostructure Stabilized at 700 Degree C, *Chem. Mater.*, 15, 4562–4570, 2003.

[30] Sakatani, Y; Grosso, D; Nicole, L; Boissie`re, C; Soler-Illia, GJAA; Sanchez, C. Optimised Photocatalytic Activity of Gride- like mesoporous TiO2 Films : Effect of Crystalinity, Pore Size Distribution, and Pore Radius, *J. Mater. Chem.*, 16, 77–82, 2006.

[31] Li, X-Y; Chen, Li-H; Rooke, JCC; Deng, Z; Hu, Z-Y; Wang, S-Z; Wang, L; Li, Y; Krief, A; Su, B-L. Mesoporous titanium Dioxide (TiO_2) with hierarchically 3D dendrimeric architectures: Formation mechanism and highly enhanced photocatalytic activity, *Journal of colloid and interface Science*, 394, 252-262, 2013.

[32] Chevallier, L; Bauer, A; Cavaliere, S; Hui, R; Rozière, J; Jones, DJ. *ACS Applied Materials and Interfaces*, 4, 1752-1759, 2012.

[33] Bauer, A; Chevallier, L; Hui, R; Cavaliere, S; Zhang, J; Jones, D; Rozière, J. Synthesis and characterization of Nb-TiO_2 mesoporous microsphere and nanofiber supported Pt catalysts for high temperature PEM fuel cells, *Electrochimica Acta*, 77, 1– 7, 2012

[34] Zhao, B; Cai, R; Jiang, S; Sha, Y; Shao, Z. Highly flexible self-standing film electrode composed of mesoporous rutile TiO_2/C nanofibers for lithium-ion batteries, *Electrochimica Acta*, 85, 636– 643, 2012.

[35] Duan, Y; Fu, N; Fang, Y; Li, X; Liu, Q; Zhou, X; Lin, Y. Synthesis and formation mechanism of mesoporous TiO_2 microspheres for scattering layer in dye-sensitized solar cells, *Electrochimica Acta*, 113, 109– 116, 2013.

[36] Gao, L; Li, X; Hu, H; Li, G; Liu, H; Yu, Y. TiO_2 mesoporous microspheres with nanorod structure: facile synthesis and superior electrochemical performance, *Electrochimica Acta* 120 231–239, 2014.

[37] Zhou, R; Zhang, Q; Uchaker, E; Yang, L; Yin, N; Chen, Y; Yin, M; Cao, G. Photoanodes with mesoporous TiO_2 beads and nanoparticles for enhanced performance of CdS/CdSe quantum dot co sensitized solar cells, accepted manuscript, *Electrochimica Acta*, xx, xx-xx, 2014.

[38] Li, C; Luo, Y; Guo, X; Li, D; Mi, J; Sq, L; Hald, P; Meng, Q; Iversen, B. Mesoporous TiO_2 aggregate photoanode with high specific surface area and strong light scattering for dye-sensitized solar cells, *Journal of Solid State Chemistry*, 196, 504–510, 2012.

[39] Park, S-B; Chung, IJ; Woo, JW; Kim, TH; Li, Z; Jin, M; Lee, DJ; Kim, JM. Improvement of dye-sensitized solar cell performance through infiltration of TiO_2 nanoparticles between mesoporous TiO_2 particles, http://dx.doi.org/10.1016/j.materresbull.2014.04.062, *Materials Research Bulletin*, xx xxx–xxx, 2014

[40] Kim, H-J; Jeon, J-D; Kwak, S-Y. Highly dispersed mesoporous TiO_2 spheres via acid treatment and its application for dye-sensitized solar cells, *Powder Technology*, 243, 130–138, 2013.

[41] Xiong, Z; Ma, J; Ng, WJ; Waite, TD; Zhao, XS. Silver-modified mesoporous TiO_2 photocatalyst for water Purification, *water research*, 45, 2095 -2103, 2011.

[42] Di, K; Zhu, Y; Yang, X; Li, C. Electrorheological behavior of urea-doped mesoporous TiO_2 suspensions, Colloids and Surfaces A: Physicochem. *Eng. Aspects*, 280, 71–75, 2006.

[43] Faisal, M; Ismail, AA; Harraz, FA; Bouzid, H; Al-Sayari, SA; Al-Hajry, A. Mesoporous TiO_2 based optical sensor for highly sensitive and selective detection and preconcentration of Bi(III) ions, *Chemical Engineering Journal*, 243, 509–516, 2014.

[44] Liu, F; Liu, C-L; Hu, B; Kong, W-P; Qi, C-Z. High-temperature hydrothermal synthesis of crystalline mesoporous TiO2 with superior photo catalytic activities, *Applied Surface Science*, 258, 7448– 7454, 2012.

[45] Wang, W; Lu, C; Ni, Y; Peng, F; Xu, Z. Enhanced performance of {0 0 1}facets dominated mesoporous TiO2photocatalyst composed of high-reactive nanocrystals and mesoporous spheres, *Applied Surface Science*, 265, 438– 442, 2013.

[46] Qiana, S; Caiyunb, J; Yupinga, W; Weibena, Y; Chuna, Y. Preparation and characterization of PVA–I complex doped, mesoporous TiO_2 by hydrothermal method, *Applied Surface Science*, 273, 769– 775, 2013.

[47] Wang, R; Cai, X; Shen, F. TiO_2hollow microspheres with mesoporous surface: Superior adsorption performance for dye removal, *Applied Surface Science*, 305, 352–358, 2014.

[48] Chen, C-Y; Cheng, M-C; Chen, A-H. Photocatalytic decolorization of Remazol Black 5 and Remazol Brilliant Orange 3R by mesoporous TiO_2, *Journal of Environmental Management*, 102,125-133, 2012

[49] Asuha, S; Zhou, XG; Zhao, S. Adsorption of methyl orange and Cr(VI) on mesoporous TiO_2 prepared by hydrothermal method, *Journal of Hazardous Materials*, 181, 204–210, 2010

[50] Onsuratoom, S; Chavadej, S; Sreethawong, T. Hydrogen production from water splitting under UV light irradiation over Ag-loaded mesoporous-assembled TiO_2-ZrO_2 mixed oxide nanocrystal photocatalysts, *international journal of hydrogen energy*, 36, 5246-5261, 2011

[51] Jantawasu, P; Sreethawong, T; Chavadej, S. Photocatalytic activity of nanocrystalline mesoporous-assembled TiO2 photocatalyst for degradation of Methyl Orange monoazo dye in aqueous wastewater, *Chem. Eng. J.*, 15, 223–233. 2009.

[52] Sreethawong, T; Suzuki, Y; Yoshikawa, S. Synthesis, characterization, and photocatalytic activity for hydrogen evolution of nanocrystalline mesoporous titania prepared by surfactant-assisted templating sol–gel process, *J.Solid State Chem.* 178, 329–338, 2005

[53] Sakulkhaemaruethai, S; Sreethawong, T. Synthesis of mesoporous-assembled TiO_2 nanocrystals by a modified urea-aided sol-gel process and their outstanding photocatalytic H_2 production activity, *international journal of hydrogen energy*, 36, 6553 -6559, 2011.

[54] Zhu, L; LIV, K; Li, H; Sun, Y; Qiu, M. Solvothermal synthesis of mesoporous TiO_2 microspheres and their excellent photocatalytic performance under simulated sunlight irradiation, *Solid state Silences*, 20, 8-14, 2013.

[55] Lakshmi, BB; Patrissi, CJ; Martin, CR. “Sol-gel template synthesis of semiconductor oxide micro- and nanostructures, *Chemistry of Materials*, 11, 2544–2550, 1997.

[56] Zhang, J; Li, M; Feng, Z; Chen, J; Li, CUV. Ramman Spectroscopic Study on TiO2. 1. Phase Transformation at the Surface and in the Bulk., *Journal of Physical Chemistry B*, 110, 927–935, 2006.

[57] Cao, T; Li, Y; Wang, C; Shao, C; Liu, Y. One Step Nonaqeous Synthesis of Pure Phase TiO2 Nanocrystals from $TiCl_4$ in Butanol and Their Photocatalytic Properties, *Journal of Nanomaterials*. 2011, Article ID 267415, 6, 2011.

[58] Nikkanen, JP; Kanerva, T; M¨antyl¨a, T. The Effect of Acidity in Low-Temperature Synthesis of Titanium Dioxide, *Journal of Crystal Growth*, 304, 179–183, 2007.

[59] Fujishima, A; Rao, TN; Tryk, DA. Intrermolecular and Supramolecular Electron Transfer Processes of Fullerene-Prophyrin/phthalocyanine Systems, *Journal of Photochemistry and Photobiology C*, 1, 1–21, 2000.

[60] Chong, MN; Jin, B; Chow, CWK; Saint, C. Recent Developments in Photocatalytic Water Treatment Technology: A review, *Water Research*, 44, 2997–3027, 2010.

[61] Yu, J; Wang, G; Cheng, B; Zhou, M. Effect of hydrothermal Temperature and Time on Photocatalytic Activity and Microstructures of Bimodal Mesoporous TiO_2 Powders, *Applied Catalysis B*, 69, 171–180, 2007.

[62] Li, Y; Li, X; Li, J; Yin, J. Photocatalytic Degradation of Methyl Orange by TiO_2-Coated Activated Carbon and Kinetic Study, *Water Research*, 40, 1119–1126, 2006.

[63] IUPAC reporting physisorption data for gas/solid system, *Pure and Applied Chemistry*, 57 1723, 1985.

[64] Phonthammachat, N; Chairassameewong, T; Gualri, E; Jamieson, AM; Wongkasemjit, S. Structural and Rheological aspect of Mesoporous Nanocrystalline TiO2 Synthesized via Sol-gel Process, *Micropor. Mesopor. Mater.* 66, 261,2003.

[65] Khalil, KMS; Baird, T; Zaki, MI; El-Samahy, AA; Awad, AM. Synthesis and Characterization of Catalytic Titania via Hydrolysis of Titanium (IV) Isopropoxide, Colloids and Surfaces A : *Physicochemical and Engineering Aspects*, 132, 31-44, 1998.

[66] Balachandran, U; Eror, NG. "Raman spectra of titanium dioxide", *J. Solid State Chem.*, 42, 276– 282, 1982.

[67] Guidotti, M; Ravasio, N; Psaro, R; Ferraris, G; Moretti, GJ. *Catal.*, 214, 242. 2003.

[68] Zhao, XP; Yin, JB. Preparation and Electrorheological Characteristic of Rare-Earth Doped TiO_2 Suspention, *Chem. Mater.*, 14, 2258-2263, 2002.

[69] Yu, JG; Xiujian, Z; Qingnan, Z. Effect of Surface Structure on Photocatalytic Activity of TiO_2 Thin Films Prepared by Sol-gel Method, *Thin Solids Films*, 379 7-14, 2000.

[70] XYLi, Quan, X; kutal, C. Synthesis and Photocatalytic Properties Quantum Confined titanium Dioxide Nanoparticle, *Scripta Mater.* 50, 499-505, 2004.

[71] Janik, MJ; Campbell, KA; Bardin, BB; Davis, RJ; Neurock, M. *Appl. Catal. A. general*, 256, 51-68, 2003.

[72] Janik, MJ; Davis, RJ; Neurock, M. Anhydrous and water-assisted proton mobility in phosphotungstic acid, *J. Am. Chem. Soc.*, 127, 5238–5245, 2005.

[73] Tanev, PT; Pinnavaia, TJ. A Neutral Templating Route to Mesoporous Molecular Sieves, *Science*, 267, 865–867, 1995.

[74] Boujday, S; W¨unsch, F; Portes, P; Bocquet, J-F; Justin, CC. Photocatalytic and electronic Properties of TiO2 Powders Elaborated by Sol-gel, *Solar Energy Materials and Solar Cells*, 83, 421–433,2004.

[75] Zhang, X; Zhang, F; Chan, KY. Synthesis of Titania-Silica Mixed Oxide Mesoporous Materials, Characterization and Photocatalytic Properties, *Applied Catalysis A*, 284, 193–198, 2005.

[76] Ding, Z; Lu, GQ; Greenfield, PF. Role of the Crystallite Phase of TiO_2 in Heterogeneous Photocatalysis for Phenol Oxidation in Water, *Journal of Physical Chemistry B*, 104, 4815–4820, 2000.

[77] Topkaya, E; Konyar, M; Yatmaz, HC; Öturk, K. Pure ZnO and composite ZnO/TiO_2 Catalyst Plates: A Comparative Study for the Degradation of Azo dye, Pesticides and

Antibiotic in Aqueous Solutions, *Journal of Colloid and interface Science*, Accepted manuscript, xx, 2014, xx.

[78] Guo, N; Ziang, Y; Lan, S; Liu, L; Ji, G; Gan, S; Feng, H; Xu, X. Uniform TiO_2-SiO_2 hollow nanospheres: Synthesis, Characterization and enhanced adsorption-photodegradation of azo dyes and phenol, *Applied Surface Science*, 305, 562-574 .

[79] Shariffuddin, JH; Jones, MK; Patterson, DA. Greener photocatalysts: Hydroxyapatite derived from waste mussel shells for the photocatalytic degradation of a model azo dye wastewater. *Chemical Engineering Research and Design*, 91, 1693-1704, 2013.

[80] Kalasin, PW; Chavadej, S; Sreethawong, T. Photocatalytic degradation of mixed azo dyes in aqueous wastewater using mesoporous-assembled TiO2 nanocrystal synthesized by a modified sol–gel process, Colloids and Surface A: *Physicochemical and Engineering Aspects*, 384, 519-528, 2011

[81] Neˇsi´C, J; Manojlovi´c, DD; Andelkovi´c, I; Dojˇcinovi´c, BP; Vuli´c, PJ; Krsti´c, J; Rogli´, GM. Preparation, characterization and photocatalytic activity of lanthanumand vanadium co-doped mesoporous TiO2 for azo-dye degradation, *Journal of Molecular Catalysis A: Chemical*, 378, 67– 75, 2013

[82] Lučic´, M; Milosavljevic´, N; Radetic´, M; Šaponjic´, Z; Radoičic´, M; Krušic´, MK. The potential application of TiO2/hydrogel nanocomposite for removal of various textile azo dyes, *Separation and Purification Technology*, 122, 206–216, 2014.

[83] Ilinoiu, EC; Pode, R; Manea, F; Colar, LA; Jakab, A; Orha, C; Ratiu, C; Lazau, C; Sfarloag, P. Photocatalytic activity of a nitrogen-doped TiO2 modified zeolite in thedegradation of Reactive Yellow 125 azo dye, *Journal of the Taiwan Institute of Chemical Engineers*, 44, 270–278, 2013.

[84] Narayanasamy, L; Murugesan, T. Degradation of Alizarin Yellow R using UV/H2O2 advanced oxidation process, *Environmental Progress & Sustainable Energy*, 33, 482–489, 2014

[85] Jiao, X; Yu, H; Kong, Q; Luo, Yi; Chen, Q; Qu, J. Theoretical mechanistic studies on the degradation of alizarin yellow R initiated by hydroxyl radical, *Journal of Physical Organic Chemistry*, 27, 519–526, 2014.

[86] Gopal, M; Moberly Chan, WJ; De Jonghe, LC. Room Temperature Synthesis of Crystalline Metal Oxides, *Journal of Materials Science*, 32, 6001–6008, 1997.

[87] Kangle, Lv; Quanjum, Xiang; Jiaguo, Yu. Effect of Calcination Temperature and Photocatalytic Activity of Anatase TiO_2 Nanosheets with Exposed {0 0 1} Facets, *Applied Catalysis B*, 104, 275–281, 2011.

[88] Han, Y; H-S Kim, Kim, H. Relationship between Synthesis Conditions and Photocatalytic Activity of Nanocrystalline TiO_2 *Journal of Nanomaterials*, 2012, Article ID 427453, 2012

[89] Tian, G; Fu, H; Jing, L; Tian, C. "Synthesis and photocatalytic activity of stable nanocrystalline TiO_2 with high crystallinity and large surface area," *Journal of Hazardous Materials*, 161, 1122–1130, 2009.

In: Comprehensive Guide for Mesoporous Materials, Volume 2 ISBN: 978-1-63463-982-8
Editor: Mahmood Aliofkhazraei

Chapter 7

ORGANIC FUNCTIONALIZATION OF ORDERED MESOPOROUS SOLIDS

Bénédicte Lebeau, Magali Bonne and Claire Marichal
Université de Haute Alsace (UHA), CNRS, Equipe Matériaux à Porosité Contrôlée (MPC), Institut de Science des Matériaux de Mulhouse (IS2M), UMR 7361, F-68093 Mulhouse, France

ABSTRACT

In the early nineties, quickly after the discovery of ordered mesoporous solids, some strategies for organic functionalization were developed. Concerning the organic functionalization of ordered mesoporous silicas, the conventional post-synthesis grafting from organosilane reagents on surface silanols and the direct synthesis via co-condensation of alkoxysilane and organosilane precursors have been widely studied, compared and reviewed. The recent societal tendencies such as evolution of researches in the field of materials towards the concept of eco-conception and biologically oriented applications have provoked the emergence of new organic functionalization strategies in terms of processes, reactants and controlled functionalization. The characterization of these organically-modified ordered mesoporous solids is of paramount importance for understanding the structure-properties relationships in order to produce tailor-made hybrid mesoporous materials. Some progresses have also been done in this area and especially with the powerful and well-adapted technique for the characterization of such solids consisting in solid state NMR spectroscopy. These evolutions in both organic functionalization strategies and functionalization characterization by NMR are highlighted in this book chapter.

INTRODUCTION

In the early nineties, two research groups from Japan and USA have reported the synthesis of ordered silica-based mesoporous solids. These materials are characterized by a regular porosity in both size and spatial arrangement in the few-nanometer scale. They differ from the synthetic route which is a hydrothermal transformation of 2D-aluminosilicate layered compounds into 3D-hexagonal aluminosilicate solids in the presence of alkyltrimethylammonium chloride surfactants for FSM-16 materials [1] with hexagonal pore structure. M41S materials reported by Mobil researchers were obtained from reaction of silica and alumina sources under hydrothermal conditions and in the presence of cetyltrimethylammonium bromide surfactants [2]. M41S silica-based solids have hexagonal (MCM-41), cubic (MCM-48) and lamellar (MCM-50) pore structure. The main advances in the synthesis of ordered mesoporous solids are summarized in Table 1. Organic or organometallic functionalization is one of the main features for ordered mesoporous silicas (OMSs). The walls are amorphous and rich in silanol groups, which offer several possibilities to anchor covalently organic or organometallic groups. This concept was validated in 1995 by Brunel et al. [3] and Maschmeyer et al. [4]. Starting from molecular silica source (usually SiO_4^- or $Si(OR)_4$) the idea to form hybrid framework by co-condensation with organosilane has quickly followed in 1996 simultaneously from the research groups of S. Mann [5] and D. Macquarrie [6]. These pioneer works were related to the pore functionalization with pendant organic groups. Few years later, in 1999, 3 research groups have, in the same time, demonstrated the possibility to synthesize hybrid ordered mesoporous siliceous solids with organic functions as main groups in the walls [7-9]. These periodic mesoporous organosilicas are prepared from bisilylated organosilane precursors in the presence of surfactant molecules as templating agents. In a general way, due to their large pore, with diameter in the molecular range, ordered mesoporous solids can be functionalized by different ways such as organic and/or organometallic groups anchoring, molecule and/or polymer adsorption [10], enzyme immobilization [11], drug or active principle encapsulation [12]. Many application fields are concerned and the number of publications or other scientific reports on this topic is considerable and still growing. In this chapter only the anchoring of organic or organometallic groups on the external particle and internal pore surface is considered. Siliceous ordered mesoporous materials are mainly concerned by this topic. Indeed the silanol groups present on the pore surface are very reactive and can easily react with organosilane molecules. Concerning non siliceous ordered mesoporous oxide solids, the nature of the frawework is not so easily functionalizable in terms of reactivity. Moreover, the synthesis of non siliceous ordered mesoporous oxide solids is more difficult compared to silica ones, the chemistry is very different and often much less flexible in terms of control of pore size and spatial arrangement and surface reactivity. Some strategies have been developed mainly based on the affinity metal/ligand [13,14].

Starting from the aim to achieve a simple surface passivation or to bring a monofunctionality to the bulk material, the anchoring of organic groups on the pore surface of ordered mesoporous siliceous solids has quickly evolved to well controlled and sophisticated surface decoration. Several strategies to achieve multifunctional, multistep and targeted functionalization have been developed.

This is a topic that has been extensively reviewed [15-22, and references therein]. The aim of this chapter is to give in a first part an overview of the classical routes to functionalize pore surface of OMS, their advantages and drawbacks, recent and/or original trends in the synthesis of hybrid OMS materials. A second part concerns the characterization of these materials; in particular the potential of solid state NMR for this purpose is emphasized. A third and last part summarize the main targeted applications for these hybrid OMS.

Table 1. Chronology of the main advances in the sysnthesis of ordered mesoporous solids

Key steps in the synthesis or ordered mesoporous solids	Year(s)	Ref.
Synthesis of ordered mesoporous solids	1990	[1]
	1992	[2]
Post-synthesis organic functionalization of ordered mesoporous silica	1995	[3,4]
Direct organic functionalization of ordered mesoporous silica	1996	[5,6]
Synthesis of periodic mesoporous silica	1999, 2000	[7-9]

I. CLASSICAL ROUTES FOR ORGANIC FUNCTIONALIZATION OF SILICA-BASED ORDERED MESOPOROUS MATERIALS

1. Main Synthesis Strategies for Organic Functionalization of OMS

1.1. Post synthesis grafting method

Pioneer work of Brunel and Marschmeyer in 1995 have demonstrated the possibility to anchor organic and organometallic groups on the pore of OMS by coupling silanols present on the surface and organosilane reagents (Figure 1) [3,4]. The organosilanes with a general formula $(Z)_{3-n}R'_n$-Si-**f** possesses alcoxy (OR, with R=CH_3, C_2H_5) or halogen groups as condensable groups (Z). Three, two or one condensable groups are present. In the two latter cases, some neutral functions are linked to the silicon center (ẽ.g. CH_3). The organic function **f** is linked to the silicon via a stable Si-C bond. Usually a spacer such as a propyl chain is present between the silicon atom and the functional groups. The OMS is first synthesized and after the extraction of the surfactant occluded within the mesoporosity, the silanol groups that are on the pore surface, are accessible. The reaction of coupling between silanols and organosilanes requires a nonpolar liquid medium, the most used solvent is toluene and heating is necessary (reflux). The absence of water is necessary to avoid the self-condensation of the organosilanes. However, halogeno-organosilanes are more reactive and often preferably used [23]. The number of condensable groups was also found determinant on the grafting efficiency. One condensable group is preferable because with 2 or 3 condensable groups the condensation is not complete and silanol groups are generated [24].

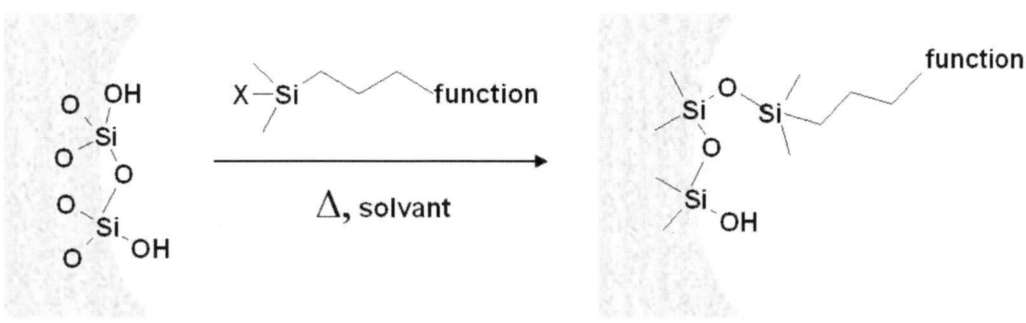

Figure 1. Post-synthesis grafting of an organic function on the pore surface of OMS.

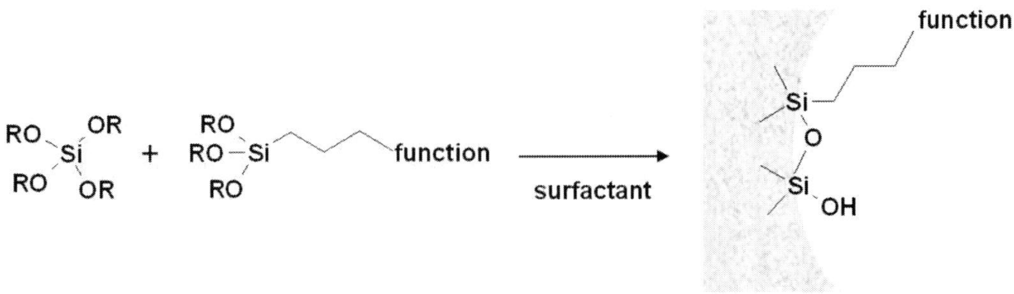

Figure 2. Direct functionalization of OMS by co-condensation of alkoxysilane and organosilane reagents.

1.2. Direct Functionalization Method

1.2.1. Functionalization with Pendant Groups

Another method for functionalizing OMS is the direct synthesis consisting in the co-condensation of the silica precursor with organosilane (Figure 2). This method has been reported for the first time by S. Mann and D. Macquarrie at the same time in 1996 [5,6]. Alkoxy-organosilanes are preferred to halogeno-organosilanes for their lower reactivity and their good compatibility with alkoxysilane, which allow promoting the hetero-condensation and thus minimizing the self-condensation. Therefore alkoxysilane are mostly used as silica precursor. In this case synthesis conditions are adapted to control the reactivity of both precursors to favor hetero-condensation.

1.2.2. Functionalization with Main Chain Groups

In 1999 the synthesis of hybrid ordered mesoporous siliceous solids incorporating the organic groups in the framework was reported by 3 research teams [7-9]. The synthesis of these materials is based on the hydrolysis - condensation of silsesquioxane precursors in the presence of surfactants as templating agents (Figure 3). The conditions are very similar to the ordered mesoporous silica solids. First Periodic Mesoporous Organosilicas (PMOs) have been synthesized with silsesquioxane or bis-silane precursor $(RO)_3Si\text{-}B\text{-}Si(OR)_3$ ($R=CH_3$, C_2H_5; B=organic bridge) with simple organic bridges such as ethane, ethylene and benzene but the use of more advanced functional bridges was rapidly explored and extended to more complexed bridges containing heteroelements, metal complexes, macrocycles,... and/or multisilylated ones. These hybrid mesoporous siliceous materials are very attractive because of the presence of a large amount of organic functional groups located on the pore surface and inside the walls which bring specific properties. Several reviews have highlighted their thermal and hydrothermal stabilities, the potential of these materials for applications such as low-k coating, catalysis, chromatography...[25-30, and references therein].

Their thermal stability depends on the nature of the organic bridge and was found quite high with benzene bridges (up to 500 - 570°C under air or N_2). Interestingly the mesostructuration of the porous network is preserved even after the thermal decomposition of the organic bridge. Pure ordered mesoporous silica can be obtained from thermal decomposition of the organic bridges of PMOs. An interesting behavior of metamorphosis was also reported for methylene-, ethane- and benzene-PMOs when thermal treatment is performed under some particular conditions. This phenomenon consists in the cleavage of the

Si-C and thus results in the transformation of the bridging organic groups into pendant ones. The consequences in surface chemistry and reactivity of PMOs are important.

Hydrophobicity is a characteristic of PMOs which improves their hydrothermal stability and makes them interesting for some applications such as catalysis. The hydrophobic character of theses solids make them stable to hydrothermal conditions and induce good mechanical properties.

Up to know the choice of organosilane and silsesquioxane is huge, most of them are commercially available but several are synthesized by organochemists. First tendencies were to functionalize OMS with simple organic groups such as alkyl, amine, thiol, cyano, halogen...Alkyl were used to control the hydrophobic character of the surface [31]. Amines are of interest for their basic properties and ability to interact with metal cations. Amino groups have also been used as coupling agent for multistep functionalization and more recently for the preparation of CO_2 sorbent materials. Thiol function is also of interest for its affinity for some metals such as Hg and Au. It can be oxidized to be transformed into sulphonate SO_3H which is interesting for acid catalysis or proton conductivity. It is also a potential coupling agent. Halogen are usually attached to the silica network to be used as a coupling agent [32].

Very quickly, more advanced functional organic or organometallic have been attached to the silica backbone of OMS: metal complexes for catalysis [33] or sensor due to their luminescent properties [34], ionic liquids for catalysis [35], macrocycles which are useful for selective metal trapping [36,37], polymers [38], dendrimers [39]....

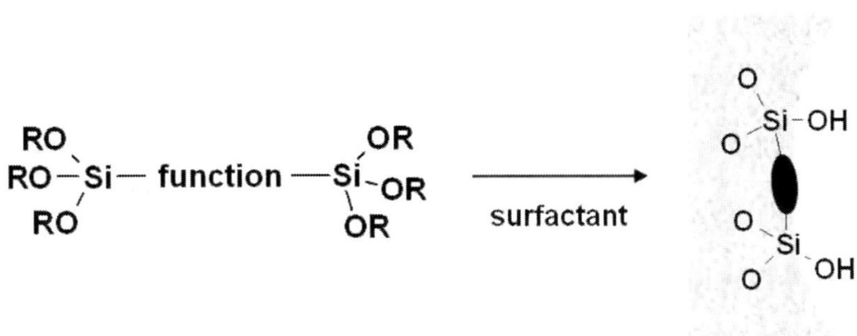

Figure 3. Synthesis of periodic mesoporous organosilica by condensation of silsesquioxane.

1.3. Anionic Surfactant Template Mesoporous Silicas

Another family of hybrid ordered mesoporous siliceous solids consists in the anionic surfactant template mesoporous silicas (AMSs), which were the subject of a recent review [40 and references therein]. The synthesis of these very interesting hybrid mesoporous solids was first reported in 2003 by Che et al. [41]. A co-structuring method using organosilane as co-structure directing agent (CSDA) was proposed. Organosilanes are chosen to promote interactions with the polar head of the surfactant in order to obtain a mesostructuration of the silica-based network at the interface of surfactant micelles. Such AMSs are synthesized in the presence of anionic surfactants which present several advantages such as low manufacturing cost and environmental friendliness. The organosilane containing amine or ammonium groups are co-condensed with tetraalcoxysilane used as silica source in the presence of anionic surfactants. Since the organosilanes play a CSDA role, their organic functions are uniformly distributed on the surface of pores with high loadings. These very promising materials have been synthesized with various structures and controlled morphology. Very interestingly, chirality has been successfully introduced in these solids by using chiral anionic surfactant molecules. The resulting chiral mesoporous solids offer several opportunities in chiral chemistry such as chiral recognition and separation, asymmetric catalysis and chiral material preparation.

Organic functionalization is not restricted to powder samples. Synthetic strategies developed for OMS processed as powder have been also applied for the preparation of mesostructured hybrid organic-inorganic thin films [42], monoliths [43], and replica of colloidal crystal [44]. These shape-modelings are mainly obtained by the EISA (Evaporation-Induced Self-Assembly) [45] or EASA (Electro-Assisted Self-Assembly) [46] processes which allow direct functionalization.

2. Advantages and Drawbacks of Functionalization Methods

Several studies based on the comparison of post-synthesis grafting and direct functionalization have been realized [47, 48]. Both organic functionalization routes have advantages and drawbacks and the choice could be made depending on the targeted application (Table 2).

2.1. Effects on Structure and Texture

Post-grafting method imposes beforehand the synthesis of the parent OMS. After post-grafting the porous structure is expected to be the same as the parent material. When post-grafting is effective, the pore surface is covered by a layer formed with silylated organic function with, in most of the cases, a spacer. As a result pore size, specific surface and pore volume decrease. For OMS with small mesopores such as MCM-41 or MCM-48, a microporous material results from the post-synthesis grafting. Pore blocking was also observed when the organic function is large in comparison to the pore size.

When co-condensation is performed the role of the organosilanes is double. They act as network modifier with their organic functional groups but also as network former with their condensable groups. A new material is synthesized. The structure will depend on the interactions of the templating molecules with the hybrid network. These interactions are often not favorable. As a consequence, the amount of organosilane is limited to obtain a regular pore structure [49]. When the interactions between surfactant and framework species are electrostatic (synthesis under basic conditions with cationic surfactants), the presence of organosilane decreases the density of charge, which can impact the micelle morphology. Hydrophobic interaction between the organic function of the organosilane and the surfactant tail could also modify the micelle morphology [50].

Some organic groups were found to direct the pore structure. The cubic *Pm3n* phase SBA-1 was obtained from co-condensation of phenyltriethoxysilane and tetraethoxysilane under acidic conditions and in the presence of cetyltrimethylammonium bromide [51]. Although hexagonal 2D phase is usually obtained with CTAB surfactant, the existence of specific interactions between the phenyl groups and the surfactant polar head group was suggested. Hybrid organic–inorganic materials exhibiting ordered mesoporous structures have been synthesized by co-condensation of tetraethoxysilane and various alkyltrimethoxysilanes with increasing length of the hydrocarbon chain (propyl, octyl, hexadecyl), in water–ethanol solution containing ammonia, in the presence of a cationic surfactant (cetyltrimethylammonium bromide) as templating agent [52]. It was shown that increasing the chain length of the organic group, from propyl to octyl and hexadecyl leads to a change of the pore structure from hexagonal *p6mm* MCM-41 type architecture to cubic *Ia3d* MCM-48 type mesostructure.

2.2. Loading and Localization of Functional Groups

For post-synthesis, the initial amount of organosilanes is usually in excess relative to the number of SiOH groups but in most cases the silylation is incomplete (residual silanol groups). The loading rate depends on the nature of organosilane reagent, the type of the condensable groups (alcoxy or halogen) and the nature and accessibility of Si-OH groups. A recent study on the post-functionalization of SBA-15 type OMS with fluoropropyl groups showed that the texture of the parent material does not really impact on the functionalization rate [24]. This latter depends more on the silanol rate, their nature and accessibility. Concerning their nature, several studies showed that isolated and germinal silanol are more favorable. When silanols are engaged into hydrogen bond with water molecules or neighbor silanols they do not react with organosilanes. The degree of organic functionalization is also dependent of the size/shape effects of the organic substituent of the silyl moiety introduced. The preconditioning conditions are of paramount importance since they can modify the nature of silanol groups [53]. The total consumption of silanol groups is never achieved. A second functionalization step is necessary for that purpose. Such post-treatment was successfully achieved with reactive silylation agent bearing small organic function such as chlorotrimethylsilane [53] or hexamethyldisilazane [54]. For post-synthesis grafting, the localization of the functional groups will depend on the diffusion of grafting reagents inside pore network, the steric effect, the affinity with already grafted groups…, which are usually not favorable for a homogeneous dispersion. It is often reported that grafting occurs preferably at the entrance of pores. However, some methods have been developed to control the distribution of the grafted functional groups within pores and on the external particle surface [55].

For direct synthesis with both pendant and main-chain groups the amount is given by the initial quantity of organosilane introduced in the reaction medium. The loading rate is therefore controlled by initial synthesis conditions. By this method a more homogeneous dispersion of functions is obtained [56]. However in the case of direct synthesis by co-condensation a good control of reactivity of precursors is required to favor hetero-condensation.

Post-synthesis grafting with alkoxy-organosilane groups is usually performed in an apolar solvent under reflux. The most used solvent is toluene, which allows a reflux at 110°C. For halogeno-organosilane a base such as pyridine is used to form a salt with chlorine ions [23]. These conditions are not eco-friendly. Few studies have been conducted to modify syntheses conditions for more environment friendly ones. The use of green template such as dendrimers was reported [57].

Direct functionalization can also be considered more ecofriendly since no organic solvent is necessary for the synthesis. Moreover it is a one-step process and thus energy saver. However the presence of organic moieties in the framework does not allow the surfactant removal by calcination and required usually a chemical extraction process. This latter is commonly realized in alcoholic medium. Depending on the nature of interactions between surfactants and organic framework, extraction is often incomplete and sometimes can provoke alcoxylation of the residual silanols [58].

Table 2. Advantages and drawbacks of functionalization methods

Post-synthesis functionalization	Ref.	Direction functionalization	Ref.
Advantages			
Structure similar to the parent material	[3,4]	One step process (energy saver)	[5,6]
		Control of the amount of grafted species	[47]
		Homogeneous localization of organic functions	[56]
		Non-hazardous solvent	[5,6]
Drawbacks			
Two steps process	[3,4]	Structure depends on the nature and amount of the organosilane	[49,50,51]
Decrease of pore size, specific surface and pore volume	[47]	Chemical extraction of the template	[58]
Bad control of the amount of grafted species	[47]		
Grafting agent (organosilane) in excess	[3,4]		
Non homogeneous localization of organic functions	[56]		
Use of hazardous solvent	[3,4, 23]		

3. Evolutions in Organic Functionalization of OMS

3.1. Multistep Functionalization

First tendencies were to graft or functionalize by co-condensation the OMS with one functional group for modifying the pore surface properties or to give a specific property to the bulk material. Amine groups have been widely used to functionalize OMS because of its metal affinity, basic properties, hydrophilic character and positive charge when protonated. More recently their ability to absorb CO_2 was extensively considered, amino groups were also used as coupling agents to anchor other organic or organometallic groups. Some other groups are used as coupling agent such as thiol, cyano, halogen…The multistep functionalization is usually realized by coupling reaction with forming covalent bond [59], click chemistry [60] and ligand exchange [61].

"Click Chemistry" is a term that was introduced by K. B. Sharpless in 2001 [62] to describe reactions that are high yielding, wide in scope, create only byproducts that can be removed without chromatography, are stereospecific, simple to perform, and can be conducted in easily removable or benign solvents. This concept was developed in parallel with the interest within the pharmaceutical, materials, and other industries in capabilities for generating large libraries of compounds for screening in research. Several types of reaction have been identified that fulfill these criteria, thermodynamically-favored reactions that lead specifically to one product, such as nucleophilic ring opening reactions of epoxides and aziridines, non-aldol type carbonyl reactions, such as formation of hydrazones and heterocycles, additions to carbon-carbon multiple bonds, such as oxidative formation of epoxides and Michael additions, and cycloaddition reactions.

3.2. Multifunctionalization

It was quickly demonstrated that both functionalization methods allowed the multifonctionalization of the pore surface of OMS. First reports were based on the use of a mixture of organosilanes for post-synthesis grafting and also for direct synthesis functionalization [63]. It is noteworthy that both post-synthesis and direct-synthesis can be coupled to give multifunctional surface. The multifunctionalization can be achieved by a multistep process. In some cases functionalization involves several steps and represents a real construction [61]. After both post-synthesis grafting and direct synthesis residual silanols are still present. These latter can undergo a subsequent post-synthesis functionalization. PMOs can be synthesized by co-condensation of polysilsesquioxanes and organosilanes to give multifunctionality. In PMOs, some silanols are present, which allow a post-synthesis functionalization. Bridges can also be used for further functionalization.

3.3. Selective Functionalization

The selective functionalization is of great interest since it is possible to decorate external particle surfaces separately and differently from internal pore surface. Functionalization of the external surface can be performed by post-synthesis grafting of the as-synthesized OMS for which porosity is occluded by surfactant molecule and thus non accessible. In 2003, Mal et al. have reported the selective functionalisation of photochromic coumarine dye on the surface of MCM-41 particles [64]. They showed that a post-synthesis grafting for a short-time procedure and on as-synthesized materials (pore filled by the surfactant) are more favorable. After surfactant removal by chemical extraction, the pores are filled with phenanthrene. The coumarin dye serves as a gate by photodimerization for the controlled release of the compound incorporated inside mesoporosity. During the last decade, the concept of outlet gate for controlled release systems has been widely developed for ordered mesoporous silica nanoparticles [65]. This approach has been widely used for drug delivery systems. Gates are grafted at the entrance of pores to control the release of drug which is inside pores. Different chemical or physical stimuli are necessary for gate opening/closing as pH, temperature, light… Several strategies have been developed to synthesize gated silica mesoporous supports for drug controlled release and signaling applications, which have been recently reviewed [65,66 and references therein]. The selective functionalization on particle surface has also been developed for other aims such as molecular recognition [67]. The grafting of 5,6-epoxyhexyltriethoxysilane onto the exterior surface of thiol-modified mesoporous silica nanoparticles-containing surfactant inside the mesopores has been realized. After surfactant removal and conversion of 5,6-epoxyhexyl into 5,6-dihydroxyhexyl groups in a one-pot method (HCl/MeOH under reflux), these latter reacted with L-lactide in presence tin(II)2-ethlyhexanoate as catalyst to form polylactic acid (PLA) coating. Finally the thiols groups were converted to *o*-phthalic hemithioacetal group which is sensitive to amine. The PLA coating serves as gatekeeper by regulating the penetration in and out of amino-containing molecules to be detected. This study also illustrates the multistep functionalization of OMS.

4. Non-Organosilane Routes for Organic Functionalization of OMS

Disilazane reagents have been suggested by Anwander et al. to functionalize the pore surface of OMS [68,69]. Used for post-synthesis functionalization, disilazane reagents can react under less drastic conditions than those of chlorosilanes and alkoxysilanes. Indeed, silylation reactions with disilazane reagents are realized under mild conditions (room-temperature) and involve easy removal of byproduct ammonia by thermal desorption. Silazanes are potentially interesting to react with silanols in vapor phase and thus leading to a total passivation of silanol groups. This strategy is applied in the following step of post-synthesis functionalization [54].

Silylamide reagents have been also reported by Anwander et al. as functionalizing agent containing metals [70]. The silylamide route allows grafting metal alkoxides on the surface of OMS. Moreover the ligand can be exchanged by post-treatment. Multilayer can be deposited and lead in some case to the formation of metal organic framework [71].

The pore surface functionalization of OMS is not limited to coupling reaction between surface silanols and silane or silazane reagents. The use of Grignard reagents has been reported to be an efficient method for organic functionalization of OMS, resulting in the formation of stable Si-C bonds. Alcoxy groups produced from incomplete hydrolysis or from alkoxylation may react with Grignard reagents and be converted to Si-R [72].

5. Processes

In the aim to develop functional materials for industrial development, eco-conception… the processes are paramount of importance. Few studies have been realized in this focus. Micro-wave heating has been used for the direct synthesis of hybrid OMS and was found to be very convenient in terms of short synthesis duration, low temperature reaction and eco-friendly solvent. [73]. Direct synthesis of hybrid OMS can be performed by using the aerosol or spray drying technique based on the EISA method [74]. This process has the advantage to integrate solid formation and drying stage in one step. There is no by-product and the production of large amounts with short-time syntheses is possible. From an industrial point of view this process is interesting but is limited to the production of spherical particles.

II. Tools for Characterization

Characterization of the porous network structure of functionalized mesoporous materials is achieved by a combination of several tools that will be briefly reviewed in the first part of this section. The second part focuses more specifically on the information provided by solid state NMR.

1. General

The pore organization of hybrid materials is studied by X-Ray Diffraction (XRD) mainly at small angle leading to the symmetry of the porous structure (hexagonal, cubic…) and the corresponding unit cell parameter [75]. The effect of post-synthesis grafting on the parent mesoporous structure can be evaluated by XRD: a decrease of the unit cell parameter is sometimes observed upon grafting [24]. In addition, Transmission Electron Microscopy also informs about the organization of the structure whereas Scanning Electron Microscopy shows the morphology of the particles.

Applications of hybrid materials are often closely related to their textural properties that are probed by nitrogen adsorption/desorption manometry. In the case of post-synthesis grafting, a decrease in BET surface area, pore volume and pore diameter with a concomitant increase of the pore thickness confirms that grafting occurs.

Fourier Transformed Infra-Red spectroscopy is also useful to evidence the vibrations characteristic of the presence of silanol groups and/or organic moieties [75].

Thermal analysis gives the amount of physisorbed water and an order of magnitude of the amount of organic matter in the sample. Indeed, during heating of the sample, the departure of organic matter occurs concomitantly with the dehydroxylation of silica distorting the results. Thus the grafting yield might be overestimated. Chemical analysis can also give an idea of the grafting yield. Nevertheless, an overestimation of the grafting yield is again possible due to the presence of solvent or surfactant in the case of incomplete extraction.

2. Solid State Nuclear Magnetic Resonance (NMR)

Whatever the functionalization strategy employed solid state NMR provides useful information on both the inorganic and the organic part of the material. Indeed ^{29}Si Magic Angle Spinning (MAS) NMR allows to quantify the amount of grafted groups whereas the influence of grafting on the inorganic framework is studied by ^{29}Si MAS and/or CPMAS (Cross Polarization Magic Angle Spinning) NMR. Indeed the local environment of framework species is reflected by the position of the ^{29}Si resonances: $Si(OSi)_n(OH)_{4-n}$ with n=0-4, namely Q_n species, are detected between -80 and -120 ppm [76 and references therein]. The presence of organosilane is confirmed by the observation of additional resonance whose position depends on the nature of the groups. Mono (M_n with n=0-1), di (D_n with n=0-2) or tri (T_n with n=0-3) functional species are detected between 15 and -80 ppm [76 and references therein]. The presence of such resonance indicates a covalent bond between the organic group and the silica framework. Decomposition of ^{29}Si MAS NMR spectrum that provides quantitative information (unlike CPMAS) will lead to the determination of the proportion of each Si species. The condensation degree of the silica framework and the grafting yield can thus be deduced from the ^{29}Si MAS NMR spectrum. Because the grafting efficiency and the condensation degree of the silica belonging to the framework are strongly dependent on the synthesis conditions, ^{29}Si NMR allows the optimization of the synthesis parameters. Such experiment also provides useful information on the nature, the number and the accessibility of the silanol before and after grafting [24].

Thanks to the possibility of studying different nuclei, NMR is a powerful tool giving different "point of view" on the material. For instance ^{1}H MAS NMR will inform on the nature and the number of silanol groups that can be correlated to data obtained from ^{29}Si MAS NMR. ^{1}H MAS NMR also evidences the presence of hydrogen bonded silanol and /or water that may influence the grafting efficiency [24]. Finally, because ^{1}H MAS NMR is very sensitive, this experiment is also useful to check the efficiency of the extraction of the surfactant [77].

In most of the cases the grafted moieties are purely organic groups that can be studied by ^{13}C NMR but sometimes they possess a specific function like phosphonic acid groups [78,79] or fluorinated groups [24]... Consequently, depending on the presence of an NMR active nucleus on the grafted groups, solid state NMR of nuclei such as ^{31}P, ^{19}F... can provide useful additional information on the sample. Concerning organic moieties, in most of the cases ^{1}H-^{13}C CPMAS NMR is performed because of the low natural abundance (~1%) of the ^{13}C nucleus. This experiment informs about the local structure or the integrity of the organic groups but can also evidence the presence of ethoxy/ methoxy groups created during the template decomposition. ^{1}H-^{13}C CPMAS NMR also checks the efficiency of the surfactant removal even if it is much less sensitive than ^{1}H MAS NMR. Information about the dynamic of grafted moieties is also accessible by solid state NMR. For example Benamor et al. [24] performed ^{19}F MAS NMR on SBA-15 type ordered mesoporous silica post-synthesis grafted with fluoropropyl groups. Differences in mobility of the fluoropropyl groups were highlighted and related to their localization in micropores (confined species) and in mesopores (more mobile).

The efficiency of further chemical reaction on the grafted groups can also be followed by solid state NMR. For example, Pan et al. [78] used ^{31}P MAS NMR to study the conversion of phosphonate groups into phosphonic acids groups upon acidification of functionalized SBA-15 type mesoporous silica.

Another interesting contribution of NMR concerns the possibility to study the conformation of the grafted species and also their interactions with the surface. Indeed a detailed description of the surface contributes to the comprehension of the material's properties. An illustration of this point is given by Bibent et al. [79] in their study of propylphosphonic acid functionalized SBA-15 mesoporous silica. Indeed, ^{31}P MAS NMR spectrum of this sample exhibits two resonances corresponding to two different environments of phosphonic acid groups suggesting the presence of hydrogen bonds with either surface silanol and/or adjacent phosphonic acid group. 2D ^{1}H-^{1}H homonuclear spin diffusion experiments have shown the absence of hydrogen bond between surface silanol and $PO(OH)_2$ groups whereas 2D ^{31}P-^{1}H-^{1}H-^{31}P correlation experiments proves the spatial proximity between phosphonic acid groups suggesting that those adjacent groups are engaged in hydrogen bonding.

More generally, the combination of 2D multinuclear NMR experiments, sometimes with theoretical calculations, leads to a detailed description of the material's surface. Nevertheless, despite the obvious interest of NMR for studying functionalized surfaces, this spectroscopy suffers from low sensitivity. Recently, Dynamic Nuclear Polarization (DNP) has been introduced to characterize functionalized mesoporous materials and looks promising for studying low concentration surface species [80]. This experiment, relying on the transfer of polarization from unpaired electron spin to the X (X= ^{29}Si or ^{13}C) nuclei, increases significantly the sensitivity in comparison with standard CPMAS NMR experiments. Lelli et al. [80] illustrated this point by studying SBA-15 type mesoporous silica prepared by direct synthesis or by post-synthesis grafting of phenol groups by ^{29}Si CPMAS NMR with and without DNP. In the second case the sensitivity is improved by a factor 21 revealing interesting structural details.

In conclusion, multinuclear 1D or 2D NMR experiments provide unique information on several aspects of functionalized mesoporous silica namely their local structure, the dynamic and the potential interaction of the grafted moieties with their environment.

III. Potential Applications

Organically modified ordered siliceous solids have been designed at the beginning for applications such as mainly catalysis, sensor and sorbents. In the domain of catalysis OMS have been identified as efficient support for catalysts with different degree of complexity; those with simple acid or basic functions, those tend to mimic enzymatic catalysts [81]. The functionalization allows covalently anchoring the catalytic functions on the pore surface and increasing its stability under conditions of use. Several studies concerning the functionalization with metal complexes by using strategies presented above have been reported [33]. Organically modified OMS were also extensively considered for applications as sensors [82]. Resulting materials for sensing are usually based on an optical or electrochemical detection. Functionalization of OMS with chromophores allows sensing surrounding environment of the chromophore. For optical detection, two conditions are necessary: 1) it is necessary that the chromophore keeps an optical activity after anchoring, and 2) any change in its surrounding environment could be detected if any modification in its optical properties occurs. Due to their large specific surface area, the loading rate of organic functions in OMS is high. OMS functionalized with organic groups able to interact with metal ions, could allow adsorbing metal ions from liquid medium and thus could be used for solid-liquid extraction and/or to detect the preconcentrated species. Thiol functions are usually anchored on the OMS surface to trap Hg species. If metal ion is electroactive such as Hg^{2+}, then voltametry can be used to detect and quantify adsorbents on the OMS surface [83,84, and references therein]. Organic functionalization of OMS has been widely developed for the synthesis of adsorbents or sensors for not only metals [85] and organic pollutants [86] in liquid media but also pollutants in gas phase [87,88].

At the beginning of 2000, the potential of hybrid OMS for medical applications such as drug/gene carrier, theranostics, biomedical imagining... has been demonstrated and researches in this direction have then exploded [89-93]. More recently a particular attention has been devoted to amine functionalized OMS for CO_2 adsorption [94,95]. Because of their large pore and due to the inertia of the amorphous silica framework, these materials were also considered as support for enzyme for use in biocatalysis [96,97, and references therein]. Functionalisation of the pore surface was found to increase stability of the entrapped enzyme or a convenient way to anchor the enzyme on the pore surface. Covalent coupling was found to prevent the enzyme leaching.

Other applications are considered such as materials for chromatography [98], electrolyte solids for fuel cells [99], photocatalysis [100].

CONCLUSION

The organic functionalization of ordered mesoporous silica is an area still in emergence. Several strategies have been developed, which allow the synthesis of tailor-made hybrid periodic mesoporous siliceous solids for a targeted application. The application potential of these materials is huge, since there are several possibilities to functionalize them. The nanoparticles for biomedical applications such as drug/gene delivery, theranostics or imaging are probably the hybrid OMS that benefit the most of the skill of chemists and biologists to realize very sophisticated functionalization.

There are still few studies devoted to the eco-conception of organically modified SMO. The present green trends for SMO concern mainly their use such as catalyst for green reactions [101].

Numerous techniques are available and easy to access for the characterization of these materials. However, the solid-state NMR remains the technique of choice for its multifunctionality in terms of accessible information (integrity, quantification, mobility, nature of surrounding environment of the organic functional groups).

REFERENCES

[1] Yanagisawa, T; Smimizu, T; Kuroda, K; Kato, C. *Bull. Chem. Soc. Jpn.*, 1990, 63, 988-992.

[2] Kresge, CT; Leonowicz, ME; Roth, WJ; Vartuli, JC; Beck JS. *Nature*, 1992, 359, 710 – 712.

[3] Brunel, D; Cauvel, A; Fajula, F; Di Renzo,F. *Stud. Surf. Sci. Catal.*, 1995, 97, 173-180.

[4] Maschmeyer, T; Rey, F; Sankar, G; Thomas, JM. *Nature* 1995, 378, 159-162.

[5] Burkett, S; Sims, SD; Mann, S. *Chem. Commun.*, 1996, 1367-1368.

[6] MacQuarrie, DJ. *Chem. Commun.*, 1996, 1961-1962.

[7] Melde, BJ; Holland, BT; Blanford, CF; Stein, A. *Chem. Mater.*, 1999, 11, 3302-3308.

[8] Inagaki, S; Guan, S; Fukushima, Y; Ohsuna, T; Terasaki, O. J. Am. Chem. Soc., *1999*, 121, 9611–9614.

[9] Asefa, T; Maclachlan, MJ; Coombs, N; Ozin, GA. *Nature*, 1999, 402, 867-871.

[10] Choi, M; Kleitz, F; Liu, DN; Lee, HY; Ahn, WS; Ryoo, R. *J. Am. Chem. Soc.*, 2005, 127, 1924-1932.

[11] Diaz, JF; Balkus, KJ. Jr. *J. Mol. Catal. B: Enzym.*, 1996, 2, 115-126.

[12] Charnay, C; Begu, S; Tourne-Peteilh, C; Nicole, L; Lerner, DA; Devoiselle, JM. *Eur J Pharm Biopharm.*, 2004, 57, 533-540.

[13] Kimura, T. *J. Nanosci. Nanotechnol.*, 2013, 13, 1461-2470.

[14] Ma, TY; Yuan, ZY. *ChemSusChem*, 2011, 4, 1407-1419.

[15] Athens, GL; Shayib, RM; Chmelka, BF. *Curr. Opin. Coll. Inter. Sci.*, 2009, 14, 281-292.

[16] Hoffmann, F; Cornelius, M; Morell, J; Fröba M. *Angew. Chem., Int. Ed.*, 2006, 45, 3216-3251.

[17] Vinu, A; Hossain, KZ; Ariga, K. *J. Nanosci. Nanotechnol.*, 2005, 5, 347-371.

[18] Brunel, D; Blanc, AC; Galarneau, A; Fajula, F. *Catal. Today* 2002, 73, 139-152.

[19] Sayari, A; Hamoudi, S. *Chem. Mater.*, 2001, 13, 3151-3168.
[20] Impens, NREN; Van der Voort, P; Vansant, EF. *Microporous Mesoporous Mater.*, 1999, 28, 217-232.
[21] Stein, A; Melde, BJ; Schroden, RC. *Adv. Mater.*, 1999, 12, 1403-1419.
[22] Maschmeyer, T. *Curr. Opin. Sol. State Mater. Sci.*, 1998, 3, 71-78.
[23] Martin, T; Galarneau, A; Brunel, D; Izard, V; Hulea, V; Blanc, AC; Abramson, S; Di Renzo, F; Fajula, F. *Stud. Surf. Sci. Catal.*, 2001, 135, 4621-4629.
[24] Benamor, T; Miesch, J; Montégut, G; Nouali, H; Marichal, C; Lebeau, B. *Eur. J. Inorg. Chem.*, 2012, 5371–5379.
[25] Van der Voort, P; Esquivel, D; De Canck, E; Goethals, F; Van Driessche, I; Romero-Salgurero, FJ. *Chem. Soc. Rev.*, 2013, 42, 3913-3955.
[26] Hoffmann, F; Fröba, M. *Chem. Soc. Rev.*, 2011, 40, 608-620.
[27] Mizoshita, N; Tani, T; Inagali, S. *Chem. Soc. Rev.*, 2011, 40, 789-800.
[28] Wang, WD; Lofgreen, JE; Ozin, GA. *Small* 2010, 6, 2634-2642.
[29] Kumar, P; Guliants, VV. *Microporous Mesoporous Mater.*, 2010, 132, 1-14.
[30] Hoffmann, F; Cornelius, M; Morell, J; Fröba, M. *J. Nanosci. Nanotechnol.*, 2006, 6, 265-288.
[31] Koyano, KA; Tatsumi, T; Tanaka, Y; Nakata, S. *J. Phys. Chem. B*, 1997, 101, 9436-9440.
[32] Alauzun, J; Mehdi, A; Reyé, C; Corriu, R. *New J. Chem.*, 2007, 31, 911-915.
[33] Brunel, D; Bellocq, N; Sutra, P; Cauvel, A; Laspéras, M; Moreau, P; Di Renzo, F; Galarneau, A; Fajula, F. Coord. *Chem. Rev.*, 1998, 178-180, 1085-1108.
[34] Kong, LL; Yan, B; Li, YJ; Li, Y. *Micorporous Mesoporous Mater.*, 2010, 135, 45-50.
[35] Zhao, H; Yu, N; Ding, Y; Tan, R; Liu, C; Yin, D; Qiu, H; Yin, D. *Microporous Mesoporous Mater.*, 2010, 136, 10-17.
[36] 36. Corriu, RJP; Mehdi, A; Reyé, C; Thieuleux, C. *Chem. Commun.*, 2002, 1382-1383.
[37] Etienne, M; Goubert-Renaudin, S; Rousselin, Y; Marichal, C; Denat, F; Lebeau, B; Walcarius, A. *Langmuir*, 2009, 25, 3137-3145.
[38] Reichhardt, NV; Guillet-Nicolas, R; Thommes, M; Klosgen, B; Nylander, T; Kleitz, F; Alfredsson, V. *Phys. Chem. Chem. Phys.*, 2012, 14, 5651-5661.
[39] Gonzalez, B; Collila, M; de Laorden, CL; Vallet-Regi, M. *J. Mater. Chem.*, 2009, 19, 9012-9024.
[40] Han, L; Che, S. *Chem. Soc. Rev.*, 2013, 42, 3740-3752.
[41] Che, S; Garcia-Bennett, AE; Yokoi, T; Sakamoto, K; Kunieda, H; Terasaki, O; Tatsumi, T. *Nat. Mater.*, 2003, 2, 801-805.
[42] Nicole, L; Boissière, C; Grosso, D; Quach, A; Sanchez, C. *J. Mater. Chem.*, 2005, 15, 3598-3627.
[43] Lebeau, B; Fowler, CE; Hall, SR; Mann, S. *J. Mater. Chem.*, 1999, 9, 2279-2281.
[44] Lebeau, B; Fowler, CE; Mann, S; Farcet, C; Charleux, B; Sanchez, C. *J. Mater. Chem.*, 2000, 10, 2105-2108.
[45] Brinker, CJ; Lu, Y; Sellinger, A; Fan, H. Adv. Mater., 1999, 11, 579-585.
[46] Etienne, M; Guillemin, Y; Grosso, D; Walcarius, A. *Anal. Bioanal. Chem.*, 2013, 405, 1497-1512.
[47] Lim, MH; Stein, A. Chem. Mater., 1999, 11, 3285-3295.
[48] Illiade, P; Miletto, I; Coluccia, S; Berlier, G. *Res. Chem. Intermediates*, 2012, 38, 785-794.

[49] Wei, Q; Nie, Z; Hao, Y; Chen, Z; Zou, J; Wang, W. *Mater. Lett.*, 2005, 59, 3611-3615.

[50] Fowler, CE; Burkett, S. L; Mann, S. *Chem Commun.*, 1997, 1769-1770.

[51] Goletto, V; Dagry, V; Babonneau, F. *Mater. Res. Soc. Symp. Proc.*, 1999, 576, 229-234.

[52] Vodolazkaya, NA; Despas, C; Lebeau, B; Marichal, C; Walcarius, A. *J. Sol Gel Sci. Technol.*, 2012, 63, 587–594.

[53] Zhao, XS; Lu, GQ. *J. Phys. Chem. B*, 1998, 102, 1556-1561.

[54] Sutra, P; Fajula, F; Brunel, D; Lentz, P; Daelen, G; Nagy, JB. *Colloid Surf., A* 1999, 158, 21-27.

[55] Bruhwiler, D. *Nanoscale*, 2010, 2, 887-892.

[56] Lim, MH; Blanford, CF; Stein, A. *J. Am. Chem. Soc.*, 1997, 119, 4090-4091.

[57] Xu, Y; Xu, S; Emmler, T; Roelofs, F; Boettcher, C; Haag, R; Buntkowsky, G. *Chem. Eur. J.*, 2008, 14, 3311-3315.

[58] Lesaint, C; Lebeau, B; Marichal, C; Patarin, J. *Microporous Mesoporous Mater.*, 2005, 83, 76–84.

[59] Mondal, J; Borah, P; Modak, A; Zhao, Y; Bhaumik, A. *Org. Process. Res. Dev.* DOI:10.1021/op4000994

[60] Kumari, S; Malvi, B; Ganai, AK; Lillai, VK; Sen Gupta, S. *J. Phys Chem. C*, 2011, 115, 17774-17781.

[61] Saito, M; Watanabe, T; Kamegawa, T; Horiuchi, Y; Matsuoka, M. *Res. Chem. Interm.*, DOI:10.1007/s11164-013-1460-1

[62] Kolb, HC; Finn, MG; Sharpless, K. B. *Angew. Chem., Int. Ed.*, 2001, 40, 2004–2021.

[63] Hall, S; Fowler, CE; Lebeau, B; Mann, S. *Chem. Commun.*, 1999, 201-202.

[64] Mal, NK; Fujiwara, M; Tanaka, Y; Taguchi, T; Matsukata, M. *Chem. Mater.*, 2003, 15, 3385-3394.

[65] Coll, C; Bernados, A; Martínez-Mañez, R; Sancenón, F. *Acc. Chem. Res.*, 2013, 46, 339-349,.

[66] Hecht, M; Climent, M; Biyikal, M; Sancenon, F; Martinez-Manez, R; Rurack, K. *Coord. Chem. Rev.*, 2013, 257, 2589-2606,.

[67] Radu, DR; Lai, CY; Wiench, JW; Pruski, M; Lin, VSY. *J. Am. Chem. Soc.*, 2004, *126, 1640-1641.*

[68] Anwander, R; Nagl, I; Widenmeyer, M; Engelhardt, G; Groeger, O; Palm, C; Röser, T. *J. Phys. Chem. B*, 2000, 104, 3532-3544.

[69] Zapilko, C; Widenmeyer, M; Nagl, I; Estler, F; Anwander, R; Raudaschl-Sieber, G; Groeger, O; Engelhardt, G. *J. Am. Chem. Soc.*, 2006, 128, 16266-16276.

[70] Anwander, R; Roesky, R. *J. Chem. Soc., Dalton Trans.*, 1997, 2, 137-138.

[71] Liang, Y; Erichsen, ES; Anwander, R. *Dalton Trans.*, 2013, 42, 6922-6935.

[72] Yamamoto, K; Tatsumi, T. *Microporous Mesoporous Mater.*, 2001, 44-45, 459-464.

[73] Procopio, A; De Luca, G; Nardi, M; Oliverio, M; Paonessa, R. *Green Chem.*, 2009, 11, 770-773.

[74] Melnyk, IV; Zub, YL; Veron, E; Massiot, D; Cacciaguerra, T; Alonso, B. *J. Mater. Chem.*, 2008, 18, 1368-1382.

[75] Lebeau, B; Gaslain, F; Fernandez-Martin, C; Babonneau F. In *Ordered Porous Solids, Recent advances and prospects*, Editor Valtchev, V; Mintova, S; Tsapatsis, M. Elsevier B.V., NL, 2009, 283-308.

[76] Alonso B; Marichal C. *Chem. Soc. Rev.*, 2013, 42, 3808-3820.

[77] Bae, YK; Han OH. *Microporous Mesoporous Mater.*, 2007, 106, 304-307.

[78] Pan, YC; Gavin Tsai, HH; Jiang, JC; Kao, CC; Sung, TS; Chiu, PJ; Saikia, D; Chang, JH; Kao, HM. *J. Phys. Chem. C*, 2012, 116, 1658-1669.

[79] Bibent, N; Charpentier, T; Devautour-Vinot, S; Mehdi, A; Gaveau, P; Henn, F; Silly, G. *Eur. J. Inorg. Chem.*, 2013, 2350-2361.

[80] Lelli, M; Gajan, D; Lesage, A; Caporini, MA; Vitzthum, V; Miéville, P; Héroguel, F; Rascon, F; Roussey, A; Thieuleux, C; Boualleg, M; Veyre, L; Bodenhausen, G; Coperet, C; Emsley, L. *J. Am. Chem. Soc.*, 2011, 133, 2104-2107.

[81] Diaz, U; Brunel, D; Corma, A. *Chem. Soc. Rev.*, 2013, 42, 4083-4097.

[82] Melde, BJ; Johnson, BJ; Charles, PT. *Sensors*, 2008, 8, 5202-5228.

[83] Sierra, I; Pérez-Quintanilla, D. *Chem. Soc. Rev.*, 2013, 42, 3792-3807.

[84] Walcairus, A. *Chem. Soc Rev.*, 2013, 42, 4098-4140.

[85] Lebed, PJ; de Souza, K; Bilodeau, F; Larivière, D; Kleitz, F. *Chem. Commun.*, 2011, 47, 11525-11527.

[86] Yoshitake, H. *J. Mater. Chem.*, 2010, 20, 4537-4550.

[87] Xu, PC; Yu, HT; Li, XX. *Anal. Chem.*, 2011, 83, 3448-3454.

[88] Ewlad-Ahmed, AM; Morris, MA; Patwardhan, SV; Gibson LT. *Environ. Sci. Technol.*, 2012, 46, 13354-13360.

[89] Li, ZX; Barnes, JC; Bosoy, A; Stoddart, JF; Zink, JI. *Chem. Soc. Rev.*, 2012, 41, 2590-2605.

[90] Vivero-Escoto, JL; Huxford-Phillips, RC; Lin, WB. *Chem. Soc. Rev.*, 2012, 41, 2673-2685.

[91] Pasqua, L; Cundari, S; Ceresa, C; Cavaletti, G. *Curr. Med. Chem.*, 2009, 16, 3054-3063.

[92] Aznar, E; Martinez-Manez, R; Sancenon, F. *Exp. Opin. Drug Deliv.*, 2009, 6, 643-655.

[93] Vallet-Regi, M; Balas, F; Arcos, D. *Angew. Chem., Int. Ed.*, 2007, 46, 7548-7558.

[94] Wei, J; Zhao S. *Adv. Mater. Res.*, 2012, 538-541, 2240-2245.

[95] Langeroudi, EG; Kleitz, F; Iliuta, MC; Larachi, F. *J. Phys. Chem. C*, 2009, 113, 21866-21876.

[96] Zhou, Z; Hartmann, M. *Chem. Soc. Rev.*, 2013, 42, 3894-3912.

[97] Magner, E. *Chem. Soc. Rev.*, 2013, 42, 6213-6222.

[98] Yang, LM; Wang, YJ; Luo, GS; Dai, YY. *Particology*, 2008, 6, 143-148.

[99] Bibent, N; Mehdi, A; Silly, G; Henn, F; Devautour-Vinot, S. Eur. *J. Inorg. Chem.*, 2011, 3214-3225.

[100] Corma, A; Navarro, MT; Rey, F; Ruiz, VR; Sabater, M. *J. Chem. Phys. Chem.*, 2009, 10, 1084-1089.

[101] Macquarrie, D. *Topics Catal.*, 2009, 52, 1640-1650.

In: Comprehensive Guide for Mesoporous Materials, Volume 2 ISBN: 978-1-63463-982-8
Editor: Mahmood Aliofkhazraei

Chapter 8

THE NANOSTRUCTURED MATERIALS CONTRIBUTION IN THE ENHANCEMENT OF THE PERFORMANCE ATTRIBUTES OF THE TEXTILE MATERIALS

***Eveline Popovici*[1], *Narcisa Vrinceanu*[1,2] *Mirela Suchea*[1,3] *and Lucian-Ionel Cioca*[2]**
[1]"Al.I.Cuza" University of Iasi, Iasi, Romania
[2]"Lucian Blaga" University of Sibiu, Faculty of Engineering, Department of Industrial Machines and Equipments, Sibiu, Romania
[3]Center of Materials Technology and Laser, School of Applied Technology, Electrical Engineering Department, Technological Educational Institute of Crete, Heraklion, Greece

ABSTRACT

The application of nanoscale materials and structures is an emerging area of nanoscience and nanotechnology, with broad scientific and commercial perspectives. The high surface energy of the inorganic materials, such as metal and metal oxides, mainly determined by size, shape, composition, crystallinity and morphology has attracted lots of attention in functional nano-finishing of textiles. The purpose of this work is to highlight the added-value technologies based on nanoparticles contribution, focusing on functional textiles with higher quality attributes, such as: antimicrobial properties, water repellence, electro-magnetic protection and thermoregulatory fabrics, antistatic, anti-infrared and flame retardant properties, dyeability, colour fastness, as well as durable, or permanent press finished garments, textiles with self-cleaning properties, UV-blocking, deodorizing, thermal regulator, memorizing shape, and so on. The use of the nanoscale materials in the textile manufacturing opens many opportunities towards the production of so-called "intelligent textiles".

Keywords: nanotechnology, nanostructured materials, tunable functional textile properties

OVERVIEW

Nanoscale science and technology have emerged over the past decades as the forefront of science and technologies and continues to draw worldwide attention because it is widely perceived as offering huge potential in a wide range of end uses.

Nobel Prize winning physicist, Richard Feynman, is recognized as one who envisioned in his well-known lecture entitled "*There is Plenty of Room at the Bottom.*" (1959) the numerous opportunities associated with the manipulation and control of particles at small level.

Since then, a number of key scientists predicted the birth of what is called today Nanotechnology.

A list of remarkable contributions in NANO-evolution in materials engineering are presented in the following table.

NANO is not a single technology, but an infinite complexity of different things. It stands for an "umbrella" term covering a wide range of technologies concerned with structures and processes on the nanometer scale, and in the same time is an unavoidable part of our future.

Nano-engineering contribution	**Definition/effect**	**Functional value-added textiles**	**References**
Norio Taniguchi (Tokyo University of Science)[i]	processing of separation, consolidation and deformation of material by a single atom or a single molecule.		[1]
	controlling atoms and molecules at nanometer scale	functional materials, devices and systems	[2]
	synergy between nanotechnology and the textile industry, exploiting the intrinsic properties of nanostructured materials	UV-blocking, antimicrobial, antistatic and self-cleaning properties, water repellence, electro-magnetic protection and thermoregulatory fabrics	[3]
Nano-Tex	started to invest in the development of nanotechnologies modifying the surface of textiles by means of nanostructured materials	fabrication of nanofibers and nanoparticle composite fibers[ii]; upgrading existing functions and performance of textile materials and	[4]
real commercial potential for the textile industry	nanostructured materials have a large surface area-to-volume ratio and high surface energy	high durability for fabrics	[5]

Nano-engineering contribution	Definition/effect	Functional value-added textiles	References
using nanosized materials and growth of nanostructures during manufacturing and finishing processes		water repellence; antiodour; soiling-resistance; antistatic, anti-infrared and flame retardant properties, dyeability, color fastness, as well as durable, or permanent press finished garments; textiles with self-cleaning properties, UV-blocking, deodorizing, thermal regulator, memorizing shape	[6]
two ways in synthesis of nanostructured materials: top-down approach and bottom up approach[ii]	classification schemes for nanostructured materials[iv]	nanostructured zero-dimensional (0D)[v], one-dimensional (1D)[vi], two-dimensional (2D)[vii] and three-dimensional (3D)[viii] nanomaterials	[7-10]

[i] 1974

[ii] developing intelligent textiles with completely new characteristics and functions

[iii] Structural elements in 0D, 1D and 2D nanomaterials can be distributed in a liquid or solid macroscopic matrix or be applied on a substrate.

[iv] The most important criterion of the classification is the dimension of the array of constituents

[v] uniform particles arrays (quantum dots), heterogeneous particles arrays, core–shell quantum dots, onions, hollow spheres and nanolenses, nanocluster materials and nanodispersions in which nanoparticles are isolated from each other

[vi] ideal systems for exploring a large number of novel phenomena at the nanoscale: nanowires, nanorods, nanotubes, nanobelts, nanoribbons, and hierarchical nanomaterials

[vii] films (coatings) with nanometer thickness

[viii] powders, fibrous, multilayer and polycrystalline materials in which the 0D, 1D and 2D structural elements are in close contact with each other and form interfaces.

Nanotechnology is the creation and utilization of materials, devices, and systems through the control of matter on the nanometer-length scale - the ability to engineer matter at the atomic, molecular, and supramolecular level - and the generation of larger structures with fundamentally new molecular organizations. The manipulation of a material at the atomic and molecular scale to create new functionality and properties is a deep scientific, productive and modern concept, based on the premise that properties can dramatically change when a dimension is reduced to the nanometer range. At nanometer scale, the surface-to-volume ratios of materials become large and their electronic energy states become discrete, leading to unique electronic, optical, magnetic, and mechanical properties of the nanomaterials.

With the intensification of global competition, textile manufacturing companies are competing for a significant share of the global market by developing *new technologies* and *new products*.

All the leading textile industries are focusing on value added applications such UV-blocking, antimicrobial, antistatic and self-cleaning properties, water repellence, electro-magnetic protection and thermoregulatory fabrics, and so on. Textile fabrics are some of the best platforms for deploying nanotechnology; the synergy between nanotechnology and the textile industry judiciously exploits the intrinsic properties of nanostructured materials [3].

The first work on nanotechnology in textiles was undertaken by Nano-Tex [4] and later, more and more textile companies started to invest in the development of nanotechnologies modifying the surface of textiles by means of nanostructured materials.

The use of nanotechnology in textiles can be subdivided as follows:

1. fabrication of nanofibers and nanoparticle composite fibers,
2. surface modification of fibers and fabrics for the purpose of upgrading existing functions and performance of textile materials and
3. developing intelligent textiles with completely new characteristics and functions.

Nanotechnology has a real commercial potential for the textile industry, due to the fact that conventional methods used to induce different properties to fabrics often do not lead to permanent effects, and will lose their functions after laundering or wearing. Nanotechnology solutions can provide high durability for fabrics, because nanostructured materials have a large surface area-to-volume ratio and high surface energy, thus presenting better affinity for fabrics and leading to an increase in durability of the function. [5].

In the textile field, the research focuses on using nanosized materials and growth of nanostructures during manufacturing and finishing processes in order to impart variable functions and applications such as:

- antimicrobial properties,
- water repellence,
- antiodour,
- soiling-resistance,
- antistatic,
- anti-infrared and flame retardant properties,
- dyeability,
- color fastness, as well as durable, or permanent press finished garments,
- textiles with self-cleaning properties, UV-blocking, deodorizing, thermal regulator, memorizing shape and other properties. [6].

There are two ways in synthesis of nanostructured materials: top-down approach and bottom up approach. Generally, the classification schemes for nanostructured materials are based on different **criteria**: their chemical composition, the technique for their manufacture, or their dimensionality [7]. The most important criterion of the classification is the dimension of the array of constituents which form the nanostructured zero-dimensional (0D), one-dimensional (1D), two-dimensional (2D) and three-dimensional (3D) nanomaterials [7-10]

Zero-dimensional nanomaterials include uniform particles arrays (quantum dots), heterogeneous particles arrays, core–shell quantum dots, onions, hollow spheres and nanolenses, nanocluster materials and nanodispersions in which nanoparticles are isolated from each other. One-dimensional nanostructured materials, such as nanowires, nanorods, nanotubes, nanobelts, nanoribbons, and hierarchical nanomaterials are ideal systems for exploring a large number of novel phenomena at the nanoscale.

Two-dimensional nanomaterials are films (coatings) with nanometer thickness. Structural elements in 0D, 1D and 2D nanomaterials can be distributed in a liquid or solid macroscopic matrix or be applied on a substrate.

Three-dimensional nanomaterials include powders, fibrous, multilayer and polycrystalline materials in which the 0D, 1D and 2D structural elements are in close contact with each other and form interfaces.

Nanostructured materials used in textile industry can be categorized in the following main groups:

- metallic and non-metallic nanostructured materials: silver, copper, gallium, gold nanoparticles, carbon nanotubes;
- oxide nanostructured materials: TiO_2, ZnO, Al_2O_3, SiO_2, nano-layered clay;
- nanocomposites;
- inorganic nanostructured materials loaded organic carriers: cyclodextrin loaded with inorganic materials, nano- and micro-capsules having inorganic nano-particles, metallic dendrimer nano-composites and inorganic nano-particles loaded in liposomes.

The aim of this chapter is to summarize the fundamentals and technical approaches in processing and behavior of nanostructured materials in textile industry, to provide a comprehensive assessment of exciting opportunities for diversification in the textile industry: upgrading the existing functions and performance of textile materials, developing smart and intelligent textiles with unprecedented functions.

1. Nanostructured Materials and Functional Values Added to Textiles

Of the many applications of nanotechnology, textile industry has been currently added as one of the most benefited sector. Application of nanotechnology in textile industry has tremendously increased the durability of fabrics; increase of comfort, hygienic properties and reduced production cost. Today, the use of nanotechnology is allowing textiles to become multifunctional.

Figure 1 presents a comprehensive representation of various properties and the added-value to fabrics that may be achieved by nanomaterials use. Most often, nanomaterials are impregnated onto textile materials without significantly affecting their texture or comfort. Chemical finishes are the addition of chemicals to natural (plant fibers, such as: cotton (*Genus Gossypium*) along with jute (*Genus Corchorus*), ramie (*Boehmeria nivea*), hemp (*Cannabis sativa L.*) and flax (*Linum usitatissimum L.*), and animal fibers, such as wool and silk) and synthetic fibers (petroleum based fibers, such as polyester, polyacrylic and polyolefine, and regenerated plant cellulosNe fibers, such as rayon), while mechanical finishes employ physical means to change properties of textiles for a particular end use.

The chemical processing of textiles covers nanofinishing, nanocoating, nanocomposite coating and nanodyeing, as well as in developing nanocomposite fibers and nanofibers with new functionalities that can find application in diverse fields [11]. A comprehensive review of various properties and the added-value to various fabrics that may be achieved by the use of nanomaterials is presented in Table 1.

1.1. Metallic and Non-Metallic Nanostructured Materials and Their Nanocomposites

The application of nanoparticles to textile materials has been the object of several studies aimed at producing finished fabrics with different functional performances.

Nanoparticles can provide high durability for treated fabrics, with respect to conventional materials, because they possess large surface area and high surface energy. These two characteristics ensure better affinity for fabrics and lead to an increase in durability of the textile functions.

Various methods have been employed for nanosilver production as: chemical reduction process, photo-catalytic reduction, photo-chemical or radiation–chemical reduction, [162], reverse micelle based methods, [163] biological methods [164], sonochemical, polyol-methods [165].

Recent, bio-preparation of silver nanoparticles have been carried out using aqueous extract of lemon leaves (Citrus limon) which acts as reducing agent and encapsulating cage for the silver nanoparticles. These silver nanoparticles have been used for durable textile finish on cotton and silk fabrics. Remarkable antifungal activity has been observed in the treated fabrics [166]. A successful procedure for the *in situ* synthesis of silver nanoparticles by chemical reduction of a silver salt on cotton and wool textiles, performed in an aqueous system, has also been reported [167]. This simple method can be applied to produce textiles for biomedical applications or with conductive properties. At present, many researchers have focused on functional and multifunctional properties of silver nanoparticles [168].

Nanosilver particles are currently the most commonly used nano-engineered material due to their outstanding bactericidal and fungicidal activity [146]. In order to fabricate antibacterial cotton fabric with excellent antibacterial property and laundering durability, amino functional silver nanoparticles were prepared by one-step reaction with amino-terminated hyperbranched polymer and grafted on the oxidized cotton fabric subsequently. The average size of silver nanoparticles was 10.8nm with narrow size distribution on the cotton fabric and 99.43% and 99.45% bacterial reduction rates were obtained for *Staphylococcus aureus* and *Escherichia coli* respectively [170]. Cross linkable polysiloxan has been used for stabilizing silver nanoparticles and ZnO nanoparticles on the textile surface, which also has many unique properties such as softness, flexibility and others [25, 169].

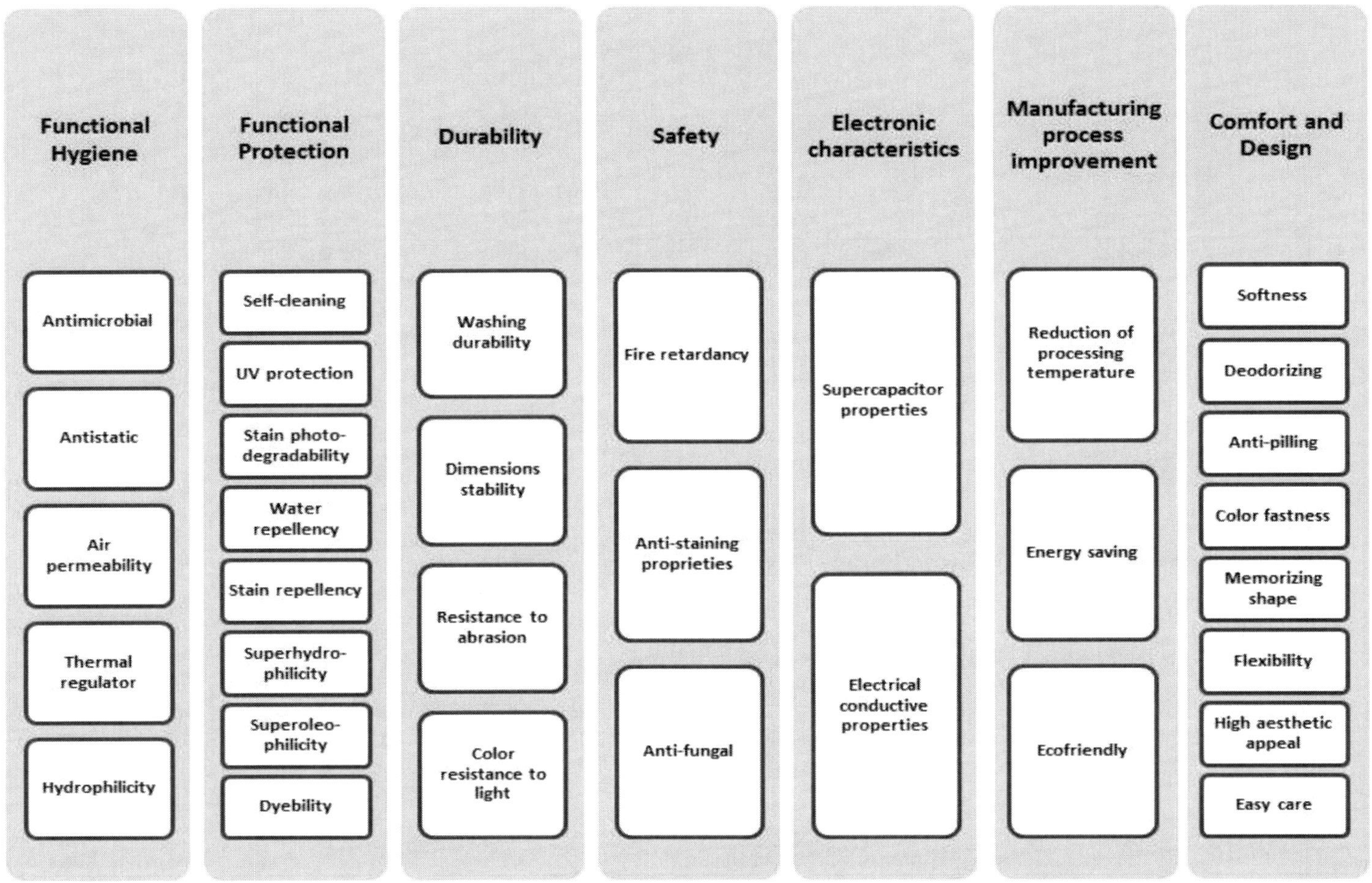

Figure 1. Nanostructured materials and their convergence with functional values –added textiles (the figure is realized by the authors; consequently, please, consider its originality).

Table 1. Nanostructured finishes and tunable functional value-added textiles

Nano-engineered finishes	Textile materials	Functional value-added textiles	References
Metallic and nonmetallic nanostructured materials			
Silver nanoparticles	Cotton fabrics Antimicrobial Superhydrophobic conductive textiles with antibacterial property Wool fabrics Grafted oxidized cotton fabrics, Antibacterial Polycaprolactam fibres Cellulose fibers Antimicrobial Jute fibers thermal stability, high surface areas: ultrafiltration, antibacterial dressing, food package, water treatment, and biomedical field PET fabrics antimicrobial membrane filter Antimicrobial; comfort performance (water repellency) Polypropylene filament yarns antibacterial	Antifungal high resistance to oxidation High thermal conductivity UV protection Antimicrobial performance, Comfort performance (water repellency)	[12-15] [16] [17] [18-21] [22] [23,24] [25-27] [28]
	Superhydrophobic conductive textiles with antibacterial property		[16]
	Wool fabrics Grafted oxidized cotton fabrics		[17]

Nano-engineered finishes	Textile materials	Functional value-added textiles	References
Metallic and nonmetallic nanostructured materials			
	Polycaprolactam fibres Cellulose fibers Antimicrobial		[18-21]
	Jute fibers thermal stability, high surface areas: ultrafiltration, antibacterial dressing, food package, water treatment, and biomedical field		[22]
	PET fabrics antimicrobial membrane filter		[23,24]
	Antimicrobial; comfort performance (water repellency)		[25-27]
	Polypropylene filament yarns antibacterial		[28]
Gold nanoparticles	Cotton fabrics - super-hydrophobicity Polycaprolactam fibres Wool fabrics	Antibacterial and antifungal properties Electrical conductive Super-hydrophobicity	[30] [33]
Copper nanoparticles	Cotton fabrics - Antibacterial Bamboo rayon Wool fabrics - antibacterial activities, textile quality and mechanical properties	Antibacterial properties UV protection, Electrical conductive	[31, 32] [33]
Gallium nanoparticles	Cotton fabrics Wool fabrics	Antibacterial	
Carbon nanotubes	Cotton fabrics - Antimicrobial, electrical conductive, fire retardant, antistatic, chemical absorber	Antimicrobial, electrical conductive, fire retardant, antistatic	[34]

Table 1. (Continued)

Nano-engineered finishes	Textile materials	Functional value-added textiles	References
Oxide nanostructured materials			
TiO_2	Cotton fabrics - Self cleaning, UV protective, antibacterial, mechanical performance (bursting, tensile and tearing strength), photicatalytic activity (self-cleaning), wash resistant, stable coating with TiO2, wrinkle recovery improvement, air permeability, whiteness value, tensile strength	Antimicrobial properties Photocatalitical properties Self -cleaning properties UV protecting Super-hydrophobicity Hydrophilicity Wicking properties	[35-73]
	Wool fabrics - Photocatalyst (self-cleaning), UV protection, surface morphology, mechanical properties, effects of silicone post-treatment on wettability, hydrophilicity, alkali solubility, antimicrobial properties, fabrics, shrinkage		[74-86]
	Wool–polyamide,		
	Polyester - Self-cleaning, crease recovery		[78, 87]
	Polyester fibers - Antibacterial, UV protective and self-cleaning properties, hydrophilicity, wicking properties, self cleaning		[88- 94]
	Polyamide fibers – self cleaning		[95-97]
	Polyester/cotton blended fabrics		

Nano-engineered finishes	Textile materials	Functional value-added textiles	References
ZnO	Cotton fabrics- Antibacterial, UV blocking, Super hydrophobic, high breathability, better strength properties, air permeability, photocatalyst self-cleaning, superhydrophobicity and ultraviolet (UV) blocking properties conductiveantistatic finishing Bamboo pulp fabric Polyester fabrics - Superhydrophilicity and superhydrophobicity	Antibacterial properties UV blocking properties Superhydrophilicity properties Superhydrophobicity properties Self-cleaning properties Conductive property Antistatic finishing properties High breathability, Better strength properties	[98] [99-101] [102-109] [110,112] [113] [114]
MnO_2 **ZnFe2O4, MnFe2O4** **Co_3O_4, Mn_3O_4** **$ZnMn_2O_4$, $CoMn_2O_4$**	Silk PET fibres – chemical recycling	UV blocking properties Increase of tensile strength	[115] [116]
SiO_2	Cotton fabric - Superhydrophobicity, resistance to ruberring Synthetic fibers - UV shields Cotton fibres – multifunctionality	Superhydrophobicity UV blocking properties Mullti-functionality	[117] [118-121] [122]
Al_2O_3	Cotton fabrics - super-hydrophobicity	Super-hydrophobicity	[126, 127]
Nano-layered clays	Cotton fabrics Polyester fabrics	Gas barrier property Increase of tensile strength Flame retardancy	

Table 1. (Continued)

Nano-engineered finishes	Textile materials	Functional value-added textiles	References
		Dimensional stability Hardness, scratch resistance and flexibility Controlled release of the incorporated species	
Nanocomposites			
TiO_2/Ag	Polypropylene fibers – Antibacterial Polyester fabrics - hydrophilicity–hydrophobicity, antibacterial, self-cleaning, stain photo-degradability, UV protection, air permeability and washing durability Cotton fabric - antimicrobial activity, self-cleaning properties and UV protection [135] Cotton/PET blend – antibacterial [136] Wool fabric – self cleaning [137] self-cleaning, stain photo-degradability, UV protection, air permeability and washing durability [138-151] Polystyrene (PS) - Packaging materials [152]	Antibacterial efficiency Antifungal effectiveness Photocatalytic activity	[130-132] [92, 133, 134]
TiO2 -apatite	Cotton	Antibacterial efficiency	

Nano-engineered finishes	Textile materials	Functional value-added textiles	References
Ag- TiO_2 **ZnO- TiO_2** **ZrO_2- TiO_2**	Linen/cotton blended fabrics - Multifunctional support Silk fibers	Multifunctional support Fastness properties to washing, rubbing and light, Antibacterial property UV-protective	[155-157]
ZnO/SiO_2	PET fibers	UV-durable super-hydrophobicity	
SiO_2/TiO_2	Cotton fabrics – self cleaning Poly-ethylene terephthalate – self-cleaning	Self cleaning	[123-125]
Ag/SiO_2	Wool fabrics - Anti-bacterial and antifungal efficiency Cotton fabrics – antibacterial	Antibacterial efficiency Antifungal effectiveness	[67, 153] [154]
TiO_2/ZrO_2	Wool fabrics - Photocatalytic activity	Photocatalytic activity	[158]
Silver–chitosan/clay	Cotton fabrics Silk fibers	Antibacterial effect	
CNT–SnO, CNT–SiO_2, CNT–ZnO, CNT–SnO_2, CNT–SiC composites	Synthetic fibres Nanofibers fibrous membranes	Electrical conductivity Fire retardancy, UV protection Increasing garment comfort High stiffness and high strength	[155]
Inorganic nanostructured materials loaded organic carriers			
Cyclodextrin loaded with inorganic and organic materials	Cotton - humidity adsorptive attribute, thermal barrier Wool	High adsorption capacity for dye removal.	[159, 161]

Table 1. (Continued)

Nano-engineered finishes	Textile materials	Functional value-added textiles	References
Microcapsules containing nanostructured inorganic and organic compounds as shell	Cellulose and polyamide fibres and fabrics Cotton, pure wool and wool blends	Controlled release of products such fragrances, essential oils, thermal regulation Antibacterial efficiency Higher resistance to washing, fresh fabrics, cosmeto-textiles,Photo chromic textile, Thermo chromic textiles	[125]
Metallic dendrimer nanocomposites	Cotton Polyamide fibres	Antibacterial efficiency Antifungal effectiveness	[155]
Inorganic nanoparticles loaded in liposomes	Cotton Silk	Reduction of processing temperature Saving energy Increasing garment comfort	[156]

Generally, gold nanoparticles were obtained via reduction of HAuCl4, stabilized with sodium (3-mercaptopropionate) [171]. Recently, gold nanostructures were synthesized by bioreduction technique of gold ions using aqueous extract of grapefruit pulp (Citrus paradisi). This eco-friendly method allows the formation in solution and support of different kinds of gold nanostructures on silk fibers, and develops the new functional fibers. [146]. Gold nanoparticles have potent antibacterial effectiveness against various gram positive, gram negative and fungi [172] and as a consequence, are known as a novel biomedical application.

The grafted fibres with Cu nanoparticles were evaluated both for the antibacterial activity against both gram positive and gram negative bacteria and the durability of their antibacterial activity after washing [173]. The product showed antibacterial activity against both types of bacteria which was found to be durable for 50 washes.

1.2. Oxide Nanostructured Materials

During the paste several decades, many efforts were focused on immobilization of various metal oxide nanoparticles onto textile substrates in order to induce multifunctional properties. Various methods and techniques were developed for this purpose. The bellow table summarizes the latest advancements in this field of research.

1.3. Nanocomposites

A wide variety of nanofillers, whiskers and nanofibers with structural modification can be used in nanocomposite coatings. In combination with modern production methods, nanocomposite coatings have made possible to provide unique solutions for all common textile coating processes in the areas of sports, medical, clothing, fashion articles, technical articles and general protective equipment.

UV illumination plays the main role in triggering the photocatalytic activities of anatase nano titanium dioxide. In order to make TiO_2 suitable for receiving solar energy efficiently in a wide range from ultraviolet to visible, many methods such as dye sensitization, metal- or nonmetal-doped TiO_2-based nanoparticles, and modification of TiO_2 by the addition of another semiconductor metal oxide- have been used [198].

It is known that the polysulfonamide fibre is a new kind of developed flame retardant material and it has excellent heat resistance and flame retardancy as well as excellent thermal stability. However, raw polysulfonamide fibre has high electrical resistance while poor ultraviolet resistance; these properties cause some difficulties in its manufacturing procedures and limit its application in the development of functional textile products.

Type of substrate	Method for applying to the textile fiber	Advantages/Drawbacks of the method	Usage/effects/results Functional properties of the textiles	References
TiO_2 nanoparticles[i]				
cotton[ii]	sol–gel method at room temperature[iii]	the occurrence of uniform continuous layer of nearly spherical TiO_2 nanoparticles[iv]/ uneven distribution of TiO_2 nanoparticles on fiber surface and poor binding efficiency[v]	UV protection[vi]	[174]
			bactericidal properties[vii] against Gram-negative bacteria *Klebsiella pneumoniae* and Gram-positive bacteria *Staphylococcus aureus,* self-cleaning attributes due to the photocatalytic activity	[42,44]
	introduction of functional groups as potential sites for binding of TiO_2 nanoparticles	enhancement the binding efficiency of TiO_2 nanoparticles to the fibers	enhanced efficiency, better stability of the textile nanocomposite, long-term durability[viii]	[37, 38], [123], [177], and [178].
	application of succinic acid, 1,2,3-propanetricarboxylic acid and 1,2,3,4-butanetetracarboxylic acid as chemical spacers		excellent self-cleaning properties	[38]
wool[ix]	coating[x]		self-cleaning, UV-protective and mechanical properties	[74]
	Introduction of carboxylic acids, including citric acid and butane tetracarboxylic acid	improves the stability of TiO_2 nanoparticles on the fabric surface		
ZnO nanoparticles				
pure cotton woven fabric	wet chemical methods (pad-dry-cure)[xi]		antibacterial activity, antistatic finishing, photoluminescence characteristics, optical properties, potential properties in antireflection coatings, UV-blocking property , photocatalytic properties	[110], [179], [180-183]

Type of substrate	Method for applying to the textile fiber	Advantages/Drawbacks of the method	Usage/effects/results Functional properties of the textiles	References
woven and knitted fabric of pure cotton and 45/55% polyester/cotton	pad-dry-cure spray gun methods[xii]	The enhancement of wash durability by manipulating the particle concentration and size	excellent antibacterial activity[xiii]	[184]
dyed polyester woven and knitted fabrics	Coating of fabric with zinc oxide nano particles dispersed in an acrylic polymer[xiv]		UV protective antibacterial activity[xv]	[183]
CuO nanoparticles				
Cotton fabrics	ultrasonic cavitation[xvi]		good antibacterial activity against *Staphylococcus aureus, Pseudomonasaeruginosa, Acinetobacter baumannii, Escherichia coli and Methicillin-Resistant Staphylococcus aureus*	[185]
Al_2O_3 or SiO_2 nanoparticles				
Cotton textiles and cotton/polyester blends			highly photocatalytically active textile materials	[185]
Cotton, silk, nylon-6, polyurethane	amino functionalized silica nanoparticles applied on epoxy functionalized cotton fabric	the nano-roughening imparted by silica nanoparticles are not durable/ strong bond between the fiber and particles.	super-hydrophobic surfaces	[186]
			wrinkle resistance	[5]
			reduce the friction and wear	[187-188]
			Flame retardant[xvii]	[189]
MgO and CaO nanoparticles[xviii]				
	introduction of unmodified clay, montmorillonite in a polymer resin based on different textiles		hardness, scratch resistance and flexibility (excellent chemical resistance and adhesion)	

(Continued)

Type of substrate	Method for applying to the textile fiber	Advantages/Drawbacks of the method	Usage/effects/results Functional properties of the textiles	References
Clay nanoparticles				
different textile materials	Coating by introduction of unmodified clay, montmorillonite in a polymer resin based	excellent chemical resistance and adhesion.	hardness, scratch resistance and flexibility	[191]
cotton or polyester	melt spinning of nanocomposites into yarns		Flame retardant	[191]
	Coating with polymer nanoclay composite[xix]		flame retardant textiles	
polyamide-fibers			biocidal agents, drugs, ethereal oils, insect repelling fragrances or other active agents and enable the controlled release of the incorporated species	[192]
	introducing of nanofillers/ nanoclays[xx]	Enhancement of dyeability through altering the fiber morphology		
Polypropylene fibers	introduction of positively charged quaternary ammonium salts in the nanoclay[xxi]	improvement the compatibility between the non-polar polypropylene and the polar layered silicate[xxii]	affinity towards acid and disperse dyes	[192-194]

Type of substrate	Method for applying to the textile fiber	Advantages/Drawbacks of the method	Usage/effects/results Functional properties of the textiles	References
Nylon fibers	melt spinning[xxiii]	the clay filled yarn dyes itself faster with disperse dyes than unfilled nylon yarn/ the dyeing rate and color value is less than that of a pure nylon fiber[xxiv]		[195]
Carbon nanomaterials (CNT-carbon nanotubes , both SWCNTs and MWCNTs single multi-walls, nano-graphite)[xxv]				
		The composite possesses the merits of light weight, little hysteresis, and low power dissipation	capacitor structure	
			improved thermal stability to oxidation (strong free-radical capture ability of nanotubes and their high thermal conductivity)	
cotton and polyester fabrics	Incorporation of functionalized multi-walled carbon nanotubes with oxygen-containing surface groups	Mimics the traditional industrial dyeing process	washing fastness, hydrophobicity and flame retardancy	[197]

[i] research on modification of TiO_2 nanoparticles which will ensure extension of the light absorption into the visible spectrum is needed [74, 176].

[ii] complete UV blocking from 332 to 280 nm has been obtained.

[iii] employing titanium tetraisopropoxide as a precursor

[iv] with an average size of about 20 nm was detected using SEM analysis [40].

[v] affect efficiency, reproducibility and durability of products [87].

[vi] preserved even after 20–55 home launderings [45]

[vii] due to the photocatalytic activity of nano-sized TiO_2

[viii] Reducing the possible environmental problems concerning the release of TiO_2 nanoparticles during exploitation and maintenance

[ix] PLUSES: excellent insulating ability, biodegradability, durability. MINUSES: low thermal and chemical resistance, poor photostability under UV light and weak resistance to microorganisms [84]

[x] The concentration of the precursor ((titanium tetraisopropoxide) strongly influenced the tendency of nanoparticles to aggregate

[xi] At low cost [169]

[xii] The average size of zinc oxide nanoparticles is 10-21 nm

[xiii] 97% against *Staphylococcus aureus* and 98% for *Klebsiella pneumonia;* the durability of antibacterial activity of nano-ZnO functionalized cotton fabric to sweat, in alkaline, acidic and inorganic salt artificial sweat medium (better salt and alkaline resistances than acid resistances)

[xiv] Zinc oxide being photoactive generates superoxide and hydroxyl radicals when irradiated with UV in the presence of oxygen and water

[xv] against two representative bacteria, *Staphytococcus aureus* and *Klebsiella pneumoniae*.

[xvi] Nanoparticles are directed by the cavitation jets (which are formed due to asymmetric bubble collapse in the presence of a solid surface) towards the surface of the textile at very high velocities and penetrate into the fibres [31].

[xvii] reduces the peak heat release of polymer matrix

[xviii] are of particular interest as they are not only stable under harsh process conditions of textile industry but also generally regarded as safe materials to human beings [190].

[xix] Nanoclay particles of montmorillonite combining with fire blocker materials (Nomex, Kevlar, glass fibre, Panox, etc)

[xx] attracts the dye molecules through Van der Val forces and possibly by hydrogen bonding.

[xxi] Resulting nanocomposites based on quaternary ammonium modified nanoclay- polypropylene fibers

[xxii] The clay reduce the polypropylene fibers crystallinity which in turn helps to increase the access of disperse dye in the fiber and the cationic dye sites from ammonium iron helps to exhaust acid dye

[xxiii] the crystallization temperature increases and the melting point of nylon decreases

[xxiv] This is because the interactions of anionic groups of nanoclay with the amino sites and dying sites of the fiber are blocked, thus clay prevents the fixation of the acid or metal complex dyes on nylon fiber

[xxv] are the most researched area for high performance nanocomposite coatings due to their unique conducting, mechanical and responsive properties, exceptional mechanical properties, very high electrical as well as thermal conductivity and electroactivity [196].

Nanocomposites	Methods (for what?)	Applications/effects/benefits to the textile	References
anatase nano titanium dioxide	dip-pad-dry-cure procedure		
TiO_2[i] and other inorganic oxides ZnO, SnO_2, SiO_2		efficient photocatalysts by irradiation with solar light	[199]
integration of silica[ii] in the TiO_2/SiO_2[iii] nanocomposites – on wool		self-cleaning function and hydrophilicity	
CNT[iv]/ polymer nanocomposites based on CNT[v]			
PSA/CNT/TiO_2		improve significantly the breaking tenacity, the electrical conductivity and photocatalitic functionality of fabrics	[201]
ZnO on cotton with nanocomposite (Poly-N-vinyl-2-pyrrolidone) AND cotton with ZnO-soluble starch nanocomposites		antibacterial activity UV-protection functions, improved reactive dyeability	[202-204]
Cotton containing ZnO (28 nm) nanoparticles[vi]and carboxymethyl chitosan (100 nm)	pad-dry-cure method	UV protection and antibacterial properties[vii] (multifunctional finishing)	
iron-nickel co-deposited nanographite		good microwave absorbent (has the potential to produce flexible radar absorbent coatings on textiles for defense applications)	[206]
Polymer/ clay nanocomposites[viii]			[207]
polylactide (PLA)-clay nanocomposites/organo-modified bentonite	direct melt compounding	Multifunctional textiles	[208]
nylon 6 nanocomposites with different nanoclay/jojoba oil contents[ix]	direct melt compounding	affects the thermal behavior and the tensile mechanical properties of polymer/clay nanocomposites, cosmetic textiles	[209]

[i] the functionality of titanium dioxide on textiles has been proved, integrating some metal oxide into the surface coating formulation has been put forth to augment the efficacy of titanium dioxide photocatalyst [200].

[ii] Increasing the concentration of silica, the TiO2/SiO2 nanocomposite showed more capability in decomposing the stains.

[iii] Increasing the concentration of silica, the TiO2/SiO2 nanocomposite showed more capability in decomposing the stains.

[iv] has excellent electrical conductivity, mechanical properties and thermal conductivity.

[v] high performance textile applications

[vi] Bionanocomposite containing ZnO (28 nm) nanoparticles

[vii] very good antibacterial properties against Gram positive and Gram negative bacteria (increased with increasing the composite concentration), and a good UV protection which increased with increasing the temperature of curing [205].

[viii] new class of composites with polymer matrix in which the dispersed phase is the silicate constituted by particles that have at least one of its dimensions in the nanometer range.

[ix] the release of the cosmetic oil is a good perspective for the application of polyamide 6 nanocomposite fibers as textiles in cosmetic field [209].

1.4. Inorganic Nanostructured Materials Loaded Loading Organic Carriers

Approaches to diversify the nanocomposite system by various inorganic or organic nanosubstances can lead to a huge number of additional functionalities which are increasingly demanded by the textile industries. Organic nanostructured materials can be prepared from a variety of materials such as proteins, polysaccharides and synthetic polymers.

In the textile area nanoparticulated chitosan and its derivates are used for the treatment of fibrous materials in classical or alternative finishing procedures [210]. Chitosan fibres are used as nonwovens for dressings and as drug carriers in the form of hollow fibres [211]. A chitosan-polypropylene nonwoven has showed antimicrobial activity against Gram-negative and Gram-positive bacteria, effective for surgical covers.

Cotton fabric treated with nanosol of carboxymethyl chitosan derivative showed good antibacterial activity against *E. coli* and *S. aureus* at 0.1 % concentration, as well as improved wrinkle recovery [212].

Chitosan based core-shell nanoparticles of poly(n-butyl acrylate) core and chitosan shell has been designed as a novel antibacterial coating for textiles [212]. Cotton fabrics treated with PBA-chitosan nanoparticles during a pad-dry-cure process demonstrate an excellent antibacterial activity against *S. aureus* with bacterial reduction over 90 %.

One of the new concepts for modification of textile substrates is based on the permanent fixation of supramolecular compounds on the surface of materials and, thus, imparts new functionality to the fabric. [213]. Szejtli published [214] a very extensive review about cyclodextrins in the textile industry. In his review he divided the application of cyclodextrins in textile sectors in the following areas: binding of cyclodextrins to fibre surfaces, cyclodextrins in textile dyeing, in textile finishing, cyclodextrins and detergents and miscellaneous applications of cyclodextrins in textile industry and textile care.

Cyclodextrin-functionalized fibrous membranes with meso- and nanostructure prepared via a sol–gel/electrospinning method [215] could provide a promising candidate for dye removal.

First, microencapsulation technology was utilized in 1980 by the US National Aeronautics and Space Administration (NASA) with the aim of managing the thermal barrier properties of garments, in particular for use in space suits. Microencapsulation processes are now used currently in textiles as a means of imparting finishes and properties on textiles that were not possible or cost-effective using other technology.

The microcapsules are functional. They have affinity and react with fibers, such as cellulose and polyamide fibers. Other type of microcapsules can even be dyed to the color of the fabric (previously or simultaneously). Microcapsules may be spherically shaped, with a continuous wall surrounding the core, while others are asymmetrically and variably shaped.

The encapsulated particles produce their required effect when their core material is released. Generally, the shells of microcapsule have been prepared from inorganic or organic nanostructured materials. The nanocapsules which are encapsulated can be used for producing special properties such as anti-microbial finishes, flame retardant finishes, fragrance finishes, phase change materials, thermoregulation, stain resistant finishes, antimicrobial fabrics, fresh fabrics, cosmeto-textiles, photochromic textile, thermochromic textiles and the very latest nano-based process of yarn formation.

Microcapsules for smart textile materials, containing an active product such as phase change materials, that are chemical bonded on the fibres, add specific functional properties to the textile materials: higher resistance to washing, controlled release of products such as fragrances, essential oils, and antibacterial functionality [216]. The major interest in microencapsulation is currently in the application of durable fragrances, antimicrobial agents, thermoregulation, phase-change materials and skin softeners. This technology in textile fibers and fabric coatings and phase-change materials capsules is now applied outdoor wear (parkas, vests, thermals, snowsuits and trousers) and in the house in blankets, duvets, mattresses and pillowcases. As well as being designed to combat cold, textiles containing phase-change materials also helps to combat overheating, so overall the effect can be described as thermoregulation. Currently, for garments and home furnishing products, microencapsulated phase change materials are incorporated into acrylic fibers. Microencapsulation of essential oil flavors has led to many novelty applications, particularly for children's garments, but it has also allowed exposure at home and in the work place to the beneficial effects of aromatherapy. The textile applications using microcapsules represents a system that is easy to apply, does not affect the existing textile properties and has a shelf-life on a garment that allows normal fabric-care processes to take place. The textile applications using microcapsules would be a system that is easy to apply, does not affect the existing textile properties and has a shelf-life on a garment that allows normal fabric-care processes to take place [217].

Dendrimers are highly hyper-branched polymers which contain internal cavities, excess branches and a number of end groups. Generally, dendrimers have various generations, sizes and molecular weights, which can be controlled during their synthesis. The dendrimers can endow a range of applications to textile materials by their unique structure.

It has been found that the dendrimers having highly branched three-dimensional architecture, could modulate the size and shape of metal nanoparticles for textile purposes. Thus, a method has been developed called "*dendrimer-assisted synthesis nanoparticles*".

It is known that the gold nanoparticles, silver nanoparticles have a tendency to aggregate through a preferred attachment of the [111] surfaces. To tune this surface interaction, polyamidoamine (PAMAM) dendrimers were used to coat the gold surface under various conditions. The adsorbed dendrimer on the nanoparticle surface strongly alters the interaction between the nanoparticles. The sizes of the synthesized Au nanoparticles decrease with the increase of the number of dendrimer generations [218, 219].

In order to fabricate antibacterial cotton fabric with excellent antibacterial property(99.43% and 99.45% bacterial reduction rates were obtained for *Staphylococcus aureus* and *Escherichia coli*, respectively) and laundering durability, amino functional silver nanoparticles of 10.8nm were prepared by one-step reaction with amino-terminated hyperbranched dendrimer and grafted on the oxidized cotton fabric subsequently [170].

It is known that cellulosic fibers like cotton is hydrophilic in nature and therefore, it can absorb sweat from the human body and can release onto the surface that makes it comfortable. But it has some inherent limitations such as it is prone to wrinkles, shrinkage, low dye uptake in absence of salt/electrolyte and microbial degradation.

Dendrimers serve the purpose of reducing some of those problems when applied on natural cellulosic fibers like cotton, and jute [220].

It has been found that the polypropylenimine dendrimer (PPI) grafted cotton fabrics shown antimicrobial activity against Staphylococcus aureus (Gram-positive bacterium), Escherichia coli (Gram-negative bacterium), Pseudomonas aeruginosa (Gram-negative bacterium) and Candida albicans (fungus). The antimicrobial activity shown by the PPI dendrimer grafted cotton fabrics is directly related to the number of free amino groups in the samples. The antimicrobial activity is attributed to the interaction between polycationic structures of PPI dendrimer and anionic cell membranes of bacteria [221].

The dendrimers have proven to be feasible for the dyeing of hydrophobic fibers such as polypropylene fibers which cannot be dyed in conventional dyeing processes. In this case hyperbranched polymer is blended with polypropylene and the blend is subsequently spun in to fiber. Addition of a hyperbranched polymer to polypropylene in spinning, increase significantly the dye uptake due to making the polar groups on the hydrophobic fibers [222].

Among all textile fibres, cotton is one of the most flammable which do not meet the flammability standards of clothing textiles. It is important to find ways to render the flame retardant textiles in an economically and ecofriendly manner. Different approaches have been proposed to solve this inconvenient, and thermal properties of cotton textiles have been improved through various additives having flame retardant functionality, such as multifunctional carboxylic acids, silica coatings, nano sized alumina, and different others nanofillers. Recently, the thermal stability and flame retardant properties of cotton fabrics modified with poly (propylene imine) dendrimer using cross linking agents have been reported [223].

Liposomes are the assembling structures formed by surface-active biological lipids that arrange themselves by exposing their polar head groups toward the water phase and hydrophobic hydrocarbon moieties that adhere together in the bilayer. They form closed vesicles with an aqueous core and an internal aqueous domain that is trapped between the lipid bilayers. The size of liposomes changes in the range of nanometers to micrometers. Liposomes can be prepared according to the textile application. This means that it can be produced as a negative or positive charge developing polyfunctional properties in preparation of textile materials Because of their amphiphilic structures; they can trap and carry both hydrophobic and hydrophilic materials. Nanoencapsulation and microencapsulation with liposomes may take place in the hydrophobic bi-layer membrane or in the aqueous core of liposomes [224].

In textiles, the major interest in liposomes is currently in the application of wool dyeing [225], durable fragrance and protective textile [226, 227], insect repellents, antimicrobial agents, and phase-change materials [227, 228]. According to the encapsulating and sustaining release properties of liposomes, they can be used in cosmetotextile, smart textile, and medical textiles that allow active agent delivery across cell membranes. Liposomes are used for dyeing with several types of dyes and fibers for the reduction of processing temperature [225] and saving energy [224]. It has been shown that liposomes can be used as an alternative-vehicle to various types of synthetic auxiliaries commonly used in dyeing. Microencapsulation with liposomes as a new efficient technology for dyeing process prevents environmental pollution. In addition, the presence of liposomes in the textile process can improve the mechanical properties of textile products, resulting in better wash fastness properties and leveling effect and handle properties [229].

2. Nanocoating the Surface of Textiles and Clothing

The conventional coating procedure has several problems like strength loss, improper adhesion, poor abrasion resistance and less durability. Researchers focus their interest to overcome these problems by minimizing coat to weight ratio using nanocoating procedure to cover with a layer with thickness on nanometer scale or to cover a surface with a nanoscale entity.

Nanocoating the surface of textiles and clothing is the new approach to the production of highly active surfaces to have UV-blocking, antimicrobial and self-cleaning properties with exceptional durability and minimum coat weight.

The most important techniques considering suitability of application on the uneven surface of a textile substrate such as self-assembly, sol–gel, plasma polymerization, layer-by-layer, and magnetron sputtered coating.

The self-assembly technique allows to control the state of molecular arrangement and/or assembly of functional materials on each of the monofilaments in nanoscale sizes and leads to the development of new functionalities as well as remarkable improvements in the existing functions of the synthetic fibers: quality, durability, feel, without losing the texture of fabric.

Low temperature sol-gel based nanocoatings have been used to deposit layers of metal oxide nanoparticles on textile fabric surfaces to impact specific functionality. Nano sol-gel method has been applied for preparing self-cleaning [230], antibacterial [231] and water repellent finishes [232] on fabrics.

Anatase titanium dioxide nanoparticles and TiO2/Ag nanocomposite were prepared by sol-gel method for coating the cotton fabric by a dip-pad-dry-cure process. It was shown that the treated fabric had a remarkable photocatalytic self-cleaning performance and it had ability to decompose a colorant and degrade red wine and coffee stains as well as proper protection against UV radiation [230].

Electrostatic interactions, hydrogen bonding, and covalent bonding are some of the associating forces between the coating and the material being coated. For a good adhesion of nanotitania particles, the cotton surface is first activated by using plasma and UV irradiation before TiO_2 colloidal solution is applied to develop sol-gel nanocoating on the fabric surface. It has been found that suitable chemical spacers like polycarboxylic acids can also impart long term stable performance of photocatalytic activity of TiO_2 on cotton [37, 38].

Also, silica nanoparticles achieved by sol-gel process has the water repellent property.

In textiles, plasma is used mainly for the following basic processes: cleaning and activation of the fabric surface respectively, removing organic contaminates from surface followed by anchoring functional groups; grafting of monomers and thin polymeric films deposition. Characteristics that can be improved through plasma processes include wettability, flame resistance, adhesive bonding, printability, electromagnetic radiation reflection, surface hardness, hydrophilic-hydrophobic tendency, dirt-repellent and antistatic properties [233].

Furthermore plasma methods including radio frequency plasma (RF-plasma) and microwave plasma (MW-plasma) have been applied as a pre-treatment to improve deposition process, and then the loading of TiO_2 on the textile surface is done by wet chemical technique that leads to form a controllable and uniform modification of the textile surfaces with increasing the polarity and hydrophilicity of the textile surface. These modified coated textiles show a remarkable photo-oxidative activity under the visible light [78].

Layer-by-layer nanocoating technique can be utilized to fabricate thin film coatings in order to create multifunctional textile surface. Cotton fibers offer unique challenges to the deposition of nanolayers because of a unique cross-section as well as chemical and physical heterogeneity of its surfaces. Thus, this procedure of nanocoating has been carried out on cotton fabric using chitosan as the cationic polyelectrolyte and poly-sodium-4-styrene sulfonate as the anionic polyelectrolyte. The process is assisted with ultrasonic treatment. The produced fabric has good antimicrobial properties and in the same time, the feel, flexibility and breathability of the fabric are not affected [234].

Antimicrobial silver nanoparticles can be immobilized on nylon and silk fibers by this method.

The multilayer Al_2O_3 nanoparticles film on the cationized cotton fabric and silver nanoparticles were deposited on the nylon and the silk fabric using layer-by-layer process. The result showed that this process causes to enable modifying multicomposite molecular assemblies with a control on the molecular structure and a high degree of control over the thickness.

Nanocoatings by physical treatments methods have attracted a great deal of attention because of the environmentally friendly technology. Metals and metal oxides can be deposited on the textile fabrics using magnetron sputter coating. For example, zinc oxide (ZnO) [235] and copper and silver nanoparticles [236] were deposited on the nonwoven fabrics by sputter coating. Magnetron sputter coating has also been used for deposing nanostructure thin film on the surface of nonwovens.

Nano size Fe_2O_3 is coated on the surface of polyamide 6 (PA 6) nanofibers and it was shown Fe_2O_3 improved the thermal stability of the composite nanofibers [237]. Also, a transparent nanofilm was deposited on the synthetic PA 6 nanofibers by tin-doped indium oxide leading to the remarkable improvement in the electrical properties of the PA 6 nanofibers.

Conclusion and Outlook

The synergy between nanotechnology and the textile industry judiciously exploits the intrinsic properties of nanostructured materials and facilitates the opportunity to overcome the limitations of applying conventional methods to impart certain properties to textile materials offering the key which will not only influence technological development in the near future, but will also have economic, social and ecological implications.

Certainly finishing of textile materials using nanostructured materials adds many valuable benefits in a view of end-use properties: functional hygiene (antimicrobial, antistatic, air permeability,thermal regulator, hydrophilicity), functional protection (self-cleaning, UV protection, stain photo-degradability, water repellency, stain repellency, superhydrophilicity, superoleophilicity, dyebility), durability, (washing durability, color fastness, dimensions stability, resistance to abrasion, color resistance to light), comfort and design (softness, deodorizing, anti-pilling, memorizing shape, flexibility, high aesthetic appeal, easy care), safety (fire retardancy, anti-staining proprieties, anti-fungal), electronic characteristics (supercapacitor properties, electrical conductive properties), manufacturing process improvement (reduction of processing temperature, energy saving, ecofriendly).

The wide range of nanostructured materials, in continuous development, will capture the attention of all the leading textile industries which are focusing on value added applications and there is no doubt that in the next few years, nanotechnology will penetrate into every area of the textile industry.

The desire for a healthier and more productive lifestyle will continue to generate a market for textiles that promotes the new added values obtained by incorporating nanostructured materials. The future holds ample opportunities for the contribution of nanostructured materials in the enhancement of the performance attributes of the textile materials.

ACKNOWLEDGMENTS

This work was partially supported by a grant of the Romanian National Authority for Scientific Research, CNCS – UEFISCDI, project number PN-II-RU-TE-2012-3-0202.

REFERENCES

[1] Taniguchi, N. (1974). On the Basic Concept of Nano-Technology. *Proceedings of the International Conference on Products Engineering Tokyo*, Part II, Japan Society of Precision Engineering.

[2] Drexler, E. K. (1986). *Engines of Creation: The Coming Era of Nanotechnology;* Anchor Books: New York, US.

[3] Samal, S. S., Jeyaraman, P. & Vishwakarma, V. (2010). *Journal of Minerals & Materials Characterization & Engineering*, *9*, 519-525.

[4] Russell, E. (2002). Nanotechnologies and the shrinking world of textiles. *Textile Horizons*, *9*, 7-9.

[5] Wong, Y. W. H., Yuen, C. W. M., Leung, M. Y. S., Ku, M. Y. S. & Lam, H. L. I. (2006). Selected applications of nanotechnology in textiles. *Autex Research Journal*, *6*, 1- 10.

[6] Jeevani, J. (2011). Nanomedicine & Nanotechnolology. http://omicsonline.org/ nanotextiles-a-broader-perspective-2157-7439.1000124.pdf.

[7] Popovici, E. & Dvininov, E. (2007). *Nanostructured materials. Present and future*; *Nanoparticles*; Demiurg Publishing House: Iasi, RO, Vol.1.

[8] Cao, G. (2008). *Nanostructures & Nanomaterials: Synthesis, Properties & Applications*, Imperial College Press.

[9] Popovici, E. & Seftel, E. (20070). *Nanostructured materials. Present and future*; *Anionic Clays*; Demiurg Publishing House; Iasi, RO, Vol. *2*.

[10] Popovici, E. & Ignat Rudei, M. (2009). *Nanostructured materials. Present and future. Nanotubes*; Demiurg Publishing House, Iasi, RO, Vol. *4*.

[11] Joshi, M. & Bhattacharyya, A. (2011). Nanotechnology—a new route to high-performance functional textiles. *Textile Progress,* 43*,* 155–233.

[12] Sataev, M. S., Koshkarbaeva, S. T., Tleuova, A. B., Aidarova, S. B., Prokopovich, P. & Perni, S. (2013). Novel process for coating textile materials withsilver to prepare antimicrobial fabrics. *Colloids and Surfaces A: Physicochemical and Engineering Aspects*, *442*, 146-151.

[13] El-Rafie, H. M., El-Rafie, M. H. & Zahran, M. K. (2013). Green synthesis of silver anoparticles using extracted from marine macro algae. *Carbohydrate Polymers*, *96*, 403–410.

[14] Parikh, D. V., Fink, T., Rajasekharan, K., Sachinvala, N. D., Sawhney, A. P. S., Calamari, T. A. & Parikh, A. D. (2005). Antimicrobial silver/sodium carboxymethyl cotton dressings for burn wounds. *Textile Research Journal*, *75*, 134–138.

[15] Edwards, J. V. & Vigo, T. (2001). *Bioactive Fibers & Polymers*, American Chemical Society: Washington DC, US.

[16] Chao-Hua, X., Jia, C., Wei, Y., Shun-Tian, J., Jian-Zhong, M. (2012). Superhydrophobic conductive textiles with antibacterial property by coating fibers with silver nanoparticles, *Applied Surface Science*, *258*, 2468– 2472.

[17] Desuo, Z., Ling, C., Chuanfeng, Z., Yuyue, C., Hong, L. (2013). Antibacterial cotton fabric grafted with silver nanoparticles and its excellent laundering durability. *Carbohydrate Polymers*, *92*, 2088–2094.

[18] Teixeira, E.D. M., Bondancia, T. J., Teodoro, K. B. R., Corrêa, A. C., Marconcini, J. M. & Mattoso, L. H. C. (2011). Sugarcane bagasse whiskers: Extraction and characterizations. *Industrial Crops and Products*, *33*, 63–66.

[19] Breitwieser, D., Mirhosseini, M., Moghaddam, Spirk, S, Baghbanzadeh, M., Pivec, T., Fasl, H., Ribitsch, V. & Kappe, C. O. (2013). In situ preparation of silver nanocomposites on cellulosic fibers Microwave vs. conventional heating, *Carbohydrate Polymers 94*, 677– 686.

[20] Malegowd, G. R., Jayaramudu, T., Varaprasad, K., Sadiku, R., Sinha, S. & Raju, K. M. (2013). Cellulose–polymer–Ag nanocomposite fibers for antibacterial fabrics/skin scaffolds. *Carbohydrate Polymers*, *93*, 553– 560.

[21] Emam, H. E., Manian, A. P., Siroká, B., Duelli, H., Pipal, A., Bechtold, T. & Redl, B. (2013). Treatments to impart antimicrobial activity to clothing and household cellulosic-textiles. Why "Nano"-silver? *Journal of Cleaner Production*, *39*, 17-23.

[22] Cao, X., Ding, B., Yu, J., Al-Deyab, S. S., Cao, X., Ding, B., Yu, J. & Al-Deyab, S. S. (2012). In situ growth of silver nanoparticles on TEMPO-oxidized jute fibers by microwave heating. Cellulose nanowhiskers extracted from TEMPO-oxidized jute fibers. *Carbohydrate Polymers*, *90*, 1075–1080.

[23] Dang Van Phu, Le Anh Quoc, Nguyen Ngoc Duy, Nguyen Quoc Hien. (2013). Study of incorporation of silver nanoparticles onto PE-g-PAAc nonwoven fabric by γ-irradiation for water treatment. *Radiation Physics and Chemistry*, *88*, 90–94.

[24] Said, H. M., Sokker, H. H. & Ali, A. E. H. (2010). Acrylation of pre-irradiated polypropylene and its application for removal of organic pollutants. *Radiation Physics and Chemistry*, *79*, 534–539.

[25] Dastjerdi, R., Montazer, M. & Shahsavan, M. (2009). A new method to stabilize nanoparticles on textile surfaces, *Colloids and Surfaces A: Physicochemical Engineering Aspects*, *345*, 202–210.

[26] Fasl, H., Stana, J., Stropnik, D., Strnad, S., Stana-Kleinschek, K. & Ribitsch, V. (2010). Improvement of the hemocompatibility of PET surfaces using different sulphated polysaccharides as coating materials. *Biomacromolecules*, *11*, 377–381.

[27] Bagherzadeh, R., Montazer, M., Latifi, M., Sheikhzadeh, M. & Sattari, M. (2007). Evaluation of comfort properties of polyester knitted spacer fabrics finished with water repellent and antimicrobial agents, *Fibers and Polymers*, *8*, 386–392.

[28] Dastjerdi, R. (2006). *Investigating the production and some properties of antimicrobial polypropylene filament yarns*, M.Sc. thesis, Amirkabir University of Technology, Textile Engineering Department.

[29] Wang, T., Hu, X. G. & Dong, S. J. (2007). A General Route to Transform Normal Hydrophilic Cloths into Superhydrophobic Surface. *Chemical Communications*, *18* 1849-1851.

[30] Wang, C. F., Wang, Y. T., Tung, P. H.,. Kuo, S. W., Lin, C. H., Sheen, Y. C. & Chang, F. C. (2006). Silver and Gold Glyconanoparticles for Colorimetric Bioassays. *Langmuir*, *22*, 82-89.

[31] Abramova, A., Gedanken, A. V., Popov, V., Ooi, E. H., Mason, T. J., Joyce, E. M., Beddow, J., Perelshtein, I. & Bayazitov, V. (2013). A sonochemical technology for coating of textiles with antibacterial nanoparticles and equipment for its implementation, *Materials Letters*, *96*, 121–124.

[32] Abramov, O. V. Gedanken, A., Koltypin, Y., Perkas, N., Perelshtein, I., Joyce, E. & Mason, T. J. (2009). Pilot scale sonoche mical coating of nanoparticles onto textiles to produce biocidal fabrics, *Surface & Coatings Technology*, *204*, 718–722

[33] Heliopoulos, N. S., Papageorgiou, S. K., Galeou, A., Favvas, E. P., Katsaros, F. K. & Stamatakis, K. (2013). Effect of copper and copper alginate treatment on wool fabric. Study of textile and antibacterial properties. *Surface and Coatings Technology*, *235*, 24-31.

[34] Liu, Y. Y., Tang, J., Wang, R. H., Lu, H. F., Liu, L., Kong, Y. Y., Qi, K. H. & Xin, J. H. (2007). Recent developments in bio-inspired special wettability. *Journal of Materials Chemistry*, *17*, 1071.

[35] Daoud, W. A. & Xin, J. H. (2004). Nucleation and growth of anatase crystallites on cotton fabrics at low temperature, *Journal of American Ceramic Society*, *87*, 953–955.

[36] Daoud, W. A. & Xin, J. H. (2005). Synthesis of single-phase anatase nanocrystallites at near room temperatures, *Chemical Communications*, *16*, 2110–2112.

[37] Bozzi, A., Yuranova, T. & Kiwi, J. (2005). Self-cleaning of cotton textiles modified with TiO_2 at low temperatures under daylight irradiation, *Journal of Photochemistry and Photobiology A*, *174*, 156–164.

[38] Meilert, K. T., Laub, D. & Kiwi, J. (2005). Photocatalytic self-cleaning of modified cotton textiles by TiO clusters attached by chemical spacers, *Journal of Molecular Catalysis A: Chemistry*, *237*, 101–103.

[39] Xin, H., Daoud, W. A. & Kong, Y. Y. (2004). A new approach to UV-blocking treatment for cotton fabrics. *Textile Research Journal*, *74*, 97-100.

[40] Daoud, W. A. & Xin, J. H. (2004). Low-temperature processed photocatalytic titania coating. *Journal of Sol-Gel Science & Technology, 29*, 25-29,

[41] Daoud, W. A. & Xin, J. H. (2004). Nucleation and growth of anatase nanocrystallites on cotton fabrics at low temperatures. *Journal of American Ceramic Society*, *87*, 953.

[42] Daoud, W. A., Xin, J. H. & Zhang, Y. H. (2005). Surface functionalization of cellulose fibers with titanium dioxide nanoparticles and their combined bactericidal activities. *Surface Science*, *599*, 69-75.

[43] Qi, K., Daoud, W. A., Xin, J. H., Mak, C. L., Tang, W. & Cheung, W. P. (2006). Self-cleaning cotton. *Journal of Materials Chemistry*, *16*, 4567-4574.

[44] Qi, K., Chen, X., Liu, Y., Xin, J. H., Mak, C. L. & Daoud, W. A. (2007). Facile preparation of anatase/SiO_2 spherical nanocomposites and theor application in self-cleaning textiles. *Journal of Materials Chemistry*, *17*, 3504-3508.

[45] Qi, K., Wang, X. & Xin, J. H. (2011). Photocatalytic self-cleaning textiles based on nanocrystalline titanium dioxide. *Textile Research Journal*, *81*, 101-110.

[46] Fei, B., Deng, Z., Xin, J. H., Zhang, Y. & Pang, G. (2006). Room temperature synthesis of rutile nanorods and their applications on cloth. *Nanotechnology*, *17*, 1927-1931.

[47] Liuxue, Z., Xiulian, W., Peng, L. & Zhixing, S. (2007). Photocatalytic activity of anatase thin films coated cotton fibers prepared via a microwave assisted liquid phase deposition process. *Surface Coatings Technology*, *201*, 7607-7614.

[48] Uddin, M. J., Cesano, F., Bonino, F., Bordiga, S., Spoto, G., Scarano, D. & Zecchina, A. (2007). Photoactive TiO2 films on cellulose fibres: synthesis and characterization. *Journal of Photochemistry Photobiology A*, *189*, 286-294.

[49] Wu, D., Long, M., Zhou, J., Cai,W., Zhu, X., Chen, C. & Wu, Y. (2009). Synthesis and characterization of self-cleaning cotton fabrics modified by TiO_2 through a facile approach. *Surface Coatings Technology*, *203*, 3728-3733.

[50] Nazari, A., Montazer, M., Rashidi, A., Yazdanshenas, M. & Moghadam, M. B. (2010). Optimization of cotton crosslinking with polycarboxylic acids and nano TiO_2 using central composite design. *Journal of Applied Polymer Science*, *117*, 2740-2748.

[51] Nazari, A., Montazer, M., Moghadam, M. B. & Anary-Abbasinejed, M. (2011). Self-cleaning Properties of Bleached and CationizedCotton using nano TiO2: A statistical approach. *Carbohydrate Polymers*, *83*, 1119-1127.

[52] Lessan, F., Montazer, M. & Moghadam,M. B. (2011). A novel durable flame-retardant cotton fabric using sodium hypophosphite, nano TiO_2 and maleic acid. *Thermochimica Acta*, *520*, 48-54.

[53] Montazer, M., Lessan, F. & Moghadam, M. B. (2012). Nano-TiO2/maleic acid/triethanol amine/sodium hypophosphite colloid on cotton to produce cross-linking and self-cleaning properties. *Journal of Textile Institute*, *103*, 795-805.

[54] Nazari, A., Montazer, M., Rashidi, A., Yazdanshenas, M. & Anary-Abbasinejad, M. (2009). Nano TiO 2 photo-catalyst and sodium hypophosphite for cross-linking cotton

with poly carboxylic acids under UV and high temperature. *Applied Catalysis A*, *371*, 10-16.

[55] Lam, J. L., Kan, C. W., Yuen, C. W. & Au, C. H. (2011). Low stress mechanical properties of plasma-treated cotton fabric subjected to titanium dioxide coating. *Textile Research Journal*, *81*, 1008-1013.

[56] Mihailovic, D., Saponjic, Z., Radoicic, M., Lazovic, S., Baily, C. J., Jovancic, P., Nedeljkovic, P. & Radetic, M. (2011). Functionalization of cotton fabrics with corona/air RF plasma and colloidal TiO_2 nanoparticles. *Cellulose*, *18*, 811-825.

[57] Ugur, S. S., Sariisik, M. & Aktas, A. H. (2011). Nano-TiO_2 based multilayer film deposition on cotton fabrics for UV-protection. *Fibers and Polymers*, *12*, 190-196.

[58] Chen, X., Liu, Y. H., Shi, X., Wang, K., Qi, X., Zhou, J. H. & Xin, M. (2011). Carboxymethyl chitosan coating to block photocatalytic activity of TiO_2 nanoparticles,*Textile Research Journal*, *80*, 2214-222,

[59] Hashimoto, K., Irie, H. & Fujishima, A. (2005). TiO_2 photocatalysis: a historical overview and future prospects, *Japanese Journal of Applied Physics*, *44*, 8269–8285.

[60] Fujishima, A. & Zhang, X. (2006). Titanium dioxide photocatalysis: present situation and future approaches, *Comptes Rendus Chimie*, *9*, 750–760.

[61] Parkin, I. P. & Palgrave, R. G. (2005). Self-cleaning coatings. *Journal of Materials Chemistry*, *15*, 1689–1695.

[62] Kikuchi, Y., Sunada, K., Iyoda, T., Hashimoto, K. & Fujishima, A. (1995). Photocatalytic bactericidal effect of TiO2 thin films: dynamic view of the active oxygen species responsible for the effect. *Journal of Photochemistry Photobiology A: Chem.*, *106*, 51–56.

[63] Sopyan, I., Watanabe, M., Murasawa, S., Hashimoto, K. & Fujishima, A. (1996). A film-type photocatalyst incorporating highly active TiO powder and fluororesin binder: photocatalytic activity and long-term stability. *Journal of Electroanalytical Chemistry*, *415*, 183–186.

[64] Dong, Y., Bai, Z., Zhang, L., Liu, R. & Zhu, T. (2006). Finishing of cotton fabrics with aqueous nano-titanium dioxide dispersion and the decomposition of gaseous ammonia by ultraviolet irradiation. *Journal of Applied Polymer Science*, *99*, 286–291.

[65] Yu, H., Lee S. C., Yu, J. & Ao, C. H. (2006). Photocatalytic activity of dispersed TiO_2 particles deposited on glass fibers. *Journal of Molecular Catalysis A*: *Chemistry*, *246*, 206–211.

[66] Jia, B. Y., Duan, L. Y., Ma, C. L. & Wang, C. M. (2007). Characterization of TiO_2 loaded on activated carbon fibers and its photocatalytic reactivity, *Chinese Journal of Chemistry*, *25*, 553–557.

[67] Wang, S., Hou, W., Wei, L., Jia, H., Liu, X. & Xu, B. (2004). Antibacterial activity of nano-SiO2 antibacterial agent grafted on wool surface. *Surface Coating Technolology*, *202*, 460–465.

[68] Peplow, M. (2004). Self-cleaning clothes, *Nature*, *429*, 620–621.

[69] Daoud, W. A., Xin, J. H., Zhang, Y. H. & Qi, K. H. (2005). Surface characterization of thin titania films prepared at low temperatures, *Journal of NonCrystalline Solids*, *351*, 1486–1490.

[70] Qiao-Zhen Yu, & Ai-Ai Shen (2008). Anti-ultraviolet Treatment for Cotton Fabrics by Dyeing and Finishing in one Bath and Two Steps. *Journal of Fiber Bioengineering and Informatics*, *1*, 65-72.

[71] Qiao-Zhen, Yu, & Ai-Ai Shen. (2001). UV-blocking effect, as well Low Temperature Sol-Gel Processed Photocatalytic Titania Coating, *Catalysis Today*, *68*, 89-97.

[72] Bach, U., Lupo, D., Comte, P., Moser, J.E., Weissortel, F., Salbeck, J., Spreitzer, H. & Gratzel, M. (1998). Solid-state dye-sensitized mesoporous TiO_2 solar cells with high photon-to-electron conversion efficiencies. *Nature*, *395*, 583-585.

[73] Haga, Y., An, H. & Yosomiya, R. (1997). Photoconductive properties of TiO2 films prepared by the sol–gel method and its application *Journal of Materials Science*, *32*, 3183-3188.

[74] Tung,W. S. & Daoud,W. A. (2008). Photocatalytic formulations for protein fibers: Experimental analysis of the effect of preparation on compatibility and photocatalytic activities, *Journal of Colloid Interface Science*, *326*, 283-288.

[75] Daoud, W. A., Leung, W S. K., Tung, S., Xin, J. H., Cheuk, K. & Qi, K. (2008). Self-Cleaning Keratins. *Chemistry of Materials*, *20* 1242- 1244.

[76] Qi, S. K.W., Daoud, J. H., Xin, C. I., Mak, W. & Tang, W. P. Cheung, (2006). Sel cleaning cotton. *Journal of Materials Chemistry*, *16* , 4567-4574.

[77] Tung, W. S. & Daoud, W. A. (2009). Photocatalytic self-cleaning keratins: a feasibility study. *Acta Biomaterialia*, *5*, 50-56.

[78] Bozzi, A., Yuranova, T. & Kiwi, J. (2005). Self-cleaning of wool-polyamide and polyester textiles by TiO_2-rutile modification under daylight irradiation at ambient temperature. *Journal of Photochemistry and Photobiology A*: *Chemistry*, *172*, 27-34

[79] Tung, W. S. & Daoud, W. A. (2009). Effect of wettability and silicone surface modification on the self-cleaning functionalization of wool. *Journal of Applied Polymer Science*, *112*, 235-243.

[80] Montazer, M. & Pakdel, E. (2010). Reducing Photoyellowing of Wool Using Nano TiO_2. *Photochemistry and Photobiology A*, *86*, 255-260.

[81] Montazer, M. & Pakdel, E. (2011). Self-cleaning and color reduction in wool fabric by nano-titanium dioxide. *Journal of Textile Institute*, *102*, 343-352.

[82] Montazer, M., Pakdel, E. & Moghadam, M. B. (2010). Nano titanium dioxide on wool keratin as UV absorber stabilized by butane tetra carboxylic acid (BTCA): a statistical prospect. *Fibers and Polymers*, *11*, 967-975.

[83] Montazer, M., Pakdel, E. & Bameni Moghadam, M. (2011). Nano titanium dioxide on wool keratin as UV absorber stabilized by butane tetra carboxylic acid (BTCA): a statistical prospect. *Colloids and Surfaces A*, *375*, 1-11.

[84] Montazer, E. Pakdel, & Behzadnia, A. (2011). Novel Feature of Nano-Tit anium Dioxide on Textiles:Antife lting and Antibac terial Wool. *Journal of Applied Polymer Science*, *21*, 3407–3413.

[85] Montazer, M., Behzadnia, A., Pakdel, E., Rahimid, M. K. & Bameni Moghadam, M. (2011). Photo induced silver on nano titanium dioxide as an enhanced antimicrobial agent for wool. *Journal of Photochemistry and Photobiology B*, *103*, 207–214,

[86] Zhang, H., Millington, K. R. & Wang, X. (2009). The photostability of wool doped with photocatalytic titanium dioxide nanoparticles. *Polymer Degradation Stability*, *94*, 278-283.

[87] Montazer, M. & Seifollahzadeh, S. (2011). Enhanced self—cleaning, antibacterial and UV protection properties of nano TiO_2 treated textile through enzymatic pretreatement. *Journal of Photochemistry Photobiology A*, *84*, 877–883.
[88] Qi, K., Xin, J. H. & Daoud, W. A. (2007). Functionalizing Polyester Fiber with a Self-Cleaning Property Using Anatase TiO_2 and Low-Temperature Plasma Treatment. *International Journal of Applied Ceramic Technology*, *4*, (2007) 554-563.
[89] Mihailovic, D., Saponjic, Z., Radoicic, M., Radetic, T., Jovancic, P., Nedeljkovic, J. & Radetic, M. (2010). Functionalization of Polyester Fabrics with Alginates and TiO_2 Nanoparticles. *Carbohydrate Polymers*, *79*, 526-532.
[90] Mihailovic, D., Saponjic, Z., Molina, R., Puac, N., Jovancic, P., Nedeljkovic J. & Radetic, M. (2010). Improved properties of oxygen and argon RF plasma-activated polyester fabrics loaded with TiO2 nanoparticles. *Applied Materials & Interfaces*, *2*, 1700-1706.
[91] Mihailovic, D., Saponjic, Z., Molina, R., Radoicic, M., Esquena, J., Jovancic, P., Nedeljkovic, J. & Radetic, M. (2011). Multifunctional properties of polyester fabrics modified by corona discharge/air RF plasma and colloidal TiO_2 nanoparticles. *Polymer Composites*, *32*, 390-397.
[92] Mihailovic, D., Saponjic, Z., Radoicic, M., Molina, R., Radetic, T., Jovancic, P., Nedeljkovic, J. & Radetic, M. (2011). Novel properties of PES fabrics modified by corona discharge and colloidal TiO_2 nanoparticles *Polymer for Advanced Technology*, *22*, 703-709.
[93] Hashemizad, S., Montazer, M. & Rashidi, A. (2012). Influence of the surface hydrolysis on the functionality of poly(ethylene terephthalate) fabric treated with nanotitanium dioxide. *Journal of Applied Polymer Science*, *125*, 1176-1180.
[94] Martinkova, L., Sima, O., Akrman, J. (2012). Enzymatic preactivation of PET substrate for improved interaction with photocatalytic systems and stability.*Biocatalysis and Biotransformations 30*, 71-77.
[95] Mejía, M. I., Marín, J. M., Restropo, G., Pulgarín, C., Mielczarski, E., Mielczarski, J., Stolitchnov, I. & Kiwi, J. (2009). Innovative UVC Light (185 nm) and Radio-Frequency-Plasma Pretreatment of Nylon Surfaces at Atmospheric Pressure and Their Implications in Photocatalytic Processes. *Applied Materials & Interfaces*, *1* 2190-2198.
[96] Zhang, H. & Yang, L. (2012). Immobilization of nanoparticle titanium dioxide membrane on polyamide fabric by low temperature hydrothermal method. *Thin Solid Films*, *520*, 5922-5927
[97] Zhang, H., Zhu, H. & Sun, R. (2012). Fabrication of photocatalytic TiO2 nanoparticle film on PET fabric by hydrothermal method. *Textile Research Journal*, *82*, 747-754.
[98] Cakir, B. A., Budama, L., Topel, Önder. & Hoda, N. (2012). Synthesis of ZnO nanoparticles using PS-b-PAA reverse micelle cores for UV protective, self-cleaning and antibacterial textile applications, *Colloids and Surfaces A: Physicochemical and Engineering Aspects*, *414*, 132– 139.
[99] Ates, E. S. & Unalan, H. E. (2012). Zinc oxide nanowire enhanced multifunctional coatings for cotton fabrics. *Thin Solid Films*, *520*, 4658–4661.
[100] Xu, B. & Cai, Z. (2008). Fabrication of a superhydrophobic ZnO nanorod array film on cotton fabrics via a wet chemical route and hydrophobic modification, *Applied Surface Science*, *254*, 5899-5904.

[101] Daoud, W. A., Xin, J. H., Zhang, Y. H. & Mark, C. L. (2008). *Thin Solid Films*, *515*, 835.

[102] Lim, Z. H., Chia, Z. X., Kevin, M., Wong, A. S. W. & Ho, G. W. (2010). A facile approach towards ZnO nanorods conductive textile for room temperature multifunctional sensors *Sensors and Actuators B*, *151*, 121–126.

[103] Lupan, O., Ursaki, V. V., Chai, G., Chow, L., Emelchenko, G. A., Tiginyanu, I. M., Gruzintsev, A. N. & Redkin, A. N. (2010). Selective hydrogen gas nanosensor using individual ZnO nanowire with fast response at room temperature. *Sensors and Actuators B*, *144*, 56–66.

[104] Wang, J. X., Sun, X. W., Yang, Y., Huang, H., Lee, Y. C., Tan, O. K. & Vayssieres, L. (2006). Hydrothermally grown oriented ZnO nanorod arrays for gas sensing applications, *Nanotechnology*, *17*, 4995–4998.

[105] Wang, H. T., Kang, B. S., Ren, F., Tien, L. C., Sadik, P. W., Norton, D. P., Pearton, J. & Lin, S. J. (2005). Detection of hydrogen at room temperature with catalyst-coated multiple ZnO nanorods. *Applied Physics A*, *81*, 1117–1119.

[106] Wang, C., Chu, X. & Wu, M. (2006). Detection of H S down to ppb levels at room temperature using sensors based on ZnO nanorods, *Sensors and Actuators B: Chemical*, *113*, 320–323.

[107] Wan, Q., Li, Q. H., Chen, Y. J., Wang, T. H., He, X. L., Li, J. P. & Lin, C. L. (2004). Fabrication and ethanol sensing characteristics of ZnO nanowire gas sensors, *Applied Physical Letters*, *84*, 3654–3656.

[108] Yeh, P. H., Li, Z. & Wang, Z. L. (2009). Schotthy-gated probe-free ZnO nanowire biosensor. *Advanced Materials 21*, 4975–4978.

[109] Yang, K., She, G. N., Wang, H., Ou, X. M., Zhang, X. H., Lee, C. S. & Lee, S. T. (2009). ZnO nanotubes array as biosensors for glucose. *The Journal of Physical Chemistry C, 113*, 20169–20172.

[110] Zhang, F. & Yang, F. (2009). Preparation of Nano-ZnO and Its Application to the Textile on Antistatic Finishing, *International Journal of Chemistry*, *1*, 18-22,

[111] Zhang, J., France, P., Radomyselskiy, A., Datta, S. & Zhao, J. G. (2003). Hydrophobic Cotton Fabric Coated by a Thin Nanoparticulate Plasma Film. *Journal of Applied Polymer Science*, *88*, 1473.

[112] Vigneshwaran, N., Kumar, S., Kathe, A. A., Varadarajan, P. V. & Prasad, V. (2006). Functional finishing of cotton fabrics using zinc oxide–soluble starch nanocomposites. *Nanotechnology*, *1*, 5087–5095.

[113] Zhang, G., Morikawa, H., Chen, Y. & Miura, M. (2013). In-situ synthesis of ZnO nanoparticles on bamboo pulp fabric, *Materials Letters*, *97*, 184–186.

[114] Ashraf, M., Campagne, C., Perwuelz, A., Champagne, P., Leriche, A. & Courtois, C. (2013). Development of superhydrophilic and superhydrophobic polyester fabric by growing Zinc Oxide nanorods. *Journal of Colloid and Interface Science*, *394*, 545–553.

[115] Khanjani, S. & Morsali A. (2013). Ultrasound-assisted coating of silk yarn with sphere-like Mn, *Ultrasonics Sonochemistry*, *20*, 413–417.

[116] Imran, M., Do Hyun, K., Waheed A. Al-Masry, Asif, M., Sajjad, H. & Shahid, M. R. (2013). Manganese-, cobalt-, and zinc-based mixed-oxide spinels as novel catalysts for the chemical recycling of poly(ethylene terephthalate) via glycolysis. *Polymer Degradation and Stability*, *98*, 904-915.

[117] Yu, M., Gu, G., Meng, W. D. & Qing, F. L. (2007). Superhydrophobic cotton fabric coating based on a complex layer of silica nanoparticles and perfluoro-octylated quaternary ammonium silane coupling agent. *Applied Surface Science*, *253*, 3669-3673.
[118] Su, C. & Li, J. (2010). The friction property of super-hydrophobic cotton textiles. *Applied Surface Science*, *256*, 4220-4225.
[119] Correa-Duarte, M. A., Giersig, M. & Liz-Marzán, L. M. (1998). Stabilization of CDS semiconductor nanoparticles against photodegradation by a silica coating procedure. *Applied Physics Letters*, *286*, 497-501.
[120] Zhao, J. X. (1998). The situation of development antiultraviolet fabric in Japan. *Textile Science Research*, *9*, 43- 47,
[121] Ren, X. M., Zhao, G. L. & Li, X. N. (2003). New development of research on anti - UV fiber. *Polyester Industry*, *16*, 4-6.
[122] Wedler, M. & Hirthe, B. *UV-absorbing micro additives for synthetic fibers*, Sachtleben Chemie Cmb H; International Textile Reader: Duisburg, GE, 2000, Vol. *2*, 48- 49.
[123] Gashti, M. P. & Elahi, A. (2013). UV radiation inducing succinic acid/silica–kaolinite network on to improve the functionality, *Composites: Part B*, *48*, 158–166
[124] Yuranova, T., Mosteo, R., Bandara, J., Laub, D. & Kiwi, J. (2006). Self-cleaning cotton textiles surfaces modified by photoactive SiO2/TiO2 coating, *Journal of Molecular Catalysis A: Chemical*, *224*, 160–167.
[125] Qi, K. H., Daoud, W. A., Xin, J. H., Mak, C. L., Tang, W. S. & Cheung, W. P. (2006). Self-cleaning cotton, *Journal of Materials Chemistry*, *16*, 4567–4574.
[126] Zhou, L., Yan, S., Tian, B., Zhang, J. & Anpo, M. (2006). Preparation of TiO2-SiO2 film with high photocatalytic activity on PET substrate, *Materials Letters*, *60*, 396-399.
[127] Guo, Z., Zhou, F., Hao, J. & Liu, W. M. (2005). Stable biomimetic super-hydrophobic engineering materials. *Journal of the American Chemical Society*, *127*, 15670-15671.
[128] Wang, S., Feng, L. & Jiang, L. (2006). One-Step Solution-Immersion Process for the Fabrication of Stable Bionic Superhydrophobic Surfaces. *AdvancedMaterials*, *18*, 767-770,
[129] Cui, Z., Wang, Q.J., Xiao, Y., Su, C. H. & Chen, Q. M. (2008). The stability of superhydrophobic surfaces tested by high speed current scouring. *Applied Surface Science*, *254*, 2911-2916.
[130] Cui, Z., Yin, L., Wang, Q. J., Ding, J. F., & Chen, Q. M. (2009). A Facile Dip-Coating Process for Preparing Highly Durable Sper-hydrophobic Surface with Multiscale Structures on Paint Films. *Journal of Colloid Interface Sciences*, *337*, 531-537.
[131] Dastjerdi, R., Mojtahedi, M. R. M., Shoshtari, A. M. & Khosroshahi, A. (2010). Investigating the production and properties of Ag/TiO_2/PP antibacterial nanocomposite filament yarns, *Journal of the Textile Institute*, *101*, 204–213, 2010.
[132] Dastjerdi, R., Mojtahedi, M. R. M. & Shoshtari, A. M. (2008). Investigating the effect of various blend ratios of prepared masterbatch containing Ag/TiO nanocomposite on the properties of bioactive continuous filament yarns, *Fibers and Polymers*, *9*, 727–734.
[133] Dastjerdi, R., Mojtahedi, M. R. M. & Shoshtari, A. M. (2009). Comparing the effect of three processing methods for modification of filament yarns with inorganic nanocomposite filler and their bioactivity against staphylococcus aureus. *Macromolecular Research*, *17*, 378–387.

[134] Dastjerdi, R., Montazer, M. & Shahsavan, S. (2010). A novel technique for producing durable multifunctional textiles using nanocomposite coating, *Colloids and Surfaces B: Biointerfaces*, *81*, 32–41.
[135] Dastjerdi, R. & Montazer, M. (2011). Nano-colloidal functionalization of textiles based on polysiloxane novel photo-catalyst assistant: Processing design. *Colloids and Surfaces B: Biointerfaces*, *88*, 381–388.
[136] Yuranova, T., Rincon, G., Pulgarin, C., Laub, D., Xantopoulos, N., Mathieu, H. J. & Kiwi, J. (2013). Performance and characterization of Ag-cotton and Ag/TiO_2 loaded textiles during the abatement of *E. Coli*. *Journal of Photochemistry and Photobiology A: Chemistry*, *181*, 363-369.
[137] Hebeish, A. A., Abdelhady, M. M. & Youssef, A. M. (2013). TiO_2 nanowire and TiO_2 nanowire doped Ag-PVP nanocomposite for antimicrobial and self-cleaning cotton textile. *Carbohydrate Polymers*, *91*, 549– 559.
[138] Messaoud, M., Chadeau, E., Brunon, C., Ballet, T., Rappenne, L., Roussel, F., Leonard, D., Oulahal, N. & Langlet, M. (2010). Photocatalytic generation of silver nanoparticles and application to the antibacterial functionalization of textile fabrics. *Journal of Photochemistry and Photobiology A: Chemistry*, *215*, 147-156.
[139] Montazer, M., Behzadnia, A., Moghadam, M.B. (2012). Superior self-cleaning features on wool fabric using TiO_2/Ag nanocomposite optimized by response surface methodology. *Journal of Applied Polymer Science*, *125*, E356-E363.
[140] Rolland, J. P., Dam, R. M. V., Schorzman, D. A., Quake, S. R. & DeSimone, J. M. (2004). Solvent resistant photocurable liquid teflon for microfluidic device fabrication, *Journal of American Chemical Society*, *126*, 2322.
[141] Shah, K., Shin, W.C., Besser, R.S. (2004). A PDMS micro proton exchange membrane fuel cell by conventional and non-conventional microfabrication techniques. *Sensors and Actuators B Chemical*, *97*, 157.
[142] Kim, J. S. & Knapp, D. R. (2001). Microfabricated PDMS multichannel emitter for electrospray ionization mass spectrometry. *Journal of the American Society for Mass Spectrometry*, *12*, 463.
[143] Oh, B., Hyung, Y. E., Vissers, D. R. & Amine, K. (2002). New interpenetrating network-type siloxane, polymer electrolyte. *Electrochemical and Solid-State Letters*, *5*, 59.
[144] Kweon, D. K. (1998). Preparation and characteristics of Chitosan-g-PDMS copolymer. *Polymer Bulletin*, *41*, 645.
[145] Heywood, D. (2003). *Textile Finishing*, Society of Dyers and Colourists.
[146] Chung, D. W. & Lim, J. C. (2009). Study on the effect of structure of polydimethylsiloxane grafted with polyethyleneoxide on surface activities, *Colloids and Surfaces A: Physicochemical Engineering Aspects*, *336*, 35-40.
[147] Kim, D. W., Noh S. T. & Jo, B. W. (2006). Effect of salt and pH on surface active properties of comb rake-type polysiloxane surfactants. *Colloids and Surfaces A: Physicochemical Engineering Aspects*, *287*, 106.
[148] Montazer, M. & Joulaie, M. M. (2010). Cyclodextrin stabilized on 3-D polyester fabric using different cross-linking agents. *Journal of Applied Polymer Science*, *116*, 210-212.
[149] Dastjerdi, R. & Montazer, M. (2010). A Review on the application of inorganic nanostructured materials in modification of textiles: focus on anti-microbial. *Colloids and Surfaces B: Biointerfaces*, *79*, 5-18.

[150] Dastjerdia, R., Montazera, M., Shahsavan, S., Zhang, L. & Yu, J. C. (2005). A simple approach to reactivate silver-coated titanium dioxide photocatalyst, *Catalysis Communications*, *6*, 684-995.

[151] Martinez, N. N., Castanon, G. A. M., Pina, A. A., Gutierrez, F. M., Mondoza, J. R. M. & Ruiz, F. (2008). Characterization of silver nanoparticles synthesized on titanium dioxide fine particles, *Nanotechnology*, *19*, 5711-5719.

[152] Chen, C. C., Wang, C. C. & Yeh, J. T. (2010). Improvement of odor elimination and antibacterial activity of polyester fabrics finished with composite emulsions of nanometer titanium dioxide-silver particles-water-borne polyurethane, *Textile Research Journal*, *80*, 291-300.

[153] Awazu, K., Fujimaki, M., Rockstuhl, C., Tominaga, J., Murakami, H., Ohki, Y., Yoshida, N. & Watanabe, T. (2008). A plasmonic photocatalyst consisting of silver nanoparticles embedded in titanium dioxide, *Journal of the American Chemical Society*, *130*, 1676-`1680.

[154] Linsebigler, A. L., Lu, G. & Yates, J. T. (1995). Photocatalysis on TiO_2 surfaces: principles mechanisms, and selected results, *Chemical Reviews*, *95*, 735-758.

[155] Youssef, A., Kamel, S. & El-Samahy, M. A. (2013). Morphological and antibacterial properties of modified paper by nanocomposites for packaging Applications, *Carbohydrate Polymers*, *98*, 1166– 1172.

[156] Oh, S. D., Lee, S., Choi, S. H., Lee, I. S., Lee, Y. M., Chun, J. H. & Park, H. J. (2006). Synthesis of Ag and Ag–SiO2 nanoparticles by irradiation and their antibacterial and antifungal efficiency against Salmonella enterica serovar Typhimurium and Botrytis cinerea, *Colloids and Surfaces A: Physicochemical Engineering Aspects*, *275*, 228–233.

[157] Nischala, K., Rao, T. N. & Hebalkar, N. (2011). Silica–silver core–shell particles for antibacterial textile application, *Colloids and Surfaces B: Biointerfaces*, *82*, 203–208.

[158] Ibahim, N. A., Eid, B. M., Abd El-Aziz, E. & Abou Elmaaty, T. M. (2013). Functionalization of linen/cotton pigment prints using inorganic nano structure materials, *Carbohydrate Polymers*, *97*, 537– 545.

[159] Dastjerdi, R., Mojtahedi, A M. R. M., Shoshtari, M., Khosroshahi, A. & Moayed, A. J. (2009). Fiber to fabric processability of silver/zinc-loaded nanocomposite yarns, *Textile Research Journal*, *79*, 1099–1107.

[160] Dastjerdi, R., Mojtahedi, M. R. M. & Shoshtari, A. M. (2008). Processing, properties of nanocomposite filament yarns with various filler concentrations from two different modification methods, *Macromolecular Symposia*, *274*, 154–165.

[161] Moafi, H. F., Shojaie, A. F. & Zanjanchi, M. A. The comparison of photocatalytic activity of synthesized TiO_2 and ZrO_2 nanosize onto wool fibers. http://research.guilan.ac.ir/icc12/data/Book.pdf,

[162] Tanasa, D., Vrinceanu, N., Nistor, A., Hristodor, C. M., Popovici, E., Bistricianu, I. L., Brinza, F., Chicet, D. L., Coman, D., Pui, A., Grigoriu, A. M. & Broasca, G. (2012). Zinc oxide-linen fibrous composites: morphological, structural, chemical and humidity adsorptive attribute, *Textile Research Journal*, *82*, 832-844.

[163] Vrinceanu, N., Tanasa, D., Hristodor, C. M. & Brinza, F. (2013). Synthesis and characterization of zinc oxide nanoparticles: application to textiles as thermal barriers *Journal of Thermal Analysis and Calorimetry*, *111*, 1107-1119

[164] Grigoriu, A. M., Luca, C., Vrînceanu, N. & Ciolacu, F. (2011). Inclusion compounds of monochlorotriazinyl-β-cyclodextrin grafted on a paper supports. *Cellulose Chemistry and Technology*, *45*, 177-183.

[165] Yu, D. G. (2007). Formation of colloidal silver nanoparticles stabilized by Na^+–poly(γ-glutamic acid)–silver nitrate complex via chemical reduction process. *Colloids and Surfaces B: Biointerfaces*, *59*, 171–178.

[166] Courrol, L. C., Silva, F. R. D. O. & Gomes, L. (2007). A simple method to synthesize silver nanoparticles by photo-reduction. *Colloids and Surfaces A: Physicochemical Engineering Aspects*, *305*, 54–57.

[167] Xie, Y., Ye, R. & Liu, H. (2006). Synthesis of silver nanoparticles in reverse micelles stabilized by natural biosurfactant. *Colloids and Surfaces A: Physicochemical Engineering Aspects*, *279*, 175–178.

[168] Sathishkumar, M., Sneha, K., Won, S. W., Cho, C. W., Kim, S. & Yun, Y. S. (2009). Cinnamon zeylanicum bark extract and powder mediated green synthesis of nano-crystalline silver particles and its bactericidal activity, *Colloids and Surfaces B: Biointerfaces*, *73*, 332–338.

[169] Nersisyan, H. H., Lee, J. H., Son, H. T., Won, C. W. & Maeng, D. Y. (2003). A new and effective chemical reduction method for preparation of nanosized silver powder and colloid dispersion. *Materials Research Bulletin*, *38*, 949–956.

[170] Vankar, P. S. & Shukla, D. (2012). Biosynthesis of silver nanoparticles using lemon leaves extract and its application for antimicrobial finish on fabric. *Applied Nanoscience*, *2*, 163-168.

[171] Osório, I., Igreja, R., Franco, R. & Cortez, J. (2012). Incorporation of silver nanoparticles on textile materials by an aqueous procedure. *Materials Letters*, *75*, 200-203.

[172] Rai, M., Yadav, A. & Gade, A. (2009). Silver nanoparticles as a new generation of antimicrobials. *Biotechnology Advances*, *27*, 76–83.

[173] Yadav, A., Prasad, V., Kathe, A. A., Raj, S., Yadav, D., Sundaramoorthy, C. & Vigneshwaran, N. (2006). Functional finishing in cotton fabrics using zinc oxide nanopartickes, *Bulletin of Materials Science*, *29*, 641–645.

[174] Zhang, D., Zang, C., Chen, C. & Lin, H. (2013). Antibacterial cotton fabric grafted with silver nanoparticles and its excellent laundering durability, *Carbohydrate Polymers*, *92*, 2088-2094.

[175] Yonezawa,T. & Kunitake, T. (1999). Practical preparation of anionic mercapto ligand-stabilized gold nanoparticles and their immobilization. *Colloids and Surfaces A: Physicochemical and Engineering Aspects*, *149*, 193–199.

[176] Zhang, Y., Peng, H., Huanga, W., Zhou, Y. & Yan, D. (2008). Facile preparation and characterization of highly antimicrobial colloid Ag or Au nanoparticles. *Journal of Colloid and Interface Science*, *325*, 371–376.

[177] Teli, M. D. & Sheikh, J. (2013). Modified bamboo rayon–copper nanoparticle composites as antibacterial textiles. *International Journal of Biological Macromolecules*, *61*, 302– 307.

[178] Radetic, M. (2013). Functionalization of textile materials with TiO_2 nanoparticles. *Journal of Photochemistry and Photobiology C: Photochemistry Reviews*, *16*, 62– 76.

[179] Tung, W. S., Daoud, W. A. & Leung, S. K. (2008). Photocatalytic formulations for proptein fibers: Experimmental analysis of the effect of preparation on compatibility and photocatalytic activities. *Journal of Colloid Interface Sciences*, *326*, 283-288.
[180] Montazer, M., Pakdel, E. & Moghadam, M. B. (2011). The role of nano colloid of TiO_2 and butane tetra carboxylic acid on the alkali solubility and hydrophilicity of proteinous fibers. *Colloids and Surfaces A-physicochemical and Engineering Aspects - Colloid Surface A*, *375*, 1-11.
[181] Montazer, M. & Pakdel, E. (2011). Self-cleaning and color reduction in wool fabric by nano titanium dioxide. *Journal of Textile Institute*, *102*, 343-352.
[182] Mejia, M. I., Marin, J. M., Restrepo, G., Polgarin, C., Mielczarski, E., Mielczarski, J., Arroyo, Y., Lavanchy, J. C. & Kiwi, J. (2009). Self-cleaning modified TiO_2–cotton pretreated by UVC-light (185 nm) and RF-plasma in vacuum and also under atmospheric pressure. *Applied Catalysis B: Environmental*, *91*, 481.
[183] Kiwi, J. & Pulgain, C. (2010). Innovative self-cleaning and bactericide textiles. *Catalysis Today*, 151, 2-7.
[184] Xin, J. H., Yeo, S. Y., Lee, H. J. & Jeong, S. H. (2003). Preparation of nano compsite fibres for permanent antibacterial effect. *Journal of Material Science*, *38*, 2143-2147
[185] Parthasarathi, V. & Thilagavathi, G. (2011). Synthesis and characterization Zinc oxide nano particles and its application on fabrics for microbe resistant defense clothing. *International Journal of Pharmacy and Pharmaceutical Sciences*, *3*, 392-398.
[186] Becheri, A., Durr, M., Lonostro, P. & Bagloini, P. J. (2008). Synthesis and characterization of zinc oxide nanoparticles: application to textiles as UV-absorbers, *Journal of Nanoparticle Research*, *2*, 679-689.
[187] Zhiping, M., Qiuping, S., Linping, Z. & Huantian, C. (2009). The formation and UV-blocking property of needle-shaped ZnO nanorod on cotton fabric. *Thin Solid Films*, *517*, 2681–2686.
[188] Lu, S., John, A., Peter, G., Xungai, W., King, K. (2008). Nano zinc oxide for UV protection of Textiles, *International Journal of Technology*, *7*, 224-235.
[189] Li, Q., Chen, S. L. & Jiang, W. C. (2007). Durability of nano ZnO antibacterial cotton fabric to sweat, *Journal of Applied Polymer Science*, *103*, 412–416.
[190] Selishchev, D. S., Karaseva, I. P., Uvaev, V. V., Kozlov, D. V. & Parmon, V. N. (2013). Effect of preparation method of functionalized textile materials on their photocatalytic activity and stability under UV irradiation. *Chemical Engineering Journal*, *224*, 114–120.
[191] Xue, C. H., Jia, S. T., Zhang, J. & Tian, L. Q. (2009). Superhydrophobic surfaces on cottin textiles by complex coating of silica nanoparticles and hydrophobization. *Thin Solid Films 517*, 4593–4598.
[192] Garcıa, M., Rooij, M de, Winnubst, L., Zyl, W. E. & van, Verweij, H. (2004). *Friction and wear studies on nylon-6/SiO_2 nanocomposites. Journal of Applied Polymer Science*, *92*, 1855–1862.
[193] Song, H. J., Zhang, Z. Z. & Men, X. H. (2008). The Tribological Behaviors of the Polyurethane Coating Filled with Nano-SiO_2 under Different Lubrication Conditions. *Composites Part A: Applied Science and Manufacturing*, *39*, 188–194.
[194] Kashiwagi, T., Morgan, A. B., Antonucci, J. M., VanLandingham, M. R., Harris, R. H. Jr., Awad, W. H. & Shields, J. R. (2003). Thermal and flammability properties of a

silica–poly(methylmethacrylate) nanocomposite. *Journal of Applied Polymer Science*, *89*, 2072–2078.
[195] Stoimenov, P. K., Klinger, R. L., Marchin, G. L. & Klabunde, K. J. (2002). Metal oxide nanoparticles as bactericidal agents, *Langmuir*, *18*, 6679–6686.
[196] Ghosh, A. (2011). Nano-Clay Particle as Textile Coating, *International Journal of Engineering & Technology*, *11*, 40-43.
[197] Bahrami, S. H. & Mirzaie, Z. (2011). Polypropylene/Modified Nanoclay Composite-Processing and Dyeability Properties. *World Applied Sciences Journal*, *13*, 493-501.
[198] Kulshreshtha, A. K., Maiti, M. S. & Rao, K. V. (2006). Nano-addition of raw bentonite enhances polypropylene (PP) properties. *Journal of applied Polymer Science*, *99*, 1004-1009.
[199] Fogelström, L., Malmström, E., Johansson, M. & Hult, A. Hard and Flexible Nanocomposite Coatings using Nanoclay-filled Hyperbranched Polymers, *ACS Applied Materials and Interfaces*, Vol. *2*, No. 6, 1679-1684.
[200] Razafimahefa, L., Chlebicki, S., Vroman, I. & Devaux, E. (2005). Effect of nanoclay on the dyeing ability of PA6 nanocomposite fibers. *Dyes and Pigments*, *66*, 55–60.
[201] Sahoo, N. G., Jung, Y. C., Yoo, H. J. & Cho, J. W. (2007). Influence of carbon nanotubes and polypyrrole on the thermal, mechanical and electroactive shape-memory properties of polyurethane nanocomposites. *Sci. Tech.*, *67*, 1920–1929.
[202] Gonçalves, A. G., Jarrais, B., Pereira, C., Morgado, J. & Freire, C. (2012). Functionalization of textiles with multi-walled carbon nanotubes by a novel dyeing-like process. *Journal of Materials Science*, *47*, 5263-5275.
[203] Dastjerdi, R. & Mojtahedi, M. R. M. (2013). Multifunctional melt-mixed Ag/Ti_{O2} nanocomposite PP fabrics: Water vapour permeability, UV resistance, UV protection and wear properties. *Fibers and Polymers*, *14*, 298-303.
[204] Habib, M. A., Shahadat, M. T., Bahadur, N. M., Iqbal, M. I. I. & Mahmood, A. J. (2013). Synthesis and characterization of ZnO-TiO_2 nanocomposites and their application as photocatalysts, *International Nano Letters*, *3*, 7-8.
[205] Esfandiar, P., Walid, A. & Xungai D. W. (2013). Self-cleaning and superhydrophilic wool by TiO_2/SiO_2 nanocomposite, *Applied Surface Science*, *275*, 397– 402.
[206] Xin, B., Chen, Z. & Wu, X. (2013). Preparation and Characterisation of PSA/CNT/TiO2 Nano Composites and Fibres. *Fibres & Textiles in Eastern Europe*, *21*, 24-28.
[207] Yeum, J. H., Park, J. H., Kim, I. K. & Cheong, I. W. (2010). *Advances in Nanocomposites - Synthesis, Characterization and Industrial Applications*, Dr. Boreddy Reddy.
[208] Selvam, V. & Sundrarajan, M. (2012). Functionalization of cotton fabric with PVP/ZnO nanoparticles for improved reactive dyeability and antibacterial activity. *Carbohydrate Polymers*, *87*, 1419–1424.
[209] Kumar, S., Kathe, A. A., Varadarajan, P. V. & Prasad, V. (2006). Functional Finishing of Cotton Fabrics using Zinc Oxidesoluble Starch Nanocomposites. *Nanotechnology*, *17*, 5087-5094.
[210] El Shafei, A. & Abou-Okeil, A. (2011). ZnO/carboxymethyl chitosan bionano-composite to impart antibacterial and UV protection for cotton fabric. *Carbohydrate Polymers*, *83*, 920–925.

[211] Bhattacharyya, A. & Joshi, M. (2010). Microwave Absorbent Nanocomposite Films of Iron-Nickel Nanographite in Thermoplastic Polyurethane Matrix. *Journal of Nanostructured Polymers and Nanocomposites*, *6*, 73–78.
[212] Choi, S. & Sankar, B. V. (2008). Gas permeability of various graphite/epoxy composite laminates for cryogenic storage systems, *Composites Part B*, *39*, 782–791.
[213] Solarski, S., Ferreira, M., Devaux, E., Fontaine, G., Bachelet, P., Bourbigot, S., Delobel, R., Coszach, P., Murariu, M., Ferreira, A. Da Silva, Alexandre, M., Degee, P. & Dubois, P. (2008). Designing polylactide/clay nanocomposites for textile applications: Effect of processing conditions, spinning, and characterization, *Journal of Applied Polymer Science*, *109*, 841–851.
[214] Persico, P., Carfagna, C., Musto, P. (2006). Nanocomposite Fibers for Cosmetotextile Applications. *Macromolecular Symposia*, *234*, 147–155.
[215] Gupta D. & Haile A. (2007). Multifunctional properties of cotton fabric treated with chitosan and carboxymethyl chitosan. *Carbohydrate Polymers*. *69*, 164-171.
[216] Niekraszewicz A. (2005). Chitosan Medical Dressings. *Fibers & Textiles in Eastern Europe*, *13*, 16-18.
[217] Ye, W.J., Leung, M. F., Xin, J., Kwong, T. L., Lee, D. K. L. & Li, P. (2005). Novel core-shell particles with poly(n-butyl acrylate) cores and chitosan shells as an antibacterial coating for textiles. *Polymers*, *46*, 10538-10543.
[218] Voncina, B. & Vivod, V. *Eco-Friendly Textile Dyeing and Finishing*, Cyclodextrins in Textile Finishing, InTech Opened. by Melih Günay, 2013.
[219] Szejtli, J. (2003). Cyclodextrins in the Textile Industry. *Starch*, *55*, 191-196.
[220] Teng, M., Li, F., Zhang, B. & Taha, A.A. (2011) Electrospun cyclodextrin-functionalized mesoporous polyvinyl alcohol/SiO2 nanofiber membranes as a highly efficient adsorbent for indigo carmine dye. *Colloids and Surfaces A: Physicochemical and Engineering Aspects*, *385*, 229-234.
[221] United States Patent 8404345.
[222] Zhang, X. X., Fan, Y. F., Tao, X. M. & Yick K. L. (2004). Fabrication and properties of microcapsules and nanocapsules containing n-octadecane, *Materials Chemistry and Physics*, *88*, 300–307.
[223] Mandal, T., Dasgupta, C. Prabal, K. & Maiti, E. (2013). Engineering Gold Nanoparticle Interaction by PAMAM Dendrimer, *The Journal of Physical Chemistry C*, *117*, 13627–13636.
[224] Castonguay, A. & Kakkar, A. K. (2010). Dendrimer templated construction of silver nanoparticles, *Advances in Colloid Interface Science*, *15*, 76-87.
[225] Zolriasatein, A. A., Yazdanshenas, M. E., Khajavi, R., Rashidi, A. & Najafi, F. (2013). The use of poly(amidoamine) dendrimer in modification of jute for improving dyeing properties of reactive dyes. *Journal of Applied Polymer Science*, *127*, 4203-4210.
[226] Abkenar, S. S. & Malek R. M. A. (2012). Preparation, characterization and antimicrobial property of cotton cellulose fabric grafted with poly(propylene imine) dendrimer. *Cellulose*, *19*, 1701-1714.
[227] Colleoni, C., Massafra, M. R., Migani, V. & Rosace, G. (2011). Dendrimer finishing influence on CO/PES blended fabrics color assessment, *Journal of Applied Polymer Science*, *120*, 2122–2129.
[228] Abkenar, S. S., Malek, R. M. A. & Mazaheri, F. (2013). Thermal properties of cotton fabric modified with poly (propylene imine) dendrimers, *Cellulose*, *20*.

[229] Barani, H. & Montazer, M. (2008). A Review on applications of liposomes in textile processing, *Journal of Liposome Research*, *18*, 249–262.
[230] Martí, M. & Barsukov, L. (2004). Physicochemical aspect of the liposome-wool internal in wool dyeing. *Langmuir*, *13*, 3068–73.
[231] Vincent, J. E. & Goheen, S. C. (2006). Performance of bioactive molecules on cotton and other textiles. *RJTA*, *10*, 19–32.
[232] Langenhove, H. (Ed.). *Smart Textiles for Medicine and Healthcare*, Woodhead Publishers in Textiles, CRC: Boca Raton, FL, 2010, 74–88.
[233] Mondal, S. (2008). Phase change materials for smart textiles – An overview. *Applied Thermal Engineering*, *28*, 1536–50.
[234] EL-Zawahry, M., EL-Shami, S. et al. (2007). Optimizing a wool dyeing process with reactive dye by liposome microencapsulation. *Dyes and Pigments*, *74*, 684–691.
[235] Qi, K. & Xin, J. H. (2010). Room-temperature synthesis of single-phase anatase TiO_2 by aging and its self-cleaning properties. *ACS Applied material and interfaces*, *2*, 3479-3485.
[236] Vilcnik, A., Jerman, I., SurcaVuk, A., Kozelj, M., Orel, B., Tomsic, B., Simoncic, B. & Kovac, J. (2009). Structural properties and antimicrobial effects of hydrophobic and oleophobic sol-gel coatings for cotton fabrics. *Langmuir*, *25*, 5869–5880.
[237] Textor, T. & Mahltig, B. (2010). A sol-gel based surface treatment for preparation of water repellent antistatic textiles. *Applied Surface Science*, *256*, 1668–1674.
[238] Höcker, H. (2002). Plasma treatment of textile fibers. *Pure and Applied Chemistry*, *74*, 423–427.
[239] Joshi, M., Khanna, R., Shekhar, R. & Jha, K. (2011). Chitosan nanocoating on cotton textile substrate using layer-by-layer self-assembly technique. *Journal of Applied Polymer Science*, *119*, 2793–2499.
[240] Wei, Q., Yu, L., Hou, D. & Huang, F. (2008). Surface characterization and properties of functionalized nonwoven. *Journal of Applied Polymer Science*, *107*, 132–137.
[241] Wei, Q., Xiao, X., Hou, D., Ye, H. & Huang, F. (2008). Characterization of nonwoven material functionalized by sputter coating of copper. *Surface & Coatings Technology*, *202*, 2535–2539.
[242] Cai, Y., Huang, F., Wei, Q., Wu, E. & Gao, W. (2008). Surface functionalization, morphology and thermal properties of polyamide6/O-MMT composite nanofibers by Fe_2O_3 sputter coating. *Applied Surface Science*, *254*, 5501–5505.

In: Comprehensive Guide for Mesoporous Materials, Volume 2 ISBN: 978-1-63463-982-8
Editor: Mahmood Aliofkhazraei

Chapter 9

STUDY OF MESOPOROUS SILICON NANOWIRES

A. Najar[1] *and J. Charrier*[2]
[1]NTT Basic Research Laboratories, NTT Corporation, Atsugi, Kanagawa, Japan
[2]Universite européenne de Bretagne, CNRS FOTON, UMR 6082, ENSSAT BP80518, F-22305 Lannion Cedex, France

ABSTRACT

We report on the structural and optical properties of mesoporous silicon nanowires (PSiNWs) fabricated using silver ions assisted electroless etching method. The strongest photoluminescence signal has been measured from samples etched with 4.8M of HF. Reflectance were measured versus nanowire length and were inferior to 0.1%. Models based on cone shape of nanowires located in circular and rectangular bases were used to calculate the reflectance using the Transfert Matrix Formalism of *P*SiNWs layer. The modeling of the reflectance permits to explain this value by taking account into the shape of the nanowires and its porosity. Optical absorbance and transmission were also theoretically studied. The absorbance was superior to that obtained with silicon nanowires and the ultimate efficiency was about equal to 25% for normal incidence angle. These results could be applied to the potential application in low-cost and high efficiency *P*SiNWs based solar cells.

1. INTRODUCTION

Nanoscale silicon has been intensively investigated and explored for its applications in microelectronics, photonics, and biomedical sensors [1–3]. Specific efforts have been concentrated in the development of new silicon nanostructures, including quantum dots, nanowires, or porous silicon (PS). Porous silicon has attracted much attention, especially in enhancing photo-emission. Much research efforts have been invested to realize an optical device with porous silicon [4–9], but the inefficiency [10] and instability [11] of optical characteristic in PS still remain. In addition, silicon nanowires (SiNWs) are also ideal candidates for the study of physics of low dimensional systems. It has potential impact in realizing nanoscale interconnects and functional device elements in future nanoscale electronic and optoelectronic devices [12, 13]. The field of SiNWs synthesis represents an exciting and rapidly expanding research area. Considerable efforts have been devoted to the development of versatile and controllable methods for the synthesis of SiNW (Table I.). These methods can be broadly classified as: (i) bottom-up, and (ii) top-down approaches. The bottom-up approach involves the construction of desirable nanostructures from the basic components, i.e., from the atomic level to the nano- or micro-scale wires.

This method is useful for the fabrication of low-dimensional heterostructure based devices in large quantities [14-18]. Using bottom-up, SiNWs were first obtained by vapor–liquid–solid (VLS) method [19], followed by an etching step to create nanowires. The VLS method has been implemented in a variety of techniques, such as pulsed laser deposition (PLD) [20, 21], gas-phase molecular beam epitaxy (GS-MBE) [22], chemical vapor deposition (CVD) [23, 24], laser ablation [25, 26], and oxide assisted growth techniques [27].

Top-down approach seeks to fabricate SiNWs from high quality single crystal silicon wafer or thin film. Silicon nanowires have also been realized using lithographically defined patterns, or spin-coating of nanospheres as etched mask [28] followed by etching of the nanowires using plasma processing technique. MacEtch is a simple, cost-effective, and powerful semiconductor etching technique that is capable of producing high aspect ratio semiconductor nanostructures. By combining with metal patterning lithography or non-lithographic patterning methods, accurate control of the nanowire orientation (vertical slanted), size (nano vs microscale), shape, architecture, density, length, doping characteristics can be achieved readily at waferscale.

The fabrication of silicon nanowires using the metal-assisted electroless etching method has also been adopted [29-33]. The silver (Ag) ions in an ionic solution of hydrofluoric acid (HF) and hydrogen peroxide (H_2O_2) have been used to prepare the arrays of SiNWs from single crystal wafers [34, 35] The effects of various process parameters such as the etchant concentration of H_2O_2, etching time and post-etch treatment on the morphology and optical properties of the SiNWs have also been investigated [36] The fabrication of nanowires using this method does not require complex sample preparation steps. Furthermore, this technique is effective, having high throughput and low cost.

Table 1. Bottom-up and Top-down growing technics

	Growth Technics
Bottom-up	VLS [19], PLD [20, 21], GS-MBE [22], CVD [23, 24], Laser ablation [25, 26], Oxide assisted growth techniques [27].
Top-down	Lithographically defined patterns, or spin-coating of nanospheres as etched mask followed by etching [28]

2. Fabrication of Porous Silicon Nanowires(PSiNWs)

PSiNWs were fabricated by Ag assisted electroless etching method from n-type Si wafer (100) with a resistivity of 0.01–0.02 ohm cm. The Si wafers were cleaned using acetone followed by ethanol for 5min in an ultrasonic bath. Next, the wafers were immersed in a piranha solution H_2SO_4/H_2O_2 (3:1) for 20 min to remove the organic deposits from the surface. The cleaned wafers were transferred into $HF/AgNO_3$ solution with a concentration of

4.8 M/0.005M for Ag-deposition (Figure 1), followed by rinsing with de-ionized water. Then, the treated Si samples were etched in the HF/H_2O_2 solution for 45 min. Finally, samples were rinsed again for10 min with HNO_3 solution to dissolve the excessive Ag nanoparticles (Ag NPs), leaving behind traces of Ag for catalyzing the etching reaction. Four samples were etched using HF concentration of 1.8, 2.8, 4.8, and 5.8 M, respectively, with a fixed H_2O_2 concentration of 0.5 M.

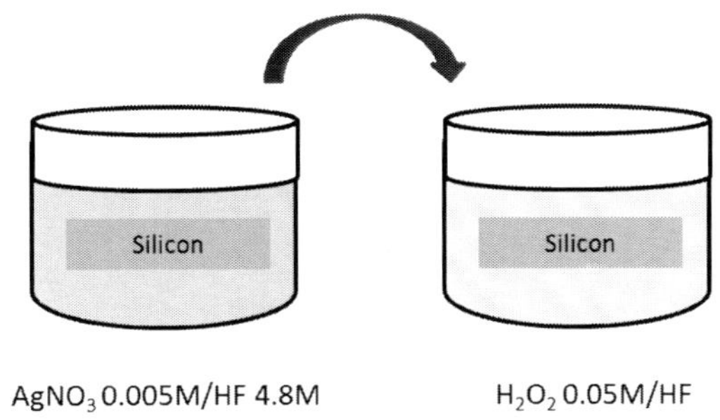

Figure 1. Fabrication of PSiNWs using Ag assisted electroless etching method with different HF concentration.

3. Physical and Optical Properties

3.1. Effect of Etching Time

Figure 2 (a) shows the scanning electron microcopy image of the vertically oriented PSiNW shaving an average length of 13 μm after 75 min wet etching. The nanowires distribute uniformly on the sample. Figure 2 (b) shows the top view with tips of NWs congregates together and the diameter of the congregated bundles is several micrometers. These congregated bundles are uniformly distributed on the entire samples and confirmed from the cross-section images. The transmission electron microscopy (TEM) image of an individual nanowire is shown Figure 2 (c), which shows that the nanowire has a diameter of ~150 nm.

A typical 2D bright-field TEM image used for volume reconstruction of the sample is presented in Figure 3 (a)-(c). The three successive longitudinal slices obtained from the reconstructed volume revealed a distribution of irregular size/shape vacuum spaces and silicon frame, constituting the PSiNWs.

A typical 3D-tomography observation was conducted using a TEM and 3D reconstructions were achieved using a simultaneous iterative reconstruction algorithm of consecutive 2D slices in Figure 3 (d). The pore sizes present a distribution from 10 to 50 nm, with an estimated measurement error of 10%, and these pores go inside the nanowire showing similar structure to porous silicon. An average distance between two neighboring pores of lower than5 nm has been observed. These mesoporous structures are expected to show strong quantum confinement effects. Our results based on TEM and 2D reconstructions were compared to the results in ref [29] and presented in Table II.

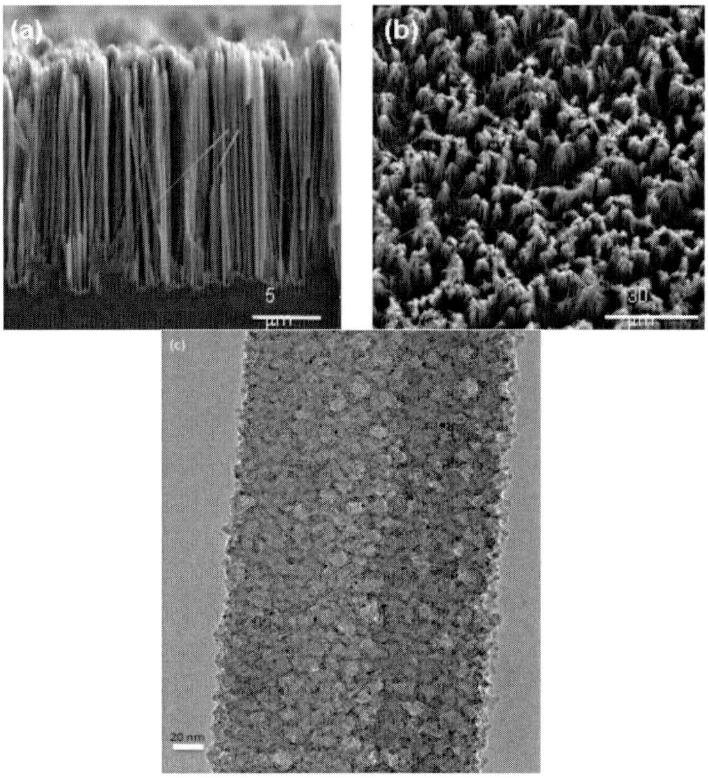

Figure 2. (a) cross-section SEM micrograph of PSiNWs etched for 75 min, (b) top view micrograph of PSiNWs, (c) TEM image of individual nanowire and inset the FFT image.

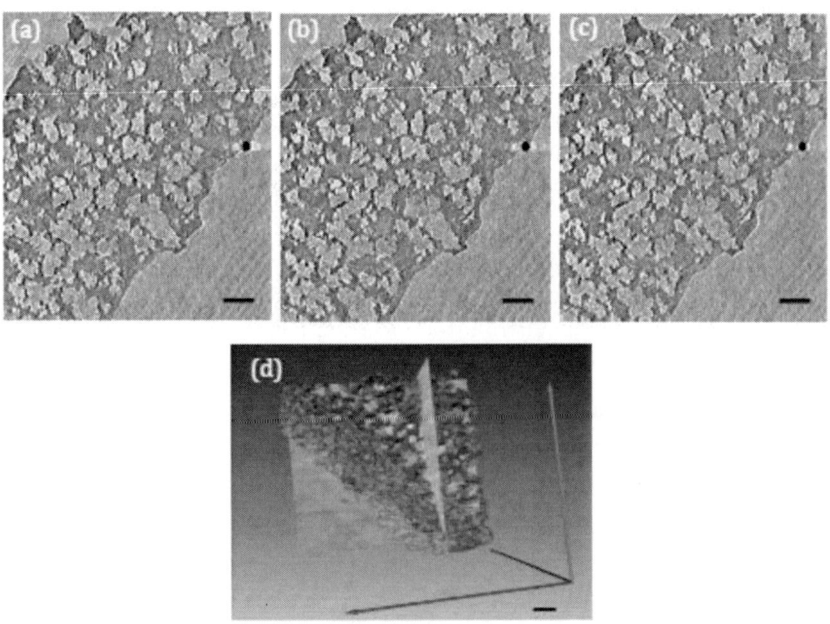

Figure 3. (a), (b), and (c) represent three longitudinal slices extracted from the volume reconstruction of the PSiNW nanostructures revealing an area of highly porous medium. The bare scale is 50 nm. (d) 3D tomography TEM image of nanowires.

Table 2. Comparison of pores properties to ref [29]

	Si wafer type & resistivity (Ω.cm)	Pore diameters (nm)	Main pore diameter (nm)	Mean surface area (m^2g^{-1})	Total pore volume (cm^3g^{-1})
Our work	n-type, 0.01–0.02	10-50	25	-	-
Ref [29]	n-type, <0.005	2-20	9.7	342	0.88

In the two-step chemical etching for highly doped Si wafers (n-type), [36-38] the silver nanoparticle catalysts are predeposited on the clean silicon surface. Since H_2O_2 itself cannot etch the silicon, the formation of porous structures needs to be facilitated by the metal nanoparticles. Previous studies suggested that the silver nanoparticles could be oxidized into Ag^+ ion [34] during the etching process as follows:

$$2Ag + H_2O_2 + 2H^+ \rightarrow 2Ag^+ + 2H_2O$$

In a typical etch process, the silver particles are partially oxidized by H_2O_2 to create alocalized Ag^+ cloud in the close proximity of the silver particles. Silver ions can quickly react with silicon and take electrons from silicon near the Ag/Si interface and be recovered into original silver particles again. In this way, the etching is localized around the silver nanoparticles and the silver nanoparticles are trapped in the nanopits created by themselves, leading to continued etching in the vertical direction and the formation of vertical silicon nanowire array. However, as the H_2O_2 concentration increases, the concentration of Ag^+ ion increases. In this case, the Ag^+ ions may not be 100% recovered into the original silver nanoparticles and some of them may diffuse out. When the amount of the out-diffused Ag^+ ions reach a certain threshold, they may nucleate on the side wall near defective sites (e.g., around the dopants) by extracting electrons from the silicon nanowires and forming new silver nanoparticles as a new etching pathway along the lateral direction of the nanowires.

In this way, the higher H_2O_2 concentration increases the amount of out-diffused Ag^+, and the increase of doping concentration increases the amount of weak defective points in the silicon lattice; both are important factors for the initiation of additional etching pathways in addition to the vertical etch, which explains the observation of the increased porosity with increasingH_2O_2 concentration during the etching of highly doped n-Si wafers.

Figure 4 shows the length of the NWs versus the etching time. The length of PSiNWs is linear as a function of the etching time up to 190 min with an etching rate of 0.12 μm.min^{-1}. This clearly indicates the rapidity of the electroless etching process. On the other hand, beyond 190 min, again revealed a linear relationship, with a change in the etching rate of the nanowires lower than the expected for 460 min etching time. The etching rate of the nanowires between 190 and 460 was 0.01 μm.min^{-1}. This clearly indicates that the reaction slows down due to the decrease in the reaction kinetics. We attribute this to the inhibition of the sinking mechanism of Ag^+ions to the bottom of the nanowire arrays, leading to a decrease in Ag^+ concentration. The drop in Ag concentration results in the deceleration of etching rate.

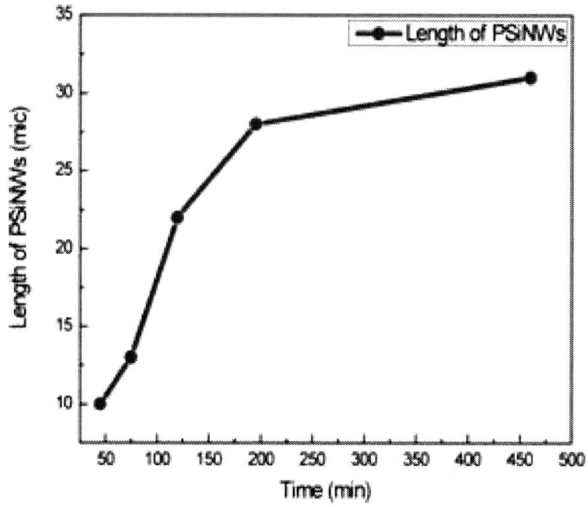

Figure 4. Variation of the nanowires length as a function of etching time 45 min, 75 min, 120 min, 195 min and 460 min.

3.2. Effect of HF Concentrations

In the presence of Ag catalyst, an increase in HF orH_2O_2concentration in electroless etching method is analogous to an increase in the current density in electrochemical based methods [39, 40]. In both cases, increasing the H_2O_2or HF concentration increases the oxidation rate and dissolution rate, respectively, resulting in nanostructures with varying optical properties [38]. Figures 5 (a)-(h) show the cross section SEM micrographs and TEM images acquired from the NW samples grown with HF concentration of 1.8 M,2.8 M, 4.8 M, and 5.8M in a fixed H_2O_2concentration of0.5 M.

The variation of NWs length, and NWs diameter with increasing HF concentrations were measured and analyzed using SEM and TEM, and the results are tabulated in Table I. The NWs length shows significant changes, while the NWs diameter varies from 80 nm to 210 nm for samples indicating the dominant effect of vertical etching enhanced by Ag catalyst with preferential etching along the low atomic density plane in[001]. An increase in aspect ratio from about 9 to 12 and 27has been calculated for NWs samples etched with 1.8 M,2.8 M, and 4.8M HF concentration (Figs. 5 (a), 5 (c), and 5 (e)). However, the aspect ratio reduces to about 9 for NWs etched with (Figure 5 (g)) signifying a change in etching mechanism with faster lateral chemical etch.

To identify the causes of the transition point between thc vertical-dominant to lateral dominant etching in the chemical process, the pore-density of the nanowires have been examined thoroughly. The TEM images show clearly that samples prepared with 4.8M of HF concentration present relatively high pore density compared to samples prepared with 5.8M of HF. The monotonic increase in the pore-density evidently showed that lateral oxidation process at a fixed H_2O_2concentration dominates at low HF concentration of 4.8M due to a limited HF dissolution rate. As HF concentration increases, the HF dissolution rate increases to the extent that the PSiNWs apexes were effectively dissolved at the highest HF concentration of 5.8 M, forming conical PSiNWs. Hence, the reaction (dissolution) rate dominates the lateral PSiNWs chemical etching process at high HF concentration, while the diffusion of HF etching species and byproducts dominates at low HF concentration. In the following paragraphs, we further explain the role of Ag nanopartiules and ionic Ag^+ in the formation of conical PSiNWs, the transport of etchant and other by products in the nanowires.

Figure 5. (a)–(h) SEM and TEM micrographs of PSiNWs fabricated with increasing HF concentration from 1.8, 2.8, 4.8, up to 5.8 M.

Table 3. Effect of HF concentration on PSiNWs length, NWs diameter, and aspect ratio analyzed from SEM micrographs

PSiNWs physical parameters	HF concentration			
	1.8 M	2.8 M	4.8 M	5.8 M
NWs length (μm)	1.3-1.8	2-2.2	4.8-5.3	1.6-1.8
NWs diameter (nm)	160 – 210	100 – 180	80 – 200	150 -200
Aspect ratio (longest length /largest diameter)	~9	~12	~27	~9

It is noted that mesopores are mostly found near the top of the nanowires with an average pore size between 10 and40 nm. Ag NPs were detected at the bottom section of the nanowires by EDS analysis in STEM mode. Based on the SEM and TEM analysis, the following qualitative model for the formation of PSiNWs is proposed.

In HF/H_2O_2 solution, part of the Ag is oxidized into Ag^+ by H_2O_2 [38]. The Ag^+ catalyzes the dissolution of Si through the formation of SiF_6^{2-}.

It is apparent that the transport of Ag^+ is isotropic while the AgNPs merely drop downwards. Hence, the AgNPs facilitated the nanowire formation through vertical chemical etching, while the isotropic diffusion of Ag^+ promoted lateral chemical etching and pore formation as illustrated in Figure 6 (a-i and a-ii).

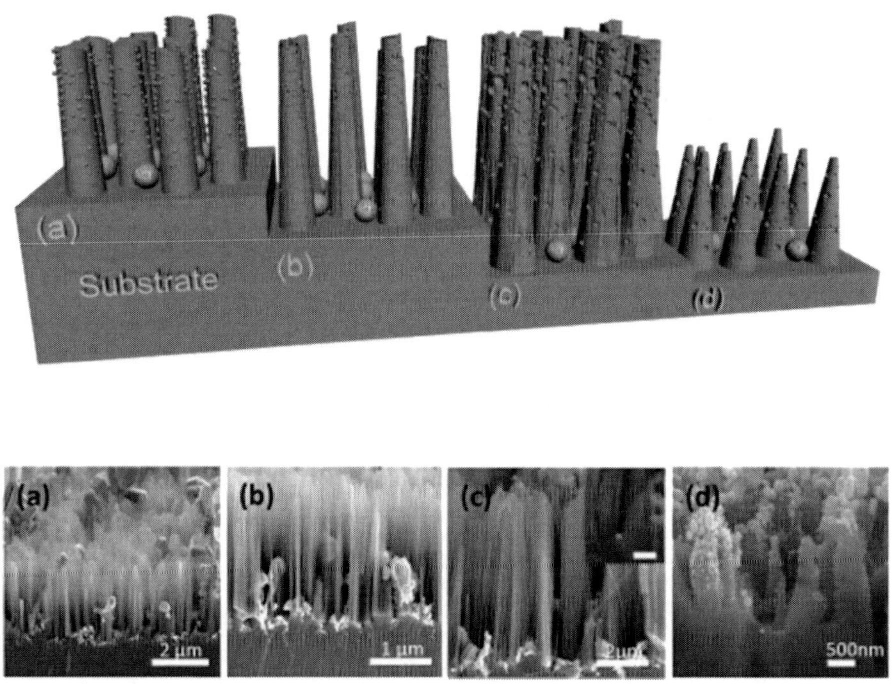

Figure 6. (a-i)–(d-i) The schematic illustration of the proposed qualitative etching model for the formation of PSiNWs using Ag-assisted HF/H_2O_2 electroless etching technique. (a-ii)–(d-ii) the corresponding SEM micrographs, prepared using various HF concentrations.

The Ag^+ may be chemically reduced by the n-type silicon nanowires [41] forming Ag particle that re-deposited onto the sidewalls forming new Ag nucleation sites, and therefore the new localized lateral etching pathway. Furthermore, the nucleation of the AgNPs on the side-walls would also reduce the Ag^+ concentration locally, and accelerates the Ag^+ diffusion in the lateral direction. The laterally diffused Ag^+ will continue to catalyze the chemical etching of the sidewalls of Si. Since the upper Si nanowires are exposed longer to the solutions, larger Si volume dissolves at the top (and creating more mesopores) than the bottom of the nanowires (Figures 6 (b-i and b-ii) and 6 (c-i and c-ii)), consistent with the TEM image shown in Figure 5 (e). Furthermore, increasing the HF concentration accelerates the process described in our qualitative model; with the Ag <→ Ag^++ e^-reversible reaction continues to penetrate the nanowires laterally to catalyze the HF/H_2O_2 chemical reaction forming mesopores. Eventually, the mesopores merged leading to a reduction in length and pore-density, with the formation of conical PSiNWs in Figure 6 (d-i and d-ii).

3.3. XPS Study

The XPS measurements have been conducted to study the role of SiOx/Si interface defects and/or defect states in surface oxide to the broadband PL emissions in PSiNWs. Figure 6exhibits the Si 2p core-level spectra having two peaks around100.0 eV and 103.5 eV, which correspond to the silicon and silicon dioxide peaks. The lower binding energies near100.0 eV, i.e., the Si $2p_{3/2}$(99.7 eV) and Si $2p_{1/2}$(100.1 eV) [42] spin-orbit splitting are well resolved. The inset of Figure 7shows the XPS spectra of the PSiNW etched using 4.8Mconcentration of HF. The XPS spectra was fitted using 1doublet (Si 2p3/2-Si 2p1/2) (with a fixed area ratio equal to 2:1and doublet separation of 0.6 eV) for each of the 5 XPS signatures for Si^0(Si), Si^{4+}(SiO_2), Si^{1+}(Si_2O), Si^{2+}(SiO), andSi^{3+}(Si_2O_3) [43–46], indicating the absence of +2 Si oxidation state. The SiO_2peak was relatively broad compared to the Si peak due to phonon broadening [47]. Also, as HF concentration increases, the SiO_2peak increases while the Si peak intensity decreases (Figures 6 (a)–6 (d)).

This is expected based on the qualitative etching model, where Ag^+ penetrates the nanowire to oxidize and dissolve Si materials forming mesopores, and tapered nanowires. The formation of mesopores further increases the surface to volume ratio of the nanowires, thereby accelerating the Si oxidation and dissolution rates.

3.4. Optical Properties

The refractive index of the PSiNW samples was measured at 1550 nm by m-lines technique. It was equal to 2.18 ± 0.2 whatever the used sample. By knowing this value and by using a classical model of effective media approximation, the Bruggeman effective medium approximation [48, 49], the porosity of layer was deduced. For this purpose, a system composed of void pore embedded in a Si matrix was considered and the porosity was about of 48 % ± 2. The evolution of the refractive index of sample was then calculated as a function of the wavelength (Figure 8). Then the experimental spectra of PSiNW samples were measured for different thicknesses (10 μm, 13 μm, 28 μm, and 31 μm) as shown in Figure 9 as well as in table VI. These spectra reveal the low reflectance of PSiNW layers and permit to use these layers as anti-reflection coating of solar cells.

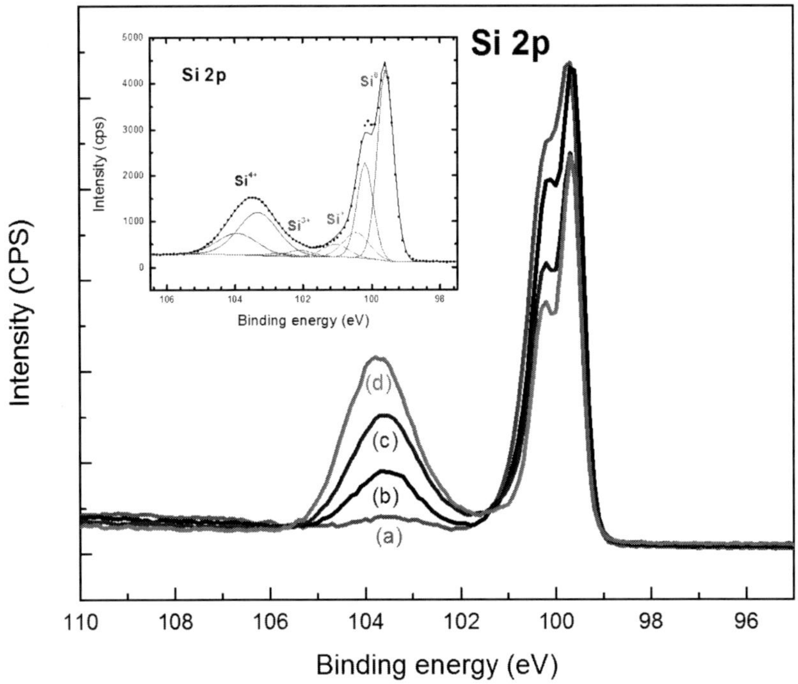

Figure 7. XPS spectra of PSiNWs as a function of HF concentration etched for 45 min: (a) 1.8 M, (b) 2.8 M, (c) 4.8 M, and (d) 5.8 M. Inset: the XPS spectra showing the 5 sets of doublet-fitted signatures of Si^0 (Si), Si^{4+} (SiO_2), Si^{1+} (Si_2O), Si^{2+} (SiO), and Si^{3+} (Si_2O_3) for the PSiNWs sample etched using 4.8M concentration of HF for 45 min.

Table 6. Experimental parameters used in the simulation

Parameters				
Refractive index @ 1550nm	2.18 ± 0.2			
NWs length (μm)	10	13	28	31
Porosity (%)	48 ± 2			
Optical reflection (%)	~1 < R < ~8			

The generic structure under investigation is schematically depicted in Figure 10. The PSiNWs were simulated by Si cones [50]. The z-axis is perpendicular to the Si substrate surface. The bottom of the PSiNW is at $z = 0$ and the top at $z = h$. The radius of the PSiNW depends on the z-position by the following equation $\rho = \frac{h-z}{h} R$ where R is the radius of the cone circle and h the thickness of layer. The PSiNW is divided into n horizontal layers with the same thickness denoted d equal to 1 nm. In the simulation model, each PSiNW is supposed to be located in a rectangular of edge equal to a or in a circle of diameter equal to a. According to the porosity value near of 50%, a is chosen as $2R$. This permits to simulate the reflectance for two different volume shape and to take the complex shape of PSiNWs into account.

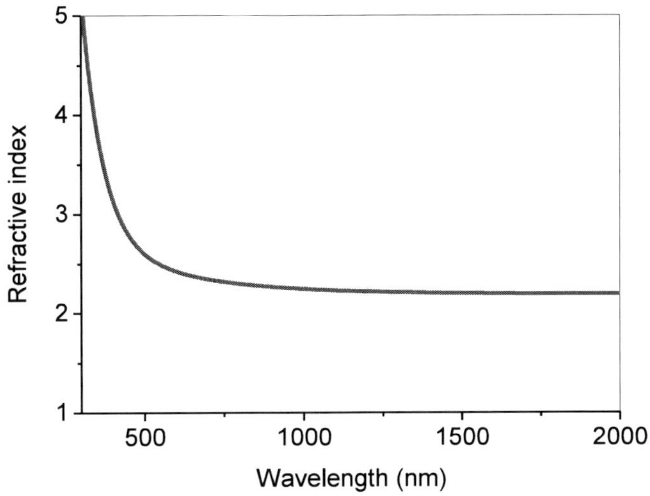

Figure 8. Evolution of refractive index of SiNW layer as a function of the wavelength. The refractive index measured by m-lines at 1550 nm was equal to 2.18 ±0.2 corresponding to a porosity of 48 % by using Bruggeman model.

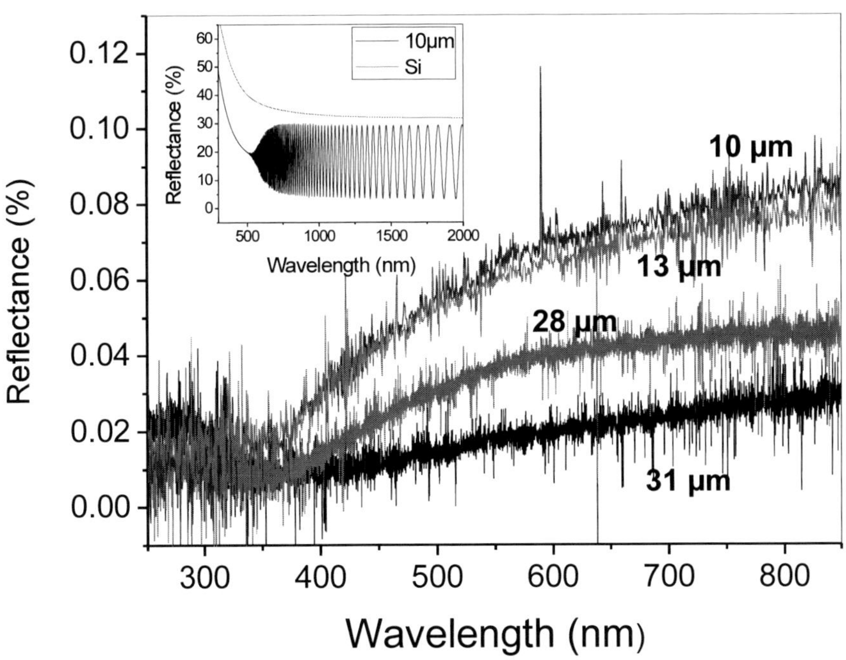

Figure 9. Experimental spectra of PSiNW samples of different thickness (10 µm, 13 µm, 28 µm, and 31 µm). Inset: theoretical reflectance spectra of porous silicon layer (thickness equal to 10 µm) on Si substrate and Si substrate.

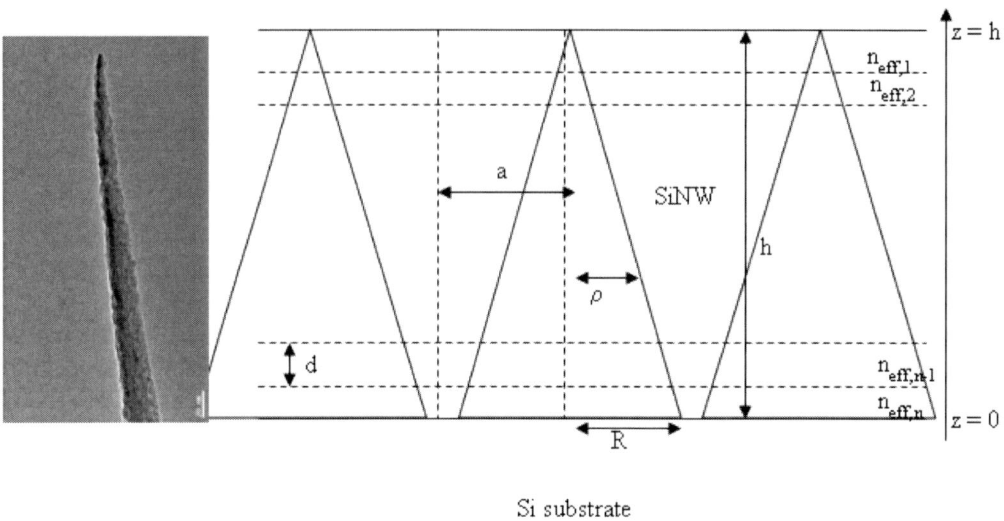

Figure 10. Schematic representation of SiNWs and effective mulitlayer structure with cone shape and TEM image showing the cone shape.

Transfert Matrix Formalism (TMF) was applied to calculate the reflectance of electromagnic wave of through the multilayer structure simulating the PSiNWs layer. In each zone, the wave-function is completely determined by two scalar components and can therefore be represented as a bi-dimensional (2D) column vector. The linearity of the Electromagnetism equations makes possible to relate the components between any two given abscissas z under a (2x2) matrix form. The transfer matrix of the whole structure is readily expressed as the matrix product:

$$[T]=\left[I_{air,n_{eff},1}\right]\left(\prod_{i=1}^{n-1}\left[M_{n_{eff},i}\right]\left[I_{n_{eff},i;n_{eff},i+1}\right]\right)\left[M_{n_{eff},n}\right]\left[I_{n_{eff},n;Si}\right] \quad (1)$$

$$[T]=\begin{bmatrix} T_{11} & T_{12} \\ T_{21} & T_{22} \end{bmatrix}. \quad (2)$$

Where the notation (n_{eff},i) corresponds to the i horizontal layer with the n_{eff} refractive index. $\left[I_{i;i+1}\right]$ is the matrix corresponding to thc waves propagation across the interface between i and $i+1$media and $\left[M_i\right]$ is the propagation matrix in the i medium according to Figure 9.

Reflectance is given by the above equation.

$$R_{PSiNW}=\left|\frac{T_{21}}{T_{11}}\right|^2. \quad (3)$$

With this TMF, the Bruggeman model was used to calculate the refractive index of PSiNWs layer. From the Bruggeman model [48], knowing the volume fraction of Si and void, the refractive index of PSiNWs layer was deduced. This model takes into account the two media of the PSiNWs material according to the following equation:

$$f_{Si}\frac{\varepsilon_{Si}-\varepsilon}{\varepsilon_{Si}+2\varepsilon}+f_{air}\frac{\varepsilon_{air}-\varepsilon}{\varepsilon_{air}+2\varepsilon}=0. \qquad (4)$$

where ε_{Si}, ε_{air} and ε are the dielectric constants of silicon, air and PSiNWs layer respectively; f_{Si} and f_{air} are the weighting coefficients and $f_{Si}+f_{air}=1$ with $\varepsilon_i = n_i^2$ where ni is the complex refractive index of Si or air.

The refractive index of bulk silicon is used in the Bruggeman model without considering the confinement effects of silicon nanostructures on the refractive index [51]. The aim of this modeling is to show the principal trends of reflectance evolution as a function of the SiNW length and the chosen base between circular and rectangular to explain experimental values of reflectance. Lastly, $f_{Si,j}=\frac{\pi\rho^2}{a^2}$ or $f_{Si,j}=\frac{\pi\rho^2}{\pi(a/2)^2}$ respectively if the SiNW is located in a rectangular or a circle and $f_{air,j}=1-f_{Si,j}$ according to Figure 10.

Firstly, the reflectance spectrum is calculated from TMF for a porous silicon layer with 50% porosity and a thickness equal to 10 μm on Si substrate layer in order to compare the reflectance spectra between PSi layer and PSiNW layer with the same porosity. Classical interfering fringes of the PSi layer and reflectance spectrum of Si substrate are observed in inset of Figure 9 and the reflectance value is inferior to that of Si substrate. The comparison between the spectra of PSi and PSiNW layers shows the low obtained reflectance of PSiNW to use this layer as anti-reflection coating of solar cells.

Figures11(a)-(b) and 12(a)-(b) show theoretical reflectance spectra of PSiNWs on silicon substrate for different layer thicknesses (0.5, 1, 5 and 10 μm) by considering cone shape according to Figure 10 in the case where the PSiNW is located in a circle of radius equal to R or in a rectangle of edge equal to 2R. The reflectance of PSiNWs is strongly lower than that of porous silicon layer and the interfering fringes have low amplitudes notably on the case where the PSiNW is located in a circle. The constructive interference disappears when the thickness increases. Whatever the considered base, the reflectance decreases when the thickness of Si increases. The rectangular base permits to take into the porosity between the PSiNW account. The circular base permits to enhance artificially the porosity to take into the porosity between the SiNW and the porosity of the SiNW account. Figure11(c) and 12(c) show the evolution of the porosity as a function of thickness by using cone shape according to different thickness of layers. The porosity drops with the increasing in layer thickness because of the shape of SiNW. Indeed, the width of PSiNW increases with the thickness so the porosity decreases. The porosity is more important in the case of the circular base. This permits to better simulate the reflectance spectrum of PSiNW layer by taking the porosity between the nanowires and the porosity of each nanowire into account. The fringes also disappear and the reflectance drops brutally for the low wavelengths due to the strong absorption of the silicon material in the simulations as we can also see in Figure 9 in the case of porous silicon layer.

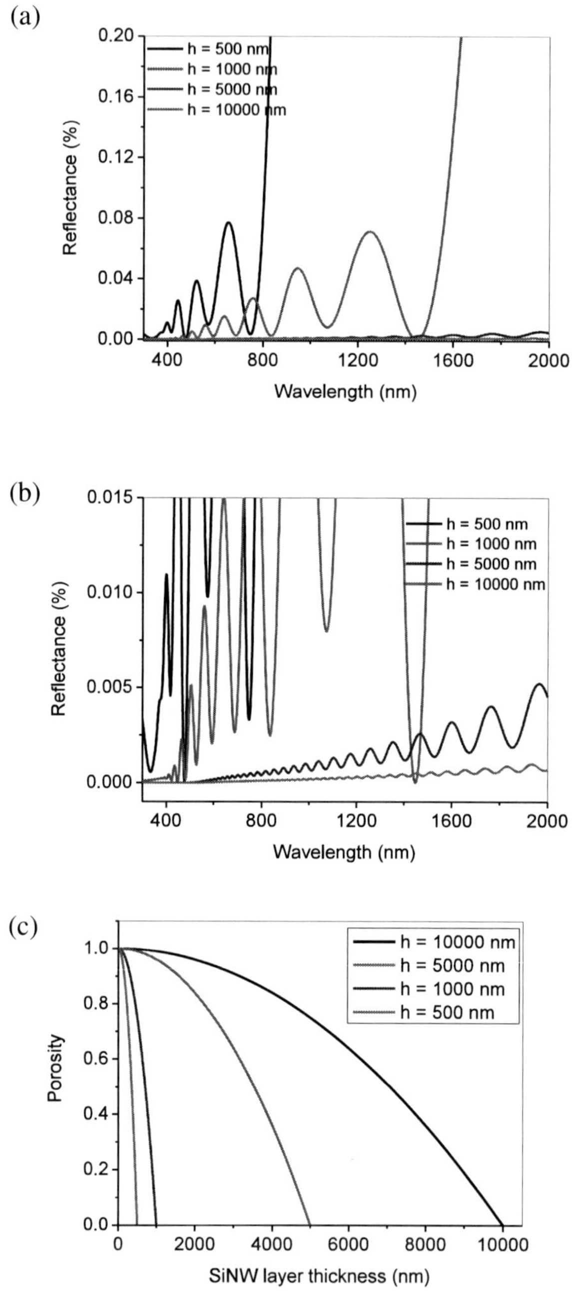

Figure 11. (a) Theoretical reflectance spectra of SiNWs on silicon substrate for different layer thicknesses (0.5, 1, 5 and 10 μm) by considering cone shape and circular base, (b) Zoom of theoretical reflectance spectra of SiNWs, (c) Evolution of porosity as a function of thickness by using cone shape and circular base.

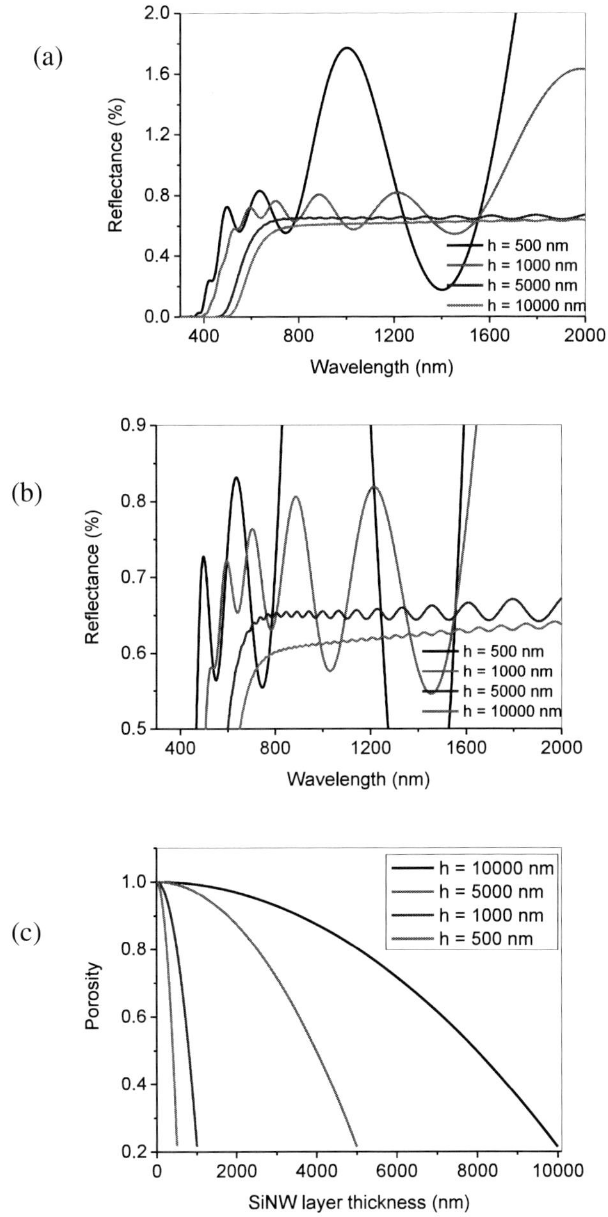

Figure 12. (a) Theoretical reflectance spectra of SiNWs on silicon substrate for different layer thicknesses (0.5, 1, 5 and 10 μm) by considering cone shape and rectangular base, (b) Zoom of theoretical reflectance spectra of SiNWs, (c) Evolution of porosity as a function of thickness by using cone shape and rectangular base.

Figure 13 represents experimental and theoretical reflectance spectra of PSiNWs on silicon substrate corresponding to different samples with their thickness by considering cone shape. The interferences are clearly observed in the calculated reflectance for the low lengths of PSiNWs whereas the experimental reflectance shows no interference. However, the same trends are observed on experimental and theoretical spectra: the experimental value of reflectance is between those obtained by the two models. The reflectance decreases strongly for the low wavelength and increases slightly towards the higher wavelengths. According to the used model, the porosity is majored or minored with a circle base or rectangular base respectively that is why the reflectance values are different.

In the Bruggeman effective medium approximation model, we do not take into account the fraction of SiO_2 on the nanowires surface and the congregate and bundles of nanowires. For these reasons we have a small difference between the theoretical and experimental spectra. So the experimental reflectance appears as the pondered sum between the two theoretical reflectance spectra due to the PSiNW complex shape.

In summary, high density and high length PSiNWs exhibiting extremely low reflectance over a wavelength range from 400 to 850 nm were elaborated. The average reflectance varies from 0.04 to 0.2 % for this wavelength range.

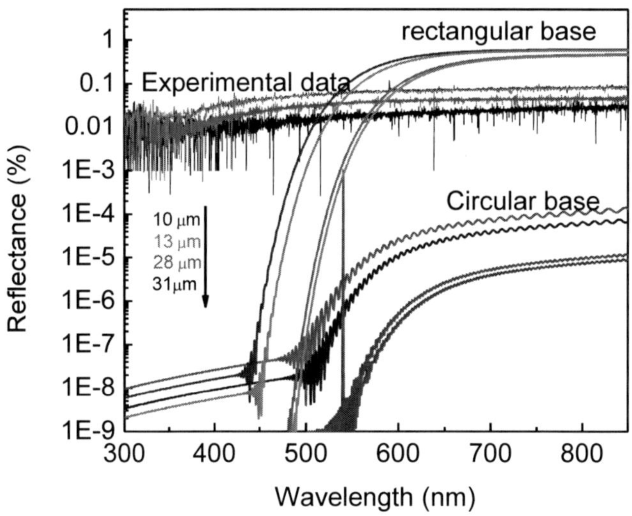

Figure 13. Experimental reflectance spectra and theoretical reflectance spectra using the conic model with circular base and rectangular base for the different thicknesses of samples (10 μm, 13 μm, 28 μm, and 31 μm).

3.5. Effect of HF Concentration on PL

Optical properties of the PSiNWs samples have been investigated using PL spectroscopy. It is known that the mesoporous silicon nanostructure, gives rise to strong visible emission [52]. Figure 14 (a) shows the PL spectra of the samples etched with HF concentrations of 1.8, 2.8, 4.8, and 5.8Mwith a constant H_2O_2concentration of 0.5 M. It is observed that the asymmetric PL spectra have two dominant peaks around 700 and 760 nm. The sample etched with 4.8M HF concentration has an additional hump at 670 nm. The highest PL peak at 760 nm is related to the emission from Si–O bond [52,53].

In our case, we observe in Figure 14 (a) that the increase in integrated intensities of the broadband PL emissions correlates well with the increase in pore densities for samples etched using 1.8–4.8M HF concentrations. With a HF concentration of 5.8 M, a reduction in PL integrated intensity is again consistent with a reduction in the pore-density of PSiNW.

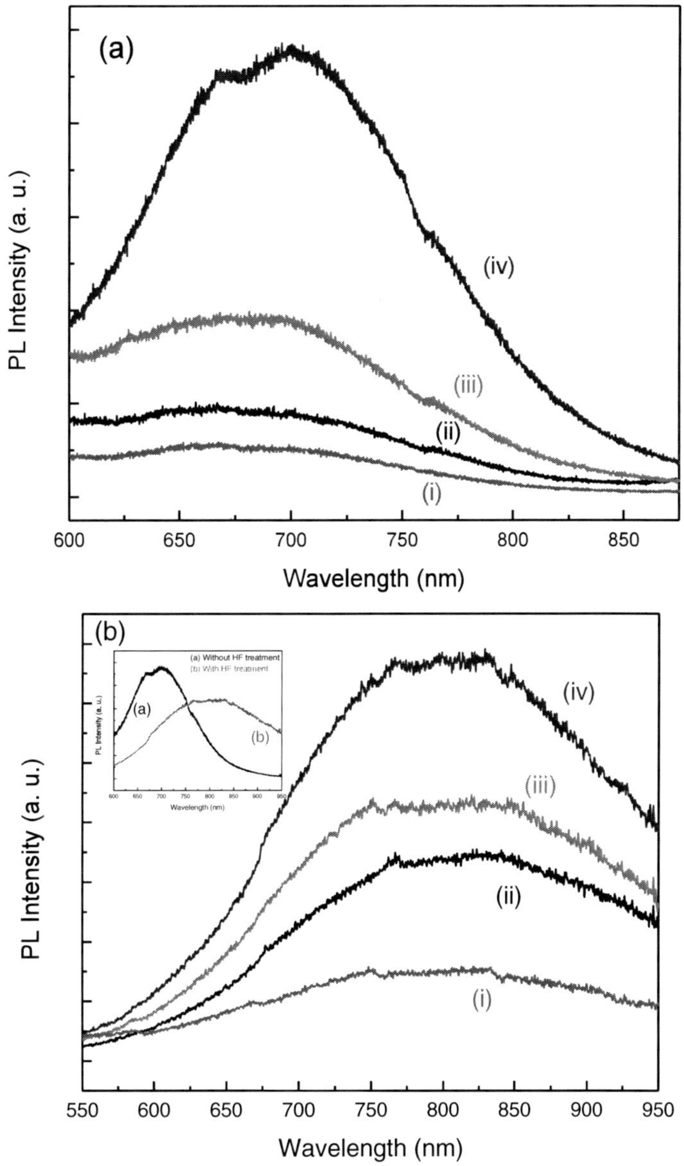

Figure 14. PL spectra of PSiNWs as a function of HF concentration etched for 45 min (a): (i) 1.8 M, (ii) 2.8 M, (iii) 4.8 M, and (iv) 5.8 M. (b) after HNO3 treatment followed by HF (5%).

After HNO_3 and HF treatments for 5min to remove the oxide, all samples show similar PL peak wavelength at800 nm (inset of Fig 14 (b)). However, without samples with HF treatment show a significant blue-shift with PL wavelength peak at 700 nm (Figure 14 (a)). This blue-shift and increase of PL intensity have been attributed to the presence of defect states in Si-Ox that has been confirmed by XPS.

This has been correlated to the red shift of PL spectrum that suggests the interfacial between Si and surface oxide, as well as the quantum confinement effect, plays a critical role in the light emission process (i.e., the presence of nanocrystallites with size <5 nm has been confirmed by HRTEM).

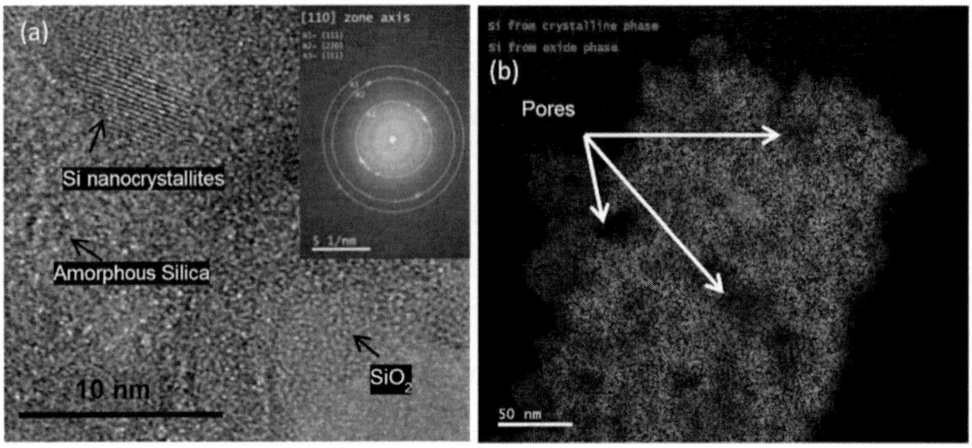

Figure 15. (a) HRTEM and (b) EFTEM analysis of the top section of a PSiNW etched using 4.8M concentration of HF.

Several mechanisms have been proposed to explain the origin of strong PL emission for the indirect band gap silicon material. These include: (i) the quantum confinement effects of free excitons within the Si-nanocrystallites, and (ii) SiOx/Si interface defects and/or defect states in the surface oxide, related to Si-O bond [29, 52, 54, 55]. Comparing the PL spectra before and after HF treatment, we believe that existence of oxide plays a significant role in changing the emission property of the nanowires.

The quantum confinement effect dictates that the characteristic size of the Si-nanocrystallites should be less than the Bohr radius of the free exciton of bulk Si, which is around 5 nm [56]. From the TEM analysis, Si-nanocrystallites of<5 nm has been observed (Figure 15 (a)). The inset of Figure 15 (a)shows the calculated fast-Fourier transform (FFT) of PSiNWs prepared using 4.8M HF concentration. The FFT revealed the spatial frequencies corresponding to {111}, {220}, and {311} d-spacing of Si crystal structure. The presence of these d-spacing implies that the PSiNW was oriented along the <110> direction. This was expected as the (001) Si wafer was used to fabricate the NWs and etching was vertical with respect to the substrate surface. Also, the energy filtered TEM analysis of the top region of the PSiNW, shown in Figure 15 (b), confirms the existence of Si nanocrystallites and silica amorphous structures indicating the possible strong PL emission from the highly quantum-confined Si nanocrytallite structures. This prediction correlates positively with our previous discussion on the effect of pore-density on PL enhancement. The above argument further confirms our observation that samples etched with HF concentration of5.8M (i.e., with low pore and Si-nanocrystallites densities) results in a decrease in PL intensity. The evolution of PLpeak intensity with HF concentration is hence consistent with the physical changes in the PSiNWs developed in our etching model.

However, SiO_2alone should not give rise to the PL emission. Rather, defect states would contribute to PL emission. Indeed, we observed $Si^{1+}(Si_2O)$ and$Si^{3+}(Si_2O_3)$ XPS signatures in the inset of Figure 7, which may originate from dangling bonds and volumetric stress that distorted the PSiNWs forming localized defect states.

Therefore, the emissions from surface/defect states in oxide may contribute to the red PL emission, in agreement with the results obtained in Refs. 29 and 57, in addition to the contribution from quantum confinement effect.

4. APPLICATIONS IN PHOTOVOLTAIC: ABSORBANCE AND EFFICIENCY

To study the theoretically spectra, the generic structure under investigation is schematically depicted in Figure 10. The transfer matrix formalism (TMF) was also used to calculate the reflectance, transmittance and absorbance of PSiNWs arrays. The absorbance was given by $A(\omega)=1-R(\omega)-T(\omega)$.

Figures 16 compare the reflectance, transmittance, and absorbance of PSiNWs by using the two models with rectangular or circular base. Three lengths were selected to show the thickness dependent absorption. The spectra of silicon thin film with the same thickness are also plotted in the same figures as references. As the thickness increases, the PSiNW absorbance increases as for the bulk silicon thin films. However, when wavelength decreases, absorbance of PSiNWs rises and reaches a plateau region shared by all lengths close to 100%. At these wavelengths, the absorbance for nanowires whatever the considered thickness is much higher than that of the Si bulk. The width of plateau region and the absorbance value increase as the PSiNWs thickness increases. We note that these absorbance values are close to 100 % and are higher than those obtained from references [58, 59] for silicon nanowires. While in the higher wavelengths, the absorption in the film is more efficient. To understand the trends of absorption in PSiNWs, reflectance and transmittance are plotted for different lengths of PSiNWs and for Si thin film for the same thickness. The reflectance is much lower than those obtained of Si thin film throughout the entire spectral region. When the wavelength decreases, the reflectance tends toward a low value near to 0 corresponding to the plateau observed for the absorbance spectrum and for each film thickness. The reflectance values are lower than those measured on silicon nanowires due to the shape of nanowires [32, 58] and the porosity of the wires. We notice that for higher wavelength, interference effects exist as observed in Figure 13because the extinction coefficient of silicon is small. Furthermore, it is the same trend for transmittance spectra by zooming at high wavelengths. Meanwhile, PSiNWs transmittance is higher than Si thin film in the considered wavelength range and approaches to zero in the low wavelengths. Therefore, in low wavelength range, the very low reflectance and the zero transmittance of PSiNWs lead to higher absorbance compared to Si thin film and also SiNWs, near to 100% at these wavelengths. Lastly, the transmittance and reflectance values decrease as the thickness increases.

To evaluate the overall absorption performance of PSiNWs, the ultimate efficiency using the Air Mass 1.5 direct normal and circumsolar spectrum [60] was calculated. It is given by the following equation [61]; where λ is wavelength, $I(\lambda)$ is the spectral irradiance (power density) of the ASTM AM1.5 direct normal and circumsolar spectrum, $A(\lambda)$ is the absorbance and λ_g=1127 nm is the wavelength corresponding to the 1.1 eV band gap of bulk silicon.

Figure 17 (a) shows the absorbance of *P*SiNWs as a function of incidence angle. The absorbance increases with the incidence angle due to the increasing to optical path. In Figure 17 (b), the efficiency is plotted versus incidence angle for *P*SiNWs with 30 μm thickness compared to Si thin film with the same thickness. The efficiency of *P*SiNWs tends to be higher than that obtained by Si thin film. Furthermore, the efficiency can be improved by using light trapping to increase optical path[58]. We can notice that the efficiency values are higher than those obtained by SiNWs arrays[58].

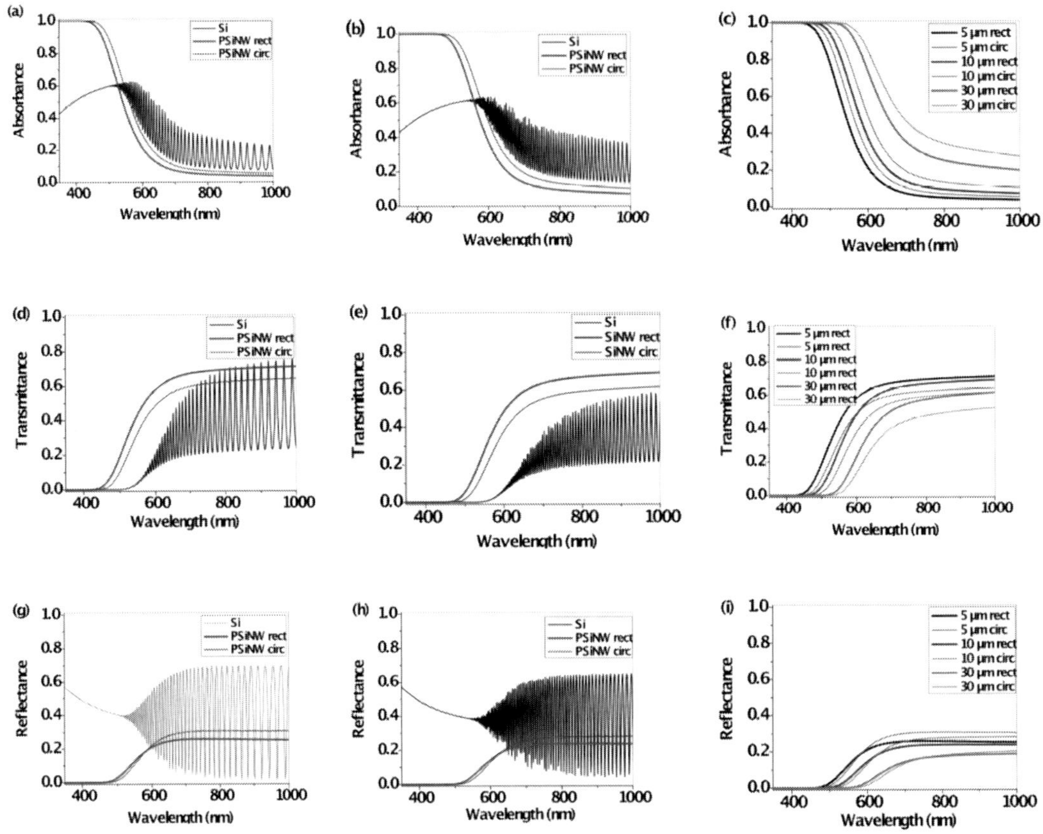

Figure 16. (a)-(b), (d)-(e), and (g)-(h) are the absorbance, reflectance and transmittance spectra of PSiNW for 5 and 10 μm respectively by considering circular base and rectangular base. The optical properties of silicon thin film are plotted for comparison. (c), (f) and (i) are the absorbance, reflectance and transmittance spectra of PSiNW for different thicknesses (5, 10 and 31μm) respectively by considering circular base and rectangular base.

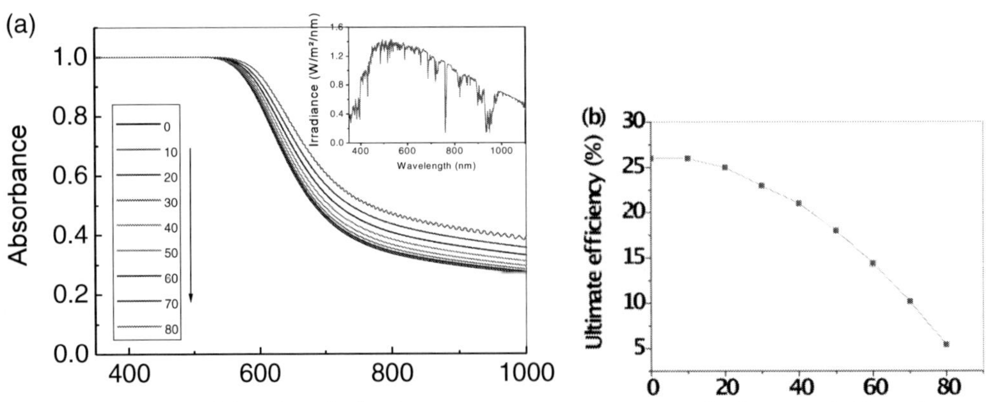

Figure 17. (a) Absorbance spectra of PSiNW with a thickness equal to 30 μm for different incidence angle (in degree). Inset: The ASTM Air Mass 1.5 direct and circumsolar spectrum from ref. 31. (b) Ultimate efficiency as a function of the incidence angle for TE polarization and for a 30-μm thick PSiNW film.

CONCLUSION

In summary, this chapter reports the structural and optical properties of porous silicon nanowires (PSiNWs) fabricated using silver (Ag) ions assisted electroless etching method. A qualitative model is proposed for the formation of PSiNWs in the presence of Ag catalyst as a function of HF concentration and was approved by microscopy observations. The optical properties of this nanostructure based on the reflection measurements were carried out and the effect of length on the reflectivity were studied (R< 0.1%). Models based on cone shape of nanowires located in circular and rectangular bases were used to calculate reflectance employing the Transfert Matrix Formalism (TMF) of the PSiNWs layer. Using TMF, the Bruggeman model was used to calculate the refractive index of PSiNWs layer. The calculated reflectance using circular cone shape fits better the measured reflectance for PSiNWs. The enhancement in PL signals is associated with the formation of PSiNWs, and to quantum confinement of electron in the Si nanocrystallites, observed in transmission electron microscopy (TEM). Furthermore, the visible emission of PSiNWs was discussed in this chapter.

Optical absorbance and transmission were also theoretically presented. The absorbance was superior to that obtained with silicon nanowires and the ultimate efficiency for solar cell application was determined (25% for normal incidence angle). These results could be applied to the potential application in low-cost and high efficiency PSiNWs based solar cells.

REFERENCES

[1] Zhao, X. Y.; Wei, C. M.; Yang, L.; Chou, M. Y. *Phys. Rev. Lett.* 2004, 92,236805.

[2] Cui,Y.; Lieber, C. M. *Science* 2001, 291, 851.

[3] Lieber,C. M.; Wang, Z. L. MRS Bull. 2007, 32, 99.

[4] Najar, A.; Charrier, J.; Ajlani, H.; Lorrain, N.; Haesaert, S.; Oueslati, M.; Haji, L. *Mater. Sci. Eng.* B 2008, 146, 260.

[5] Najar, A.; Lorrain, N.; Ajlani, H., Charrier, J.; Oueslati, M. ; Haji, H. *Appl. Surf. Sci.* 2009, 256, 581.

[6] Najar, A.; Ajlani, H.; Charrier, J.; Lorrain, N.; Haesaert, S.; Oueslati, M.; Haji, L. *Physica* B.2007, 396, 145.

[7] Najar, A.; Charrier, J.; Lorrain, N.; Haji, L.; Oueslati. M. *Appl. Phys. Lett.* 2007, 91, 12, 121120-121120.

[8] Najar, A.; Charrier, J.; Ajlani, H.; Lorrain, N.; Elhouichet, H.; Oueslati, M.; Haji, L. *J. of lumin.* 2006, 121, 2, 245-248.

[9] Najar, A.; Elhouichet, H.; Lorrain, N.; Oueslati, M. *Appl. Surf. Scien.*2006, 252, 16, 5808-5813.

[10] Koyama, H.; Nakagawa, T.; Ozaki, T.; Koshida, N. *Appl. Phys. Lett.* 1994, 65,1656.

[11] Canham, L. T.; Houlton, M. R.; Leong, W. Y.; Pickering, C.; Keen, *J. M. J. Appl. Phys.* 1991, 70, 422.

[12] Huang, Y.; Duan, X.; Lieber, C. M. Small 2005, 1, 142.

[13] Bohr, M. T.; *IEEE Trans. Nanotechnol.* 2002, 1, 56.

[14] Lieber, C. M.; *Solid State Commun.* 1998, 107, 607.

[15] Xia, Y. N.; Yang, P. D.; Sun, Y. G.; Wu, Y.; Mayers, N.; Gates, B.; Yin, Y.; Kim, F.; Yan, H.; *Adv. Mater.* 2003, 15, 353.
[16] Najar, A.; Gerland, M.; Jouiad,M.; *J. of Appl. Phys.* 2012, 111 ,9, 093513-093513.
[17] Slimane, A. B.; Najar, A.; Elafandy, R.; San-Román-Alerigi, D.; Anjum, D.; Ng, T. K.; Ooi, B. S. *Nano. Res. Lett.* 2013, 8, 342-348.
[18] Slimane, A.B.; Najar, A.; Ng, T. K.; OoiB. S. *Photonics Conference (IPC) 2012 IEEE,* 921 - 922
[19] Wagner,R. S.; Ellis, W. C.; *Appl. Phys. Lett.* 1964, 4, 89.
[20] Morales,A. M.; Lieber, C. M.; *Science* 1998, 279, 208.
[21] Zhang, Y. F.; Tang, Y. H.; Wang, N.; Yu, D. P.; Lee, C. S.; Bello, I.; Lee, S. T.; *Appl. Phys. Lett.* 1998, 72, 1835.
[22] Schubert, L.; Werner, P.; Zakharov, N. D.; Gerth, G.; Kolb, F. M.; Long,L.; Geosele, U.; Tan, T. Y. *Appl. Phys. Lett.* 2004, 84, 4968.
[23] Stelzner, T.; Pietsch, M.; Andrea, G.; Falk, F.; Ose, E.; Christiansen,S.; *Nanotechnology* 2008, 19, 295203.
[24] Lu, M.; Li, M. K.; Kong, L. B.; Guo, X. Y.; Li, H. L.; *Chem. Phys. Lett.* 2003, 374, 542.
[25] Au, F. C. K.; Wong, K. W.; Tang, Y. H.; Zhang, Y. F.; Bello, I.; Lee, S. T.; *Appl. Phys. Lett.* 1999, 75, 1700.
[26] Yu, D. P.; Lee, C. S.; Bello, I.; Sun, X. S.; Tang, Y. H.; Zhou, G. W.; Bai, Z. G.; Zhang, Z.; Feng, S. Q. *Solid State Commun.* 1997, 105, 403.
[27] Wang, N.; Tang, Y. H.; Zhang, Y. F.; Lee, C. S.; Bello, I.; Lee,S. T. *Chem. Phys. Lett.* 1999, 299, 237.
[28] Juhasz, R.; Kylmeanen, K.; Galeckas, A.; Linnros, J. *Mater. Sci. Eng.* 2005, 25, 733.
[29] Hochbaum, A. I.; Gargas, D.; Hwang, Y. J.; Yang, P.; *Nano Lett.* 2009, 9, 3550.
[30] Kumar, D.; Srivastava, S. K.; Singh, P. K.; Sood, K. N.; Singh, V. N.; Dilawar, N.; Husain, M.; *J. Nanopart. Res.* 2010, 12, 2267.
[31] Srivastava, S. K.; Kumar, D.; Singh, P. K.; Kar, M.; Kumar, V.; Husain, M. *Sol. Energy Mater. Sol. Cells* 2010, 94, 1506.
[32] Najar, A.; Charrier, J.; Pirasteh, P.; Sougrat, R. *Optics Express* 2012, 20, 16861-16870.
[33] Najar, A.; Slimane, A. B.; Hedhili, M. N.; Anjum, D.; Sougrat, R.; Ng, TK.; Ooi, B.S. *J. of Appl. Phys.* 2012, 112, 033502-033502.
[34] Peng, K. Q.; Yan, Y. J.; Gao, S. P.; Zhu, J.; *Adv. Mater.* 2002, 14, 1164.
[35] Qiu, T.; Wu, X. L.; Mei, Y. F.; Wan, G. J.; Chu, P. K.; Siu, G. G.; *J. Cryst. Growth* 2005, 277, 143.
[36] Zhong, X.; Qu, Y.; Lin, Y. C.; Liao, L.; Duan, X. *ACS Appl. Mater. Interfaces* 2011, 3, 261–270.
[37] Qu, Y. Q.; Liao, L. ; Li, Y. J.; Zhang, H.; Huang, Y.; Duan, X. F.*; Nano Lett.* 2009, 9, 4539.
[38] Qu, Y.; Zhou, H.; Duan, X.; *Nanoscale* 2011, 3, 4060.
[39] Kooij, E. S.; Butter, K.; Kelly, J. *J. Electrochem. Solid State Lett.* 1999, 2,178.
[40] Turner, D. R. *J. Electrochem. Soc.* 1960, 107, 810.
[41] M. L. Zhang, K. P. Peng, X. Fan, J. S. Jie, R. Q. Zhang, S. T. Lee, andN. B. Wong, *J. Phys. Chem.* C 112, 4444 (2008).
[42] M. F. Beaux, I. I. N. J. Bridges, M. DeHart, T. F. Bitterwolf, and D. N. McIlroy, *Appl. Surf. Sci.* 257, 5766 (2011).

[43] Himpsel, F. J.; Meyerson, B. S.; McFeely, F. R.; Morar, J. F.; Taleb-Ibrahimi,A.; Yarmoff, J. A.; in *Photoemission and Absorption Spectroscopyof Solids and Interfaces with Synchrotron Radiation,* edited by M. Campagnaand R. Rosei -NorthHolland, Amsterdam, 1990, 203.
[44] Barbagiovanni, E. G.; Goncharova, L. V.; Simpson, P. J.; *Phys. Rev.* B2011, 83, 035112.
[45] Green, M. L.; Gusev, E. P.; Degraeve, R.; Garfunkel, E. L.; *J. Appl. Phys.* 2001, 90, 2057.
[46] Renault, O.; Marlier, R.; Gely, M.; Salvo, B. D.; Baron, T.; Hansson, M.; Barrett, N. T. *Appl. Phys. Lett.* 2005, 87, 163119.
[47] Hollinger,G.; Himpsel, F. J.; *Appl. Phys. Lett.* 1984, 44, 93.
[48] Aspnes, D. E. *Thin Solid Films* 1982, 89,249–262.
[49] Theiss, W. *Surf. Sci. Reports* 1997, 29, 91–93 and 95–192.
[50] Pei, T-H., Tiyagu, S.; Pei, Z. *Appl. Phys. Lett.* 2011, 99, 153108-1153111.
[51] Diebold, A. C.; Price, *J. Phys. Stat. Sol.* (a) 2008, 205, 896-900.
[52] Cullis, A. G.; Canham, L. T.; Calcott, P. D. J.; *J. Appl. Phys.* 1997, 82, 909.
[53] Lin, L.; Guo, S.; Sun, X.; Feng, J.; Wang, Y. *Nano. Res. Lett.* 2010, 5, 1822.
[54] Kontkiewicz, A. J.; Kontkiewicz, A. M.; Siejka, J.; Sen, S.; Nowak, G.; Hoff, A. M.; Sakthivel, P.; Ahmed, K.; Mukherjee, P.; Witanachchi, S.; Lagowski, *J. Appl. Phys. Lett.* 1994, 65, 1436.
[55] Zhao, Y.; Yang, D.; Sang, W.; Li, D.; Jiang, M. *Mater. Sci. Eng.* B 2005, 116,95.
[56] Bai, Z. G.; Yu, D. P.; Wang, J. J.; Zou, Y. H.; Qian, W.; Fu, J. S.; Feng,S. Q.; Xu, J.; You, L. P. *Mater. Sci. Eng.,* B 2000, 72, 117.
[57] Lin, L. H.; Sun, X. Z.; Tao, R.; Li, Z. C.; Feng, J. Y.; Zhang, Z. J. *J. Appl. Phys.* 2011, 110, 073109.
[58] Hu L.; Chen, G. *Nano Lett.*2007, **7**, 3249.
[59] Pei, T-H. ; Tiyagu, S.; Pei, Z. *Appl. Phys. Lett.*2011, 99, 153108.
[60] ASTM, "Reference Solar Spectral Irradiance: Air Mass 1.5 Spectra," *http://rredc.nreal.gov/solar/spectra/am1.5.*
[61] Shockley, W.; Queisser, H. J. *J. Appl. Phys.*1961, 32, 510.

In: Comprehensive Guide for Mesoporous Materials, Volume 2 ISBN: 978-1-63463-982-8
Editor: Mahmood Aliofkhazraei

Chapter 10

EFFECT OF SYNTHESIS PARAMETERS ON THE MICROSTRUCTURE OF MESOPOROUS TiO_2 FILMS

Jinshu Wang *
School of Materials Science and Engineering,
Beijing University of Technology, Beijing, P. R. China

ABSTRACT

Mesoporous TiO_2 thin films had been prepared through sol-gel combined with evaporation-induced self-assembly (EISA) by using F127 and $Ti(OBu^n)_4$ as templating agent and inorganic precursor respectively. It was found that the calcinations temperature, relative humility and the mass ratio of templating agent to inorganic precursor could vary the TiO_2 mesoporous structure. The thermodynamic study for the sample had been carried out and the critical crystallite size expression of the mesoporous film based on our established four-coordinate channel model was presented for the prediction of the thermal stability of the mesoporous structure at high temperature. Varying the amount of templating agent to inorganic precursor could adjust the pore size and structure of mesoporous TiO_2. When the mass ratio of F127/$Ti(OBu^n)_4$ varied from 0.1 to 0.5, the pore sizes of mesoporous TiO_2 changed from 5.4nm to 9.1nm. The pore size regulation mechanism had been discussed. N, S co-doped TiO_2 exhibited framework with a high porosity. Both pure mesoporous TiO_2 and N,S co-doped TiO_2 exhibited higher photocatalytic activity, especially for the N,S co-doped samples.

1. INTRODUCTION

Porous solids can be classified into three groups based on their pore diameter according to the definition of IUPAC, namely microporous (5–20A°), mesoporous (20–500A°), and

* Jinshu Wang: School of Materials Science and Engineering, Beijing University of Technology, Beijing 100124, P. R. China. E-mail: wangjsh@bjut.edu.cn.

macroporous (>500A°) materials. Due to their large surface area and tailored pore structure, mesoporous transition metal oxides can be used as adsorbents, catalysts, advanced detectors, gas sensors, inorganic membranes, photovoltaic cells, lithium batteries, and so on [1-8]. Since the discovery of periodically organized silica-based mesostructural materials [9], similar templating approaches have been extended to fabricate mesoporous transition metal oxides [10-14]. These mesoporous materials show higher surface areas and much more uniform and controllable pore sizes and morphologies, thus display unique electronic, magnetic, and catalytic properties. Hard - and soft -templating syntheses are the two most widely used methods to prepare this class of porous materials.

Many great progresses have been made in this area, as shown in Table 1. Hard-template assisted route involves in the use of colloidal particles (e.g., polymeric beads, silica balls, anodic alumina membranes). The nanostructures are formed on the inner or outer surfaces of the templates by using sol–gel casting or other particle adsorption methods, and then removal of these templates generates these unique porous materials. Soft-template assisted route involves in the use of ionic organic surfactants or nonionic polymeric surfactants, which self-assemble into a diversity of supermolecular structures including spherical micelles, hexagonal rods, lamellar liquid crystals and other assemblies in solution, driven by non-covalent weak interactions such as hydrogen bonding, van der Waals forces, and electrostatic interaction between the surfactants and the building blocks. These assemblies then are used as soft templates to tune pore structure and size of porous materials.

The evaporation-induced self-assembly (EISA) method was proposed by Brinker et al. [21, 22], which allows for regulating the inorganic condensation rate with the formation of an organized liquid-crystal template, and the application of Pluronic-type amphiphilic triblcok copolymers as templates developed by Stucky's group [23-25] offered excellent opportunities for the construction of well-defined mesoporous materials.

However, there are some main hurdles to be taken into account during the preparation of highly crystalline mesoporous TiO_2 films depending on EISA route [26]: (1) chemical diversity of metal oxides as well as the availability of an appropriate precursor (a synthetic approach to creating a mesoporous metal oxide often strongly depends upon the nature of the metal precursor and targeted metal oxide. (2) The high reactivity of most of titania inorganic precursors toward hydrolysis and condensation.

Table 1. Different parameters of various materials prepared with hard and soft templates

Template		Product	Structure	Pore size (nm)	S_{BET} [a] (m^2/g)	References
Soft	Ionic liquids $[C_4mim]^+BF_4^-$	SiO_2	Worm-like	2.5	801	[15]
	Cationic surfactants (CTAB, DOTAB, OTAB)	TiO_2	Sphere–like	5.7-7.0	90-135	[16]
	F127	C	Hexagonal	5.8	651	[17]
Hard	Silica nanoparticles	C_3N_4	Graphitic	12	86-440	[18]
	KIT-6	$LiCoO_2$	mesoporous	3.67	92	[19]
	Carbon aerogel	Al_2O_3	granules	10.4-16.6	365	[20]

[a]Brunauer–Emmet–Teller (BET) surface area.

One possible strategy of controlling the instantaneous hydrolysis-condensation of the precursor is to perform the cooperative assembly process in non-aqueous media [27, 28]. (3) Adding suitable chelating agent into the reaction solution [29-32] or carrying out the reaction under strong acidic conditions ($pH < 1$) [33-36] can also control the reactivity of titania precursor.

EISA offers excellent opportunities for the reproducible synthesis of mesoporous metal oxide films. During the preparation, selection of appropriate starting materials, especially the Ti source and block copolymer, is vital in the successful preparation of highly organized mesoporous structure of TiO_2 films. The suitable Ti source is quite limited. Available titanium alkoxides include titanium ethoxide ($Ti(OEt)_4$), titanium isopropoxide ($Ti(O\text{-}iPr)_4$), and titanium butoxide ($Ti(OBu)_4$), listed in an order of increasing reactivity in hydrolysis. Compared with other Ti sources such as TiF_4, they have predominant performances in controlling the homogeneity of initial solution and final size of TiO_2 crystallites on the nanoscale [37].

Block copolymers are molecules with a polar hydrophilic head which attracts to water and a hydrophobic hydrocarbon chain which attracts to oil. In water, the hydrocarbon chains tend to self-associate to minimize contact with the water molecules, leading to the formation of various aggregates. Organized soft-template systems such as micelles, reverse micelles, vesicles, liquid droplets, etc. have been widely exploited to provide effective routes to grow porous structures [38, 39]. For the suitable block copolymers, commercially available amphiphilic poly(ethylene oxide)–block–poly(propylene oxide)–block–poly(ethylene oxide) (PEO–PPO–PEO) triblock copolymers, typically Pluronic P123 ($PEO_{20}PPO_70PEO_{20}$, EO=ethylene oxide, PO=propylene oxide, AMW = 5800) and F127 (($PEO_{106}PPO_{70}PEO_{106}$, AMV = 12,600), consisting of a central hydrophobic PPO chain and two hydrophilic PEO tails, have been widely used as the structure-directing agents for the construction of highly organized mesoporous structure for TiO_2, where P and F stand for the physical form of paste and flake, respectively [40]. Rankin et al. combined P123 surfactant templating and poly (propylene glycol) (PPG) phase separation to generate hierarchically porous titania thin films [41]. However, the calcinations at the temperature over 450°C would destroy the mesostructure prepared using P123 as the surfactant templating.

In this parper, we introduced our recent work on the synthesis of homogeneous msoporous TiO_2 thin films with highly thermal stability (>500°C) by carefully adjusting the hydrolysis/condensation conditions, utilizing the F127 as the structure-directing agent and tetrabutyle titanate as the inorganic precursor [42-46]. The effects of relative humility, mass ratio of F127, calcinations temperatures, and introduction of foreign ions such as sulfur and nitrogen on the microstructure of the thin films had been studied. The photocatalytic activity of the mesoporous TiO_2 was also investigated.

2. Synthesis and Thermal Stability of Mesoporous TiO_2

2.1. Synthesis

Triblock copolymer $HO(CH_2CH_2O)_{106}(CH_2CH(CH_3)O)_{70}(CH_2CH_2O)_{106}H$ ($EO_{106}PO_{70}$ EO_{106}, Pluronic F127, MW=12600) was purchased from BASF Corp. and used as received without further purification. F127 and $Ti(OBu^n)_4$ were used as the templating agent and inorganic precursor, respectively. In a typical synthesis process, F127 was dissolved in 50 ml absolute ethanol (EtOH) and concentrated HCl solution at room temperature under magnetic stirring for 30 min, obtaining the mixed F127/EtOH/HCl solution. Separately, 50 ml of absolute ethanol was added to the solution of $Ti(OBu^n)_4$ and acetylacetone (AcAc) under vigorous stirring for 20 min, obtaining the mixed $Ti(OBu^n)_4$/EtOH/AcAc solution. Then, the mixed F127/EtOH/HCl solution was added dropwise to the mixed $Ti(OBu^n)_4$/EtOH/AcAc solution under vigorous stirring for 30 min, and the obtained sol was stirred for another 24 h at room temperature at different relative humility. The F127/$Ti(OBu^n)_4$ /HCl/EtOH/AcAc molar ratio was 0.0008/0.0726/0.5036/19.928/0.1476.

In order to study the effect of the amount of the block copolymers on the pore size of TiO_2 mesostructure, the mass ratio of F127 to TBT, EtOH and AcAc was designed as Xg: 5g: 100g: 2.5g, where X=0.5g,1.0g, 1.5g, 2.0g, 2.5g, which were assigned as S-1, S-2, S-3, S-4 and S-5, respectively. Uniform and transparent thin films were produced by spin-coating on the silicon wafer or indium tin oxide (ITO) coated glass substrates at the constant spinning rate of 2500 rpm. The deposited films were aged at 28°C and different relative humidity (RH) for 24 h. The as-prepared thin films were subsequently calcined at a heating rate of $1°C\cdot min^{-1}$ in oxygen atmosphere from room temperature to 350, 450, 500, 600, or 700°C and held at this temperature for 2.5 h. The corresponding powdered mesoporous TiO_2 sample was prepared by scrapping the films from the substrate.

2.2. Effect of Relative Humility on the Structure of Porous TiO_2 Films

Figure 1 shows the TiO_2 films aged at 28°C with different RH followed by calcinations at 500°C. As shown in Figure 1a, the film aged at 20%RH was composed of wormlike particles and some cracks existed in the film.

The formation of particles was caused by the insufficient separation of hydrophilic phase and hydrophobic phase due to the rapid evaporation of solvent [47].

Furthermore, the rapid evaporation of the solvents under low RH led to the uneven shrinkage of the films with the result of formation of cracks shown in Figure 1a. The films aged at 45%RH had uniform mesoporous structure (Figure 1b).

Increasing the RH to 60%, although mesopores existed in the film, the uniformity of the film still need to be improved, as shown in Figure 1c. The high RH such as 70% was not favorable for the formation of mesoporous films, and the film consisted of particles. Ti-oxo species combined with hydrophilic PEO of micelles through weak coordination bonds.

Under the environment with high RH, Ti-oxo species formed bonds with water molecules, leading to the breakup of coordination bonds formed between Ti-oxo species and PEO, therefore, only TiO_2 nanoparticles were formed. Therefore, in the following work, we synthesized the films at 40-50% RH.

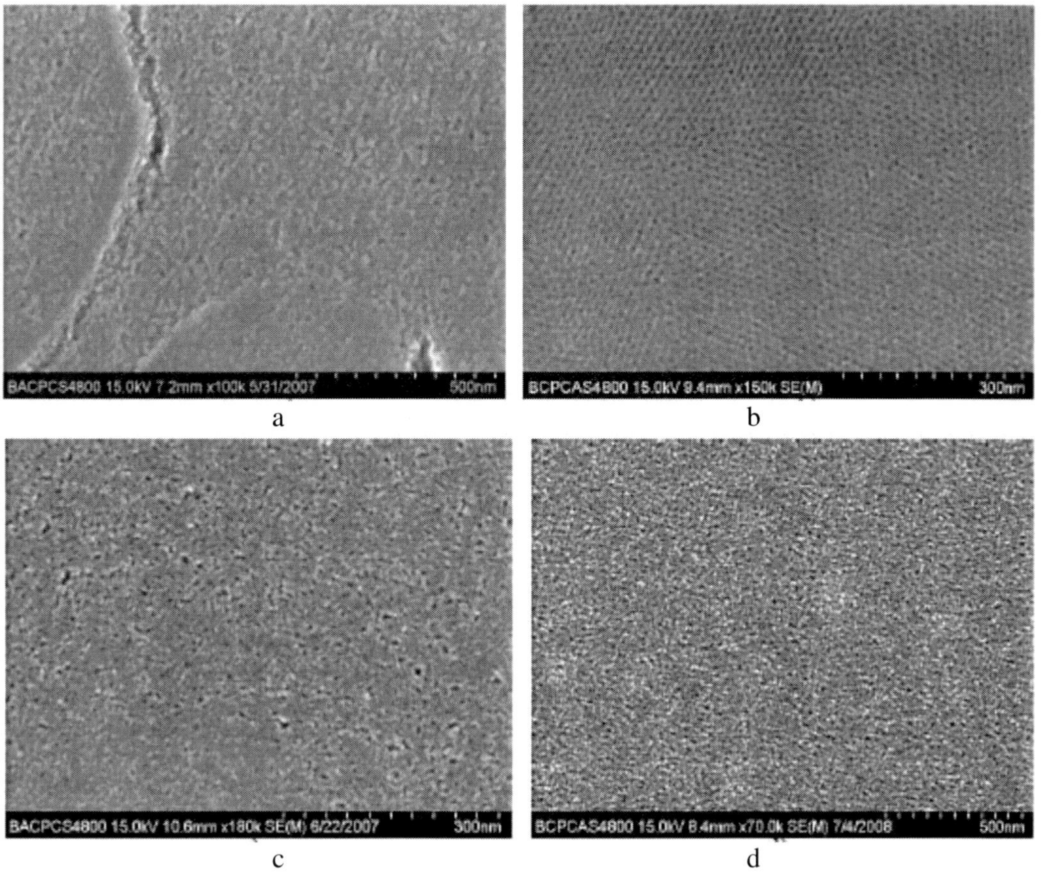

Figure 1. Films aged at different relative humidity (RH) following calcined at 500°C. a) 20%RH, b) 45%RH, c) 60%RH and d) 70%RH.

2.3. Effect of the Amount of the Block Copolymers on the Structure of Porous TiO_2 Films

The XRD patterns of mesoporous TiO_2 samples synthesized using F127 as the templating agent with different mass ratios followed by calcinations at 500°C are shown in Figure 2.

The small angel XRD results indicated that the thin films had highly organized mesoporous framework. Table 2 shows the pore size and crystallite size of the mesoporous film prepared with different mass ratio.

As presented in Table 2, the crystallite size decreased and the pore size increased with the increase of mass ratio of F127/Ti(OBu^n)$_4$, i.e., when the F127/Ti(OBu^n)$_4$ mass ratio increased from 0.1 to 0.5, the crystallite size decreased from 14.3nm to 8.9nm and pore size increased from 5.4nm to 9.1nm, respectively, indicating that the wall thickness decreased.

The above results show that the pore size of the samples could be regulated by the mass ratio of F127/Ti(OBu^n)$_4$. The pore size regulation mechanism could be explained as follows. Taking triblock-copolymer Pluronic F127 as the templating agent was the premise for the formation of the mesoporous TiO_2 with different pore size.

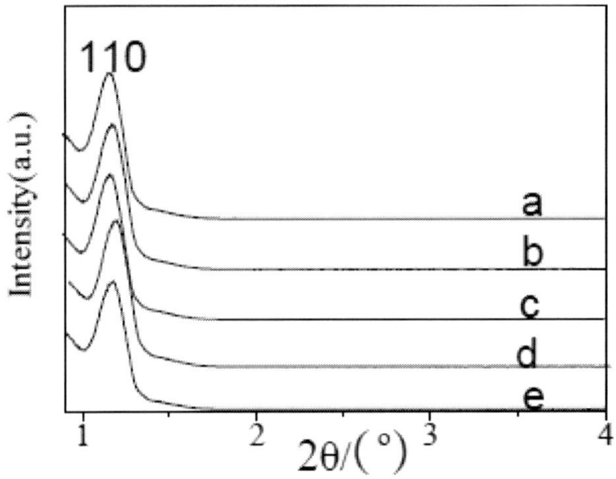

Figure 2. The small angel XRD patterns of the synthesized thin films. (a) S-1; (b) S-2; (c) S-3; (d) S-4; (e) S-5.

Table 2. The textural parameters of the synthesized mesoporous TiO_2 materials prepared with different mass ratios of F127/Ti(OBu^n)$_4$

Sample	F127/ Ti(OBu^n)$_4$	$D_p{}^a$/nm	D^b/nm
S-1	0.1	5.4	14.3
S-2	0.2	6.2	13.6
S-3	0.3	7.1	11.2
S-4	0.4	7.4	10.2
S-5	0.5	9.1	8.9

a. BJH average pore size, *b*. Average crystallite size.

F127 is a unique surfactant owing to its high molecular weight with long hydrophilic PEO and hydrophobic PPO segments, which offers the possibilities of the formation of hydrophobic core (PPO) of the micelles and hydrophilic corona with a certain thickness. According to the "three-phase model" which was presented for silica mesoporous structure [48], Ti-oxo species might react with micelles to form a three phase structure. Figure 3 shows the schematic diagram of the formation of the synthesized mesoporous TiO_2 materials with different pore size. As shown in Figure 3, hydrophobic phase I (PPO) was located in the center, and the PEO chains II existed in two parts, namely, embedding in the Ti-oxo matrix and situating between the hydrophobic core (phase I) and the Ti-oxo (this part of PEO is called as phase II). In other words, a part of the PEO chains might form a separate "phase' between the hydrophobic block and the Ti-oxo species. This part of PEO also contributed to the mesoporosity. In other words, since the templating agent could form micelles with a fixed thickness of PEO under a certain condition, the change of the organic- inorganic hybrid material (Phase III) would change the thickness of the rest of PEO (II).

Therefore, the thickness of the organic- inorganic hybrid material could be changed by adjusting the mass ratio of templating agent to inorganic precursor, resulting in the change of pore size. It is known that the pore size is, to a large extent, dependent on the effective volume of the hydrophobic core (PPO) of the micelles.

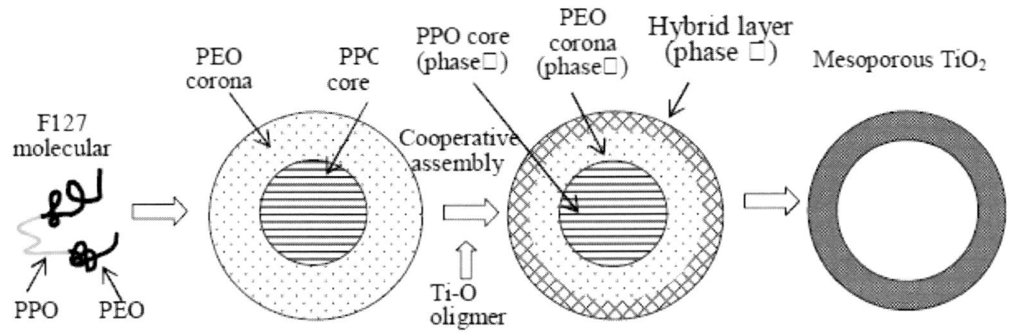

Figure 3. Schematic diagram of the formation of the synthesized mesoporous TiO_2 materials with different pore size.

As shown in Table 2, when the concentration of templating agent F127 was low, more number of oligomeric Ti-oxo species could assembly with PEO chains to form inorganic-organic hybrid (phase III), therefore, only a small number of PEO chains (phase II) contributed to the microposity with the result of small pores in the framework. Since the concentration of precursor was fixed, the number of Ti-oxo species which could assembly with ethylene oxide decreased with the increase of the concentration of F127. Therefore, more number of PEO chains would contribute to the microposity, resulting in the formation of large pores and the decrease of the wall thickness of mesoporous TiO_2. Since the pore size of the mesoporous network was primarily dependent on the hydrophobic core diameter, the *n*-Butanol released by $Ti(OBu^n)_4$ inclined to penetrate the interface between the hydrophilic and hydrophobic domains of micelles, and located at the hydrophobic core due to the nonpolar nature, therefore, *n*-Butanol in situ also played a pronounced role in the construction of large pore mesostructured TiO_2 framework.

2.4. Effect of Calcinations Temperature on the Porous Structure of TiO_2 Films

The mesostructural evolution of the synthesized TiO_2 thin films with temperature is shown in Figure 4. As shown in Figure 4, ordered mesoporous structure was found for the TiO_2 films except for the sample prepared by calcinations at 650°C. When the sample was calcined at 350°C, as shown in Figure 4a, many organic compounds still remained on the surface of the films, resulting in parts of pores had been blocked. When the calcinations temperature was increased to 450 and 500°C, as shown in Figure 4b,c, uniform and ordered mesoporous structure was clearly shown in the sample without any substance absorbed on the film surface. Increasing the temperature to 600°C, uniformly ordered mesoporous structure was still maintained without the collapse of the mesoporous structure (Figure 4d). Further increasing the temperature to 650°C, the sample lost its initial well-defined mesostructure and wormlike particles existed in the film because of the grains' overgrowth, as shown in Figure 4e. In order to investigate the mesoporous structure clearly, high resolution Transmission Electron Microscope (HRTEM) observation was carried out for the films. It could be seen from the HRTEM micrographs (Figure 4f), a well-organized cylindrical pore structure was formed in the sample calcined at 350°C. Its pore size estimated from high resolution TEM (HRTEM) (Figure 4f) was about 4-6 nm. In the HRTEM micrographs, almost no nanocrystallite was found in the channel walls, implying a low crystallinity of the sample. During the calcination process at 350°C, the initial amorphous *Im3m* cubic structure transformed into the two-dimensional grid-like structure. At this temperature, anisotropic crystallites grew within the geometry of the framework without obvious deformation or destruction of the cylindrical pore structure. A HRTEM image obtained from the sample calcined at 600°C (Figure 4g) also displayed a well-organized mesostructure. But when the calcination temperature increased from 350°C to 600°C, the pore shape was transformed into elliptical shape due to the growth of crystallites at high temperature. The excessive growth of the crystallites which exceeded the geometry of the inorganic framework resulted in the pore walls distortion in some degree, i.e., the randomly orientated anatase particles occupied some space and led to the deformation of the pores. The thick walls of the framework were attributed to the excellent thermal stability of the framework.

Figure 5 shows the Atomic Force Microscope (AFM) of mesoporous TiO_2. It could be seen that the grain size increased with the increase of calcination temperature since the longitudinal depth increased when the temperature increased from 350°C to 650°C.

Figure 4. (Continued).

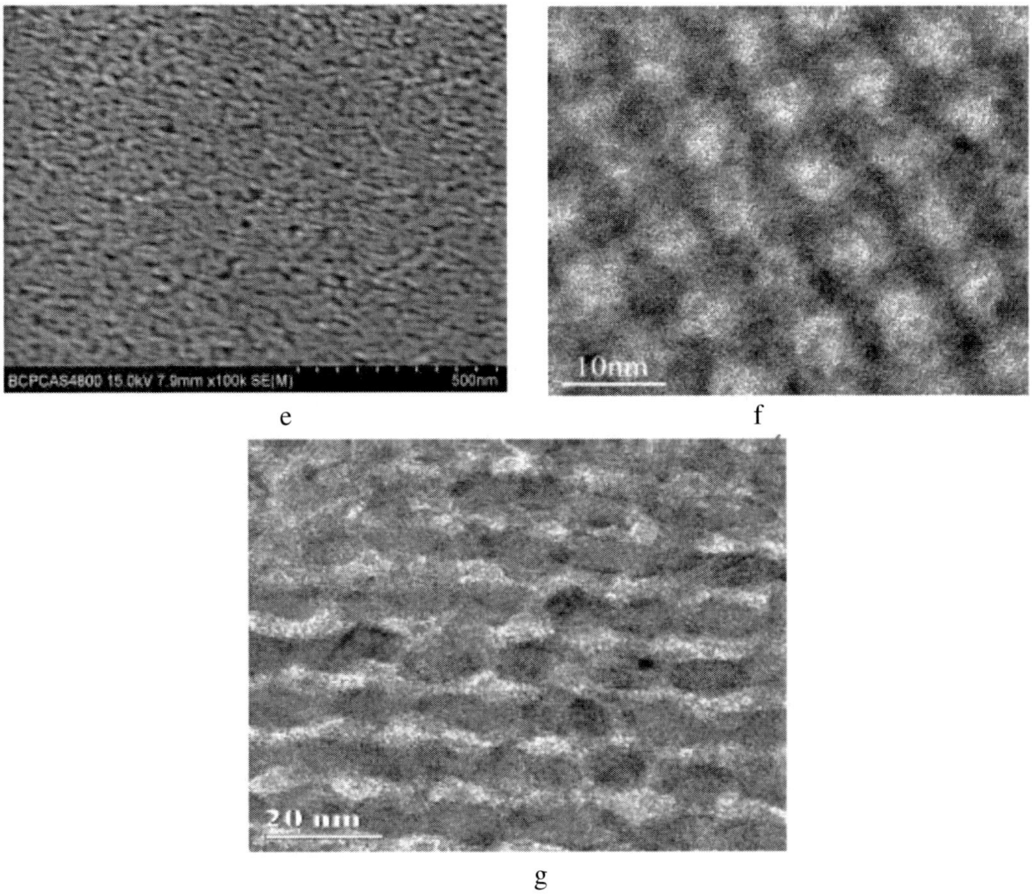

Figure 4. SEM images (a-e) and TEM (f,g) images of prepared mesoporous samples calcined at different temperatures (a, f) 350 °C, (b) 450 °C, (c) 500 °C, (d, g) 600 °C, (e) 650°C.

The root-mean-square surface roughness (Rms) was in the range of 1.0-3.0nm in the temperature range of 350-500°C. However, the Rms increased to 3.75nm when the temperature increased to 600°C. Further increasing the temperature to 650°C, the crystallite size increased to over 20nm and the Rms increased rapidly to 17.9nm, resulting in the collapse of the mesoporous structure. The AFM results showed that TiO_2 kept its mesoporous structure to 600°C, exhibiting highly thermal stability, which was consistent with the Scanning electron microscope (SEM) observation results.

The Barrett-Joyner-Halenda (BJH) pore size analysis (Figure 6) performed on the adsorption branches indicated that the samples calcined below 600°C had a narrow pore size distribution. The structural parameters of the films at different calcinations temperatures were listed in Table 3. The sample calcined at 350°C possessed a small mean pore size of 4.8 nm, whereas the sample calcined at 450°C was 7.8 nm.

Furthermore, despite the increase in the calcination temperatures from 450°C to 500°C and then 600°C, no significant change in the pore diameter could be observed, suggesting a domain of high thermal stability of the films. The BET specific surface area was lower than those reported for mesoporous TiO_2 materials synthesized by templating F127 [49, 50].

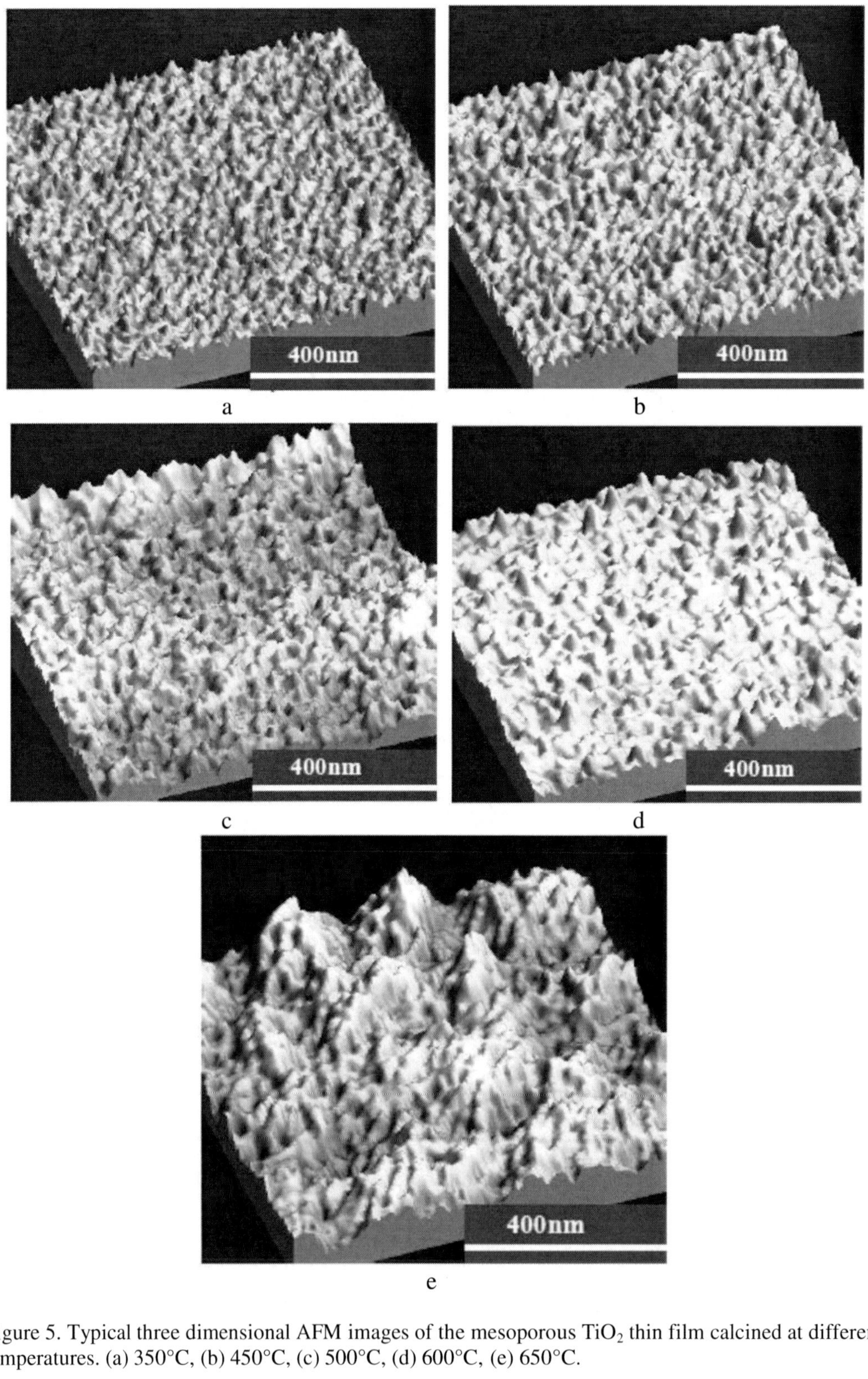

Figure 5. Typical three dimensional AFM images of the mesoporous TiO_2 thin film calcined at different temperatures. (a) 350°C, (b) 450°C, (c) 500°C, (d) 600°C, (e) 650°C.

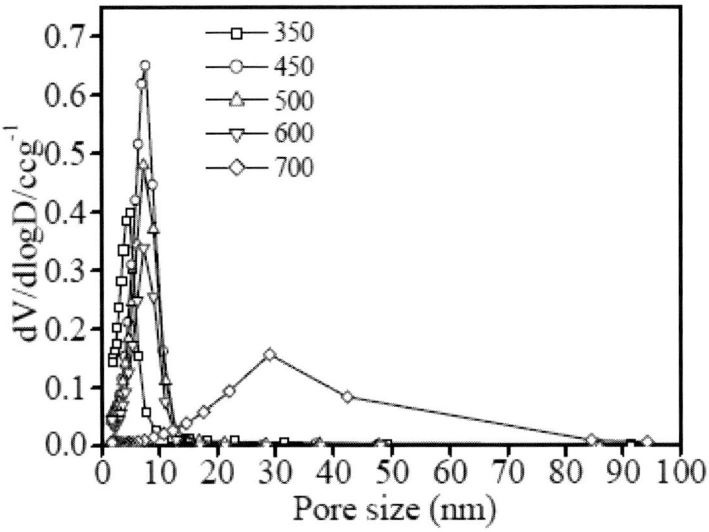

Figure 6. The pore size distribution cures for the synthesized samples calcined at 350, 450, 500, 600, and 700°C.

This might be attributed to the thick inorganic walls of our samples. From the adsorption isotherms measured at 700°C, it could be seen that the sample had a broad pore size distribution due to the collapse of most pore walls.

2.5. Thermal Stability Prediction of Mesoporus TiO_2 Film

The thermal stability of the film is correlated with the temperature. In order to find out the mechanism on thermal stability of the mesoporous structure, the thermodynamic study of the collapse of the mesoporous structure was studied. Taking the grid mesopores as orderly arranged pores surrounded by crystallites, and four crystallites, which had the cylindrical morphology with the base circle diameter of D and height D, connected to form a pore with the radius of R. The schematic diagram of this model is shown in Figure 7. The stress imposed on the pores could be expressed as:

$$\sigma = \sigma_{comp} - \sigma_{tension} \tag{1}$$

where, σ_{comp} and $\sigma_{tension}$ are the compression stress and tension stress applied on the pores, respectively.

Based on tension stress calculated by the surface energy of crystallites and the compression calculated by interface energy between the crystallites, the stress imposed on the pores could be expressed as

$$\sigma = \sigma_{comp} - \sigma_{tension} = \gamma_{interf.} \frac{nL}{\pi D_p} - \gamma_{surf.} \tag{2}$$

where, γ_{interf} is the interface energy, γ_{surf} is the specific surface energy, n is the coordinate number of the pore, L is the contact length between two crystallites and D_p is the pore diameter.

Based on the TEM observation results shown in Figure 4f, four crystallites connected to form one pore, indicating that coordinate number of the pore was 4.

After calculation according to the geometrical rule, equation (2) could be written as

$$\sigma = \sigma_{comp} - \sigma_{tension} = \gamma_{interf} \frac{2\left(-D_p + \sqrt{2D^2 - D_p^{\ 2}}\right)}{\pi D_p} - \gamma_{surf} \tag{3}$$

The change of pore volume was equal to the diffusion flux of the particles (atoms, molecules) moving to these vacancies. The diffusion flux was a function of the effective diffusion coefficient and diffusion driving force (chemical potential gradient), which could be described as

$$|J| = \left[\frac{D_{eff}}{\Omega_\alpha kT}\right] |\nabla u| \tag{4}$$

In which, $|J|$ is absolute value of the particle flux across of the unit cross-sectional area, D_{eff} is the effective coefficient, Ω_α is the volume of diffusion particles, k is Boltzmann coefficient, T is the absolute temperature and $|\nabla u|$ is chemical potential gradient.

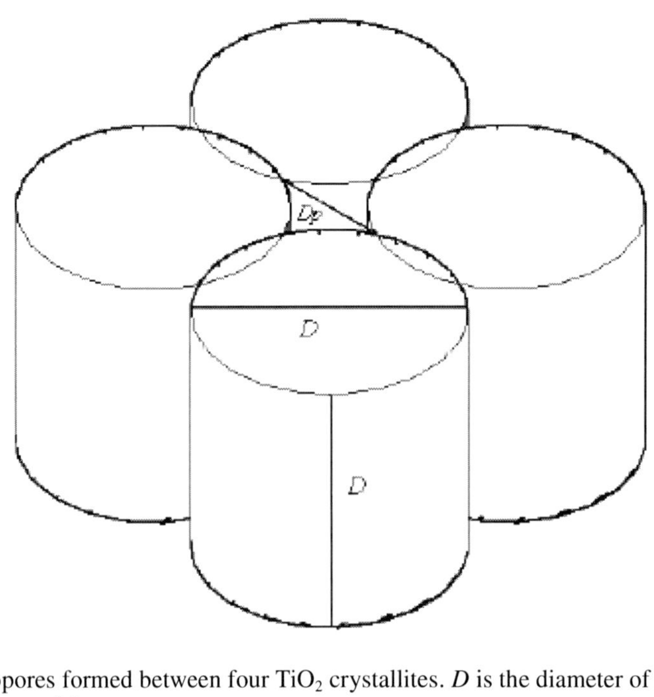

Figure 7. The mesopores formed between four TiO_2 crystallites. D is the diameter of TiO_2 crystallites and Dp is the diameter of the mesopore.

Chemical potential gradient $|\nabla u|$ could be approximately defined as the chemical potential divided by the particle diffusion distance which was equal to the distance between the pore center and crystallite center. Therefore, $|\nabla u|$ could be expressed as:

$$|\nabla u| = 2\frac{\Delta u}{D_p + D} \quad (5)$$

where, Δu is the chemical potential, and it could be calculated by the stress multiplied by the volume of diffusion particles.

$$\Delta u = \sigma\Omega_\alpha \quad (6)$$

Combining expressions (5), (6) and (4), and taking the diffusion direction of the particles towards the pores as the positive direction, equation (4) can be written as

$$J = \frac{2\sigma D_{eff}}{kT(D_p + D)} \quad (7)$$

It was apparent that when J>0, the particles would diffuse toward the pores, leading to the shrinkage of the pores. Substitution of equation (3) into expression (7) yield

$$J = \frac{4D_{eff}}{kT(D_p + D)}\left[\gamma_{interf}\frac{\left(-D_p + \sqrt{2D^2 - D_p^{\,2}}\right)}{\pi \mathrm{D_p}} - \gamma_{surf}\right] \quad (8)$$

In the calculation, the morphology of crystallite was taken as a cylinder, while spherical crystallite was usually adopted for the crystallite size calculation. Although the morphologies of the crystallite were different, the volume of these two kinds of crystallites should be constant, namely,

$$\frac{4\pi(D_e/2)^3}{3} = \pi(\frac{D}{2})^2 D \quad (9)$$

Therefore,

$$D = 0.87D_e \quad (10)$$

In which De is the diameter of spherical crystallite.

Substitution of the expression above into equation (8) yield

$$J = \frac{4D_{eff}}{kT(D_p + 0.87D_e)} \left[\gamma_{interf} \frac{\left(-D_p + \sqrt{0.90D_e^{\,2} - D_p^{\,2}}\right)}{\pi D_p} - \gamma_{surf} \right] \tag{11}$$

The mesoporous structure stability could be predicted by the value of J. When J>0, the particles would diffuse toward the pores, causing the collapse of the mesoporous structure. Otherwise, the mesoporous structure would be stable.

The surface energy and interface energy of TiO_2 are given by equations (12) and (13), respectively.

$$\gamma_{surf} = 1.91 - 1.48 \times 10^{-4}(T - 298) \tag{12}$$

$$\gamma_{interf} = 1.32 - 1.48 \times 10^{-4}(T - 298) \tag{13}$$

Table 3 shows the parameters and calculated results of TiO_2 mesoporous films calcined at different temperatures. As shown in Table 3, *J* was smaller than zero in the range of 450~600°C, indicating that TiO_2 had a mesoporous structure, while the mesoporous structure would collapse when the temperature was higher than 600°C. The calculation results for the stability of mesostructure were in agreement with the experimental results shown in Figure 4. Furthermore, it could be seen that equation (11) should be a universal one if the surface energy and interface energy of the mesoporous films composed of other substance could be obtained.

2.6. Formation Mechanism of Mesoporous TiO_2 with Thick Wall

The formation of thick and robust mesoporous TiO_2 can be explained according to the model proposed by Smarsly et al. [51]. Taking triblock-copolymer F127 as the structure-directing agent is the premise for the formation of the thick walls in the mesostructure.

Table 3. Physical parameters of TiO_2 mesopores and values of *J* calcined at different temperatures

T/°C	D_p^{a}/nm	D_e^{b}/nm	$S_{BET}/m^2\cdot g^{-1}$	γ_{surf}^{c}	γ_{interf}^{d}	J
350	4.8	7.6	145	1.86	1.27	<0
450	7.8	9.6	119	1.85	1.26	<0
500	7.8	10.4	101	1.84	1.25	<0
600	7.3	12.6	70	1.82	1.23	<0
650	7.6	22.4	21	1.82	1.23	>0
700	-	29.6	19	1.81	1.22	>0

a, b are the pore size, crystallite size and critical crystallite size of TiO_2 mesoporous film, respectively. c and d are the surface energy and interface energy of TiO_2, respectively.

It is well-known that F127 has a high molecular weight with long hydrophilic and hydrophobic segments, which permits thicker inorganic walls to be obtained [52, 53].

Furthermore, carefully controlling the hydrolysis/condensation conditions are pivotal during cooperative self-assembly and ageing process for the construction of solid and thick wall frameworks, which lies in the following facts: Firstly, during the synthesis of sol, the strong chelating agent accetylacetone and acid solvent permit control the reactivity by avoiding the fast hydrolysis and condensation of the precursor in solution. Additionally, nonaqueous solvent also depresses the reactivity of the precursor. Under the "dry conditions", the slow hydrolysis/condensation speed of inorganic precursor prefers the formation of oligermeric Ti-oxo species or titania clusters, which will form the skeleton of thick and solid inorganic walls architecture. These small entities preferentially associate with the polar poly (ethylene oxide) (PEO) moieties through electrostatic, hydrogen bonding and van der Waals forces. Secondly, regulating the ageing procedure of the films at proper temperature (28°C) and RH (40-50%) is also important for modulating the condensation step, and consolidating the inorganic framework before the template removal. In this quasi-anhydrous medium (water only provided by air moisture and reagents), the Ti-oxo species can be chelated by the ether functions of the PEO chains through the mediation of HCl. During ageing procedure, with the evaporation of the solvent, the bonding of the polar heads to the Ti centers becomes more predominant, as well as the microphase segregation is promoted, which results in the formation of a hybrid mesophase. Finally, appropriate ageing time (24 h) is also necessary for the full condensation and the formation of the thick and compact walls in the framework. Proper ageing time promotes the cross-linking process between the Ti-oxo species and the surfacants. At the same time, the binding proceeds completely during this procedure, which excludes the possibility of micropores formation in the framework during the organic groups removing process. This assumption can be confirmed by the pore size distribution plots and the HRTEM images, in which no micropores could be found.

The dense channel walls also benefit the thermal stability of the framework. However, excessively long ageing time is profitless for the formation of high-quality mesostructure. As an unexpected result, the relative solubility difference between the PPO and the PEO blocks is diminished, which leads to worm-like mesophase.

3. Preparation of Anion Doped Mesoporous Structure of TiO_2

3.1. Synthesis

Pluronic F127 (MW=12600, Basf) was dissolved in 50ml solution of absolute ethanol and 12 M concentrated HCl under magnetic stirring. Then the solution was added dropwise to the mixture of $Ti(OBu^n)_4$ (TBT), ethanol and acetylacetone (AcAc) at vigorous stirring under the room conditions. Then the thiourea ethanol solution was added to the above mixture, finally the obtained solution was stirred for another 24 h. The mole ratios of each component were F127/TBT/HCl/EtOH/AcAc=0.0008: 0.098: 0.11: 1.8: 0.02. The mole ratios of thiourea to TBT in the solution were 0, 2.5%, 5%, 7.5% and 10%, which were designated to SN-1, SN-2, SN-3 SN-4 and SN-5, respectively.

Uniform and transparent thin films were produced by spin-coating on the silicon wafer or indium tin oxide (ITO-) coated glass substrates at a constant spinning rate of 2500 rpm. The deposited films were aged at approximately 28°C and 30-50% relative humidity (RH) for 24 h. Equivalent powders were obtained by pouring and aging the residual sol in a large Petri dish at the same situations. The resulting thick layers were ground to the powders. The as-made thin films and powders were subsequently calcined at 450°C (ramp of $1°C \cdot min^{-1}$) for 2.5 h under the flow of oxygen to remove the organic template and other organic impurities.

3.2. Microstructure of S,N Co-Doped Mesoporous TiO_2

Figure 8 shows the morphologies of the synthesized TiO_2 mesoporous films. All the samples presented homogeneous mesostructures. TiO_2 mesoporous film (SN-1) without adding thiourea in the reactive solution exhibited a uniform pore size distribution (Figure 8a). On the other hand, although long-range order was not obvious, the mesopores arrangement in some orientations was ordered, i.e. the pores arrangement was regular in certain areas. The pore size could be estimated 6-9 nm from the image.

However, when 2.5% (mol) thiourea was added in the solution, apart from the uniform mesoporous structure, some particles were found on the surface of the film (Figure 8b). With the increase of content of thiourea in the solution, more particles appeared on the surfaces of the samples (Figure 8c and d). When the concentration of thiourea was increased to 10%, the mesoporous structure almost disappeared, see Figure 8e.

The physical parameters of these S,N co-doped samples were listed in table 4. As shown in Table 4, the N,S codoped samples had similar grain size, the pore size and porosity decreased gradually with the increase of amount of thiourea in the solution.

The hydrolysis of the Ti-alkoxide precursors in the acidic solution would result in the formation of nanometer sized Ti-oxo clusters, which was preferentially associated with the hydrophilic PEO block of F127 micelles which form the hybrid amorphous mesostructure. When thiourea was added in the acidic solution, thiourea might have a chemical reaction with the absorbed water in the air to form H_2S and NH_3. Then, H_2S and NH_3 produced would react with oxygen in the air and hydrochloric acid to produce sulfur and ammonia chloride, respectively. The reactions described above are shown as follows:

$$2H_2O\ (\text{in air}) + (NH_2)_2CS = CO_2\uparrow + 2NH_3\uparrow + H_2S\uparrow \tag{14}$$

$$2H_2S + O_2\ (\text{in air}) = 2S\downarrow + 2H_2O \tag{15}$$

$$NH_3 + HCl = NH_4Cl \tag{16}$$

Water produced by reaction (15) increased the hydrolysis and polycondensation speed of Ti-oxo species, therefore, titania particles were produced. The more amount of thiourea in the reactive solution, the larger amount of titania particles formed in the mesostructures. When the addition of the amount of thiourea exceeded a certain amount, excess amount of NH_4Cl caused the collapse of mesoporous microstructure. Figure 9 shows the schematic diagram of the effect of ions NH_4^+ on the miscellar curvature of TiO_2.

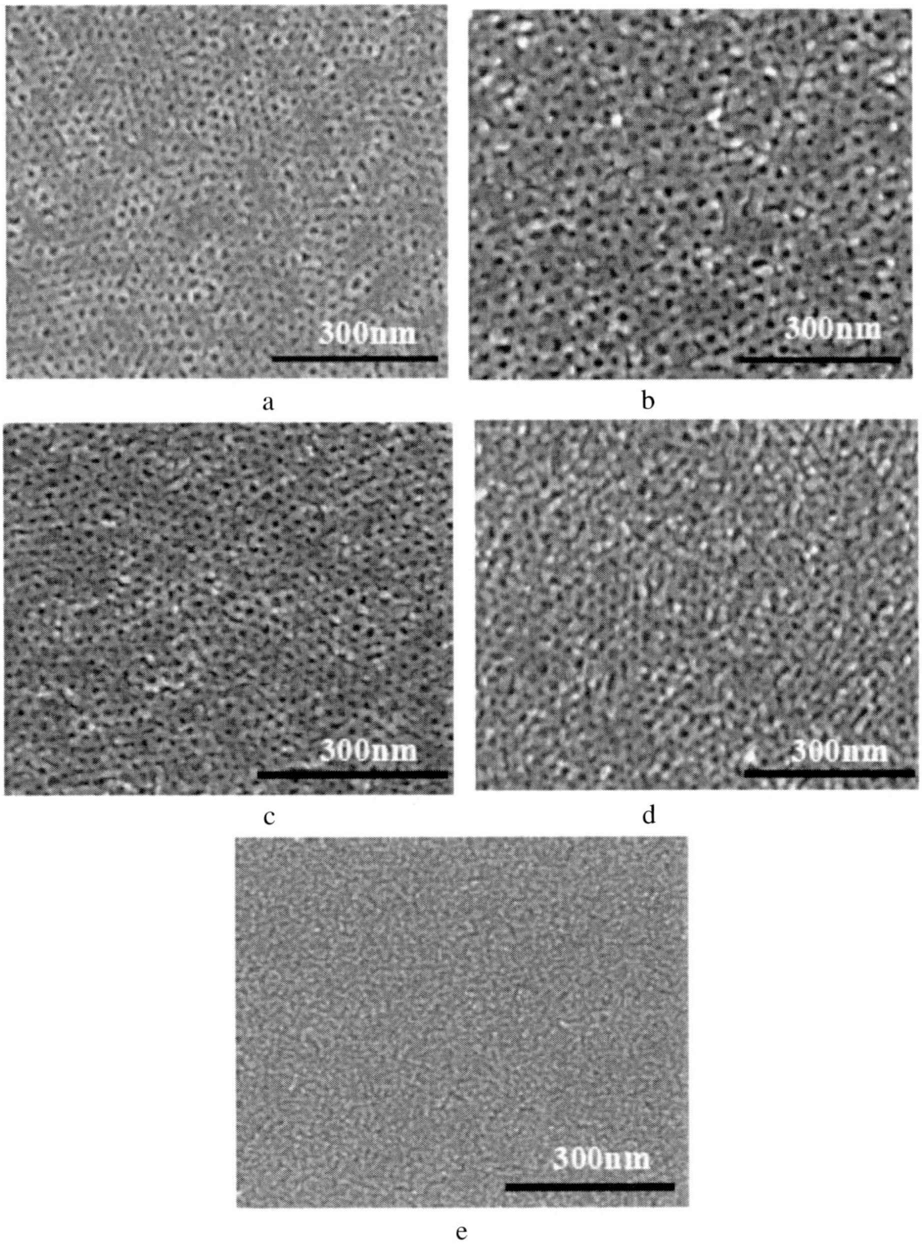

Figure 8. SEM images of the synthesized samples: (a) SN-1 (without adding thiourea), (b) SN-2 (added 2.5% mol thiourea), (c) SN-3 (added 5% mol thiourea), (d) SN-4 (added 7.5% mol thiourea), (e) SN-5 (added 10% mol thiourea).

It was known that the effect of salt addition had been attributed to the change in the curvature caused by the adsorption of ions at the micellar interface.

We deduced that ionized product NH_4^+ of NH_4Cl formed by reaction (16) adsorbed at the micellar interface and changed the micellar curvature.

The ions NH_4^+ decreased the miscellar curvature due to the electrostatic attraction, therefore, compared with pure mesoporous TiO_2 film, mesoporous TiO_2 prepared with the addition of thiourea had larger pore size.

Table 4. The physical parameters of the synthesized samples with N,S-codoping

Sample	D_e[a] /nm	D_p[b]/nm	S_{BET}/m^2·g^{-1}	Porosity (%)
SN-1	10.94	7.0	105	38.96
SN-2	11.99	12.4	76	51.28
SN-3	11.76	12.2	92	50.96
SN-4	12.58	11.2	95	50.54
SN-5	13.51	5.3	86	3.23

a and b are the average crystallite diameter and pore size, respectively.

When the concentration of thiourea exceeded a certain value, the excess amount of inorganic salt led to the dehydration of ethylene oxide (EO) groups, in other words, the aggregation number of micelles increased with the inorganic salt concentration. Under these conditions, the hydrophilic poly(ethylene oxide) (PEO) blocks became gradually more dehydrated, increasing the effective size of the hydrophobic core at the expense of the hydrophilic corona, leading to the change of the micelle shape from sphere to column. As a result, the micelles were difficult to assemble with inorganic-oxo species. The hydrolysis of Ti-oxo species and the change of miscellar curvature caused by NH_4^+ ions contributed to the change of the microstructure of mesoporous TiO_2.

4. Application of Mesoporous TiO_2

Mesoporous solids possessing large specific surfaces and easily accessible cavities have found wide applications in heterogeneous catalysis, sorption, separation, gas sensing, optoelectronics, host–guest chemistry, molecular electronic devices. In this chapter, we mainly introduced the application of mesoporous TiO_2 in photocatalysis field. Semiconductor photocatalysis, is an established, significant and still-expanding area of research which has led to numerous commercial products, such as self-cleaning glass, tiles and paint. The current commercial manifestations of photocatalysis usually utilize a layer of the semiconductor photocatalyst, invariably anatase titania [54].

The photocatalytic mechanism for the semiconductor is that upon excitation, a conduction-band electron and a valence-band hole separate, and then the hole initiated an oxidative reaction while the electron initiated a reductive reaction. Organic compounds were believed to be destroyed through direct oxidation by the trapped holes or attack by hydroxyl groups. Many methods have been adopted for assessing and characterizing the activities of such films and include: (i) the photobleaching of a dyestuff, such as methylene blue (MB) and methyl orange dissolved in aqueous solution, (ii) the photomineralisation of a thin solid film of a waxlike substance, such as stearic or palmitic acid and, (iii) the photo-oxidation of a gas phase pollutant, such as nitric oxide or acetaldehyde.

In our work, we evaluated the phtocatalytic activity of mesoporous TiO_2 by photodegradation of methyl orange (MO) solution. Figure 10 shows the decomposition of MO aqueous solution of pure mesoporous TiO_2 and N, S doped mesoporous TiO_2 prepared with thiourea under UV illumination. As shown in Figure 10, compared with the standard sample P25, both pure mesoporous TiO_2 and N, S doped TiO_2 exhibited higher photocatalytic activity, especially for the co-doped sample, MO solution was degraded completely in 3h.

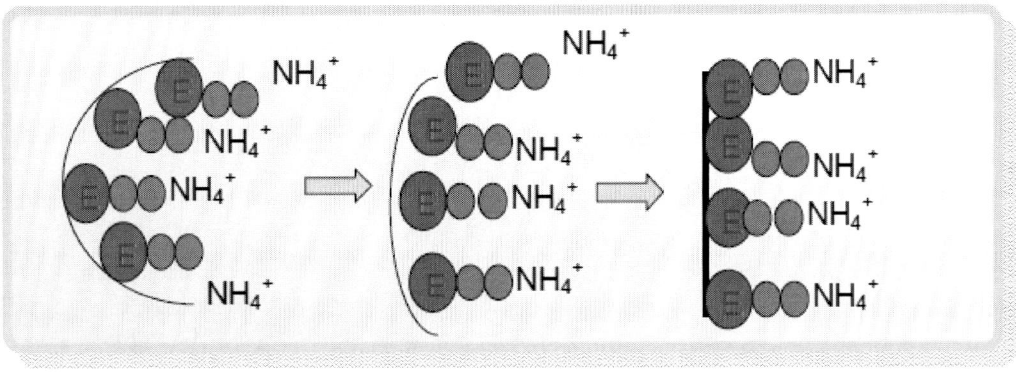

Figure 9. Schematic diagram of the effect of ions NH_4^+ on the miscellar curvature of TiO_2.

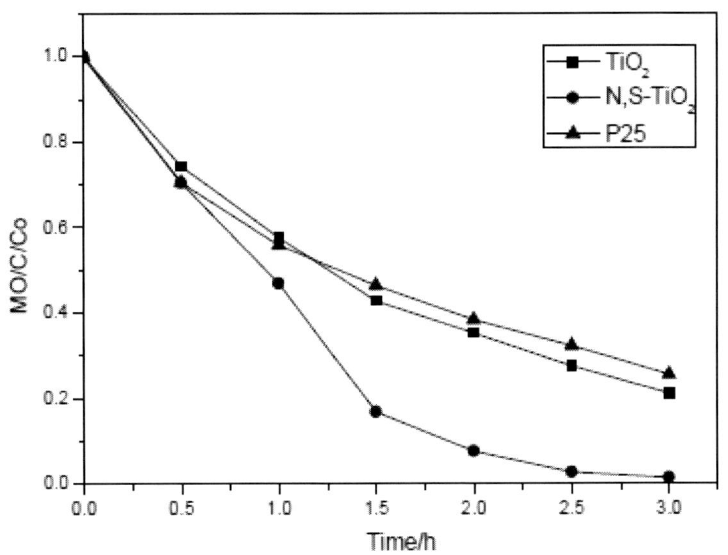

Figure 10. The photocatalytic activity of the pure and N,S co-doped mesoporous TiO_2 samples together with P25.

The high porosity, the large pore size, and high pore volume of the mesoporous TiO_2 samples which were favorable for the adsorption of the organic molecule in the mesostructure, led to improvement of the organic degradation capability. Furthermore, the UV-visible reflection result of the samples indicated that the N, S doping improved the photoabsorption of the light, as a result, the N, S doped mesoporous TiO_2 exhibited the highest photocatalytic activity among these three samples.

Conclusion

Mesoporous materials show higher surface areas and much more uniform and controllable pore sizes, thus display unique electronic, magnetic, and catalytic properties. Crystalline TiO_2 associated with the photo-induced electronic transfer behavior render their usefulness in super-hydrophilic surface, photovoltaic devices, and photocatalysts. The mesoporous structure could be adjusted by changing the calcinations temperature, relative humility, the amount of block copolymers and addition of anion solvent. The thermal stability of the mesoporous structure could be predicted by using the pore collapse criterion established by the thermodynamic calculation. Future developments will rely on facile fabrication processes and novel synthetic methods. Specifically, more work needs to be done to address the combination of sol-gel chemistry with other synthesis techniques to improve both in morphology, purity, composition, and yield control and in terms of flexibility for pore size choice.

References

[1] Gammoudi, L., Blanc, F., Morote, C., Grauby-Heywang, C., Boissiere, R., Kalfat, D., Rebiere, T., Cohen-Bouhacina, C. Dejous, *Biosens. Bioelectron.* 2014, 57, 162-170.

[2] S. Liu, Z. Y. Wang, H. R. Zhao, T. Fei, T. Zhang, *Sensor. Actuat. B-Chem.* 2014, 197, 342-349.

[3] T. Fu, B. G. Liu, Y. M. Zhou, X. M. Wu, *J. Sol-Gel Sci. Technol.* 2011, 58, 307-311.

[4] D. J. Kim, S. J. Kim, D. K. Roh, J. H. Kim, *Phys. Chem. Chem. Phys.* 2013, 15, 7345.

[5] M. B. Zakaria, N. Suzuki, N. L. Torad, M. Matsuura, K. Maekawa, H. Tanabe, Y. Yamauchi, *Eur. J. Inorg. Chem.* 2013, 2330–2335.

[6] X. Liu, Y. Gu, J. Huang, *Chem. Eur. J.* 2010, 16, 7730-7740.

[7] J. J. Choi, X. H. Yang, Z. M. Norman, S. J. L. Billinge, J. S. Owen, *Nano. Lett.* 2014, 14, 127-133.

[8] C. Y. Cho, J. H. Moon, *Adv. Mater.* 2011, 23, 2971-2975.

[9] C. T. Kresge, M. E. Leonowicz, W. J. Roth, J. C. Vartuli, J. S. Beck, *Nature,* 1992, 359, 710-712.

[10] S. Kaskel, D. Farrusseng, K. Schlichte. *Chem. Commun.* 2000, 24, 2481-2482.

[11] Z. R. Tian, W. Tong, J. Wang, N. Duan, V. V. Krishnan, S. L. Suib, *Science,* 1997, 276, 926-930.

[12] E. L. Crepaldi, G. J. D. Soler-Illia, D. Grosso, F. Cagnol, F. Ribot, C. Sanchez. *J. Am. Chem. Soc.* 2003, 125, 9770-9786.

[13] M. N. Ghazzal, H. Kebaili, M. Joseph, D. P. Debecker, P. Eloy, J. D. Coninck, E. M. Gaigneaux, *Appl. Catal. B-Environ.* 2012,115–116, 276–284.

[14] T. Katou, B. Lee, D. Lu, J. N. Kondo, M. Hara, and K. Domen, *Angew. Chem. Int. Ed.* 2003, 42, 2382-2385.

[15] Y. Zhou, J. H. Schattka, M. Antonietti, *Nano. Lett.* 2004, 4, 477-481.

[16] K. Saravanan, K. Ananthanarayanan, P. Balaya, *Energy. Environ. Sci.*, 2010, 3, 939-948.

[17] J. Jin, N. Nishiyama, Y. Egashira, K. Ueyama, *Micropor. Mesopor. Mater.* 2009, 218–223.

[18] F. Goettmann, A. Fischer, M. Antonietti, A. Thomas, *Angew. Chem. Int. Ed.* 2006, 45, 4467-4471.

[19] F. Jiao, K. M. Shaju, P. G. Bruce, *Angew. Chem. Int. Ed.* 2005, 44, 6550-6553.

[20] W. C. Li, A. H. Lu, W. Schmidt, F. Schuth, *Chem. Eur. J.* 2005, 11, 1658-1664.

[21] C. J. Brinker, Y. Lu, A. Sellinger, H. Fan, *Adv. Mater.* 1999, 11, 579–585.

[22] Y. Lu, R. Ganguli, C. A. Drewien, M. T. Anderson, C. Jeffrey Brinker, W. Gong, Y. Guo, H. Soyez, B. Dunn, M. H. Huang, J. I. Zink, *Nature*, 1997, 389, 364–368.

[23] D. Zhao, J. Feng, Q. Huo, N. Melosh, G. H. Fredrickson, B. F. Chmelka, G. D. Stucky, *Science*, 1998, 279, 548–552.

[24] D. Zhao, Q. Huo, J. Feng, B. F. Chmelka, G. D. Stucky, *J. Am. Chem. Soc.* 1998, 120, 6024–6036.

[25] D. Zhao, P. Yang, N. Melosh, J. Feng, B. F. Chmelka, G. D. Stucky, *Adv. Mater.* 1998, 10, 1380–1385.

[26] G. J. D. Soler-Illia, C. Sanchez, B. Lebeau, J. Patarin, *Chem. Rev.* 2002, 102, 4093-4138.

[27] P. D. Yang, D. Y. Zhao, D. I. Margolese, B. F. Chmelka, G. D. Stucky, *Nature*, 1998, 396, 152-155.

[28] P. D. Yang, D. Y. Zhao, D. I. Margolese, B. F. G. Chmelka, D. Stucky, *Chem. Mater.* 1999, 11, 2813-2826.

[29] D. M. Antonelli, J. Y. Ying, *Angew. Chem. Int. Ed.* 1995, 34, 2014-2017.

[30] R. L. Putnam, N. Nakagawa, K. M. McGrath, N. Yao, I. A. A. S. M. Gruner, A. Navrotsky, *Chem. Mater.* 1997, 9, 2690-2693.

[31] D. M. Antonelli, J. Y. Ying, *Chem. Mater.* 1996, 8, 874-881.

[32] S. Cabrera, J. E. Haskouri, C. Guillem, J. Latorre, A. Beltrán-Porter, D. Beltrán-Porter, M. D. Marcos, P. Amoros, *Solid Stat. Sci.* 2000, 2, 405-420.

[33] U. H. Lee, H. Lee, S. Wen, S. L. Mho, Y. U. Kwon, *Micropor. Mesopor. Mater.* 2006, 88, 48-55.

[34] E. L. Crepaldi, G. J. D. Soler-Illia, D. Grosso, F. Cagnol, F. Ribot, C. Sanchez, *J. Am. Chem. Soc.* 2003, 125, 9770-9786.

[35] B. L. Kirsch, E. K. Richman, A. E. Riley, S. H. Tolbert, *J. Phys. Chem. B,* 2004, 108, 12698-12706.

[36] J. C. Yu, X. C. Wang, X. Z. Fu, *Chem. Mater.* 2004, 16, 1523-1530.

[37] X. F. Jiang, N. Suzuki, B. P. Bastakoti, W. J. Chen, Y. T. Huang, Y. Yamauchi, *Eur. J. Inorg. Chem.* 2013, 3286–3291.

[38] Y. J. Kim, S. Y. Chai, W. I. Lee, *Langmuir*, 2007, 23, 9567-9571.

[39] L. Malfatti, M. G. Bellino, P. Innocenzi, G. J. A. A. Soler-Illia, *Chem. Mater.* 2009, 21, 2763-2769.

[40] G. J. D. Soler-Illia, E. L. Crepaldi, D. Grosso, C. Sanchez, *Curr. Opin. Colloid Interface Sci.*, 2003, 8, 109–126.

[41] Q. L. Wu, N. Subramanian, S. E. Rankin, *Langmuir*, 2011, 27, 9557-9566.

[42] J. S. Wang, H. Li, H. Y. Li, C. Zou, H. Wang, D. S. Li, *ACS Appl. Mater. Inter.* 2014, 6, 1623-1631.

[43] J. S. Wang, H. Li, H. Y. Li, H. Wang, Q. Cai, *J. Nanosci. Nanotechno.* 2013, 13, 1493-1497.

[44] J. S. Wang, H. Li, H. Y. Li, C. Zou. *Solid State Sci.* 2010, 12, 490-497.
[45] J. S. Wang, H. Li, H. Y. Li, C. Zou, *J. Phys. Chem. C.* 2012, 116, 9517-9525.
[46] H. Li, J. S. Wang, H. Y. Li, S. Yin, T. Sato. *Mater. Lett.* 2009, 63, 1583-1585.
[47] G. J. D. Soler-Illia, C. Sanchez, B. Lebeau, *J. Patarin. Chem. Rev.* 2002, 102, 4093-4138.
[48] B. Smarsly, S. Polarz, M. Antonietti. *J. Phys. Chem. B*, 2001, 105, 10473-10483.
[49] K. M. Coakley, Y. X. Liu, M. D. McGehee, K. L. Frindell, G. D. Stucky. *Adv. Funct. Mater.* 2003,13, 301-306.
[50] P. Praserthdam, P. L. Silveston, O. Mekasuwandumrong, V. Pavarajarn, J. Phungphadung, P. Somrang. *Cryst. Growth. Des.* 2004, 1, 39-43.
[51] B. Smarsly, S. Polarz, M. Antonietti, *J. Phys. Chem. B*, 2001, 105, 10473-10483.
[52] P. N. Hurter, T. A. Hatton, *Langmuir*, 1992, 8, 1291-1299.
[53] P. Alexandridis, T. A. Hatton, *Colloid surface A,* 1995, 96, 1-46.
[54] A. Mills, M. Sheik, C. O. Rourke, M. McFarlane, *Appl. Catal. B: Environ.* 2009, 89, 189–195.

In: Comprehensive Guide for Mesoporous Materials, Volume 2 ISBN: 978-1-63463-982-8
Editor: Mahmood Aliofkhazraei

Chapter 11

THE I-POINT METHOD AND THE SURFACE AREA OF POROUS MATERIALS

***A. K. Ladavos*[1] *and P. J. Pomonis*[2]**
[1]University of Patras, Agrinio, Greece
[2]University of Ioannina, Ioannina, Greece

ABSTRACT

This Chapter deals with the Inversion or Inflection point(s), termed collectively *I-point(s),* which appeared in all the N_2 adsorption isotherms $V=f(P/P_o)$ of porous solids. The *I-point(s)* correspond to the equivalent point(s) of titration of the adsorbent material (analyte or titrand) by the adsorbate N_2 (titrant or titrator) at 77K. If there is just one I-point, this corresponds to the completement of physical adsorption and is useful for the determination of the specific surface area *-ssa-* of solid (S_p, m^2g^{-1}) without any knowledge or use of the traditional BET methodology. If there is a second equivalent point, this corresponds to the completement of the condensation of N_2 into the capillary pores. The *I-point(s) method* is a precise and convenient procedure for estimating the *ssa* of porous materials based on N_2 adsorption data, $V=f(P/P_o)$. Such data if plotted in the form $V(1\text{-}P/P_o)/(P/P_o) = f(V(1\text{-}P/P_o))$ may show one I-point, if mesopores are spatially mixed with micropores (case i) or two I-points if the mesopores form separate domains in the porous solid (case ii). The projection of the single I-point (case i), or the low pressure I-point (case ii), on the $(V(1\text{-}P/P_o))$ axis corresponds precisely to the volume of monolayer V_m from which the S_p values can be easily estimated ($S_p = 4.356V_m$ for N_2). Finally the variable slope of the I-plots corresponds to the *variable equilibrium constant C of adsorption,* introduced in the BET equation, which is not attainable via the classical linear treatment.

INTRODUCTION

The determination of specific surface area *-ssa-* (S_p, m^2g^{-1}) of solids is a very important prerequisite for the physicochemical characterization of adsorbing and/or reacting capacity of powder solids, porous materials and heterogeneous catalysts. This is the reason the efforts for developing a convenient method for its determination started back in 1916 by Irving Langmuir [1] in relation to the problem of adsorption of simple inert gases on the surface of metallic filaments used in the incandescent light bulbs. That technology of electrical lights is by now gradually outdated, but the *Langmuir isotherm* (eq.1a and figure 1-left) remains a scientific and teaching *paradigm*, in line with the dictum that few things are more eternal than an equation. The symbols in eq.(1a) have their customary meaning: V-is the adsorbed volume of the gas (usually N_2), V_m-is the volume corresponding to the N_2 monolayer covering the whole surface, θ-is the fraction of the covered surface ($0<\theta<1$), K- is the thermodynamic equilibrium constant of adsorption and finally (P/P_o) –is the partial pressure at each pressure P and P_o - is the equilibrium pressure, corresponding usually to 1atm.

The non- linear form of Langmuir isotherm	$\frac{V}{V_m} = \theta = \frac{K\left(\frac{P}{P_0}\right)}{1 + K\left(\frac{P}{P_0}\right)}$	(1a)
The linear form of Langmuir isotherm	$\frac{1}{V} = \frac{1}{V_m\theta} = \frac{1}{V} = \frac{1}{KV_m\left(\frac{P}{P_0}\right)}$	(1b)

Relation (1a) is not linear and is often used in its linearized form (1b). Then plots of the form *(1/V)* vs. *1/(P/P₀)*,as shown in Figure 1-right, are used in order to determine graphically the monomolecular volume of adsorption V_m as well as the adsorption constant K. The crucial problem is the V_m value determination from which, in the case of the most common and cheap adsorbent N_2, the *ssa* can be determined via the single relationship

$$S_p\ (m^2g^{-1}) = 4.356\ V_m\ (cm^3g^{-1}) \qquad (2)$$

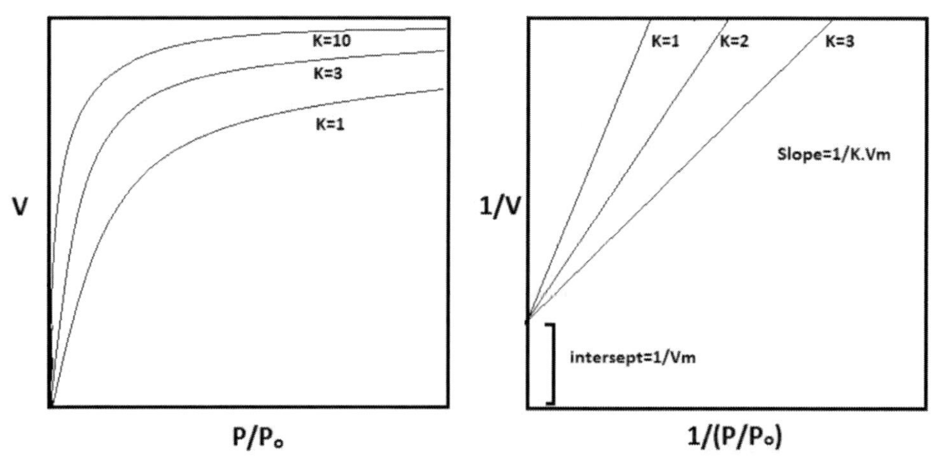

Figure 1. Left-The Langmuir isotherm in its non-linear form (Eq. 1a) for various values of K=1,3,10. Right-The Langmuir isotherm in its linear form (Eq. 1b). Plots of *(1/V)* vs. *1/(P/P₀)* have slope $s=1/KV_m$ and intercept $i=1/V_m$ which may be used for the determination of K and V_m. The V_m values are considered identical in the three cases.

Langmuir isotherm bears some well-known shortcomings the most notable been the following:

(i) It assumes that adsorption takes place in the form of monolayer. But monolayer formation is possible only under low pressure condition (see Figure 2). Under high pressure condition the assumption breaks down as gas molecules attract more and more molecules towards each other into the nanopores, resulting finally in condensation. BET theory, proposed by Brunauer, Emmett and Teller in 1938, tackled this problem and proposed a more realistic multilayer adsorption isotherm.

(ii) The adsorption parameter *K* is considered constant – *which is not*. Actually *K* is the equilibrium constant of adsorption, and if the entropy of adsorption is small, which is true for adsorption of small molecules, then *K* is related to the enthalpy, or heat, of adsorption, $(-\Delta H_{ads}) = q_{ads}$ as follows:

$$K \approx \exp(-\Delta H_{ads})/RT = \exp(q_{ads})/RT \tag{3}$$

But it is well established that the heat of adsorption decreases (linearly, exponentially or otherwise) as the surface coverage θ increases. Therefore at $\theta \to 0$, $(V << V_m)$, $q_{ads} \to$ large and $K \to$ very large. On the contrary at $\theta \to 1$, $(V \approx V_m)$, $q_{ads} \to$ zero and $K \to$ zero.

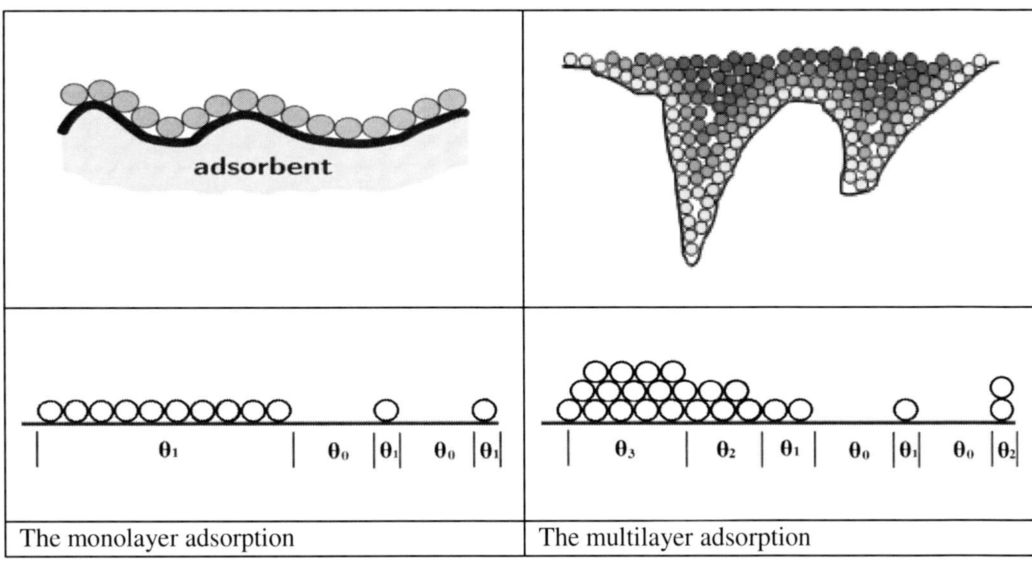

Figure 2. The basic assumption of the Langmuir isotherm is that the adsorption takes place in a monolayer (left). In reality the adsorption on solids is multilayered (right), a problem addressed by the BET isotherm. θ_i-reflects the number of adsorbed layers (i-0,1,2,3…..).

THE BET EQUATION

Aware of the problems described above, Brunauer, Emmett and Teller [2] proposed in 1938 the BET equation (4a, 4b). This relation is one of the most *classical equations* in the field of surface science and catalysis in the original meaning of this word: Every possible proposed improvement of methodology, or calculations, for the determination of *ssa* of porous materials, will use inevitably BET equation as a stick of comparison. Nowadays the determination of $S_p(m^2g^{-1})$ of various adsorbents and catalysts, in all the academic and industrial labs around the globe, is practically based on this equation which is depicted in Figure 3.

The non- linear form of BET isotherm	$$\frac{V}{V_m}=\frac{C\left(\frac{P}{P_0}\right)}{1+(C-1)\left(\frac{P}{P_0}\right)}\frac{1}{\left(1-\left(\frac{P}{P_0}\right)\right)}$$	(4a)
The linear form of BET isotherm	$$\frac{\left(\frac{P}{P_0}\right)}{V\left(1-\left(\frac{P}{P_0}\right)\right)}=\frac{1}{CV_m}+\frac{(C-1)\left(\frac{P}{P_0}\right)}{CV_m}$$	(4b)

In relations (4a) and (4b) the symbols have their customary meaning and details about its derivation and properties can be found in some standard textbooks on this subject [3-5].

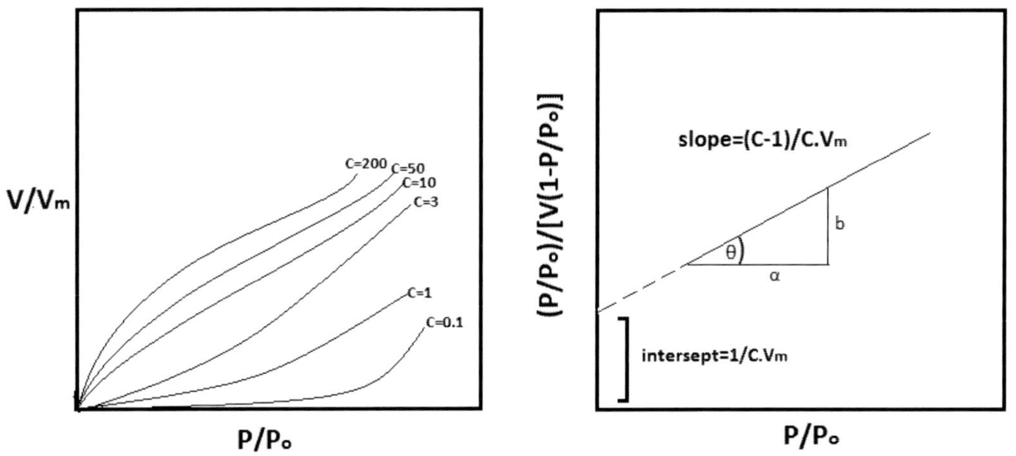

Figure 3.The BET equation in its non-linear form (left) and its linear representation (right) for various values of parameter C. Plots of $(P/P_0)/[V(1-(P/P_0)] = f(P/P_0)$ provide straight lines at low pressure range $0.05 < P/P_0 < 0.25$, with slope $s=(C-1)/CV_m$ and intercept $i=1/CV_m$, from which two unique values of C and V_m can be estimated.

Relation (4a) is not linear and for the determination of V_m is used in its linear form (4b). Then plots of the form $(P/P_0)/[V(1-(P/P_0)] = f(P/P_0)$ provide straight lines at low pressure range, usually $0.05<P/P_0<0.25$, depending on the sample. From such linear plots, with slope $s= (C-1)/CV_m$ and intercept $i=1/CV_m$, two unique values of C and V_m can be estimated (see Figure 3-right). Then, the determination of *ssa* takes place via estimation of the volume of the adsorbed monolayer V_m via relation (2).

Nevertheless there are some well-known shortcomings in the applied procedure:

First, the estimation of V_m via the classical linear treatment may be influenced *quantitatively* by the pressure range used for drawing the linear plots, which in turn influences the estimated value of surface area up to 25-30%.

Second, and perhaps more important, the estimation of a *single val*ue for the *C*- parameter conveys *the qualitative wrong impression* that this is a constant which *is not.* To be more precise the explicit relationship for *C* is given by equation (5) [3-5]

$$C = [\alpha_1 v_2/\alpha_2 v_1] \exp[(q_1 - q_L)/RT] \quad (5)$$

where α_1 and α_2 are the condensation coefficients for the first and the second layers, v_1 and v_2 are the frequencies of oscillation of the molecules in the first and the second layers normal to the surface, $q_1$is the heat of adsorption in the first layer on the bare surface and finally q_L is the heat of condensation which is equal for all of the layers except the first.

The term $(\alpha_1 v_2/\alpha_2 v_1) = m$ is often considered equal to unity although it has been suggested by Kembal and Schreiner [6] that it can varies as much as 10^6 orders (!) of magnitude in the whole range of adsorption $0< P/P_0<1$. So the *C*-constant, estimated in the traditional BET-treatment, corresponds to

$$C = \exp (q_1 - q_L)/RT \quad (6)$$

and actually approximates the value of the *C*-parameter at the point where the adsorbed monolayer is formed. *The determination of C- parameter in the whole range of adsorption is unattainable in the traditional treatment.*

The high values of *C* during the first stages of adsorption reflect the high values of the condensation coefficient α_1 in the first layer compared to the values of α_2, as well as the high values of q_1 compared to q_L. As the adsorption proceeds, the case where $\alpha_1 = \alpha_2$ is eventually reached. For $v_1 \approx v_2$, which is a realistic approximation as the pressure and the surface coverage increase, *V* tends to V_m. At this limit, equation (6) applies. To cite an example at 77K, RT=0.6kJ/mol. Then for $(q_1 - q_L)$ =4.2kJ/mol, $C \approx 1000$ whereas for $(q_1 - q_L)$ =2.0kJ/mol, $C \approx 20$.

So as far as q_1 is larger than q_L even by 1-2 kJ/mol, there should be a positive value of *C*. But if q_1 diminishes to smaller and smaller values, then a point may be reached where *q_L becomes larger than q_1.* This will lead to negative values of C, but the exact determination of P/P_0 values where this transition takes place (from $q_1>q_L$ and $C>0$ to $q_1<q_L$ and $C<0$) *is not possible via the traditional linear plots.*

In any case the difficulty of choosing the pressure range for drawing the classical BET-plots leads sometimes to negative values of C- parameter, and consequently erroneous values of V_m. But even in the case where the choice of pressure range is correct for drawing the linear BET-plots, which is usually achieved with some experience and a few trials, the estimation of the *variable equilibrium constant C of adsorption, as said, is not possible via the classical treatment.* It will be shown how this can be achieved via the I-point method.

Aware of the problems described above, Rouquerol, Rouquerol and Imelik back in 1964 [7] made a very penetrating observation, namely that the corrected choice of pressure range, providing the corrected values of *ssa,* is that where the term $[V(1-P/P_0)]$ of the BET equation increases with the increase of *pressure* P/P_0. Beyond that limit, negative values of C are obtained. In essence this is the *first precise definition of a singular Inflection point of the BET equation*, termed *B-point* in various handbooks [3-5] and used for the determination of S_p. But the potential of this observation was not appreciated for a long time by the researchers in the field.

THE I-POINT OF ADSORTION ISOTHERMS

In 2004-2005 the *I-point method* was proposed by our group in Ioannina [8-10]. The method was realized by rearranging the standard BET equation (4a) in a non-linear form as follows:

$$\frac{V(1-P/P_0)}{P/P_0} = CV_m - (C-1)\,[V(1-P/P_0)] \qquad (7)$$

Then plots of the right hand part $V(1\text{-}P/P_0)\,/\,(P/P_0)$ *vs.* $V(1\text{-}P/P_0)$ provide lines, which have the shape of an inclined V, i.e. (>), (see Figure 4).

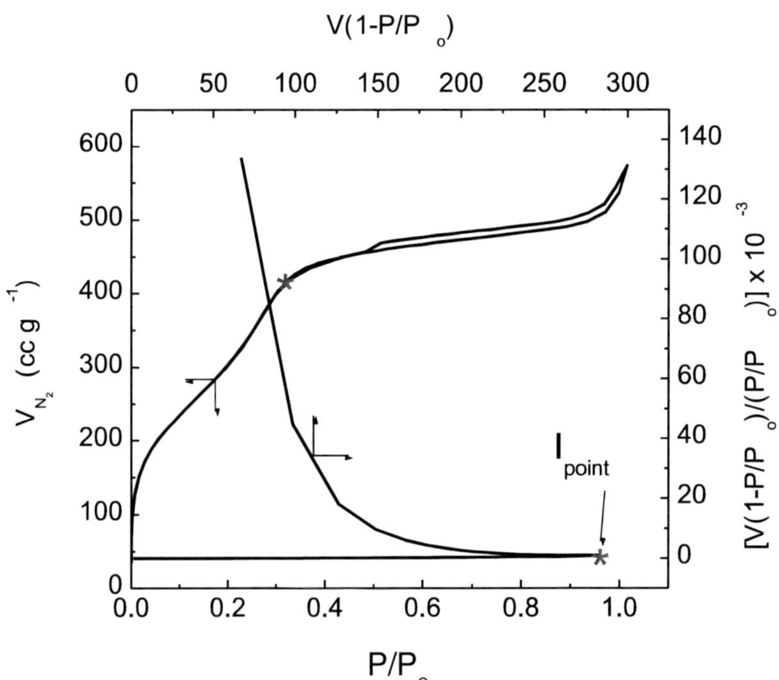

Figure 4. For appreciating the I-point method [18]. Left and down axes: The adsorption isotherm. Right and upper axes: The I-point plots. The projection of I-point on the upper $V(1\text{-}P/P_0)$ axis corresponds to V_m= 283 therefore easily S_p=4.356x283=1234 (m^2g^{-1}). The *ssa* determined by the classical BET treatment was found 1216 (m^2g^{-1}). The star (*) on the adsorption isotherm corresponds to the values P_I and V_I for which $V_m = [V_I(1\text{-}P_I)]$.

Such lines exhibit *an Inversion point, termed I-point*, which is the main subject of this chapter. The I-point has two very interesting features:

First, their slope, corresponding to {-(C-1)}, is not constant but varies from very large values at low P/P_0 to about zero at high P/P_0 as suggested indeed by Kembal and Schreiner [6]. Therefore the *values of C can be determined in the whole range of adsorption independently of the* V_m parameter. On the same token the *C*-value at the point where the formation of monolayer is completed can be also defined and corresponds, more or less, to its single value estimated during the linear BET treatment.

Second, the I-point where the slope of the inclined V-line changes from negative (upper part) to positive (lower part), projected on the horizontal $V(1-P/P_0)$ axis, (x-axis), corresponds *exactly* to the volume of monolayer V_m. So at the I-point we have

$$V_m = [V_I(1-P_I)] \tag{8}$$

where V_I and P_I is the volume and the pressure at the I-point. As a result the value of *S* is determined without any knowledge of *C* parameter with unique precision via the trivial relationship $S_p(m^2g^{-1})=4.356V_m(cm^3g^{-1})$. The above become clear in the typical example shown in Figure 4.

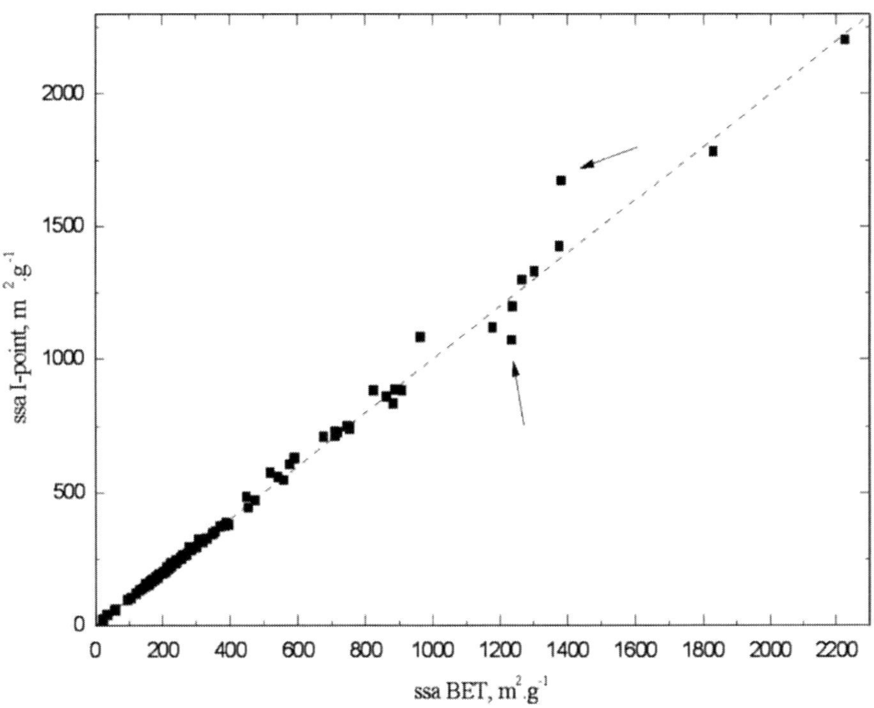

Figure 5. A comparison between the ssa determined by the BET method, (S_{BET}-horizontal axis) and the I-point method ($S_{I\text{-}point}$– vertical axis). The *ssa* of the tested materials was in the range from 60-2200 m^2g^{-1}. A perfect matching is observed with 1 or 2 exceptions, indicating by arrows that correspond to MCM–type mesoporous materials. The reasons are discussed in [8]. Reproduced with permission from [8]. Copyright 2013, Elsevier Science Publication.

In order to verify this method, in ref [8] one hundred and twenty seven (127) adsorbents were studied. Those materials belonged to quite different kind of solids, with different types of porosity, and included clays and pillared clays, mesoporous silicas, MCM's, mesoporous vanado-phosphoro-aluminates and alumino-phosphoro-ferrates with mixed micro- and meso-porosity, alumina containing heteroatoms, non-porous manganese based oxides, manganese-based mesoporous mixed oxides, commercial zeolites and finally samples of carbon cloths [8]. The *ssa* of the tested materials was in the range from 60-2200 m^2g^{-1} and a comparison between the *ssa* determined by the two methods, S_{BET} and $S_{I\text{-}point}$, is shown in Figure 5. A perfect matching is observed with 1 or 2 exceptions, which correspond to MCM–type mesoporous materials. The reasons for the mismatching, as discussed in [8], may be traced not in the I-plots but in the difficulty of drawing the straight lines in the BET treatment.

Consequently in 2006 the I-point methodology applied successfully for the determination of specific surface areas of various mesoporous materials like MCM-41, MCM-48, SBA-15 and SBA-16 containing microporosity [11] as well in studying the properties characterizing the complexity of pore structure of silicas by adsorption of nitrogen and methanol [12].

In 2007 Rouquerol, Llewellyn and Rouquerol, in order to define precisely the linear region of the BET equation, have used graphs of the form given by relation (9).

$$[V\,(1\text{–}P/P_0)]\ vs.\ P/P_0 \tag{9}$$

Such a typical graph is shown in Figure 6. Based on such graphs they proposed the following *two consistency criteria* [13]:

(i) The pressure range selected for drawing the linear plots in the classical BET treatment should have values of $V(1\text{-}P/P_0)$ increasing with P/P_0.

(ii) The *y* intercept of the linear region must be positive to yield a meaningful value for the *C*-parameter, which should be greater than zero.

Those consistency criteria are related to the original observation made by members of the same group in 1964 [7]. The next few years 2008-2010 these consistency criteria, and plots similar to (9), were applied extensively by the Snurr's group in order to study the inflections and steps in Carbon Dioxide Adsorption Isotherms in Metal-Organic Frameworks [14-15] and for evaluating the BET method in the determination of surface areas of MOFs and Zeolites that contain Ultra-Micropores [16]. These studies showed beyond any doubt that the consistency criteria are indeed a very useful guide for choosing the pressure range used in the fitting of adsorption data and obtaining correct results via the BET treatment, especially for microporous materials. Otherwise the determined *ssa* may be up to ~25% smaller.

In 2010 the group of Collins undertook the task of evaluation of the BET, I-Point, and α_s-plot procedures for the routine determination of external specific surface areas of highly dispersed and porous silicas [17]. After comparing the nitrogen adsorption isotherms of four candidate reference materials by means of shape ratio, a certain silica was chosen to be the reference material. Then twenty two (22) solids were evaluated for the effects of microporosity on the determination of their specific surface areas. The sample isotherms were then adjusted to minimize micropore adsorption and the BET plot and I-point procedures were applied to the external isotherms to give external specific areas from all three procedures. The authors conclude that "...... the I-point procedure, as proposed by Pomonis and collaborators [8-10] or the modified approach described by Rouquerol's and co-workers [7, 13] presents the advantage, for the many cases where the contribution of micropores to the specific surface area is unimportant, that it does not require prior definition of the (P/P_o) range in order to obtain the total specific surface area. However, when it is important to determine the external (i.e., non-micropore) specific surface area, the α_s-plot procedure has a significant advantage over the other procedures as no independent determination of the micropore capacity is required: both the micropore capacity and the external specific surface area are obtained from the same plot. For this, however, an appropriate reference isotherm, without micropore capacity or low P/P_o condensation, must be available".

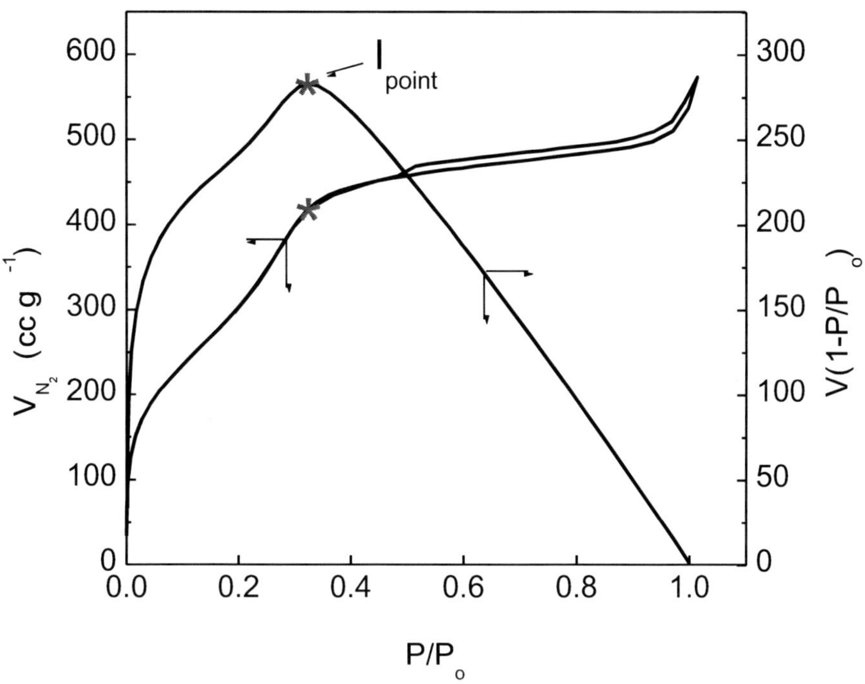

Figure 6. For appreciating the consistency criteria [13]. Left and down axes: The experimental N_2 adsorption isotherm, which is that of Figure 4. Right and upper axes: Plots $[V(1-P/P_0)]$ *vs* P/P_0 for the determination of consistency criteria. The correct *ssa*, via the classical BET treatment, is obtained if the linear part is restricted in the range up to the Inflection point of this plot. The star (*) is the I- point corresponding to the values P_I, V_I and $V_m = [V_I(1-P_I)]$. The I-point in Figure 4 and the Inflection point, shown above, are actually identical.

ADSORTION ISOTHERMS WITH TWO I-POINTS

The same year 2007 where the consistency criteria were proposed, it was also observed by our group that in some mesoporous materials of MCM-41 type exhibiting morphogenesis [18] as well of MSU type solids [19], the I-plots exhibit *not one but two (2) Inversion points/I-points* as seen in Figure 7. The appearance of two I-points was attributed to the existence of *different, discreet and distinct domains of microporosity and mesoporosity in the solids* [18, 19]. If the porosity is random and mixed such effects are masked.

The first inversion/inflection point at lower *(P/P_o), V* values, termed *"low pressure I_{LP}-point"*, corresponds to the completion of the adsorption procedure on the bare surface and the formation of V_m. This point is useful for the determination of the *ssa*.

The second inversion/inflection point at higher *(P/P_o), V* values, termed *"high pressure I_{HP}-point"*, corresponds to the completion of pore filling via condensation and may be useful for the determination of mesopore volume. If the spectrum of pore sizes is continuous and without gaps, the pores are filled at each and every pressure and no second kink appears.

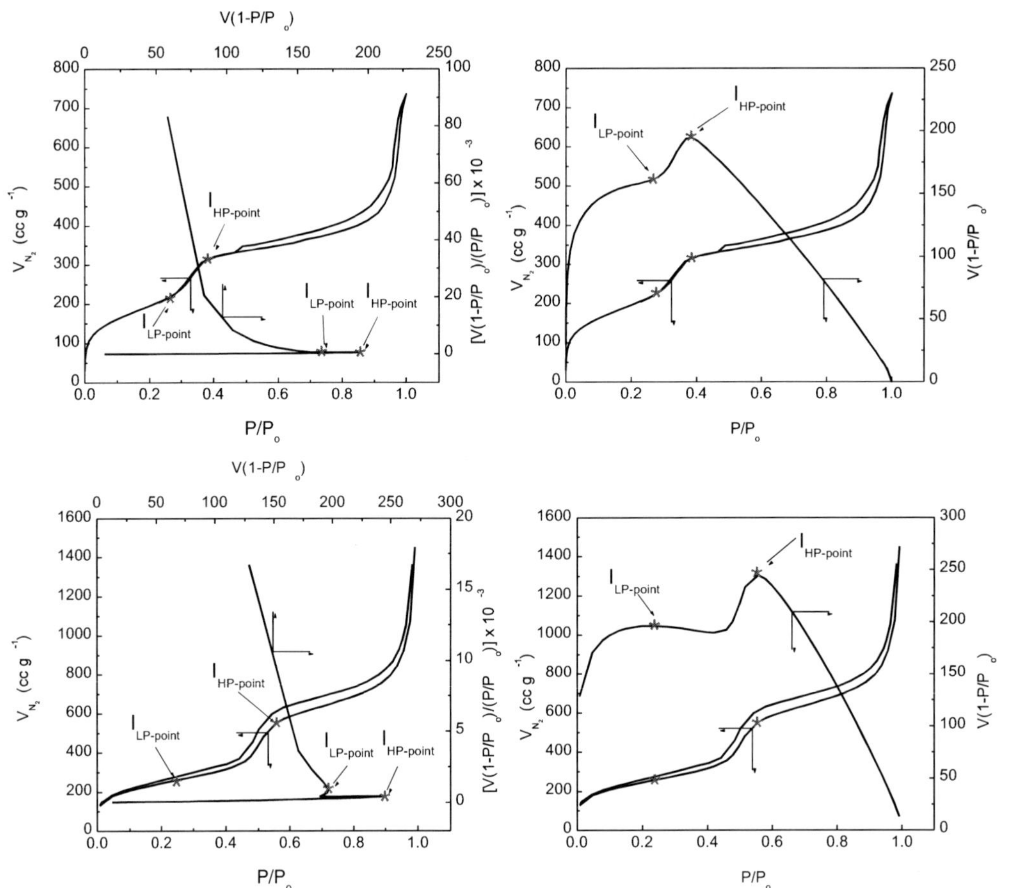

Figure 7. For appreciating the meaning of double *I-point* two examples of adsorption isotherms with two I-points are shown in the upper and lower part. Left and down axes: The original adsorption isotherms. Left figures, right and upper axes: The *I-point* plots. Right figures, right and down axes: The consistency criteria plots. The stars (*) in all graphs are the I- points (Inversion or Inflection points) which are actually identical. In the upper part example the lower pressure *I-point* (*I_{LP}-point*) is masked by the higher pressure *I-point* (*I_{HP}-point*). In the lower part example, the lower pressure *I-point* is distinct from the higher pressure *I-point*. It is the lower pressure *I_{LP}-point* which corresponds to the V_m, from which the *ssa* can be estimated [20]. Reproduced with permission from [20]. Copyright 2013, Elsevier Science Publication.

These conjunctions are elaborated in the data shown in Figure 8 taken from ref [20]. In this figure the difference *{I_{HP}-point minus I_{LP}-point} = ΔV (pore core volume)* is shown as a function of the (%) content of the purely mesoporous solid Al-MCM-41 in mechanical mixtures with purely microporous zeolite Beta. Details about the relevant data are in ref. [20].

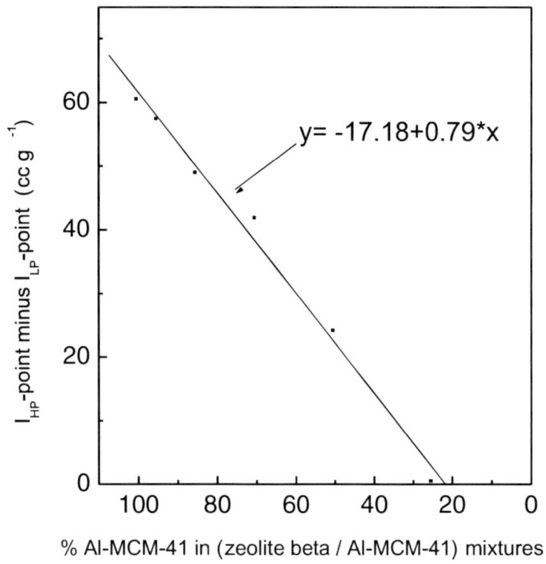

Figure 8. The difference {I_{HP}*-point minus* I_{LP}*-point*} = ΔV (pore core volume) as a function of % content of the purely mesoporous sample Al-MCM-41 in a mechanical mixtures with purely microporous zeolite Beta. Details about the relevant data are in ref [20]. Reproduced with permission from [20]. Copyright 2013, Elsevier Science Publication.

As seen in Figure 8, the difference {I_{HP}*-point minus* I_{LP}*-point*} is directly related to the (%) Al-MCM-41 content in the mechanical mixtures with zeolite-B via the relationship:

$$\Delta V = \{I_{HP}\text{-point minus } I_{LP}\text{-point}\} = 79\ (\%\ \text{Al-MCM-41}) - 17.2 \quad (10)$$

where the difference ΔV={I_{HP}*-point minus* I_{LP}*-point*} is expressed in volume units (cm^3) of adsorbed gaseous N_2. This relationship conveys the fact that the difference {I_{HP}*-point minus* I_{LP}*-point*} depends linearly on the content of mesopores in the mixture since only Al-MCM-41 contributes mesoporosity in the mixtures with zeolite. Furthermore, for the mechanical mixture (20% of Al-MCM-41 + 80% zeolite beta), then ΔV={I_{HP}*-point minus* I_{LP}*-point*} → *zero* (cm^3) which means that, at least for those materials and instrumentation, the distinction of organized mesoporous domains is not attainable in adsorption isotherms when the specific pore volume of those organized domains is less than about 20% of the rest not-organized volume.

Let us elaborate further this point: For the pure sample Al-MCM-41, the mesopore core volume corresponds to ΔV~60 cm^3 of gaseous nitrogen per gr of solid. This corresponds to geometrical volume 60x0.001518=0.091 cm^3, that tallies to the core of mesopores as seen in Figure 9. Since the total pore volume in that material was found equal to V_p= 0.96 cm^3/g [19], it follows that the mesopore core volume is only ~10% of the total pore volume. This may be the reason that in the example of Figure 8, the values of ΔV→*zero* when the content of Al-MCM-41 in the mechanical mixture with the microporous zeolite falls to ~20%. At this mass ratio, V_p= (20%)(0.96) = 0.19 cm^3g^{-1} and ΔV= (10%)(0.19) = 0.019 cm^3g^{-1} which corresponds to ~13 cm^3 of gaseous nitrogen. This value lays in the range of routine porosimetry.

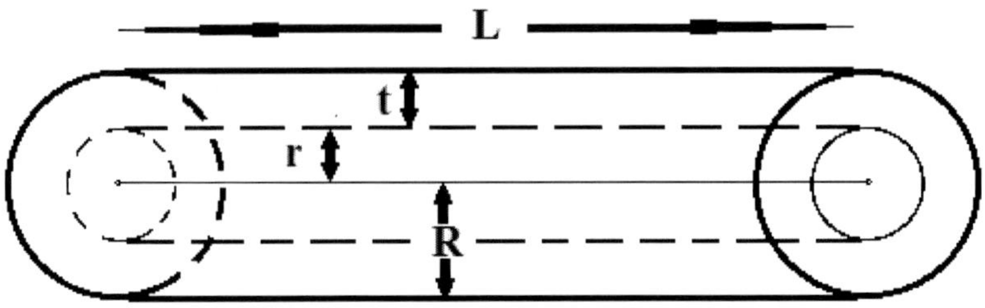

Figure 9. The core volume ΔV=*{I_{HP}-point minus I_{LP}-point}* filled during adsorption between the *I_{LP}-point* and the *I_{HP}-point* The initial pore radius is *R* and the thickness of the adsorbed layer at *I_{LP}-point* equals *t*.

Such data could be used for an educated approximation, and appreciation, of the number of filled mesopores, which is huge: Let assume that initially there are N mesopores, similar in size, with length *L* nm and radius *R* nm. As seen is Figure 8, such pores are suddenly filled-up at $P/P_0 \approx 0.5$. At this pressure, the adsorbed layer has a thickness of around *t=0.5nm (5Å)*. Then the core radius filled at $P/P_0 \approx 0.5$ has a value $r = (R-0.5)$ *nm*. Easily, the filled core volume will be $\Delta V = N\pi L(r)^2 = N\pi L(R-t)^2$. The values of R may be known from the classical pore size distribution while the relation between *L* and *R* may be estimated from the pore anisotropy *b=L/D=L/2R* [21]. For R=1.5nm and b=100, which are typical values for such MCM–type materials, the number of mesopores lays in the range $N \approx 6.10^{19}$ *per gr*. This value should be considered only indicative but shows the possibilities for mining information from such data.

The Attainability of C-Parameter

According to I-plots – see Eq. (7) - the slope of the diagrams $V(1-P/P_0)/(P/P_0)$ *vs* $V(1-P/P_0)$ is negative and corresponds to $(C-1)$ where C is the celebrated parameter of the BET equation. As evidence from the graph in Figure 4, this slope *is not constant*, especially at low P/P_0 values, except perhaps within a limited range at high pressure.

To be more precise the negative slope, at each pressure P/P_0, equals $(-s_i)=(C_i-1)$ therefore $C_i = (-s_i)+1$. Thus the C_i values can be estimated by differentiation of the plots $V(1-P/P_0)/(P/P_0)$ *vs* $V(1-P/P_0)$ and presentation of the results in the form $d[V(1-P/P_0)/(P/P_0)] / d[V(1-P/P_0)]$ *vs* P/P_0. Such a general treatment is shown in Figure 10. In the same Figure 10 the *C*–constant, calculated from the classical treatment of the BET equation, is also shown.

The graphs $C=f(P/P_0)$ provide a very clear view of the variation of *C*-parameter of BET equation as a function of partial pressure P/P_0:

(i) Initially at small P/P_0, the C-parameter is quite large, reaching values of *many thousands*. This part of the line reflects very high degree of condensation coefficient α_1 (see Eq.5) in the first layer as compared to the α_2 - value in the second layer, which does not actually exists yet.

(ii) As P/P_0 increases, the C-value decreases and between, let us say $0.05 < P/P_0 < 0.20$–0.30, it obtains values, which are almost—but not quite—constant. This region corresponds to the C-constant found from the linear BET-plots (Eq. 5). The ''flat'' part of the plots corresponds, more or less, to an ''average'' C-constant in this region of P_i, as actually suggested from the classical treatment of BET equation. In this region the condensation coefficients α_1 and α_2 coefficients should be about equal and the C-parameter should be given by Eq. (6) . Therefore

$$RT\, lnC = 0.6lnC = (q_1 - q_L) \tag{11}$$

where $RT = 0.6kJ/mol$ at $77\ K$. Then, let us say, for $20<C<1000$ we obtain $2<(q_1-q_L)<4kJ/mol$.

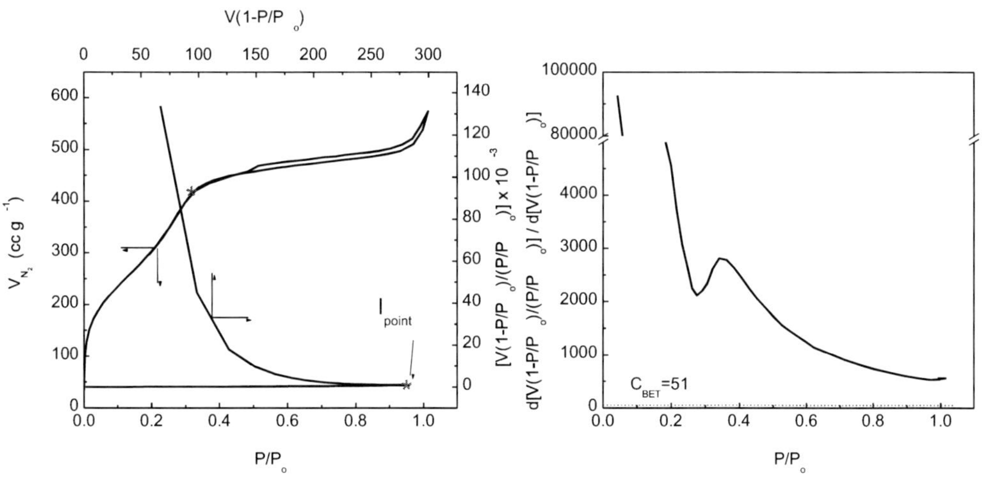

Figure 10. For the estimation of C-parameter via the I-plots: Left: The adsorption isotherm $V=f(P/P_0)$ and the I-plots $V(1-P/P_0)/(P/P_0)= f(V(1-P/P_0))$ shown in Figure 4. Right: The variation of the C-parameter as a function of P/P_0 estimated from the slope $d[V(1-P/P_0)/(P/P_0)] / d[V(1-P/P_0)]$ vs P/P_0. The discontinuity of the slope corresponds to the I-point. The C–constant, calculated from the classical treatment of the BET equation, is also shown.

Those values indicate that, at this range of adsorption, the heat of adsorption q_1 in the first layer is still slightly higher by about 2–4 kJ/mol compared to the heat of condensation q_L of nitrogen in the confine pore spaces.

(i) In the neighborhood of I-point the plots suffer an inversion and their slope alters from positive to negative. During this transition the C-values tend to infinitive. When the inversion ends, the C-values return to finite negative values. Although there is the possibility that this transition is an artifact, due to the peculiar form of I-plots, we think that this *corresponds to a real physical transition from adsorption to liquefaction,* as discussed next.

(ii) After the I-point, at higher values of P/P_0, the C parameter obtains small negative values.

$$(C\text{-}1)_{after\ the\ I\text{-}point} \approx -1.5 \text{ to } -2.5 \rightarrow C \approx -0.5 \text{ to } -1.5 \tag{12}$$

Therefore C appears to be negative and the difference $(q_I\text{-}q_L)$ cannot be calculated according to Eq. (11). A way to overcome this dilemma is to use the mathematical expansion

$$C = e^{\frac{q_1 - q_L}{RT}} = 1 + \frac{q_1 - q_L}{1!(RT)} + \frac{(q_1 - q_L)^2}{2!(RT)^2} + \cdots \tag{13}$$

After ignoring the terms equal or higher than second order in Eq. (13), and combining it with Eq. (12), we obtain

$$(C\text{-}1)_{after\ the\ I\text{-}point} = -1.5 \text{ to } -2.5 = (q_I\text{-}q_L)/RT \text{ or } (q_L\text{-}q_I) = 1 \text{ to } 1.5\ kJ/mol \tag{14}$$

In other words:

Before the I-point at lower $(P/P_0) : q_I > q_L$
After the I-point at higher $(P/P_0) : q_L > q_I$

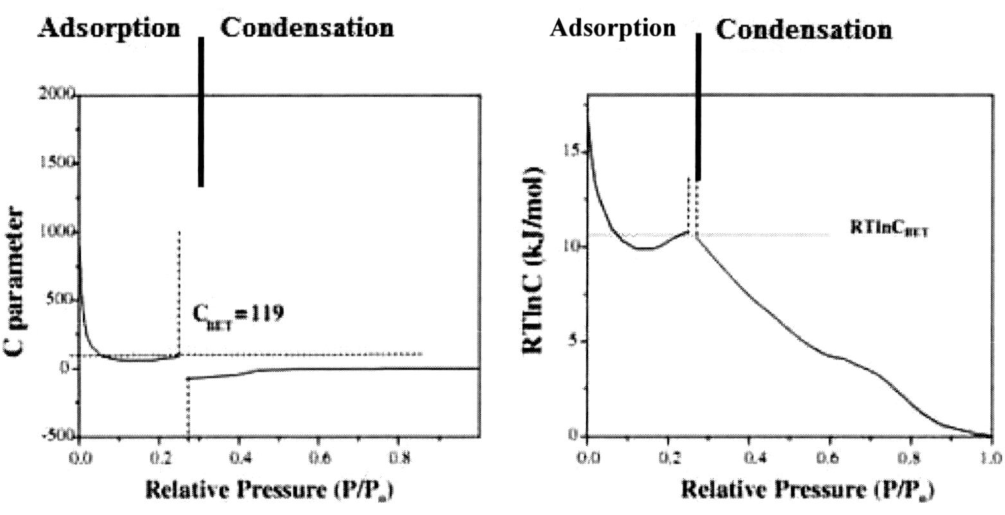

Figure 11. Left: Typical example of variation of C parameter in the whole pressure range $0<P/P_0<1$. The determined C_{BET} value is shown. Right: The corresponding chemical potentials $(q_I\text{-}q_L)= RTlnC$ in the adsorption range $(0< P/P_0< P_I)$ and $(q_L\text{-}q_I) = RTlnC$ in the condensation range $(P_I< P/P_0< 1)$. Data are from ref. [9]. Reproduced with permission from [9]. Copyright 2013, Elsevier Science Publication.

This is equivalent as to accept that the adsorbate–adsorbate interaction becomes stronger than the adsorbate–adsorbent interaction after the I-point (higher P/P_0) and therefore liquefaction takes place there. This is actually the case for the type-III isotherm according to IUPAC. A final consequence, stemming from the above discussion, is that the parameter $RTlnC = (q_I\text{-}q_L)$ expresses the *chemical potential of adsorption* before the I-point in the low pressure range $0<P/P_0<P_I$.

On the contrary at the higher pressure range $P_I< P/P_0\leq 1$, the potential is $RTlnC = (q_L\text{-}q_I)$ and corresponds to the *chemical potential of condensation* after the I-point. Both potentials, of adsorption as well of condensation, drop with the increase of partial pressure P/P_0 as shown in a typical example in Figure 11.

THE ESTIMATION OF MICROPOROSITY

The purpose of this section is to show how the information extracted from the I-plots can be used in order to distinguish between microporosity and mesoporosity of porous materials. The method proposed is based on the whole spectrum of C values estimated in the previous section, as well as on the value of V_m determined via the I-point graphs. In a sense this methodology is self-consistent without need for any additional information from theoretical isotherms or experimental reference material.

It is reminded that the most usual ways of determining microporosity in materials with mixed micro- and meso-porosity are the *t-plot* and *a_s-plot* methods [3-5]. These two approaches are based on the same basic concept, namely *the mathematical comparison of the test isotherm with another isotherm of either theoretical (standard isotherm) or experimental nature.*

We shall address the problems arising when using those treatments and, at the same time, indicate the usefulness of the proposed method.

(i) *In the t-plot method, a theoretical ''standard isotherm'' is needed.* This should be drawn according to some ''standard values'' of C-parameter of the BET equation proposed by Brunauer and his associates [22] and by Lecloux & Pirard [23]. The last two authors have suggested that the standard five isotherms are characterized by C-values in the following range: $C>300$; $300>C>100$; $100>C>40$; $40>C>30$ and $30>C>20$ and similar proposal are found in Ref. [22]. But clearly C is not constant at all, as explained in the previous section, and only as we approach the I-point its value tends to the constant value estimated from the traditional BET treatment.

(ii) *In the a_s-plot method, an experimental ''standard isotherm'' is used.* If a standard isotherm is not employed, then we need as reference the adsorption isotherm of a ''reference sample'', chemically similar to the one under test but without porosity. Then

$$\frac{ssa_{(test)}}{ssa_{(reference)}} = \frac{b_{(test)}}{b_{(reference)}} = \frac{V_{0.4(test)}}{V_{0.4(reference)}}$$

where $V_{0.4}$ is the N_2 volume adsorbed at $P/P_0 = 0.4$, where the formation of monolayer is completed under any circumstances, and $b = V_m/\sigma$ where $\sigma = 3.54\text{Å}$ is the thickness of a single monolayer [24]. The questions in this case are related to the reference sample and another to the choice of the value $(P/P_0) = 0.4$ for comparison.

It will be shown how the I-point method can be used to separate and estimate the microporosity from the mesoporosity of porous solids without resolving to the use of any standard isotherm or any reference sample or any pre-fixed and pre-determining pressure for the formation of monolayer [10].

Let us be more precise: The decrease of C parameter as a function of the surface coverage, $\theta = V/V_m$, is well known and related to the fact that the heat of adsorption decreases, either linearly or exponentially, with the surface coverage [3-6]. At $\theta = V/V_m \rightarrow 0$ the surface possesses a much higher potential to attract and bind adsorbed species compared to the case $\theta = V/V_m \rightarrow 1$, where the surface bonding potential for adsorption in nullified.

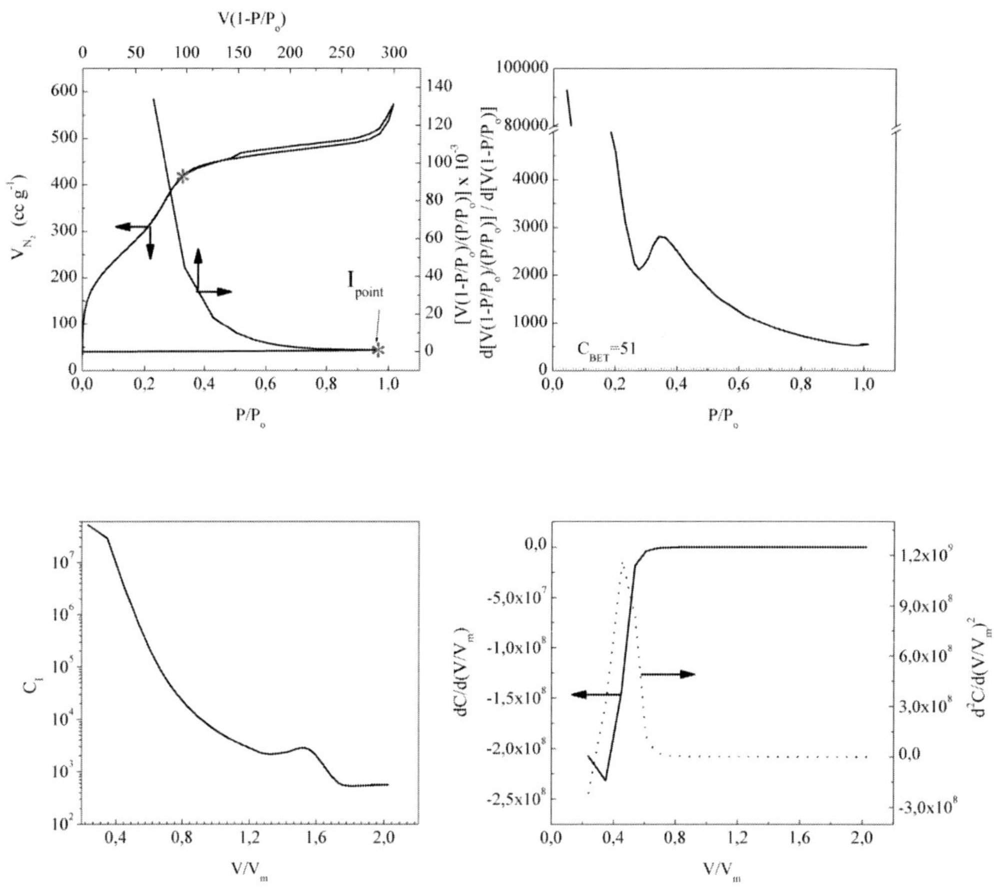

Figure 12. For the estimation of microporosity via the I-point methodology: Upper left: The adsorption isotherm $V=f(P/P_0)$ and the I-plots $V(1-P/P_0)/ P/P_0 = f(V(1-P/P_0))$ shown in Figure 4. Upper right: The variation of the C- parameter as a function of P/P_0 estimated from the slope $d[V(1-P/P_0)/P/P_0] / d[V(1-P/P_0)]$ *vs.* P/P_0. The discontinuity of the slope corresponds to the I-point. The C–constant, calculated from the classical treatment of the BET equation, is also shown. Bottom left: Plots of C *vs.* (V/V_m). Bottom right: The derivative $dC/d(V/V_m)$ *vs.* (V/V_m). The microporosity ends at the range where $dC/d(V/V_m) \rightarrow$ zero. The second derivative $d^2C/d(V/V_m)^2$ as a function *of* (V/V_m) is also shown with the dashed line.

Now a very important point is not actually the net variation of C but its relative variation $dC/d(V/V_m)$ as a function of (V/V_m). The microporosity ends at the pressure range where

$dC/d(V/V_m) \rightarrow zero$. Details of such calculations are shown in Figure 12 for the same sample shown in Figures 4 and 10.

Using this kind of treatment we have estimated the microporosity of ten (10) mesoporous silicas with different kinds of mixed micro- and mesoporosity [10] and the results were compared with those obtained via the a_s - *plot* method. As seen in Figure 13 there appears satisfactory matching with a discrepancy of (±5%): At lower % microporosity, the I-method treatment tends to yield higher values of (% microporosity) and vice versa.

It is not clear which results are closer to reality but in any case such data may be influenced by the density of adsorption points in the adsorption isotherm, especially in the lower pressure range: The more points in the adsorption isotherm, the more precise the estimation of local slopes (see Figure 12) and the more reliable the outcome.

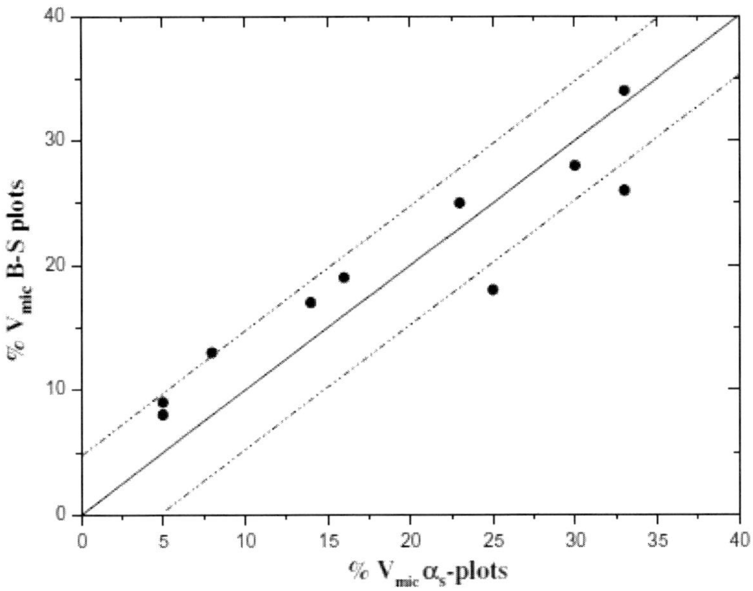

Figure 13. Estimated microporosity of ten (10) mesoporous silicas with different kinds of mixed micro- and mesoporosity from ref. [10]. Horizontal axis: Results estimated via the $a_s - plot$ method. Vertical axis: Results estimated via the *I - plot* methodology. At lower % microporosity, the I-method treatment tends to yield higher values of (%microporosity) and vice versa. Reproduced with permission from [10]. Copyright 2013, Elsevier Science Publication.

The Adsortion As a Titration

In 2011 it was recognized [20] that the derivatives of the Inflection-type graphs described by Rouquerols and co-workers [7, 13], bear remarkable similarities with the derivative of the corresponding adsorption isotherm. Namely the plots $d(V[1-(P/P_o)])/ d[(P/P_o)]$ vs. (P/P_o) on one hand and $dV/d[(P/P_o)]$ vs. (P/P_o) on the other, exhibit similar shape and their maximum appears exactly at the same (P/P_o) value. In addition the derivative $d(V[1-(P/P_o)]) / d[(P/P_o)]$ exhibits systematically lower values compared to $dV/d[(P/P_o)]$. Such derivatives are shown in Figures 14 for a typical adsorption isotherm with just one I-point and in Figure 15 for a typical adsorption isotherm with two I-points. The consequences of this distinction will be clarified in a minute.

The similarities and the differences between those derivatives were explained as follows [20]: The derivative $d(V[1-(P/P_o)])/d[(P/P_o)]$ equals $(1-P/P_o)[dV/d(P/P_o)]+V$. Therefore if each local slope $[dV/d(P/P_o)]$ is determined, and this local slope is multiply by each local value $[1-(P/P_o)]$ and then in the result the local V values are added, the outcome corresponds to $d(V[1-(P/P_o)])/d[(P/P_o)]$. At very low P,V values $(P,V \rightarrow 0)$, and the total effect , i,e. multiplication of $[dV/d(P/P_o)]$ by $[1-(P/P_o)]$ and addition of V, will be negligible. As the P, V values increase, the effect becomes considerable but does not affect the overall shape of the curve $[dV/d(P/P_o)]$. It is only at high P,V range, i.e. at $(P/P_o)>08\text{-}0.9$, where the overall effect becomes so important as to change the shape of the curves, but this range is not of use in the present study. In order to refine further the research two cases are distinguished. The 1st case corresponds to sample(s) with single Inversion or Inflection (I-) point – see Figure 14. The 2nd case corresponds to sample(s) with two Inversion or Inflection (I-) points – see Figure 15.

1st case: Samples with single I-point –see Figure 14.

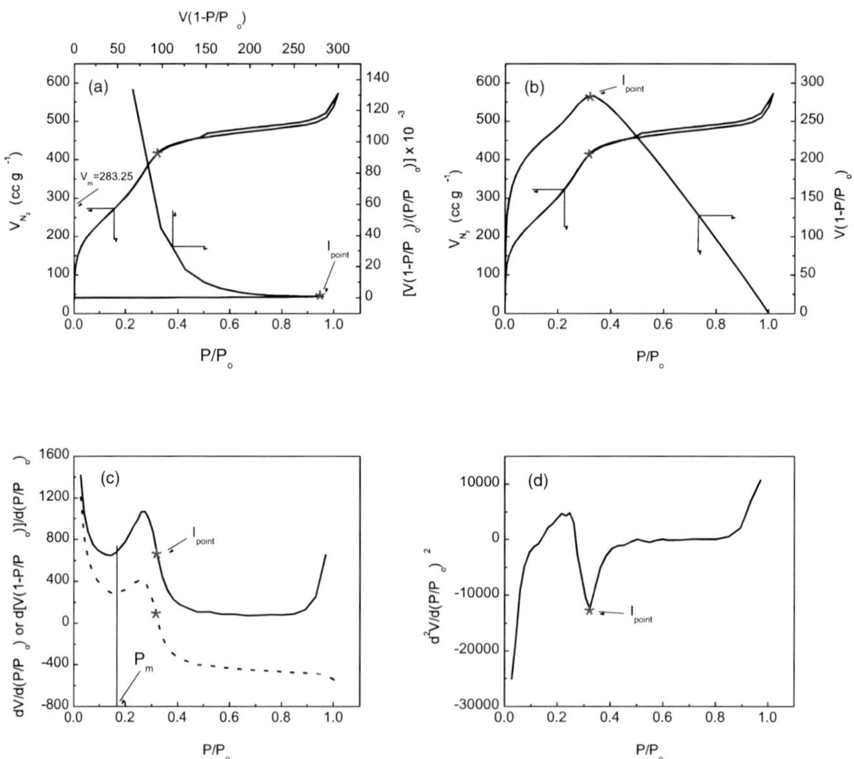

Figure 14. Graphs for OMSi–Ce–Co/1:9/A porous material (Organized Mesoporous Silicate containing Ce and Co in 1:9 ratio) [20]. (a) the N_2 adsorption –desorption isotherms as well as the I-point plot are shown in the form $(V[1-(P/P_o)]/(P/P_o)$ *vs* $V[1-(P/P_o)])$. The I-point is indicated by a star (✱). (b) plot of the form $(V[1-(P/P_o)])$ *vs.* (P/P_o). The N_2 adsorption-desorption isotherms $V=f(P/P_o)$ are shown for comparison. (c) plots in the form dV/d$[(P/P_o)]$ *vs.*(P/P_o) (solid line) and $d(V[1-(P/P_o)])/d[(P/P_o)]$ *vs.* (P/P_o) (dashed line). (d) plot in the form $d^2V/d[(P/P_o)]^2$ *vs.*(P/P_o). The star (✱) corresponds to the single I-point. Reproduced with permission from [20]. Copyright 2013, Elsevier Science Publication.

The above observations in relation to the derivatives have the following important consequences as exemplified in Figure 14 for a OMSi–Ce–Co/1:9/A porous material (Organized Mesoporous Silicate containing Ce and Co in 1:9 ratio) taken from ref. [20]: If the intercept of the maximum of the $V(1\text{-}P/P_o) = f(P/P_o)$ curve (Figure 14b), or the I-point (Figure 14a), corresponds to V_m, from which the values of specific surface area S_p can be determined ($S_p = 4.356V_m$ for N_2), then this point, at exactly the same (P/P_o) value corresponds also to the second peak, at negative values, of the second order derivative $d^2[V(1\text{-}P/P_o)]/d(P/P_o)^2$ or $d^2V/d(P/P_o)^2$ (Figure 14c). Therefore the volume of the monolayer V_m corresponds to, and can be estimated directly from, the derivative. In this case the volume (V_I) and the pressure (P_I/P_o) at the Inversion or Inflection point are the volume and the pressure that corresponds to the second peak (minimum) of the second derivative of the adsorption isotherm. *Therefore the S_p values can be estimated from the raw P, V data, from the simple second order derivative $d^2V/d(P/P_o)^2$ of the adsorption isotherm, without any knowledge, or use, or treatment of the BET equation.* As shown in Figure 14 this is exactly the case.

2nd case: Samples with two I-points –see Figure 15.

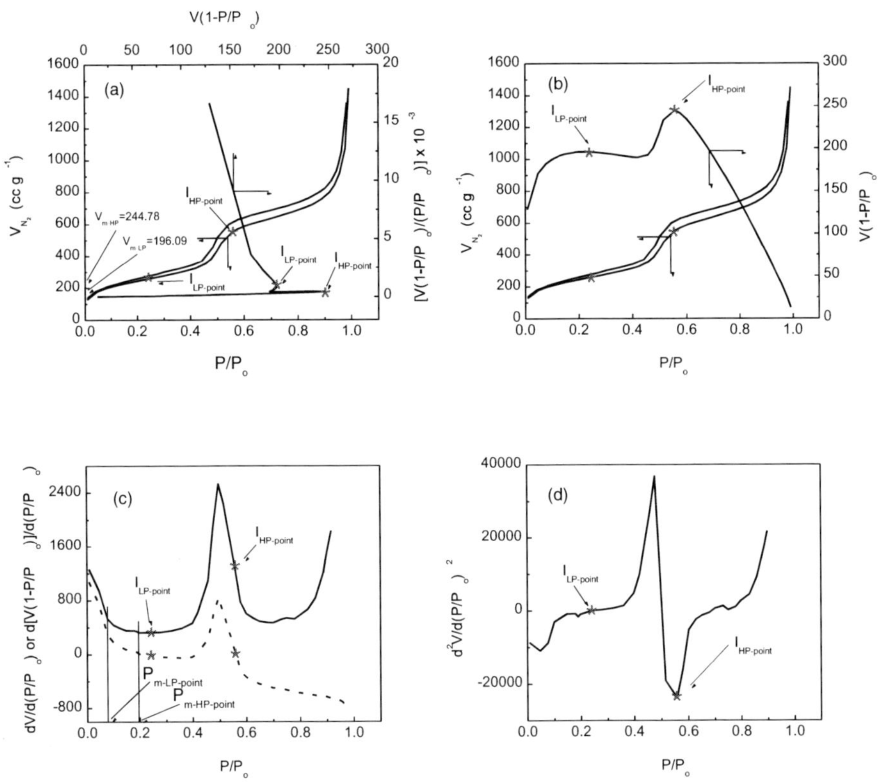

Figure 15. Graphs for 2%Al-MSU-S/W(BEA) porous sample [20]. (a) the N_2 adsorption –desorption isotherms as well as the I-point plot are shown in the form $(V[1\text{-}(P/P_o)]/(P/P_o)$ *vs* $V[1\text{-}(P/P_o)]$. I_{LP}*-point* and the I_{HP}*-point* are indicated by a star (✱). (b) plot of the form $(V[1\text{-}(P/P_o)])$ *vs* (P/P_o). The N_2 adsorption-desorption isotherms $V=f(P/P_o)$ are shown for comparison. (c) plots in the form $dV/d[(P/P_o)]$ *vs.*(P/P_o) (solid line) and $d(V[1\text{-}(P/P_o)])/d[(P/P_o)]$ *vs.* (P/P_o) (dashed line). (d) plot in the form $d^2V/d[(P/P_o)]^2$ *vs.*(P/P_o). The two stars (✱) correspond to the two distinct I-points. Reproduced with permission from [20]. Copyright 2013, Elsevier Science Publication.

As referred previously for some mesoporous materials, the I-plots show *not one but two Inversion points/I-points*. This is shown in Figure 15 for a porous sample termed 2%Al-MSU-S/W(BEA), the data taken from ref [20]. The appearance of two I-points was attributed to the existence of *different, discreet and distinct domains of microporosity and mesoporosity in the solids.* If the porosity is random such effects are masked. A careful scrutiny of the data have shown that the first inversion/inflection point at lower *(P/P$_o$), V* values, termed *"I_{LP}-point"*, corresponds to the completion of the adsorption procedure on the bare surface and the formation of monolayer and is useful for the determination of specific surface area S_p. The second inversion/inflection point at higher *(P/P$_o$), V* values, *"I_{HP}-point"*, corresponds to the knee of the adsorption isotherm or in other words to the completion of the pore filling via condensation. In that case the second derivative (Figure 15c) provides a minimum that correspond rather to the completement of the pore filling via condensation rather than the completement of monalayer via adsorption.

The above discussion indicates that in the classical porosimetry experiments for the determination of specific surface areas of solids, the adsorption of N_2 can be considered as a kind of titration of the adsorbent material (analyte or titrand) by the adsorbate N_2 (titrant or titrator). The obtained data in the form *V= f(P)* bear physicochemical similarities with the well-known acid–base titration curves. The Inversion or Inflection points of adsorption isotherms, termed I-point(s), correspond to the equivalent points of titration. After some simple mathematical manipulations they can be used for the determination of the "equivalent" points of any adsorption isotherm: The first I-point corresponds to the completion of monolayer coverage and can be used for the determination of *ssa*. The second I-point (if exists) corresponds to completion of pore filling by condensation and can be employed for extracting data referred to the mesopores.

Summary and Conclusion

In this Chapter the *I-point(s) method* for the determination of the specific surface area (*ssa*) of porous materials based on N_2 adsorption data, $V=f(P/P_o)$ is presented in depth.

This methodology, which provides unique precision and convenience, is based on the Inversion or Inflection point(s), termed collectively *I-point(s),* which appeared in all N_2 adsorption isotherms $V=f(P/P_o)$ in porous solids.

Such *P,V* data plotted in the form $V(1\text{-}P/P_o)/(P/P_o) = f(V(1\text{-}P/P_o))$ show one I-point, if mesopores are spatially mixed with micropores or two I-points if mesopores form separate domains in the porous solid. The projection of the single I-point, or the low pressure I point, on the $(V(1\text{-}P/P_o))$ axis tallies exactly to the volume of monolayer V_m from which the S_p values can be estimated with unique precision.

The *I-point(s)* correspond to the equivalent point(s) of titration of the adsorbent porous material (analyte or titrand) by the adsorbate N_2 (titrant or titrator) at 77K.

If a single I-point appears in the adsorption curve, this corresponds to the completement of physical adsorption and is useful for the determination of the specific surface area -*ssa*- of solid without any knowledge or use of the traditional BET methodology.

If a second equivalent point appears in the adsorption isotherm, this corresponds to the completement of the condensation of N_2 into the capillary pores.

The method provides also a self-consistent methodology for the determination of microporosity without resolving to reference isotherm or reference non-porous material.

Last, but not least, the variable slope of the I-plots corresponds to the *variable equilibrium constant C of adsorption,* introduced in the BET equation, which is not attainable via the classical linear treatment.

The above are summarized in the following Table 1.

Table 1. The N_2 adsorption $V=f(P/P_o)$, the Langmuir isotherm, the BET equation, the I-point method and the adsorption as a titration

Isotherms and their variables	Strong points	Observations	Refs
The Langmuir isotherm (eqs. 1a and 1b).	From $V=f(P/P_o)$ data and linear plots, the *ssa* of non- porous materials and a unique adsorption 'constant' *C* can be determined.	-Not for porous solids. -The *K* parameter is considered constant, which is not.	[1]
The BET isotherm (eqs. 4a and 4b).	From $V=f(P/P_o)$ data and linear plots, the *ssa* of porous materials and a unique adsorption 'constant' *C* can be determined.	-The *C* parameter is considered constant, which is not. -Questionable range of linearization.	[2]
The consistency criteria (eq. 9).	Precise definition of the linear part of BET plots.	-Accurate definition of *ssa*. -The *C* parameter is considered constant, which is not.	[7, 13]
The I-point plots (eq. 7) (single inversion, one I-point)	From $V=f(P/P_o)$ data and non-linear plots unique definition of ssa and definition *C* in the whole range of adsorption is achieved	-Non-linear plots. -Accurate definition of *ssa*. -The variation of *C* parameter can be followed.	[8-12]
The I-point plots (eq. 7) (double inversion, two I-points)	As for the single I-points but appears in cases of discreet domains of microporosity and mesoporocity in porous solids.	-Non-linear plots. -The lower pressure I-point corresponds to the surface coverage. -The higher pressure I-point corresponds to pore condensation.	[18,19]
Second order derivative $d^2V/d(P/P_o)^2$ of the adsorption isotherm.	The *ssa* values can be estimated from the minimum of such plots without any knowledge, or use, or treatment of the BET equation or any other model.	-No model is used. -The adsorption is considered as a titration. -The minimum of the second derivative corresponds to the I-point. -Valid for isotherms with a single I-point.	[20]

ACKNOWLEDGMENTS

This research has been co-financed by the European Union (European Social Fund – ESF) and Greek national funds through the Operational Program "Education and Lifelong Learning" of the National Strategic Reference Framework (NSRF) - Research Funding Program: THALIS- Investing in Knowledge Society through the European Social Fund.

REFERENCES

[1] Langmuir, I., *J. Am. Chem. Soc.,* 1916, 38, 2221-2295.
[2] Brunauer, S.; Emmett, P.H.; Teller, E., *J. Am. Chem. Soc.,* 1938, 60, 309-319.
[3] Lowell, S.; Shields, J.E. *Power Surface Area and Porosity*; 3rd Edition, Chapman and Hall - Kluwer, 1991.
[4] Gregg, J.S.; Sing, K.S.W., Adsorption, Surface Area and Porosity, Academic Press, 1982.
[5] Rouquerol, F.; Rouquerol, J.; Sing, K. S. W., Adsorption by Powders and Porous Solids: Principles, Methodologies and Applications, Academic Press, 1999.
[6] Kembal, C.; Schreiner, G.D.L., *J. Am. Chem. Soc.,* 1950, 72, 5605- 5607.
[7] Rouqeurol, F.; Rouquerol, J.; Imelik, B., *Bull. Soc.Chim.Fr.,* 1964, 635-639.
[8] Pomonis, P.J.; Petrakis, D.E.; Ladavos, A.K.; Kolonia, K.M.; Armatas, G.S.; Sklari, S.D.; Dragani, P.C.; Zarlaha, A.; Stathopoulos, V.N.; Sdoukos, A.T., *Microp. Mesop. Mater.,* 2004, 69, 97-107.
[9] Pomonis, P.J.; Petrakis, D.E.; Ladavos, A.K.; Kolonia, K.M.; Pantazis, C.C.; Giannakas, A.E.; Leontiou, A.A., *Catal. Commun.,* 2005, 6, 93-96.
[10] Armatas, G.S.; Petrakis, D.E.; Pomonis, P.J., *Microp. Mesop. Mater.*, 2005, 83, 251-261.
[11] Knowles, J.K.; Armatas, G.S.; Hudson, M.J.; Pomonis, P.J., *Langmuir*, 2006, 22, 410-418.
[12] Denoyel, R.; Meneses, J.M.; Armatas, G.S.; Rouquerol, J.; Unger, K.K.; Pomonis, P.J., *Langmuir*, 2006, 22, 5350-5357.
[13] Rouquerol, J.; Llewellyn, P.; Rouquerol, F., *Stud. Surf. Sci. Catal.*, 2007, 160, 49-56.
[14] Walton, K.S.; Millward, A.R.; Dubbeldam, D; Frost, H.; Low, J.J.; Yaghi, O.M.; Snurr, R.Q., *J. Am. Chem. Soc.*, 2008, 130, 406-407.
[15] Yazaydın, A.O.; Snurr, R.Q.; Park, T.H.; Koh, K.; Liu, J.; LeVan, M.D.; Benin, A.I.; Jakubczak, P.; Lanuza, M.; Galloway, D.B.; Low, J.J.; Willis, R.R. *J. Am. Chem. Soc.*, 2009, 131, 18198-18199.
[16] Bae, Y.S.; Yazaydın, A.O.; Snurr, R.Q., *Langmuir* 2010, 26, 5475-5483.
[17] Collins, K.E.; Collins, C.H.; Maroneze, C.M.; Cappovila, V.; Custodio, R., *Langmuir*, 2010, 27, 187-195.
[18] Katsoulidis, A.P.; Petrakis, D.E.; Armatas, G.S.; Pomonis, P.J., *J. Mater. Chem.*, 2007, 17, 1518-1528.
[19] Triantafyllidis, K.S.; Pinnavaia, T.J.; Iosifidis, A.; Pomonis, P.J., *J. Mater. Chem.*, 2007, 17, 3630-3638.
[20] Ladavos, A.K.; Katsoulidis, A.P.; Iosifidis, A.; Triantafyllidis, K.S.; Pinnavaia T.J.; Pomonis, P.J., *Micropor. Mesopor. Mater.* 2012, 151, 126-133.
[21] Pomonis, P. J.; Armatas, G.S., 2012. Chapter 3, "The Pore Anisotropy in Porous Solids" in "Anisotropy Research" Editor H. Lemu, Nova Science Publishers, New York, p.45.
[22] (a) Brunauer, S.; Mikhail, R.M.; Bodor, E.E., *J. Colloid Interface Sci.,*1967, 24, 451-463; (b) Hanna, K.M.; Odler, I.; Brunauer, S.; Hagymassy, J.; Bodor, E.E. *J. Colloid Interface Sci.,*1973, 45, 27-37.
[23] Lecloux, A.; Pirard, J.P., J. *Colloid Interface Sci.,* 1979, 70, 265-281.
[24] (a) Lippens, B.C.; deBoer, J.H., *J. Catal.,*1965, 4, 319-323; (b) Lippens, B.C.; Linsen, B.G.; de Boer, J.H., *J. Catal.,*1964, 3, 32-37.

In: Comprehensive Guide for Mesoporous Materials, Volume 2 ISBN: 978-1-63463-982-8
Editor: Mahmood Aliofkhazraei

Chapter 12

ADSORPTION OF TWO OR MORE MOLECULES ON ONE CATIONIC CENTER IN POROUS MATERIALS

K. Hadjiivanov, E. Ivanova, M. Mihaylov and K. Chakarova
Institute of General and Inorganic Chemistry,
Bulgarian Academy of Sciences, Sofia, Bulgaria

ABSTRACT

In most cases the chemisorption process is described in terms of formation of adsorbate complexes of 1 : 1 stoichiometry, *i.e.* one surface site binds one adsorbate molecule. However, there are two more possibilities: (i) one adsorbate molecule bridges more than one site and (ii) one site binds two or more adsorbate molecules. The second possibility is typical of cations in porous materials. In this chapter we summarize the achievements on why and when two or more identical or different molecules can be simultaneously attached to one cationic site.

INTRODUCTION

Porous crystalline solids are an important class of materials and have found a wide application in different branches in chemistry, as adsorption, catalysis, etc. The properties of these materials can be tuned in different ways, e.g. (i) through controlling the pore size or (ii) by appropriate chemical modification. In the latter case, incorporation of different cations in the materials is an often used and useful approach. These cations can act as active sites in many catalytic reactions or are sites for specific adsorption. That is why detailed knowledge of the coordination chemistry of such cations is of great importance not only from fundamental point of view but also for design of materials with pre-set properties.

Chemisorption is specific and proceeds *via* bonding of adsorbate molecules to active centers, *e.g.* coordinatively unsaturated surface ions or atoms, OH groups, etc. In most cases the chemisorption process is described in terms of formation of adsorbate complexes of 1 : 1

stoichiometry, *i.e.* one surface site binds one adsorbate molecule. However, there are two more stoichiometric possibilities:

- one adsorbate molecule bridges more than one site;
- one site binds two or more adsorbate molecules.

The first possibility is mainly realized on metals where bridging surface species are often formed [1]. There are also some particular cases (mainly with certain zeolites) where the adsorbate is bound to the so-called dual cationic sites [2,3]. In these cases the cations should be situated on a suitable distance and the adsorbate molecule normally bridges these cations with different atoms.

In this chapter we will concentrate on the second possibility, *i.e.* binding of two or more guest molecules to one adsorption site. There is no generally accepted classification system of these complexes. Hereafter we will use the classification shown of Figure 1.

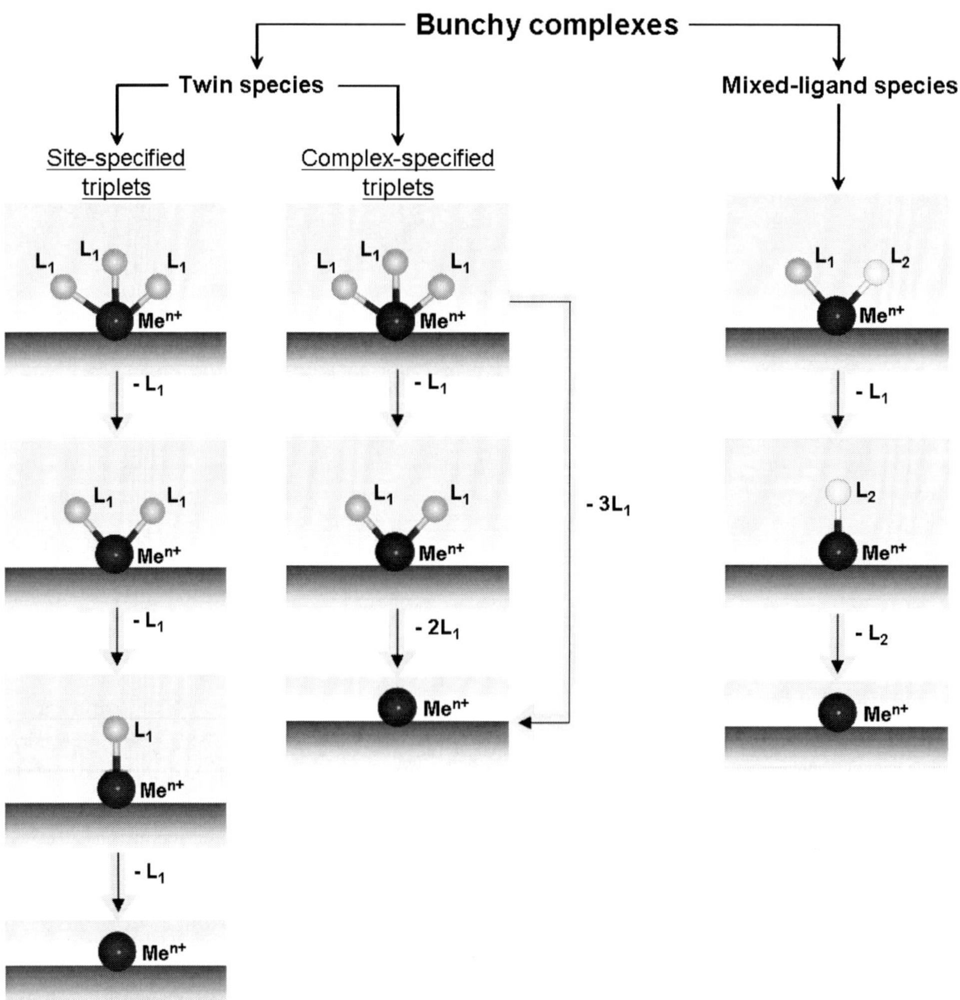

Figure 1. Schematic presentation of different kinds of bunchy complexes. The arrows show the evolution of the complexes during decomposition.

All adsorption complexes having more than one adsorbate molecule will be called *bunchy* complexes. When the adsorbed molecules are identical, the complexes are *twin*-species. Twin species with two adsorbed molecules are *geminal.* Species with three or four identical guest ligands are also often called geminal, but more accurate denotation seems to be *triplets* and *quadruplets*, respectively. Note that the proposed classification takes into account only the adsorbed molecules although the cation is also coordinated to some atoms from the support framework. Consequently, the twin complexes are denoted simply as $Me^{n+}(L)_n$, where Me^{n+} is the cation, L, the adsorbed ligand, and n, the number of adsorbed ligand molecules. When at least one of the adsorbed molecules is different from the others, the complexes formed are *mixed-ligand* species and are denoted as $Me^{n+}(L_1)_n(L_2)_m$. Some large molecules can be attached to the surface by two functional groups producing chelate complexes. Formally, these species may be regarded as *pseudo-bunchy* complexes.

The most used technique for studying bunchy complexes is infrared spectroscopy because it can selectively detect species with different number of ligands. The spectral performance of twin bunchy species in porous materials is summarized in Table 1, whereas the data for mixed-ligand species are shown in Table 2.

Table 1. Spectral performance of twin surface species formed with cations in porous materials when interacting with CO, NO and N_2.The complex-specified species are shown in grey cells

Cation	Porous material	Ligand	Characteristic IR frequencies, cm^{-1}				Ref.
			$M(L)_1$	$M(L)_2$	$Me(L)_3$	$Me(L)_4$	
Li^+	MFI		2195	2188			[4]
Na^+	FAU	CO	2175	2167-64			[5, 6]
Na^+	FAU	CO	2171.5	2159.5	2148,5		[7]
Na^+	MFI	CO	2178	2170			[8]
Na^+	ETS-10	CO	2176	2164			[9]
Na^+	FAU	$^{15}N_2$	2158,7	2255.5	.	.	[6,10]
Na^+	EMT	$^{15}N_2$	2158,5	2255	.	.	[11]
Na^+	FAU	N_2	2336.8	2333.5	.	.	[12]
Na^+	ETS-10	N_2	2133	2131			[9]
K^+	EMT	CO	2175	2164			[11]
K^+	EMT	N_2	2252	2251.5			[11]
Mg^{2+}	FER	CO	2205	2194	-	-	[13]
Ca^{2+}	FAU	CO	2201	2191	2185		[5,14,15]
Sr^{2+}	FAU	CO	2193	2187.5	2182.5		[15]
Ba^{2+}	FAU	CO	2181	2179	2177		[15]
Cr^{3+}	MFI	NO	1782	1902, 1768			[16]
Cr^{2+}	MFI	NO	-	1919, 1782			[16]
Mn^{2+}	MFI	CO	2214	2202	-	-	[17]
Fe^{2+}	FER	CO	2195	2188			[18]
Ru^{2+}	FAU	CO		2082, 2016	2152, 2091, 2086; 2154, 2081, 2049		[19]
Co^{3+}	FER	CO	2204	2194.5			[20]
Co^{2+}	MFI	NO	-	1894, 1812			[21]
Co^+	MFI	CO	-	2113, 2042	2137, 2089, 2079	2130, 2105, 2075	[22]

Table 1. (Continued)

Cation	Porous material	Ligand	Characteristic IR frequencies, cm^{-1}				Ref.
Rh^{2+}	MFI	CO		2176, 2142			[23]
Rh^{+}	MFI	CO		2114, 2048	2181, 2118, 2084	2164, 2112	[24]
Rh^{+}	FAU	CO	-	2118, 2053	2119, 2083	2151, 2135, 2124, 2111	[25]
Rh^{+}	MFI	NO		1861, 1786			[24]
Ir^{2+}	MFI	CO		2173, 2129			[26]
Ir^{+}	MFI	CO		2104, 2033	2182, 2099, 2074	2155, 2145, 2125, 2105	[26]
Ni^{+}	MFI	CO	2109	2136, 2092	2156, 2124, 2109		[27,28]
Ni^{+}	MOR	CO	2109	2138, 2094	2158, 2124, 2111		[29]
Ni^{+}	MFI	N_2	2343	2287, 2270			[30]
Pd^{3+}	MFI	CO		2221, 2200			[31]
Pd^{3+}	MFI	CO		2219, 2199			[32]
Pt^{3+}	MFI	CO	-	2112, 2176; 2196, 2155	-	-	[33]
Pt^{+}	MFI	CO	-	2122, 2092	2161, 2153, 2110	-	[33]
Cu^{3+}	MFI	NO	1960 and 1947	1915, 1863-1853			[34]
Cu^{2+}	MOF	CO					[35]
Cu^{+}	MFI	CO	2157	2178, 2151	2190, 2164		[36]
Cu^{+}	FAU	CO	2159 2143	2178, 2148; 2168, 2135	2188, 2165, (2138)		[37]
Cu^{+}	MFI	NO	1812	1827, 1734			[38]
Cu^{+}	[Al]MCM-41	CO	2159	2182, 2154	2191, 2173	-	[39]
Ag^{+}	MFI		2192	2192, 2186	-	-	[40,41]
Au^{3+}	NaFAU	CO	2205	2208	-	-	[42]

The formation of bunchy complexes is of great importance because in this way the adsorption capacity increases several times. Thus, the fundaments of gas storage and gas purification are also closely related to the process. For instance, developing of a material where two or more dihydrogen molecules could be simultaneously adsorbed on one center (as recently theoretically suggested [49]) would cause revolution in the field of hydrogen-storage materials.

Bunchy complexes are very important for catalysis. For instance, a number of catalytic reactions, *e.g.* polymerization, require initial formation of geminal species. Also, catalytic interaction of two different adsorbate molecules is facilitated when they are adsorbed at the same site.

At first sight, the formation of bunchy complexes could be easily predicted on the basis of analogy with other species, *e.g.* matrix isolated complexes. However, the surface chemistry is much more complicated: the coordination of the adsorption site to the solid (leading to partial filling of its coordination sphere) and the local environment (often responsible for steric hindrances) strongly affects the binding of additional ligands.

Table 2. Spectral features of mixed-ligand species formed with cations in porous materials when interacting with CO, NO, N_2 and H_2O.

Cation	Porous material	Mixed ligand species	Data on the spectral performance	Ref.
Na^+	FAU	$Na(CO)(^{15}N_2)$	ν(CO) at 2167 cm^{-1}; ν(^{15}N-^{15}N) at 2154 cm^{-1}	[6]
Cu^+	MFI	$Cu(CO)(H_2O)$	ν(CO) at 2130 cm^{-1}	[39,43]
Cu^+	MFI	$Cu(CO)(NH_3)$	ν(CO) at 2098 cm^{-1};	[43]
Cu^+	MFI	$Cu(CO)_2(NH_3)$	ν(CO) at 2158 and 2128 cm^{-1}	[43]
Cu^+	MFI	Cu(CO)(NO)	ν(CO) at 2137 cm^{-1}; ν(NO) at 1890 cm^{-1}	[44]
Cu^+	MFI	$Cu(CO)(CH_3COCH_3)$	ν(CO) at 2128 cm^{-1};	[45]
Rh^+	MFI	$Rh(CO)_2(H_2O)$	ν(CO) at 2108 and 2040 cm^{-1}	[24]
Pd^{2+}	MFI	$Pd(NO)(H_2O)$	ν(NO) at 1839 cm^{-1}	[32,46]
Rh^{2+}	MFI	$Rh^{2+}(CO)_2(NO)$	ν(CO) at 2181 + 2153 cm^{-1}; ν(NO) at 1890 cm^{-1}	[47]
Ni^{2+}	MFI	Ni^{2+}(CO)(NO)	ν(CO) at 2147cm^{-1}; ν(NO) at 1863 cm^{-1}	[48]
Ni^+	MFI	$Ni^+(CO)(N_2)$	ν(CO) at 2098 cm^{-1}; ν(NN) at 2305 cm^{-1}	[30]
Pd^{2+}	MFI	Pd^{2+}(CO)(NO)	ν(CO) at 2146 cm^{-1}; ν(NO) at 1837-1791 cm^{-1}	[32]

Twin Species (Coordination of Two or More Identical Molecules to One Cationic Site)

Systematic investigations of the formation of twin species are relatively recent. Based on analysis of the available data (see Table 1), the twin adsorption complexes have been divided into two main groups (see also Figure 1): *site-specified* and *complex-specified* [50, 51]. Site-specified complexes are formed as a result of the low coordination of the cation acting as adsorption site. The reasons of the formation of complex-specified twin species are multiplex and are generally related to the achievement of a stable electron configuration. As a result the latter are decomposed without passing through single-ligand species.

Site-Specified Twin Complexes

During decomposition site-specified twin complexes lose their ligands stepwise (Figure 1). Typically, they are formed with cations exchanged in zeolites. However, such species are also reported with other samples, e.g. mesoporous materials [39], bulk and supported oxides [52, 53] and metal-organic frameworks (MOFs) [35, 54]).

The background of the formation of site-specified twin complexes is easy to understand. Let us consider Y-zeolite (FAU structure) with exchanged Ca^{2+} ions. There are many investigations revealing that Ca^{2+} ions occupy preferentially S_I, S_{II} and S'_{II} positions [55-58]. Accessible for adsorption are the cations situated in S_{II} positions and facing the zeolite supercage. In dehydrated samples these cations are coordinated to three framework oxygen atoms. Because the normal coordination number of Ca^{2+} is 6, the ions in the dehydrated sample are coordinatively unsaturated and will tend to fill their coordination sphere by accepting three guest molecules. Indeed, it was established, by XRD investigations, that these ions can adsorb up to three H_2S molecules each, thus forming triplet species [59].

An important consequence of the above considerations is that the formation of site-specified twin complexes should not be sensitive to the adsorbate nature and, as a rule, the change of the ligand (if no steric hindrance occurs) should also result in formation of twin species.

In 1983 Paukshtis et al. [14] reported that low-temperature CO adsorption on CaNaY zeolite resulted in appearance, in the IR spectra, of a carbonyl band at 2197 cm^{-1}. This band disappeared when the CO equilibrium pressure increased and another band at 2186 cm^{-1} was formed on its expense. The authors assigned the band at 2197 cm^{-1} to Ca^{2+}-CO species that were converted into $Ca^{2+}(CO)_2$ dicarbonyls (2186 cm^{-1}). However, the work remained underestimated because at that time all known surface dicarbonyls were characterized by two IR bands, corresponding to the symmetric and antisymmetric C-O modes. More recent studies [5] revealed that at 85 K and in the presence of CO in the gas phase up to three CO molecules can be adsorbed on one Ca^{2+} site in CaNaY and CaNaEMT (See Figure 2). It is also reported that, when the CO - cation interaction is essentially electrostatic, the CO ligands behave as independent oscillators and no split of the CO stretching modes occurs. Thus, the possibility of coordination of up to three guest molecules to one Ca^{2+} site in S_{II} positions in Y zeolite was confirmed by various techniques (XRD and FTIR).

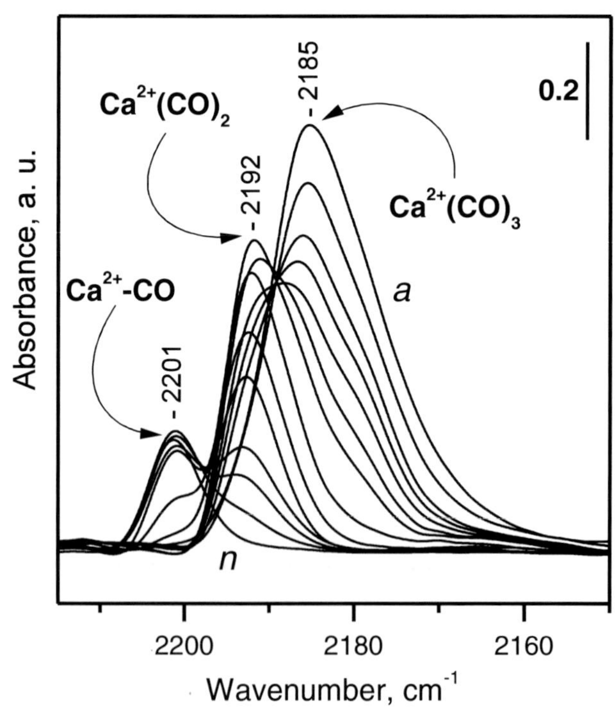

Figure 2. FTIR spectra of CO adsorbed at 85 K on CaNaEMT. Equilibrium pressure of 100 Pa CO (a) and gradual decrease of the coverage during evacuation down to 10^{-2} Pa (b-n).

An important conclusion is that the formation of these complexes is determined by the low coordination of the Ca^{2+} cations and is not (or is hardly) related to the adsorption strength.

It can be expected that other cations in S_{II} position in FAU zeolites should demonstrate the same coordination chemistry. Indeed, it was reported that three CO molecules can be bound to one Sr^{2+} or to one Ba^{2+} site in SrY and BaY, respectively [15]. Similar result has been obtained with NaY: coordination of two [6] or three CO molecules [10] or two N_2 molecules [6, 10, 60] to one Na^+ site. Up to three CO molecules were reported to be attached to one Cu^+ site in Cu-exchanged FAU samples [37].

However, studies of small cations (e.g. Li^+, [11], Mg^{2+} [15], Mn^{2+} [17, 61]) exchanged in FAU zeolites revealed that only one guest molecule can be attached to one site. For the case of MnY (Mn^{2+} ions occupying S_{II} positions) this conclusion has been confirmed by different techniques and for different guest molecules: XRD study of H_2S adsorption [61] and FTIR study of CO adsorption [17]. It has been concluded that the cationic radius is decisive for the formation of twin complexes: small cations penetrate in the plane of the three oxygen atoms to which they are coordinated and thus are able to accept only one guest molecule. This is illustrated on Figure 3 with the examples of exchanged Mg^{2+} and Ca^{2+} cations. Indeed, it is well known that the coordination number of the cation depends on its radius and small cations are characterized by a low coordination number. Therefore, in order to form twin complexes, the cation should possess a radius higher than a critical value (critical cationic radius). Thus, it appears that for S_{II} positions in FAU zeolites, the critical ionic radius is between 80 pm (Mn^{2+}) and 96 pm (Cu^+). The ionic radii used here are taken from Ref. [62]. The above picture has recently been impeached on the basis of theoretical studies: it has been proposed that, upon CO adsorption on NaY, the Na^+ cations migrate from S_{II} to vacant S_{III}' positions where form dicarbonyls [63].

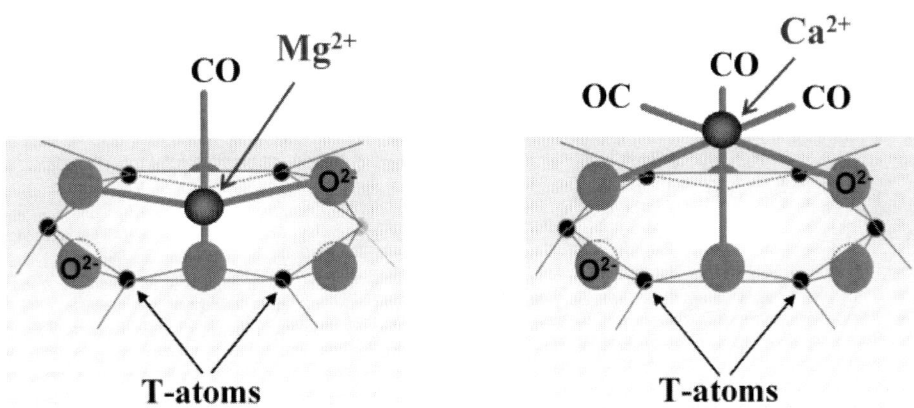

Figure 3. Coordination of CO to Mg^{2+} and Ca^{2+} cations in S_{II} positions in FAU zeolite.

When the cationic radius is around the critical one, the process of the formation of twin species could depend on the adsorption temperature. Thus, adsorption of CO on a Au^{3+}/NaY zeolite at ambient temperature leads to the appearance, among others, of a band at 2207 cm^{-1} attributed to Au^{3+}-CO species formed with the participation of gold(III) cations in S_{II} positions [42]. When the temperature was lowered to 100 K in presence of CO a second CO molecule was attached to the site, the carbonyl stretching frequency being lowered by 3 cm^{-1}. However, when CO is admitted to the sample directly at 100 K, no carbonyls of Au^{3+} are detected. The results are taken as evidence that the Au^{3+} ions (radius of 85 pm) were not accessible to CO at 100 K. However, at higher temperatures these cations have sufficient energy to migrate to the side of the aperture facing the supercage, thereby becoming accessible to CO. Once the Au^{3+}-CO species are formed, the Au^{3+} ions are able to accommodate a second CO molecule at low temperatures thus forming geminal dicarbonyls.

Due to the big variety of the zeolite structures, the local environment of exchanged cations is different. This is true not only for different zeolites but for different positions in one zeolite. Unfortunately, in many cases there are no detailed structural studies on the cation positions. This is valid above all for the high-silica zeolites where the low concentration of exchanged cations hinders the detailed investigations. Nevertheless, many theoretical studies points out the general trend cations in zeolites to be low coordinated [4, 29, 64-68]. Indeed, nowadays there are many examples of site-specified twin complexes formed with different cations and different guest molecules in various zeolites. It is not possible to describe here all findings, but we will give several examples helping for better understanding of the process.

It is now well established that cations exchanged in MFI (ZSM-5) zeolite can form, as a rule, site-specified twin complexes even with small cations [4, 17, 69, 70]. For instance, formation of dicarbonyls has been reported for Mn^{2+}-ZSM-5 [17]. Taking into account the already described facts that Mn^{2+} ions in MnY can accept only one guest molecule, it appears that the critical cationic radius is smaller for MFI than for FAU zeolites.

The ability of Cu^+ sites in Cu-ZSM-5 to form di- and tri-carbonyl complexes was reported in 1994 [36]. The phenomenon attracted a big interest because at that time Cu-ZSM-5 was (and still is) extensively studied as a catalyst for environmental purposes (NO decomposition, selective catalytic reduction of NOx with hydrocarbons, etc.). Consequently, this provoked investigations or revisiting of the interaction of CO with many supported copper systems [71-74]. Normally, it was expected that the ability of Cu^+ to form polycarbonyls was an intrinsic property of this cation. However, it appeared that formation of tricarbonyls is rather typical of zeolite systems, whereas with oxide supported copper monocarbonyls and only in some cases dicarbonyls were evidenced. This is due to the higher coordination of copper ions when they are supported on oxides as compared to the case when exchanged in zeolites [75]. It was also established [38] that Cu^+ ions in zeolites can form site-specified dinitrosyl species (these are also called true dinitrosyls because are decomposed via formation of mononitrosyls). However, Pulido and Nachtigall [76] suggested a cyclic dinitrosyl complex, $Cu^+(ON)_2$, proposed to be reaction intermediate in the NO decomposition process.

Among the zoo of the site-specified twin species formed in a ZSM-5 matrix we shall mention the following: $Li^+(CO)_2$ [4,69,70], $Na^+(CO)_n$ [8] or $Na^+(CO)_2$ [9]; $Na^+(CO_2)_2$ [77], $Ni^+(CO)_x$ (x = 1-3) [27,28]; $Ni^+(N_2)_2$ [30]; $Ag^+(CO)_2$ [40,41]. Some other examples (e.g. $Rh^+(CO)_x$) will be regarded below, after considering the so-called complex-specified twin species. Twin species are also reported with cations in other zeolites, e.g. LTA [78] or mesoporous materials, e.g. HMS [79].

The above described examples concerned *cis*-twin complexes. A peculiarity of the FER zeolite is the existence of the so-called β-cationic positions (see Figure 4). In this case the cation can be approached by two guest molecules from different zeolite cages. As a result, *trans*-twin species can be formed [13,20] the ligands being separated by the O-ring where the cation is coordinated. Obviously, the properties of *cis*- and *trans*-complexes should be very different. Thus, the formation of *trans*-species could be essential for the development of high-capacity adsorbents, but not for catalysis because the two molecules remain separated and are not able to interact chemically.

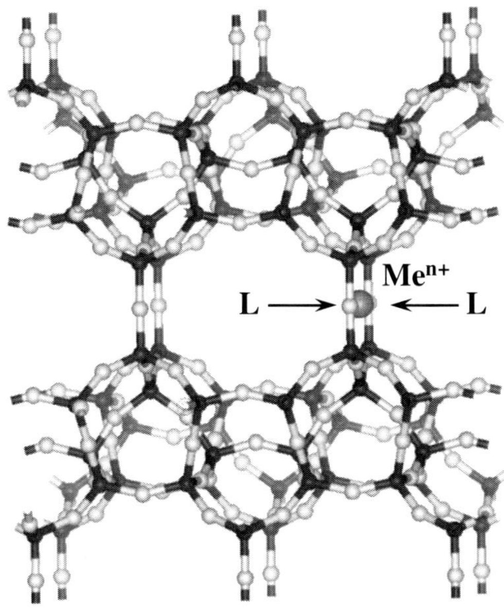

Figure 4. Formation of *trans*-geminal species with the participation of cations in β-positions in FER zeolite.

It is also evident that the conception of the critical ionic radius is not valid for *trans*-twin species because even very small cations can be approached by two molecules. However, in this case another peculiarity should be taken into account. If the interaction between the cation and the adsorbate is strong, the formation of linear complex could lead to pulling the cation out from its original position in the centre of the O-ring and, as a result, it will be not accessible from the opposite cage. Thus, no twin complexes will be formed. This can be illustrated by the following example: NO adsorption on Fe^{2+}-FER results in formation of stable Fe^{2+}-NO mononitrosyls only [18]. However, low-temperature adsorption of CO on the same sample results in formation of $Fe^{2+}(CO)_2$ dicarbonyl species because the interaction between CO and Fe^{2+} is weak (essentially electrostatic) and thus CO is not able to cause substantial changes in the cation position.

In general, the process of ion-exchange in zeolites is associated with the presence of the so-called Brønsted acid sites that represents acidic bridging hydroxyls of the Al-(OH)-Si type. Therefore, one could expect that introducing of cations into materials that are not characterized by bridging hydroxyls should not lead to the appearance of cations with two or more coordinative vacancies. Therefore, the latter should not be able to form twin complexes. Indeed, a series of investigations of cations introduced in vacant T-sites of dealuminated BEA zeolite have shown that these cations are not able to form site-specified geminal complexes [80-82].

However, this statement is not true in all cases. Consider now, Al-containing mesoporous materials, for instance [Al]MCM-41. These materials do not show in the IR spectra bands attributable to bridging hydroxyls. However, the classic ion-exchange procedure results in formation of low-coordinated cations, as in zeolites. Thus, it was reported that tricarbonyls were formed with Cu^+ ions in [Al]MCM-41 [39]. Moreover, the process was typical of the majority of Cu^+ sites. In contrast, only a part of the Cu^+ ions in Cu-ZSM-5 are able to accommodate up to three CO molecules, while an important fraction can coordinate maximum two CO molecules per cation. Therefore, mesoporous materials seem to be preferable for developing of high-capacity adsorbents. Although the reasons for the phenomena are far from well understood, it seems that the so-called pseudo-bridging silanols [83] participate in the ion-exchange process. The latter are Si-OH groups in vicinity of aluminum atom. They could be converted into bridging hydroxyls upon adsorption of probe molecules and thus demonstrate a higher acidity than the silica silanols. Because no steric hindrances are expected with cations in mesoporous materials, all of them are able to coordinate up to three guest molecules.

There are also data suggesting that highly unsaturated cation can be stabilized in porous materials even in absence of aluminum in the framework. For instance, copper tricarbonyls have been detected with Cu/silicalite-1 [84] and Cu/DAY [85]. Further investigations in this direction could lead to development of stable adsorbents with high capacity.

It should also be noted that formation of site-specified twin complexes has been reported with some metal-organic frameworks having exposed metal cations [36, 54].

Adsorption isotherms describing the formation of twin complexes have been kinetically derived [50, 86]. Here we will consider the isotherms for geminal species. For isotherms describing the formation of triple species the reader can see Ref. [50].

To derive the equations, it is considered that formally all adsorption centers consist of two sites. Thus, the part of the cations occupied by one adsorbate molecule is denoted as Θ_A, and the part occupied by two molecules, by Θ_{2A}. The following correlations are valid at equilibrium:

$$(1 - 2\Theta_A - \Theta_{2A})\,\mathrm{P} = a\,\Theta_A \qquad (1)$$

$$\Theta_A P = b\,\Theta_{2A} \qquad (2)$$

where P is the equilibrium pressure, and a and b, equilibrium constants having pressure dimensions. The isotherms are easily derived from eqs. (1) and (2):

$$\Theta_A = b\,P\,/\,(P^2 + 2\,b\,P + a\,b) \qquad (3)$$

$$\Theta_{2A} = P^2\,/\,(P^2 + 2\,b\,P + a\,b). \qquad (4)$$

A typical plot of the adsorption isotherms is presented on Figure 5. It is seen that with the increase of the partial adsorbate pressure initially the sites are preferentially occupied by one molecule but even at low pressures an important fraction of geminal species is formed. The fraction of one-ligand species (Θ_A) sharply increases with the pressure, passes through a maximum and then slowly decreases. The maximum is achieved at total coverage of 50 %.

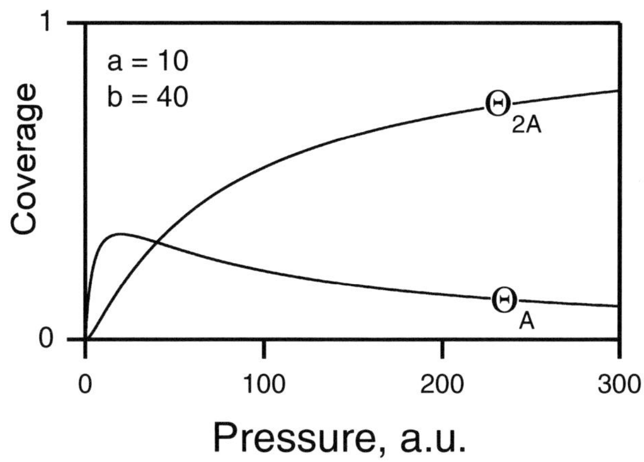

Figure 5. Adsorption isotherms describing stepwise occupation of active sites by one and two adsorbate molecules when a = 10 and b = 40 a. u. (see equations 3 and 4).

In practice the adsorption isotherms could be obtained by the use of spectral techniques because the latter can differentiate between species with different numbers of ligands. Examples of experimentally obtained isotherms can be found in Refs. [6, 14, 69, 87]. On their basis the equilibrium constants *a* and *b* can be easily obtained.

Although from the above considerations it seems that the formation of site-specified geminal complexes is easy to predict, there are still too many unclear and unresolved problems. Few examples:

- Ca^{2+} ions in S_{II} positions in CaY zeolite can accommodate up to three CO molecules [5]. Taking into account the similar cationic radii of Na^+ and Ca^{2+}, as well as the similar coordination of the two cations to the Y zeolite framework, one can expect that Na^+ ions in S_{II} positions should also be able to accept up to three CO molecules. However, although the stoichiometry in the latter case is still under debate [6, 60, 88] there are strong evidences that only two CO molecules can be attached to one Na^+ site.
- It has been established that Ag^+ and Cu^+ ions in ZSM-5 have essentially the same coordination. However, Cu^+ cations in this zeolite accommodate up to three CO molecules each, while Ag^+, up to two.
- CO easily forms mono-, di- and tricarbonyls with Cu^+ sites in Cu-ZSM-5. On this basis one could expect formation of $Cu^+(N_2)_x$ and $Cu^+(NO)_x$ ($x = 1 \div 3$) adducts. However, maximum two NO [38] and only one N_2 molecule can be attached to one Cu^+ site in this zeolite [89, 90].

It seems that, when the interaction between the cation and the adsorbed molecule is essentially electrostatic, the formation of twin complexes in porous materials is controlled mainly by the number of coordinative vacancies, the position of the cation and its radius. In these cases the nature of the ligand is not essential (providing the lack of steric hindrance). However, when the bond between the cation and the guest molecule is partly covalent (e.g. Cu^+-CO, Cu^+-N_2), additional factors should be taken into account, as the availability of suitable electron orbitals and possibility of interaction (repulsion) between the adsorbed molecules.

COMPLEX-SPECIFIED TWIN COMPLEXES

The complex-specified twin species do not leave any substantial fraction of single-ligand species upon decomposition (see Figure 1). The metal-ligand bond in these species is covalent and they are not formed with cations with a noble-gas electron configuration (alkali and alkaline earth metal cations, Al^{3+}, Ti^{4+}, etc.). There are no data of such species for cations with d^{10} electron configuration. Concerning the nature of the ligands, the complex-specified species are not formed with ligands tending to form ionic bond, e.g. H_2O. Because the complex-specified twin species are characteristic of particular metal-ligand systems, the change in the nature of one of the components may not result again in formation of twin complexes.

A typical example of complex specified twin species are the geminal dicarbonyls of Rh^+ that have attracted the interest of many researchers [23-25, 47, 91-96]. It has been established that $Rh^+(CO)_2$ species are formed after CO adsorption even on reduced samples of supported rhodium: because of the high stability of the dicarbonyls, oxidative disruption of the rhodium metal particles occurs. The respective Rh^+-CO species are much less stable and, as a result, thermal decomposition does not lead to production of mono-ligand species. It was reported that a small fraction of monocarbonyls could be isolated after flash desorption [94]. However, DFT studies indicated that in this case mixed ligand species, $Rh^+(H)(CO)$, are produced [95]. Attempts have been made to isolate tricarbonyl species. Yates et al. [92] reported the appearance of a negligible amount of tricarbonyls with Rh/Al_2O_3 in presence of CO and at low temperature. These species were evidenced only with weak IR bands in the difference spectra.

These results show that tricarbonyls are not typical of the system. However, studying the low-temperature CO adsorption on Rh/DAY, Miessner et al [25] established formation of tri- and tetracarbonyl species. Analogous results were reported more recently with Rh-ZSM-5 zeolite [24]. These results indicate that, although the dicarbonyls are complex-specified, the possibility of the insertion of a third and fourth CO molecule in the complex arises from the low coordination of the Rh^+ cations in porous materials. Evidently, when the premises for the formation of both, site-specified and complex-specified twin species are fulfilled, complexes with a high number of ligands are formed.

Complex-specified dicarbonyls are produced also with the participation of Rh^{2+} cations, the only reports concern zeolitic systems [23]. In this case, the smaller radius of Rh^{2+}, as compared to Rh^+, is probably the reason that, even at low temperature, no tricarbonyls are formed.

Consistent with the similar electronic structure of rhodium and iridium cations, the same carbonyl species were reported to appear in Ir-ZSM-5 zeolite: $Ir^+(CO)_x$ (x = 2-4) and $Ir^{2+}(CO)_2$ [26]. The coordination chemistry of Co^+ cations in Co-ZSM-5 is essentially the same [22]: CO adsorption produced $Co^+(CO)_x$ (x = 2-4) species. However, no complex-specified dicarbonyls of Co^{2+} were reported. In this case the difference of the energy of the d-electrons appears to be essential.

Another element typically forming complex-specified polycarbonyls is ruthenium [19, 97-99]. However, in this case the situation is less clear. Different authors reported on the formation of $Ru^{n+}(CO)_3$ and $Ru^{n+}(CO)_2$ polycarbonyls, the oxidation state of ruthenium (2+ or 3+) being still discussed. Miessner et al. [19] reported that CO adsorption on a Ru/DAY sample led to formation of $Ru^{2+}(CO)_3$ species with slightly distorted C_{3v} symmetry. Decarbonylation of these species (in a flow of a 1 : 1 $H_2 + N_2$ mixture at 573 K) resulted in appearance of *cis*-dicarbonyls. The latter, in presence of CO were transformed into tricarbonyls with a meridional structure. Although few studies on CO adsorption on osmium-containing samples are available, it seems that the carbonyl chemistry of ruthenium and osmium cations is very similar.

In this case again the element from the III row of the Periodic table, i.e. iron, behave in different way. No complex-specified polycarbonyls of Fe^{2+} or Fe^{3+} cations have been reported. However, similar species are observed with reduced Fe-ZSM-5 and Fe-BEA samples, where the oxidation state of iron is believed to be 1+ [100].

Only site-specified polycarbonyls have been found with nickel ions, but complex-specified $Pt^{3+}(CO)_2$, $Pt^{+}(CO)_2$ and $Pt^{+}(CO)_3$ species were reported with Pt-ZSM-5 [33]. Similarly to Pt, $Pd^{3+}(CO)_2$ species were reported in Pd-ZSM-5 [31]. Complex-specified tricarbonyls are formed also with some Cr^{n+} [101, 102] Mo^{n+} [103] and W^{n+} cations [104].

Another ligand typically forming complex-specified twin species (essentially geminal) is NO. A peculiarity in this case is that, due to the presence of an unpaired electron in the molecule, NO tends to dimmerize. Thus, the complex-specified dinitrosyls (also called non-true dinitrosyls) can also be regarded as adsorbed N_2O_2 dimmer. Indeed, these dinitrosyls could be produced even with cations with proven only one coordinative vacancy [105]. Although a simple calculation of the number of electrons in the complex seems to be an easy way to predict formation of dinitrosyls, this approach is not generally applicable. Thus, dinitrosyls are formed after adsorption of NO on Rh^{+} sites (d^8 electrons) [24] but also on Co^{2+} (d^7 electrons) containing systems [21, 105]. These species are typical of both, V^{3+} (d^2) and V^{4+} (d^1) sites [106]. Other well known complex-specified dinitrosyls are $Mo^{4+}(NO)_2$, $W^{4+}(NO)_2$, $Cr^{2+}(NO)_2$ and $Cr^{3+}(NO)_2$ [107].

It is generally accepted that the complex-specified twin complexes are stabilized as a result of reaching a stable electron configuration of the entire complex (including the ligands from the solid to which the active center is attached), but the details are far from well understood. Thus, it has been suggested that $Rh^{+}(CO)_2$ species are stabilized because of the achievement of 18-electron configuration, typical of carbonyl species. However, the stable rhodium(I) dinitrosyl complex has two more electrons. Noticeably, Rh^{2+} also forms stable dicarbonyls where the complex has one less electron as compared to $Rh^{+}(CO)_2$.

Mixed-Ligand Species (Coordination of Different Molecules to One Cationic Site)

The mixed-ligand species (see Table 2) cannot be divided into *site-* and *complex-specified* ones because (at least according to the available literature data) when decomposed they preferentially lose one of the ligands. However, mixed-ligand species demonstrate some peculiarities typical of the two types of twin species. For instance, when site-specified twin species are formed and the metal-ligand bond is essentially electrostatic, one of the ligands could be substituted by another. Thus, as already discussed, Na^+ cations in S_{II} positions in FAU zeolites can form $Na^+(CO)_2$ and $Na^+(N_2)_2$ geminal species. When CO and N_2 are simultaneously present in the gas phase mixed $Na^+(CO)(N_2)$ species were reported [6], although some authors [12] have not detected them. Note, however, that in this case a mixture of geminal and mixed ligand species is formed, the ratio between them depending (at a constant temperature) on the partial pressures of the two gases. Xu et al. [108] demonstrated that O_2 and propane can simultaneously attach to one M^{2+} site (M = Mg, Ca, Sr, Ba) in FAU zeolite to form $M^{2+}(O_2)(C_3H_8)$ mixed ligand species. The latter were proposed to be precursors in thermal propane selective oxidation.

When site-specified twin species are formed via covalent bonding to a cation in a given position the situation is more complicated. Thus, Cu^+-CO carbonyls in various zeolites can accept not only additional CO molecules, but can also form mixed-ligand species of the type $Cu^+(CO)(B)_x$, and $Cu^+(CO)_2(B)$ where B is H_2O [43,44,75,85], NH_3 [44], $(CH_3)_2CO$ [45], etc. Aqua-carbonyl complexes are also reported with other systems, for instance with Rh^+ ions in Rh-ZSM-5 [24]. Aqua-nitrosyl complexes have been found with Pd^{2+} ions in zeolites [32,46]. When the two ligands are CO and NO and they are covalently bonded, mixed ligand species are not always produced. Thus, no mixed carbonyl-nitrosyl species are formed with Rh^+ in Rh-ZSM-5, irrespective of the fact that up to four CO molecules can be attached to one Rh^+ site [24]. In contrast, Rh^{2+} ions in the same zeolite form complex-specified dicarbonyls that are not able to accept additional CO molecules and also mixed $Rh^{2+}(CO)_2(NO)$ species [47]. In the latter case, on the basis of DFT calculations, change of the coordination of the cation to the zeolite framework is proposed. Similarly to the complex-specified twin species, mixed-ligand complexes could be produced even with cations with proven only one coordinative vacancy.

The investigations on the mixed-ligand species are rather sporadic and there are too many white spots in this field. Although there are sound explanations for particular cases, no general theory able to predict the formation of mixed-ligand species is available nowadays. The complexity of the problem could be well illustrated by several selected examples with simple diatomic ligands as CO, NO and N_2:

- Ni^{2+} ions on Ni-ZSM-5 can form $Ni^{2+}(CO)(NO)$ mixed-ligand species [48]. As expected from the electronic structure, similar species should be formed with Pd^{2+} and Pt^{2+} ions in the same matrix. Indeed, $Pd^{2+}(CO)(NO)$ species are reported with Pd-ZSM-5 [32], but no carbonyl-nitrosyl complexes with Pt^{2+} ions in Pt-ZSM-5 have been registered.
- Ni^+ ions in Ni-ZSM-5 easily form tricarbonyls and can attach up to two N_2 molecules [30]. In this case $Ni^+(CO)(N_2)$ mixed ligand species are also produced. Although the carbonyl chemistry of Cu^+ ions in ZSM-5 is similar (they can accommodate up to three CO molecules [36]), only single-ligand complexes (Cu^+-N_2) are formed after adsorption of N_2 [89] and no mixed-ligand species are produced in this case.

All these examples clearly show that the formation of bunchy species depends on many factors and the present possibilities of prediction are restricted. For instance, the 16- and 18-electron rules fail to predict many *complex-specified* geminal species.

Evidently, numerous factors affect the processes and elaboration of an integral theory of bunchy complexes is still a question of the future.

Conclusion

The formation of bunchy adsorption complexes is of great importance not only for catalysis but also for other processes related to adsorption, as gas storage and gas separation. Zeolites and other porous materials are the best known systems for hosting cations able to coordinate simultaneously several small molecules during adsorption. This arises from the fact that exchanged cations in porous materials are, in general, low coordinated.

During the last two decades our knowledge on the formation of bunchy complexes significantly increased. The main achievements are related to twin complexes. It is now well established that these species are two kinds: site-specified and complex-specified. Although porous supports generally favor formation of site-specified twin complexes, they also provide the possibility to insert additional ligands when complex-specified species are initially formed. It is also well established that, for production of *cis*- twin complexes, the cationic radius should be higher than a critical value (characteristic of the cation position). Trans-complexes are also isolated. The presently known examples are restricted to cations in the FER zeolite, but similar structures are probably formed with other zeolites as well. Kinetically derived adsorption isotherms describe very well the adsorption phenomena because the adsorption sites in zeolites are normally isolated.

Irrespective of the substantial achievements, however, there are still many unresolved problems. For instance, it is not clear why, when site-specified twin complexes are formed, the maximum number of adsorbed small molecules depends on the adsorbate nature. The situation with the complex-specified species is even more complicated and theoretical investigations seem to be decisive in these case.

Concerning the mixed-ligand species, the situation seems relatively clear only when the adsorption bonds are essentially electrostatic. Presently, there are no systematic data on mixed-ligand species with covalent bonding.

Although structural techniques (*e.g.* XRD) have been used in case studies and there are also some profound theoretical investigations, most of the works dealing with bunchy complexes are based on vibrational spectroscopy. We hope that this chapter will provoke further investigations helping in the elaboration of integral theory of the formation of bunchy species.

Acknowledgments

This work was supported by the Bulgarian Scientific Fund (grant DCVP 02/2).

References

[1] Somorjai, G.; Li, Y. *Introduction to Surface Chemistry and Catalysis*; 2nd ed., John Wiley & Sons: Canada, 2010.

[2] Nachtigall, P.; Delgado, M. R.; Nachtigallova, D.; Arean, C.O. *Phys. Chem. Chem. Phys.* 2012, *14* 1552-1569.

[3] Bulánek, R.; Koudelková, E. *Microporous Mesoporous Mater.*, 2012, *151*, 149-156.

[4] Sillar, K.; Burk, P. *Phys. Chem. Chem. Phys.* 2007, *9*, 824-827.

[5] Hadjiivanov, K.; Knözinger, H. *J. Phys. Chem. B*, 2001, *105*, 4531-4534.

[6] Hadjiivanov, K.; Knözinger, H. *Chem. Phys. Lett.* 1999, *303*, 513-520.

[7] Cairon, O.; Guesmi, H.*Phys. Chem. Chem. Phys.* 2011, *13*, 11430-11437.

[8] Bordiga, S.; Escalona Platero, E.; Otero Arean, C.; Lamberti, C.; Zecchina, A. *J. Catal.* 1992, *137*, 179-185.

[9] Zecchina, A.; Otero Arean, C.; Turnes Palomino, G.; Geobaldo, F.; Lamberti, C.; Spoto, G.; Bordiga, S. *Phys. Chem. Chem. Phys.* 1999, *1*, 1649-1657.

[10] Hadjiivanov, K.; Knözinger, H. *Catal. Lett.* 1999, *58*, 21-26.

[11] Hadjiivanov, K.; Massiani, P.; Knözinger, H. *Phys. Chem. Chem. Phys.* 1999, *1*, 3831-3838.

[12] Cairon, O. *ChemPhysChem* 2013, *14*, 2744-2749.

[13] Bulánek, R.; Voleská, I.; Ivanova, E.; Hadjiivanov, K.; Nachtigall, P. *J. Phys. Chem. C* 2009, *113*, 11066-11076.

[14] Paukshtis, E.A.; Soltanov, R.; Yurchenko, E. *React. Kinet. Catal. Lett.* 1983, *22*, 147-151.

[15] Hadjiivanov, K.; Ivanova, E.; Knözinger, H. *Microporous Mesoporous Mater.* 2003, *58*, 225-236.

[16] Mihaylov, M.; Penkova, A.; Hadjiivanov, K.; Daturi, M. *J. Mol. Catal. A* 2006, *249*, 40-46.

[17] Hadjiivanov, K.; Ivanova, E.; Kantcheva, M.; Cifitikli, E.; Klissurski, D.; Dimitrov, L.; Knözinger, H. *Catal. Commun.* 2002, *3*, 313-319.

[18] Ivanova, E.; Mihaylov, M.; Hadjiivanov, K.; Blasin-Aube, V.; Marie, O.; Plesniar, A.; Daturi, M. *Appl. Catal. B* 2010, *93*, 325-338.

[19] Miessner, H.; Richter, K. *J. Mol. Catal. A* 1999, *146*, 107-115.

[20] Chakarova, K.; Hadjiivanov, K. *Microporous Mesoporous Mater.* 2009, *123*, 123-128.

[21] Campa, M.; de Rossi, S.; Ferraris, G.; Indovina, V. *Appl. Catal. B* 1996, *8*, 315-331.

[22] Hadjiivanov, K.; Tsyntsarski, B.; Venkov, Tz.; Klissurski, D.; Daturi, M.; Saussey, J.; Lavalley, J.-C. *Phys. Chem. Chem. Phys.* 2003, *5*, 1695-1702.

[23] Ivanova, E.; Mihaylov, M.; Thibault-Starzyk, F.; Daturi, M.; Hadjiivanov, K. *J. Catal.* 2005, *236*, 168-171.

[24] Ivanova, E.; Hadjiivanov, K. *Phys. Chem. Chem. Phys.* 2003, *5*, 655-661.

[25] Miessner, H.; Burkhardt, I.; Gutschick, D.; Zecchina, A.; Morterra, C.; Spoto, G. *J. Chem. Soc., Faraday Trans.* 1990, *86*, 2321-2327.

[26] Mihaylov, M.; Ivanova, E.; Thibault-Starzyk, F.; Daturi, M.; Dimitrov, L.; Hadjiivanov, K. *J. Phys. Chem. B* 2006, *110*, 10383-10389.

[27] Hadjiivanov, K.; Knözinger, H.; Mihaylov, M. *J. Phys. Chem. B* 2002, *106*, 2618-2624.

[28] Serykh, A. I.; Amiridis, M. D. *J. Phys. Chem. C*, 2007, *111*, 17020-17024.

[29] Aleksandrov, H. A.; Zdravkova, V. R.; Mihaylov, M. Y.; Petkov, P. St.; Vayssilov, G. N.; Hadjiivanov, K. I. *J. Phys. Chem. C* 2012, *116*, 22823-22831.

[30] Zdravkova, V.; Mihaylov, M.; Hadjiivanov, K. *J. Phys. Chem. C* 2012, *116*, 12706-12711.

[31] Aylor, A.; Lobree, L.; Reimer, J.; Bell, A. T. *J. Catal.* 1997, *172*, 453-462.

[32] Chakarova, K.; Ivanova, E.; Hadjiivanov, K.; Klissurski, D.; Knözinger, H. *Phys. Chem. Chem. Phys.* 2004, *6*, 3702-3709.

[33] Chakarova, K.; Mihaylov, M.; Hadjiivanov, K. *Catal. Commun.* 2005, *6*, 466-471.

[34] Tortorelli, M.; Chakarova, K.; Lisi, L.; Hadjiivanov, K. *J. Catal.* 2014, *309*, 376-385.
[35] Lieb, A.; Leclerc, H.; Devic, T.; Serre, C.; Margiolaki, I.; Mahjoubi, F.; Lee, J. S.; Vimont, A.; Daturi, M.; Chan, J.-S. *Microporous Mesoporous Mater.* 2012, *157*, 18-23.
[36] Spoto, G.; Zecchina, A.; Bordiga, S.; Ricchiardi, G.; Martra, G.; Leofanti, G.; Petrini, G. *Appl. Catal. B* 1994, *3*, 151-172.
[37] Turnes Palomino, G.; Bordiga, S.; Zecchina, A.; Marra, G. L.; Lamberti, C. *J. Phys. Chem. B*, 2000, *104*, 8641-8651.
[38] Spoto, G.; Bordiga, S.; Scarano, D.; Zecchina, A. *Catal. Lett.* 1992, *13*, 39-44.
[39] Chakarova, K.; Petrova, G.; Dimitrov, M.; Dimitrov, L.; Vayssilov, G.; Tsoncheva, T.; Hadjiivanov, K. *Appl. Catal. B* 2011, *106*, 186-194.
[40] Hadjiivanov, K.; Knözinger, H. *J. Phys. Chem. B* 1998, *102*, 10936-10940.
[41] Hadjiivanov, K. *Microporous Mesoporous Mater.* 1998, *24*, 41-49.
[42] Mihaylov, M.; Fierro-Gonzalez, J. C.; Knözinger, H.; Gates, B. C.; Hadjiivanov, K. *J. Phys. Chem. B* 2006, *110*, 7695-7701.
[43] Zecchina, A.; Bordiga, S.; Turnes Palomino, G.; Scarano, D.; Lamberti, C.; Salvalaggio, M. *J. Phys. Chem.* 1999, *103*, 3833-3844.
[44] Góra-Marek, K.; Datka, J. *Vibr. S pectrosc.* 2011, *58*, 104-108.
[45] Szanyi, J.; Paffett, M.T. *Catal. Lett.* 1997, *43*, 37-44.
[46] Descorme, F.; Gerlin, P.; Primet, M.; Lercuyer, C. *Catal. Lett.* 1996, *41*, 133-138.
[47] Ivanova, E.; Mihaylov, M.; Aleksandrov, H.; Daturi, M.; Thibault-Starzyk, F.; Vayssilov, G.; Rösch, N.; Hadjiivanov, K. *J. Phys. Chem. C* 2007, *111*, 10412-10418.
[48] Mihaylov, M.; Hadjiivanov, K. *Langmuir* 2002, *18*, 4376-4383.
[49] Georgiev, P.; Albinati, A. *Chem. Phys. Lett.* 2012, *545*, 66-70.
[50] Hadjiivanov, K.; Ivanova, E.; Klissurski, D. *Catal. Today* 2001, *70*, 73-82.
[51] Hadjiivanov, K.; Knözinger, H. *Surf. Sci.* 2009, *603*, 1629-1636.
[52] Coluccia, S.; Baricco, M.; Marchese, L.; Martra, G.; Zecchina, A. *Spectrochim. Acta A* 1993, *49*, 1289-1298.
[53] Scarano, V.; Bordiga, S.; Lamberti, C.; Spoto, G.; Richiardi, G.; Zecchina A.; Arean, C. O. *Surf. Sci.* 1998, *411*, 272-285.
[54] Drenchev, N.; Ivanova, E.; Mihaylov, M. ; Hadjiivanov, K. *Phys. Chem. Chem. Phys.* 2010, *12*, 6423-6427.
[55] Frising, T.; Leflaive, P. *Microporous Mesoporous Mater.* 2008, *114*, 27-63.
[56] Bennett, J. M.; Smith, J. V. *Mater. Res. Bull.* 1968, *3*, 633-642.
[57] Yeom, Y. H.; Jang, S. B.; Kim, Y.; Song, S. H.; Seff, K. *J. Phys. Chem. B* 1997, *101*, 6914-6920.
[58] Pluth, J. J.; Smith, J. V. *Mater. Res. Bull.* 1972, *7*, 1311-1322.
[59] Jang, S. B.; Jeong, M. S.; Kim, Y.; Han, Y. W.; Seff, K. *J. Phys. Chem. B* 1998, *102*, 33-44.
[60] Cairon, O. *Phys. Chem. Chem. Phys.* 2010, *12*, 14217-14226.
[61] Bae, M. N.; Song, M. K.; Kim, Y.; Seff, K. *Microporoud Mesoporous Mater.* 2003, *63*, 21-31.
[62] Shannon, R.D. *Acta Cryst. A* 1976, *32*, 751-767.
[63] Nour, Z.; Berthomieu, D.; Yang, Q.; Maurin, G. *J. Phys. Chem. C* 2012, *116*, 24512−24521.

[64] Aleksandrov, H. A.; Petkov, P. S.; Vayssilov, G. N. *Energy Environ. Sci.* 2011, *4*, 1879-1885.
[65] Rice, M. J.; Chakraborty, A. K.; Bell, A. T. *J. Catal.* 2000, *194*, 278-285.
[66] Vayssilov, G.; Hu, A.; Birkenheuer, U.; Rösch, N. *J. Mol. Catal. A* 2002, *162*, 135-145.
[67] Göltl, F.; Hafner, J. *J. Chem. Phys.* 2012, *136*, 064503.
[68] Pulido, A.; Nachtigall, P.; Delgado, M. R.; Areán, C .O. *ChemPhysChem* 2009, *10*, 1058-1065.
[69] Otero Arean, C.; Manoilova, O. V.; Rodrigues Delgado, M.; Tsyganenko, A. A.; Garrone, E. *Phys. Chem. Chem. Phys.* 2001, *3*, 4187-4188.
[70] Bonelli, B.; Garrone, E.; Fubini, B; Onida, B.; Rodriguez Delgado, M.; Otero Arean, C. *Phys. Chem. Chem. Phys.* 2003, *5*, 2900-2905.
[71] Lamberti, C.; Bordiga, S.; Zecchina, A.; Salvalaggio, M.; Geobaldo, F.; Otero Areán, C. *J. Chem. Soc., Faraday Trans.* 1998, *94*, 1519-1525.
[72] Kefirov, R.; Penkova, A.; Hadjiivanov, K.; Dzwigaj, S.; Che, M. *Microporous Mesoporous Mater.* 2008, *116*, 180-187.
[73] Bulánek, R.; Drobná, H.; Nachtigall, P.; Rubešb, M.; Bludský, O. *Phys. Chem. Chem. Phys.* 2006, *8*, 5535-5542.
[74] Giordanino, F.; Vennestrøm, P. N. R.; Lundegaard, L. F.; Stappen, F. N.; Mossin, S.; Beato, P.; Bordiga, S.; Lamberti, C. *Dalton Trans.* 2013, *42*, 12741-12761.
[75] Hadjiivanov, K.; Kantcheva, M.; Klissurski, D. *J. Chem. Soc., Faraday Trans.* 1996, *92*, 4595-4600.
[76] Pulido, A.; Nachtigall, P. *Phys. Chem. Chem. Phys.* 2009, *11*, 1447-1458.
[77] Bonelli, B.; Onida, B.; Fubini, B.; Otero Arean, C.; Garrone E. *Langmuir* 2000, *16*, 4976-4983.
[78] Pulido, A.; Nachtigall, P.; Rodriguez Delgado, M.; Otero Arean C. *ChemPhysChem* 2009, *10*, 1058-1060.
[79] Bonelli, B.; Onida, B; Chen, J.D.; Galarneau, A.; Di Renzo F.; Fajula, F.; Garrone E. *J. Porous Mater.* 2007, *14*, 299-304.
[80] Mihaylova, A.; Hadjiivanov, K.; Dzwigaj, S.; Che, M. *J. Phys. Chem. B* 2006, *110*, 19530-19536.
[81] Penkova, A.; Dzwigaj, S.; Kefirov, R.; Hadjiivanov, K.; Che, M. *J. Phys. Chem. C* 2007, *111*, 8623-8631.
[82] Hadjiivanov, K.; Ivanova, E.; Kefirov, R.; Janas, J.; Plesniar, A.; Dzwigaj, S.; Che, M. *Microporous Mesoporous Mater.* 2010, *131*, 1-12.
[83] Busca, G. *Chem. Rev.* 2007, *107*, 5366-5410.
[84] Hadjiivanov, K.; Knözinger, H.; Milushev, A. *Catal. Commun.* 2002, *3*, 37-44.
[85] Miessner, H.; Landmesser, H.; Jaeger, N.; Richter, K. *J. Chem. Soc., Faraday Trans.* 1997, *93*, 3417-3422.
[86] Hadjiivanov, K.; Klissurski, D. *React. Kinet. Catal. Lett.* 1991, *44*, 229-235.
[87] Cairon, O. *Phys. Chem. Chem. Phys.* 2012, *14*, 12083-12085.
[88] Cairon, O.; Bellat, J.-P. *J. Phys. Chem. C* 2012, *116*, 11195-11999.
[89] Recchia, S.; Dossi, C.; Psaro, R.; Fusi, A.; Ugo, R.; Moretti, G. *J. Phys. Chem. B* 2002, *106*, 13326-13332.
[90] Bordiga, S.; Lamberti, C.; Turnes Palomino, G.; Geobaldo, G.; Arduino, D.; Zecchina, A. *Microporous Mesoporous Mater.* 1999, *30*, 129-135.

[91] Van't Blik, H. F. J.; van Zon J. B. A. D.; Huizinga, T.; Vis, J. C.; Koningsberger, D. C.; Prins, R. *J. Am. Chem. Soc.* 1985, *107*, 3139-3147.
[92] Wang, H.; Yates, J. T. Jr. *J. Catal.* 1984, *89*, 79-92.
[93] Basu, P.; Panayotov, D.; Yates, J. T. Jr. *J. Am. Chem. Soc.* 1988, *110*, 2074-2081.
[94] Hadjiivanov, K.; Ivanova, E.; Dimitrov, L.; Knözinger, H. *J. Mol. Struct.* 2003, *661-662*, 459-463.
[95] Vayssilov, G. N.; Rösch, N. *J. Am. Chem. Soc.* 2002, *124*, 3783-3786.
[96] Ogino, I.; Gates, B. C. *J. Phys. Chem. C* 2010, *114*, 8405-8413.
[97] Mihaylov, M.; Lagunov, O.; Ivanova, E.; Hadjiivanov, K. *J. Phys. Chem. C* 2011, *115*, 13860-13867.
[98] Álvarez-Rodríguez, J.; Rodríguez-Ramos, I.; Guerrero-Ruiz, A.; Arcoya, A. *Appl. Catal. A* 2009, *366*, 114-121.
[99] Yokomizo, G. H.; Louis, C.; Bell, A. T. *J. Catal.* 1989, *120*, 1-14.
[100] Mihaylov, M.; Ivanova, E.; Chakarova, K.; Novachka, P.; Hadjiivanov, K. *Appl. Catal. A* 2011, *391*, 3-10.
[101] Ghiotti, G.; Garrone, E.; Zecchina, A. *J. Mol. Catal.* 1991, *65*, 73-83.
[102] Kohler, S. D.; Ekerdt, J. G. *J. Phys. Chem.* 1994, *98*, 4336-4342.
[103] Williams, C. C.; Ekerdt, J. G. *J. Phys. Chem.* 1993, *97*, 6843-6852.
[104] Kohler, S. D.; Ekerdt, J. G. *J. Phys. Chem.* 1994, *98*, 1276-1281.
[105] Djonev, B.; Tsuntsarski, B.; Klissurski, G.; Hadjiivanov, K. *J. Chem. Soc., Faraday Trans.* 1997, *93*, 4055-4063.
[106] Bulánek, R.; Drenchev, N.; Čičmanec, P.; Setnička, M. *Adsorption* 2013, *19*, 339-347.
[107] Hadjiivanov, K. *Catal. Rev. - Sci. Eng.* 2000, *42*, 71-144.
[108] Xu, J.; Mojet, B. L.; van Ommen, J. G.; Lefferts, L. *J. Phys. Chem. B* 2005, *109*, 18361-18368.

In: Comprehensive Guide for Mesoporous Materials, Volume 2 ISBN: 978-1-63463-982-8
Editor: Mahmood Aliofkhazraei

Chapter 13

FUNCTIONALIZED TWO-DIMENSIONAL MESOPOROUS MATERIALS: LAYERED MATERIALS GO 3D

Heloise O. Pastore and Hipassia Marcondes de Moura
Institute of Chemistry, University of Campinas
Campinas, SP, Brazil

ABSTRACT

Crystalline layered silicates, which possess two-dimensional nano-spaces between the layers, have been utilized for various applications, as adsorbents, catalysts/catalyst supports, ion exchangers etc. These uses are possible due to the covalent modifications on the silicate surface $SiOH/SiO^-$ groups. This chapter will examine the feasibility of synthesizing organo–inorganic composites based on layered silicates. It aims to explore the basic and advanced aspects of preparation and characterization of silylated and/or pillared interlayered clays by quaternary ammonium salts pre-swelling step, which is responsible for the controllable and designed pore structure and accessibility.

INTRODUCTION

This chapter focuses on the main strategies for modifying the surfaces of layered silicates. First, an overview on the composition of the lamella and its reactive groups is shown. The covalent modifications are generally preceded by a swelling step aiming at separating the lamella apart and making the interlayer region hydrophilic; it is detailed along the second section. Finally, the grafting and pillaring strategies are shown in separated sections. All these key steps can be envisaged in Table 1.

Materials with lamellar structure are materials with a two-dimensional (2D) oriented condensed structure. Clay minerals, phyllosilicates, lamellar hydrated silicates such as silicic acids, graphite, oxides and hydroxides of transition metals present lamellar structural arrangement resulting from stacking the sheets along a specific crystallographic axis. [1, 2]

Table 1. Key steps on modifying layered silicates

Topics	References
Layered Silicates Structures	[1] – [40]
Pre-swelling step with organic cations	[76] – [82]
Grafting processes with different functional groups	[83] – [112]
Cross-Linking Lamella: The pillaring process	[41] – [75] and [113] – [137]

Stacking the lamella defines a region between them called 'interlamellar space' [3], while the distance between the outer surface and an inner surface of the adjacent lamella is called interlayer distance (Scheme 1).

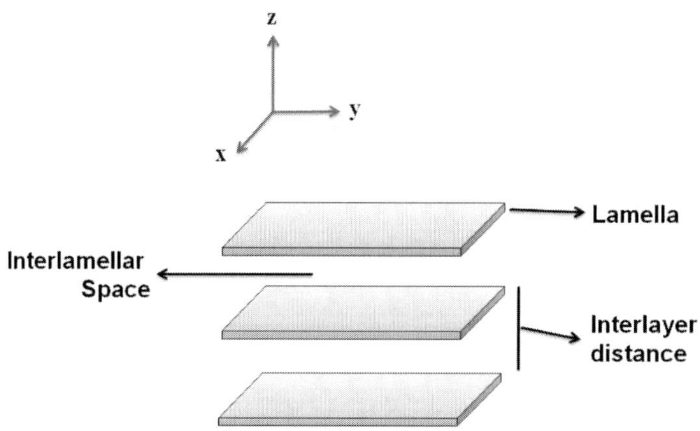

Scheme 1. Schematic representation of a lamellar structure.

When the sheets are electrically neutral, as in graphite or FeOCl, the interlamellar region is empty, adjacent sheets interact through van der Waals forces, and neutral guest species can be incorporated between the host sheets. This incorporation is accompanied by an electron transfer reaction between the molecule and the sheet. [3]

Many types of layered solids are electrically charged due to isomorphic substitutions into the layer framework. Therefore, it is possible to synthesize different structures with varied compositions and from them to generate molecular sieves with different composition. The ones made exclusively of SiO_4 tetrahedra may be modified by the insertion of transition metals such as iron [4, 5, 6,], gallium [7, 8, 9], titanium [10, 11], germanium [12, 13], indium [14], tin [15], vanadium [16], chromium [17, 18], boron [19, 20, 21], and beryllium [22, 23]. To achieve electronic neutrality in the solid, counter ions, typically solvated by water or other polar molecules, occupy the interlamellar region.

Surface properties of layered materials, such as their polarity and ionicity, are important for intercalation process. From the viewpoint of ionicity, inorganic layered materials can be classified into three categories: i) cation-exchange layered materials, which have negative charge in the layers, such as certain layered clay minerals, metal oxides, metallosilicates and silicates; ii) anion-exchange layered materials, which have positive charge in the layers, such as layered double hydroxides (LDH); and iii) neutral layered materials, such as graphite, layered chalcogenides and certain clays. There are numerous studies on layered clay minerals, in particular, organically modified clay minerals by ion exchange with cationic organic substances are extensively studied and many recent excellent literature are available. [24, 25, 26] Natural sodium layered silicates, for example magadiite [27, 28, 29], kenyaite [27, 30] makatite [31] and kanemite [32, 33] are widely studied. Magadiite and kenyaite crystal structures have not been determined yet but from its family, the silicate frameworks of both kanemite and makatite are composed of only six-membered rings with Q^3 $((O^-/OH)Si(OSi)_3)$ environmental silica species.

The important driving force for any intercalation reaction is the surface silanol groups, Si-OH, and silanolates, SiO^- responsible for the negative charge compensated by hydrated metal ion [34, 35, 36, 37] or organic ions [38] present in interlayer space of the solid. Thus, the lamella of such materials are composed exclusively of Q^3 units, $([SiO_4]_3Si\text{-}OH)$ or $([SiO_4]_3Si\text{-}O^-)$, and Q^4 units, $([SiO_4]_4\text{–}Si))$. Also, it is possible that these solids present Q^2 units, $([O^-/HO]_2Si(OSi)_2)$, as observed in $K_2Ca_4Si_8O_{21}$.[39] Intercalation reactions in layered silicates (and the corresponding layered silicic acids) are driven by hydrogen bonding, cation exchange, dipole–dipole interaction, van der Waals force, and acid–base reaction. [40] Since these intercalation reactions do not involve covalent bonds, they are denoted as non-covalent modifications. Layered silicates can also be modified covalently, as in reaction such as condensation, silylation, esterification of $SiOH/SiO^-$ groups and pillaring. In this chapter, aspects of preparation and characterization of silylated and pillared interlayered clays will be explored in the interest to show the richness of these solids in terms of options to create interesting chemical modifications in the interlayer spaces and in the surfaces of the lamella themselves.

3D Structures from 2D Silicates

Layered structures have been used as catalysts for a very long time. For example, Gurvich [41] in 1915 used attapulgite and montmorillonites to polymerize pinene. The most widely known application of clay catalysis is the French Houndry cracking process developed in the 1930s. [42] By the 1960s, clays were replaced by synthetic zeolites because they show better activity and selectivity for cracking. [43]

The oil embargo of 1973 and the subsequent rise in oil prices acted as fuel to the search for novel catalysts to process heavy crude oil. General techniques, such as coking and hydroprocessing employed with these crudes are relatively expensive. Thus, the favored method is fluid catalytic cracking (FCC), which utilizes zeolites in the cracking step. [44] The pores of the zeolites are in the range of 8 Å, although the cavities may be 12-13 Å in diameter so that a large amount of nonselective pre-cracking, to reduce the molecular sizes of the crude, is required. Also, the limited mesopore size of zeolites catalysts presents a concrete limit to the size of molecules that can undergo the very selective intrapore reactions. In this period of time, an 18-ring aluminum phosphate with a ~12 Å pore has been synthesized [45] but its framework is neutral and hence lacks cracking capability.

In this sense, Pillared InterLayered Clays (PILCs) parallel zeolites as potential materials for the processing of heavy crude oils to fuels. The concept behind the original work [46] was to obtain clay minerals with larger non-schrinkable spacings between the clay sheets to accommodate large molecules. The solid consists of an inorganic oxide from the group consisting of alumina, zirconia, silica, magnesia, and mixtures of them, intercalated to the smectite clay (hectorite, chlorite, bentonite, montmorillonite, beidellite, and their mixtures), having greater than 50 percent of its surface area in pores of less than 30 Å in diameter, after dehydration. [47]

Since then, their use in petroleum cracking exceeds that of other catalysts. Casuscelli et al. [48] prepared pillared bentonites with Al and Al/Ga polyoxycations and characterized in order to obtain materials suitable as active components in catalysts for heavy oil cracking. The X-ray diffraction results and the textural properties at different calcination temperatures revealed a stabilizing effect of Ga in the pillars. The pillared clays with Al/Ga oligomers retained a higher surface area than calcined Al-PILC. The Al/Ga catalyst showed higher activity due to its higher acidity. With respect to selectivity, the Al/Ga-PILC sample produced less gasoline and more gases and coke than the Al-PILC sample in all conversion range. This may be due to the dehydrogenating effect of Ga, which gives a higher amount of alkenes. Other authors have observed this effect in the cumene conversion and dehydrocyclodimerization reactions with catalysts containing Ga (zeolite and pillared clays). [49, 50] The gasoline obtained with the Al/Ga-PILC catalyst, although in smaller quantity, may represent a superior quality due to higher olefin content. Table 2 gives the ratio of gases as well as the percentage of olefins for these catalysts at 50% conversion. Indeed, the ratio $C_4^{=}/C_4$ (butenes/butanes) which is related to the octane number of gasoline, was higher for the Al/Ga-PILC catalyst, as was the percentage of olefins.

Table 2. Ratio of gases and percentages of olefins of the samples at 50% conversion. Reprinted from Casuscelli, S. et al., Applied Catal. 1999, 181, 71–76. Copyright 1999, with permission from Elsevier

Catalyst	$C_4^{=}/C_4$	$iC_4/C_3^{=}$	$(C_1+C_2)/iC_4$	% olefins
ZCO-7	0.50	1.00	1.40	44
Al-PILC	1.35	0.50	2.50	52
Al/Ga-PILC	1.85	0.29	6.60	59

ZCO-7 = commercial zeolitic catalyst.

PILC also found applications in environmental protection and as molecular sieves, selective adsorbents, cracking of vegetable oils to produce biofuels, electrochemical and optical devices and others that will appear along this chapter. [51]

Research into the synthesis of PILC originates from the pioneering work by Barrer and McLeod [52] approximately 60 years ago. They obtained microporous materials by replacing the interlayer exchangeable cations in smectites with tetralkylammonium ions. By varying the charge on the host layer, and hence the separation of the pillars, one can design the pore structure to differentiate adsorbates of different molecular sizes. In most of the pillared lamellar solids reported until then, the gallery height is comparable to the van der Walls thickness of the host layers. However, it should be possible, through an accurate choice of the pillaring agent, to engineer the solids in which the interlayer gallery is larger than the thickness of the host layers. [53] The synthesis of the first organic pillared clay minerals, with an interlayer space of 0.5-0.6 nm and a permanent porosity, was reported by Mortland and Berkheiser. [54] However, organic pillared clays of this type are thermally unstable at high temperatures (>250°C) when interlayer collapse as in the unmodified clay. Below this temperature, organic pillared clays can usefully serve as molecular sieves, selective adsorbents and catalysts.

Layered structures intercalated with inorganic species retain their micro and mesoporosity after heating at temperatures >300°C. Although polymeric hydroxy and/or oxymetal cations were used as pillaring agents for clay minerals since the late 1960s [55], the paper by Brindley and Sempels [56] is considered the first on the preparation of an inorganic pillared clay. The procedure consists of exchanging the Na^+ ions in the smectite with oligomeric hydroxyaluminum cations, which are converted, after a heating step, into aluminum oxide. The species formed act as a pillar in the interlayer space.

In addition to the increase in surface area, the pillaring can generate moderate acidity on the material or introduce redox sites. [57] Applications of Na-RUB-18 pillared with metal oxides (Ta, Nb and Si) in the conversion of cyclohexanone oxime to ε-caprolactam, the raw material for Nylon-6, has been quite promising. [58] Mochizuki et al. [59] prepared microporous, crystalline inorganic-organic nano-hybrid by pillaring Na-RUB-18 with 1,4-bis (trichlorosilyl)benzene or 1,4-bis (dichloromethylsilyl)benzene (Figure 1). The materials presented surface areas of ~500 $m^2.g^{-1}$ and uniform micropores distribution. The micropores from the silicate modified with 1,4-bis (trichlorosilyl)benzene exhibited hydrophilic nature due to the remaining Si-OH groups on the lamella while those modified with 1,4-bis (dichloromethylsilyl)benzene shown hydrophobic properties.

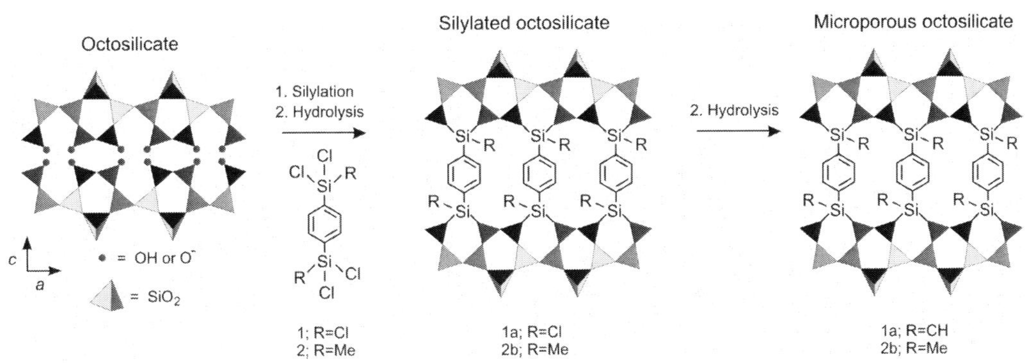

Scheme 2. Silylation of layered octosilicate with 1,4-bis(trichloro- and dichloromethyl-silyl)benzenes. Reprinted with permission from Mochizuki, D. et al., Chem. Mater. 2006, 18, 5223-5229. Copyright 2006, American Chemical Society.

Bergaya [60] investigated different metal ions as pillaring agents. Among them: Be, Mg, Y, La, Ti, Zr, V, Nb, Ta, Ce, Cr, Mo, Pr, Mn, Nd, U, Fe, Ru, Os, Co, Rh, Ni, Cu, Zn, B, Al, Ga, Si, Ge, Sn, P and Bi. Al-pillared clay minerals were the first to be prepared and extensively documented [61]. Although the hydrolysis behaviour of metal cations is generally understood, that of aluminum is profoundly known.

In the early 90s, the interest in pillared clays was renewed, because of varied sizes of pillars or the spacing between them, or both, were necessary since one may adjust the pore size to suit a particular application. Galarneau et al. [62] reported the successful synthesis of mesoporous solids from layered clays using quaternary ammonium surfactants as templating agents based on the templated synthesis of mesoporous molecular sieves of silica or aluminosilicate, first reported by Kresge et al. [63] and Beck et al. [64]

The introduction of surfactants as spacing agents in lamellar silicates can result in significant variations in the interlayer distance. Surfactants - quartenary ammonium salts - in aqueous medium, self-organize in micelles, and thus can direct the formation of a well-defined mesoporous structure from the condensation of silicate groups around them. The silicon source, i.e, tetraethyl orthosilicate (TEOS), condenses around the micelles to form a silica matrix embedded with the organic templates, which are typically removed by calcination, creating a continuous network of pores with controlled pores sizes and distribution. [65] This system is schematically shown in Figure 1. [66, 67] The advantages of this method are that the pore volume is controlled by the volume fraction of the template on the reaction mixture, and pore size is controlled by the size of the surfactant tails. [By this method, solids can achieve surface areas of the order of 1000 $m^2.g^{-1}$. [62, 68, 69, 70, 71]

Thus, the choice of spacing director allows obtaining materials with controllable interlayer distances, enabling applications as diverse as adsorbents, reinforcement for polymers, catalyst supports, chemical sensors and ion exchangers according to the interest. [72]

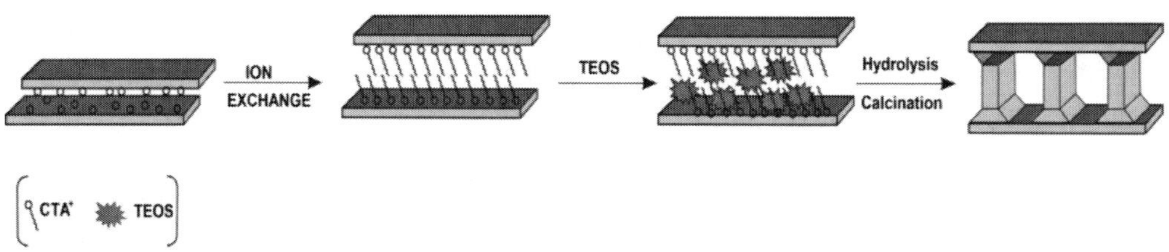

Figure 1. The PILCs (Pillared Interlayered Clays) preparation. In this example, montmorillonite undergoing a process of ion exchange with CTA^+ ions, after which condensation of silicic groups of tetraethyl orthosilicate occurs (TEOS). After calcination, an open-framework of silica is formed in the galleries with clay layers making up part of the pore walls. Reprinted from Liu, Z. et al., J. Coll. Int. Sci. 2009, 336, 244–249. Copyright 2009, with permission from Elsevier.

Tetraethylorthosilicate (TEOS) is the silicon source largely used, however, other silanes containing functional groups as amine, fenil or methyl can be applied, leading to functional solids. [70] The development of microporous organic–inorganic hybrid nanocomposites by reaction with 4,4'-biphenyl-bridged alkoxysilane compounds, which contain triethoxysilyl, methyldiethoxysilyl and dimethylethoxysilyl groups at each end of the 4,4'-biphenylene unit in ilerite [73] is well known as is the preparation of heterostructures by the bis(triethoxysilyl)benzene and TEOS condensation applied in adsorption of volatile organic compounds. [74, 75]

The Pre-Swelling Step

The replacement of natural inorganic cations such as Na^+ or K^+, for example, by various alkylammonium cations (AACs) converts the layered silicate surface from hydrophilic to hydrophobic. During this process, the metal cations and water molecules are released into solution and the intercalated organic cations sustain the lamellas apart.

The negative charged layers attract organic cations by electrostatic forces. Cations are incorporated by the cation exchange mechanism, in which the inorganic cations initially present in the silicate are replaced by organic cations. Many of the organic salts are water-soluble therefore, when possible, cation exchange reactions are performed in aqueous suspensions. For example, the exchange reaction between an inorganic metallic cation, M^{m+}, initially in the silicate structure, M^{m+}-Silicate, and an aqueous solution of an aliphatic ammonium salt ($RNH_3^+Cl^-$) can be formulated by Scheme 3.

$$m\,RNH_3^+Cl^-_{(aq)} + M^{m+}\text{-Silicate}_{(s)} \rightarrow MCl_{m(aq)} + (RNH_3^+)_m\text{-Silicate}_{(s)}$$

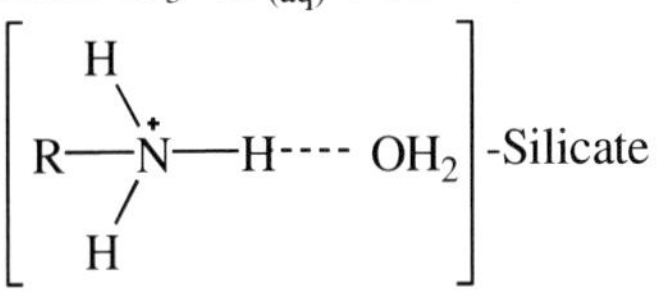

Scheme 3. Alkylammonium ions interacting via H-bond to water molecules.

For exchange of primary, secondary and tertiary AACs aqueous solutions with pH within 5-7 are employed. At a higher pH the adsorption of the cation is accompanied by the adsorption of free amine, whereas at a lower pH AACs compete with the protons for the exchange sites. The AACs compensate the negative charge in the framework but not only this. In the interlayer space, the adsorbed organic cation is not free of water. In this case, N-H^+ groups of alkylammonium ions are interacting via H-bond to water molecules by donating protons to the water molecules [76] as indicated in Scheme 3.

A primary ammonium group may form up to three hydrogen bonds with three water molecules in the first hydration sphere. The induction effect of the ammonium group increases the acid strength of the hydrating water molecules, and may donate protons to additional water molecules, which may form a second hydration sphere around the AACs. The same occurs to secondary and tertiary salts. Quaternary ammonium cations are not proton donors and consequently are not hydrophilic, i.e; they are disorganizers of the water structure as well the hydrophobic alkyl chains. Their presence into the silicates reduces their hydration ability. [77]

For this reason, in the preparation of silylated or pillared silicates it is common to replace the metallic cations by quaternary ammonium cations of the type $[(CH_3)_3RN]^+$ or $[(CH_3)_2RR'N]^+$, where R and R' are aliphatic or aromatic substituents. The introduction of a quaternary ammonium cation by a cation exchange mechanism is independent of the pH of the system unlike the primary, secondary and tertiary salts. In addition, the quaternary ammonium cation does not show acidic or basic properties and does not form H bonds with proton donors or acceptors. Thus, the choice of the quaternary AAC is determined by their chain size and concentration is controlled with the aim to obtain the desired final porous solid.

If layered silicates are exchanged with ions large enough to hold the layers permanently apart from each other by filling all the interlayer space, a notable range of shape-selective microporous sorbents are obtained. Microporosity was first demonstrated in 1955 [78, 52] with Me_4N^+ and Et_4N^+ as the interlayer cations. The reasoning is seen in Figure 2. It is d_1 and d_3, each of molecular dimensions, which control access to the interlayer regions of the crystal. [79] In the Me_4N^+ and Et_4N^+ -montmorillonites, d_1 was, respectively, 4.1 and 4.5 Å. The expanded crystals freely intercalated many permanent gases and non-polar molecules that were not intercalated in the parent montmorillonite.

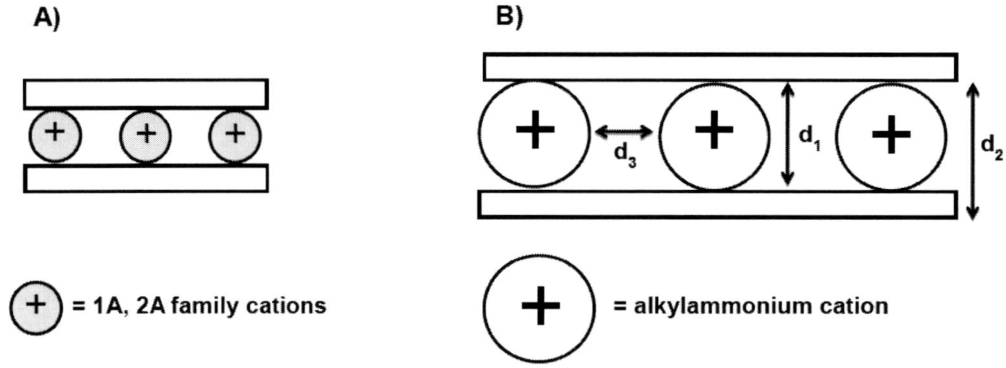

Figure 2. Representation of an unexpanded (A) and a permanently expanded layered silicate (B), showing the important free distances d_1 and d_3. Adapted from Barrer, R. M., *Clays Clay Miner.* 1989, 37, 385-395, with kind permission of The Clay Minerals Society, publisher of *Clays and Clay Minerals*.

Barrer [79] obtained pillared smectites with micropores of varying size and shape (Table 3) by applying different ammonium cations with different sizes and smectites with diferents charge densities.

The resulting smectites were used for shape selecting sorption and molecular sieving.

Table 3. Calculated interlayer free areas and porosities of some organomontmorillonites and organohectorites. Reproduced from Barrer, R. M., *Clays Clay Miner.* 1989, 37, 385-395, with kind permission of The Clay Minerals Society, publisher of *Clays and Clay Minerals*

Cation	Ion area (Å^2)	d_1 (Å)	Free area (m^2/g)	Porosity (cm^3/g)
Alkylammonium montmorillonites (CEC= 85 meq/100g)				
$CH_3NH_3^+$	21.6	2.2	294	0.065
$C_2H_5NH_3^+$	29.8	3.4	214	0.073
n-$C_3H_7NH_3^+$	35.6	3.7	184	0.068
n-$C_4H_9NH_3^+$	41.4	3.9	155	0.060
n-$C_5H_{11}NH_3^+$	47.1	4.0	126	0.050
n-$C_6H_{13}NH_3^+$	52.9	4.0	98	0.039
α, ω-alkylammonium montmorillonites (CEC= 85 meq/100g)				
$^+NH_3(CH_2)_2NH_3^+$	34.8	2.8	278	0.078
$^+NH_3(CH_2)_3NH_3^+$	40.6	3.5	262	0.092
$^+NH_3(CH_2)_4NH_3^+$	46.4	4.0	246	0.098
$^+NH_3(CH_2)_5NH_3^+$	52.2	3.9	231	0.090
$^+NH_3(CH_2)_6NH_3^+$	58.0	3.8	216	0.082
$^+NH_3(CH_2)_7NH_3^+$	63.7	3.9	201	0.078
$^+NH_3(CH_2)_8NH_3^+$	69.5	3.9	187	0.073

Cation	Ion area (Å^2)	d_l (Å)	Free area (m^2/g)	Porosity (cm^3/g)
$^+NH_3(CH_2)_9NH_3^+$	74.3	3.9	172	0.067
α, ω-alkyldiammonium hectorites (CEC= 91 meq/100g)				
$^+NH_3(CH_2)_2NH_3^+$	34.8	2.8	240	0.067
$^+NH_3(CH_2)_3NH_3^+$	40.6	3.6	224	0.080
$^+NH_3(CH_2)_4NH_3^+$	46.4	3.9	208	0.081
$^+NH_3(CH_2)_5NH_3^+$	52.2	3.9	192	0.075
$^+NH_3(CH_2)_9NH_3^+$	75.3	3.9	128	0.050

CEC= cation exchange capacity; d_l= the free distance between adjacent siliceous layers (Figure 3).

Most investigations have focused on the behavior of the intercalated surfactant arrangement in the fully loaded state using magadiite and its acidic crystalline form. In general, a linear correlation between the chain length of intercalated organic cations and the observed basal spacing of magadiite after intercalation has been obtained. Based on these results different models for the interlayer structure related to the type of long-chain cations have been developed.

Usually, partially loading with long-chain organic cations yields a non-uniform distribution of the surfactant molecules in the interlamellar space of the layered silicates. The resulting samples contain fully expanded domains within the crystals besides unreacted crystals or domains. Wuttig et al. [80] described the formation of partly loaded surfactant-magadiite complexes and presented a method to rearrange the organic cations (n-cetylpyridinium (CP)) in the interlayer space of magadiite.

All changes in the X-ray patterns in the range from 2 to 7.5° 2θ for the sample of magadiite containing 72% of CP (CP-mag5) for the different preparation steps are shown in Figure 3. The basal spacing of CP-Mag5 (40.2 Å) is nearly the same as that of the fully loaded CP-Mag7 (131%) at 39.3 Å. After rearrangement reaction of the as-synthesized magadiite the 001 reflexion disappear and the basal spacing shifts from 40.2 to 33.4 Å [CP-Mag5-re (72%)].

This investigation envisages the arrangement of the organic cation into the interlayer space of layered silicates. In a first step, the intercalation conditions have to be carefully adjusted to ensure that only the desired number of long-chain cations is taken up by the solid. However, their effect results in non-homogeneous samples. In a second step, suspending solid in deionized water at 333K, leads to the rearrangement of the surfactant molecules which have already been intercalated, induced by an equilibration process carried in salt-free aqueous solution.

It is sufficient to yield a homogeneous distribution of the intercalated organic cations (Figure 4). Lagaly et al. [81] developed a model for the incorporation of the long-chain surfactant in magadiite. The organic cation binds to the silicate layer and the chains take up a tilt angle of 35°. This orientation corresponds to the energetically favorable all-*trans* conformation of long alkyl chains.

Figure 4 shows two possibilities for arrangement of the cations in partly loaded samples before the rearrangement at 333K. It reveals that intercalating organic cations, organized, ordered and variable voids in the interlayer space are obtained which allow one to covalently modify the interlayer region.

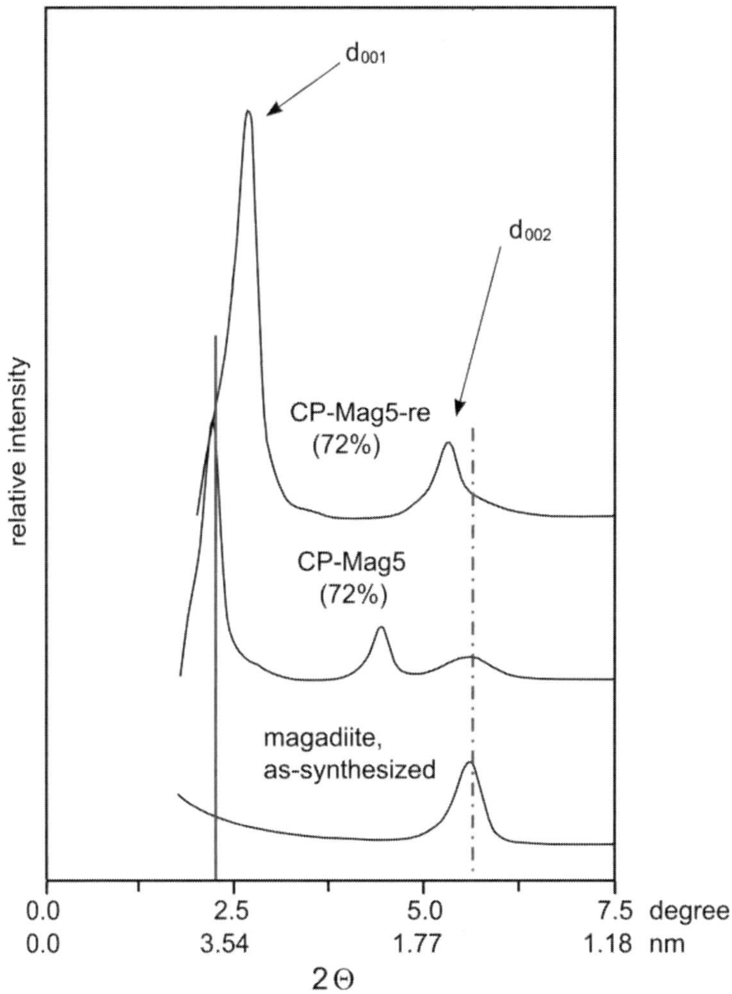

Figure 3. X-ray powder diffraction patterns of CP-Mag5 after different reaction stages. Adapted from reference [80].

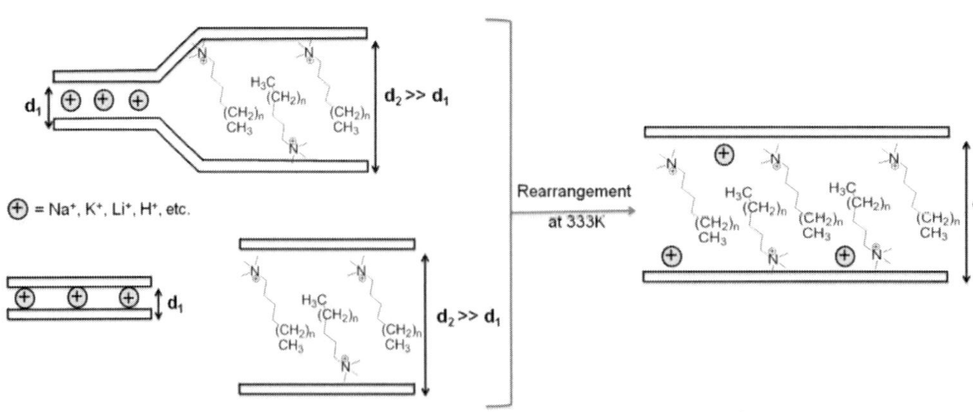

Figure 4. Representation of the arrangement of the surface cations inside the interlayer spaces: non-uniform distribution of organic and inorganic cation after ion exchange (lefthand side) and their uniform distribution after equilibration in water at 333 K (righthand side). Adapted from reference [80].

Organic cations incorporated into a layered silicate, sustain the lamellas apart and convert the layered silicate surface from hydrophilic to hydrophobic, allowing post-swelling modification on its surface. The size of the organic chains can control shape selective sorption and molecular sieving on the final solid as it was discussed in this section. The most used organic cations are summarized in Table 4.

Table 4. Organic cations widely used in the pre-swelling step

Ion exchanger	Reference	
$(CH_3)_4N^+$	[78]	Barrer, R. M. (1986)
$(CH_2CH_3)_4N^+$	[78]	Barrer, R. M. (1986)
$C_nH_{2n+1}NH_3^+$ (n=1-6)	[79]	Barrer, R. M. (1989)
$^+NH_3(CH_2)_nNH_3^+$ (n=2-9)	[79]	Barrer, R. M. (1989)
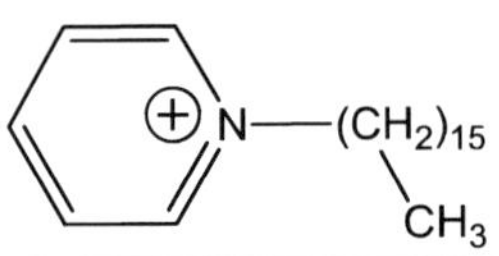	[80]	Brenn, U. et al. (1981)

The Grafting Process

The controlable voids in the interlayer space generated by the alkylammonium cations intercalation are perfect for the incorporation of silylation agents. In general, SiOH groups on silica surfaces can be covalently modified with alkyl, chlorosilyl or alkoxysilyl silylation reagents. [82] Organochlorosilanes react with silanol groups as in Reaction 1.

$\equiv Si–OH + Cl–SiR_3 \rightarrow \equiv Si–O–SiR_3 + HCl$ Reaction 1

Because of the high reactivity of chlorosilyl groups, this reaction proceeds at room temperature. Silylation reaction is also achieved with organoalkoxysilanes (typically organomethoxysilane). The advantage of the use of organotrialkoxysilanes is their relatively high stability. However, higher temperature and/or the addition of acid are needed for the desired high degree of silylation (Reaction 2).

$\equiv Si–OH + MeO–SiR_3 \rightarrow \equiv Si–O–SiR_3 + MeOH$ Reaction 2

Silylation reagents with three reactive sites (e.g. triethoxysilyl and trichlorosilyl groups) are immobilized onto the silica surfaces through one to three covalent bonds, and the Si atoms become T^1 ($RSi(OSi)(OR)_2$), T^2($RSi(OSi)_2(OR)$) and T^3 ($RSi(OSi)_3$) species. In addition, residual silanol groups of silylation reagents can be condensed with other reagents.

The first study of silylation reaction onto layered silicates was reported by Ruiz-Hitzky and Rojo in 1980. [83] Protonated layered silicates (magadiite) were treated with polar organic solvents, such as dimethylsulfoxide (DMSO), for the expansion of the interlayer spaces. Trimethylchlorosilane and hexamethyldisilazane were reacted with silanol groups. The success of reaction indicated that interlayer silanol groups of layered silicates can be used as reactive groups for this kind of reactions. Ruiz-Hitzky, Rojo and Lagaly [84] studied the mechanism of silylation reactions by using various organosilanes, such as (chloromethyl) dimethylchlorosilane, dimethylphenylchlorosilane, triphenylchlorosilane, and hexaethyldisilazane. According to their proposal, it is necessary for a successful silylation reaction that i) hydrogen bonds between interlayer surfaces be minimized by intercalation of polar organic molecules, and ii) diffusion of silylation reagents.

Dodecyltrimethylammonium cation ($C_{12}TMA^+$)-intercalated layered silicates, such as magadiite kenyaite, and octosilicate, can be used for silylation reactions with trimethylchlorosilane, diphenylmethylchlorosilane, triethylchlorosilane, triiso-propylchlorosilane, butyldimethylchlorosilane, octyldimethylchlorosilane, octadecyl-dimethylchlorosilane [85, 86] and octyltriethoxysilane. [87] Alkyltrimethylammonium-intercalated layered silicates are stable and the grafted moieties cannot be eliminated by washing with polar solvents. In addition, the basal spacings of intercalated compounds are larger than 2 nm and adjustable by choosing the length of alkyl chain. Bulky silylation reagents, such as octadecyldimethylchlorosilane, can be immobilized into interlayers. [86] Nowadays, intermediates for reactions with organochlorosilanes are usually layered silicates intercalated with alkyltrimethylammoniums or alkylamines.

Ogawa et al. [88] reported that the surface properties of layered silicates could be designed to bind guest species by controlling the degree of silylation as well as the use of organochlorosilanes with different functions. Magadiite silylated with octyltrichlorosilane adsorbs n-octanol in the interlayer space while the product with octyldimethylchlorosilane does not adsorb n-octanol. When the degree of silylation is decreased, both solids adsorb n-octanol. Therefore, one can envisage the criteria of designed shape/size selective solids by simply choosing the most adequate silylating agent.

Furthermore, the adsorption properties of alcohols in silylated layered silicates were investigated in detail, by using magadiite silylated with octyltrichlorosilane. [89, 90] Larger amounts of alcohols were adsorbed onto the derivatives with lower degree of silylation, indicating a possible control of the interlayer nanospace by varying the amount of grafting alkylsilyl groups. [90] In addition, 1-hexanol was adsorbed preferentially from a 1-hexanol/1-butanol aqueous mixture under limited conditions.[91] This concept of selective adsorption was applied to molecule-specific adsorbents by using layered titanate. Ide and Ogawa [91] reported that layered titanates immobilized with two different organic functional units (alkyl and phenyl groups) specifically adsorb 4-nonylphenol (NPh) which is a known contaminant of water. Ţoriya et al. [92] reported that trimethylsilylated kanemite adsorbed benzene rather than water because of hydrophobic interlayer region. Magadiite immobilized with [2-(perfluorohexyl)-ethyl]dimethylchlorosilane is advantageous for preparation of the films and showed high thermal stability. [93]

The defined crystal structure of layered silicates is essential for the discussion of structural details of the silylated layered silicates. Silylation of kanemite (structure determined by Gies et al. [94]) with mono-, di-, and trichloro(alkyl)silanes was performed by Kuroda's group. [95] Kanemite is composed of single layered silicate sheets whose structure is decribed as fusion of single six-membered rings. The result of the silylation reaction with dichloro- and trichloro-(alkyl)silanes is the formation of additional five- and six-membered rings (Figure 5).

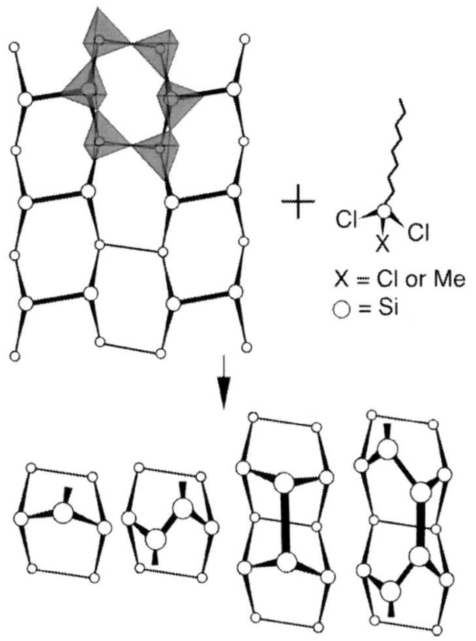

Figure 5. Possible silicate structures formed by silylation of kanemite with di- and trichloro(alkyl)silanes. Reprinted with permission from Kuroda, K. et al., Chem. Mater. 2001, 13, 3603–3609. Copyright 2001, American Chemical Society.

The ^{29}Si MAS NMR spectra of alkyltrichlorosilylated kanemite products showed signals assigned to T^2 ($RSi(OSi)_2(OR)$), T^3 ($RSi(OSi)_3$), Q^3 and Q^4 environments due to the attached alkylsilyl groups. The degree of silylation was varied from 63 to 73% calculated from the $Q^4/(Q^3+Q^4)$ ratio. Well-ordered complete silylation has not been achieved with kanemite. Kanemite has high density of surface SiOH/SiO^- groups (5.6 groups/nm^2) because all silicon atoms are Q^3 environment. The authors concluded that SiOH/SiO^- groups are too close for well-ordered and complete silylation.

The main application of hybrids obtained by alkylsilylation is as selective adsorbents. This selectivity depends on the quantities and on the geometrical ordering and/or packing of silylated groups in the two-dimensional confinement of the interlayer spaces. This effect cannot be observed for silica-based materials that do not display confined nanospaces.

A wide variety of silylation reagents, including imidazolium salts [96] and azobenzene groups, [97] have been used for silylation. When allyldimethylchlorosilane is immobilized onto magadiite and kenyaite, [98] HCl (formed from reaction between –SiOH and –SiCl) reacts with allyl group and creates new $-Si-O-Si(CH_3)_2OH$ groups (Reaction 3 - 5). Such immobilized functional groups can be active for additional covalent modifications.

$$CH_2CHCH_2(CH_3)_2SiCl + \equiv Si-OH \rightarrow \equiv Si-O-Si(CH_3)_2CH_2CHCH_2 + HCl$$

Reaction 3

$$\equiv Si-O-Si(CH_3)_2CH_2CHCH_2 + HCl \rightarrow \equiv Si-O-Si(CH_3)_2Cl + CH_3CHCH_2$$

Reaction 4

$$\equiv Si-O-Si(CH_3)_2Cl + H_2O \rightarrow \equiv Si-O-Si(CH_3)_2OH + HCl$$

Reaction 5

Actually, immobilized functional groups on layered silicates are further reacted with other reagents by using multi-step immobilization. γ-Methacryloxypropyltrimethoxysilane is immobilized onto magadiite and used for copolymerization with methylmethacrylate monomers, [99] which provides a nanocomposite with covalent bonding, because normally clay-polymer nanocomposites are prepared by the use of organoammonium exchanged clays with various polymers with non-covalent interactions.

Shindachi et al. [100, 101] reported the multi-step immobilization of 4-(bromomethyl) phenylchlorodimethylsilane onto C_{16}TMA-magadiite and the use for further reaction with trialkylamine. Covalently immobilized diarylethene onto magadiite exhibits an improvement of photochromic reversibility if compared with diarylethene intercalated into LDHs. Matsuo et al. [102] reported three-step immobilization of pyrene chromophores as follows: i) immobilization of [2-(perfluorohexyl)-ethyl]trichlorosilane onto C_{12}TMA-magadiite, ii) immobilization of APTES to SiOH groups, iii) reaction between the pyrene chromophore with amino groups of APTES (Figure 6).

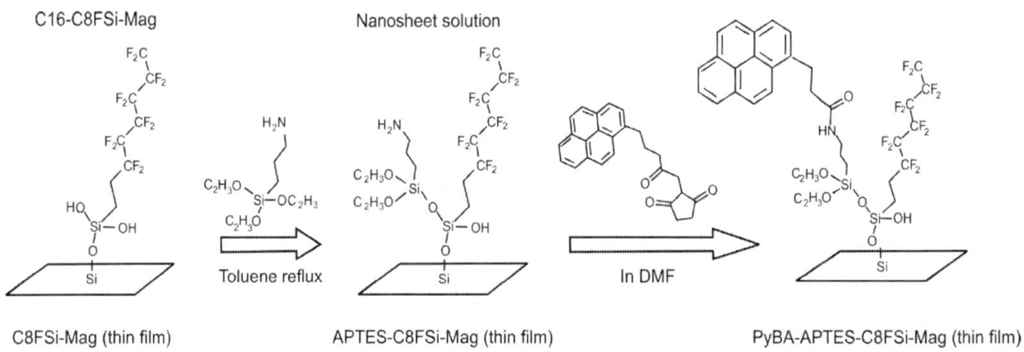

Figure 6. Immobilization of pyrene chromophores in magadiite. Adapted from Matsuo, Y. et al., J. Fluorine Chem. 2008, 129, 1150–1155. Copyright 2008, with permission from Elsevier.

Specht et al. [103] achieved immobilization by using hydrosilylation (Scheme 4). Apophyllite is reacted with phenylmethylchlorosilane (HMP) which contain a Si–H group. The Si–H groups of silylated apophyllite are reacted with octane, 3-buten-1-ol, and but-3-enyl-β-D-glucoside through hydrosilylation.

Aminopropyltriethoxysilane (APTES) is one silylation agent frequently used because of its wide spectrum of reaction. Immobilization of APTES onto the interlayer surfaces of magadiite and kenyaite are reported. [104, 105] Leu et al. [106] reported that APTES-immobilized kenyaite was used as an intermediate for the preparation of layered silicate/polyimide nanocomposites with higher thermal stability. The nanocomposite displays a maximum increase of 36°C in the degradation temperature and a maximum reduction of 54% in moisture absorption.

Zhang et al. [107] studied the utility of amino-functionalized magadiite for selective adsorption of metal cations. A silylation reagent including two amino groups (3-(2-aminoethylamino) propyltrimethoxysilane, denoted as AAPTMS) is immobilized onto magadiite. Silylation reagents with one (APTES), two (AAPTMS) and three (*N*-3-trimethoxysilylpropyldiethylenetriamine) amino group(s) were immobilized onto kenyaite or layered octosilicate (Figure 7).

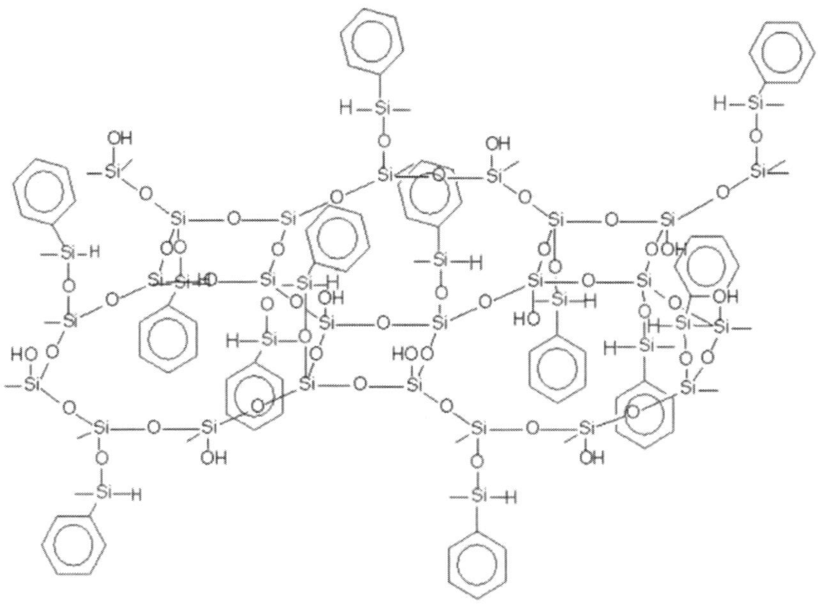

Scheme 4. Synthesis and schematic representation of HPM-apophyllite. Reprinted from Specht, K. M. et al., Appl. Clay Sci. 2010, 47, 212–216. Copyright 2010, with permission from Elsevier.

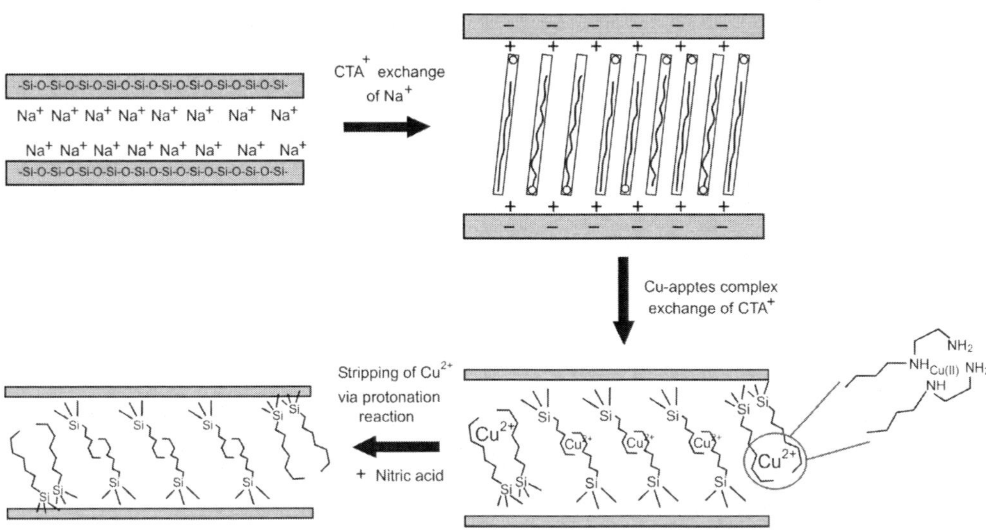

Figure 7. Adsorption of Cu^{2+} in amino-functionalized magadiite. Reprinted with permission from Zhang, Z. et al., Chem. Mater. 2003, 15, 2921–2925. Copyright 2003, American Chemical Society.

The layered hybrid is applicable for selective adsorption of Cu^{2+} in diluted aqueous solutions and the process is reversible by nitric acid addition to the system. Ishii et al. [108] reported the silylation of n-hexylamine-intercalated octosilicate with *p*-aminophenyltrimethoxysilane (APhTES). The obtained product contains both n-hexylamine and APhTES in the interlayer space. The subsequent ethanol extraction leads to the removal of n-hexylamine and complete immobilization of APhTES, however, these products have no microporosity. An acid treatment eliminates excess APhTES molecules and oligomers yelding a microporous solid. The surfaces of magadiite and [Al]-magadiite were functionalized with aminopropyl

(APTES) pending groups to understand the effect of the organofunctionalization. [109] The control of the interlayer nanospace and variation of the amount of grafting alkylsilyl groups were achieved by investigating the effect of CTA^+ surface density on the magadiite capacity to bind the aminopropyl and chloropropyl [110] functional groups. The introduction of the CTA^+ in different CTA/Na ratios before the addition of APTES in the interlayer space of [Al]-magadiite allowed the formation of materials with variable amounts of amino functional groups. Could be observed in the ^{29}Si NMR spectra the preferencial diminishing of Q^3 species (around -100 ppm) in relation to Q^4 (around -110 ppm) as commented before for magadiite, indicating the silylation dependence with the silanol group at the surface. Thus, the signal of T^2 and T^3 (around -60 ppm) increases considerably at lower concentrations of surfactant.

The grafting process is accompanied by the loss of part of CTA^+ cations as proposed by Ogawa et al. [111] In this model (Figure 8), the free silanol groups react with the organoalkoxysilane (in the case of Figure 8, APTES) to form grafted silyl groups in the interlayer space with the generation of methanol (1). Methanol can react with neighboring silanol groups to form alkoxy groups and water (2). In (3), the generated water can interact with $CTA^+OSi\equiv$sites to regenerate a silanol group. (4) The free silanol groups formed in step 3 react with the organoalkoxysilane as in the step 1. After this process, a covalent bond is formed between the organoalkoxysilane and magadiite and a large part of CTA^+ is eliminated from the solid.

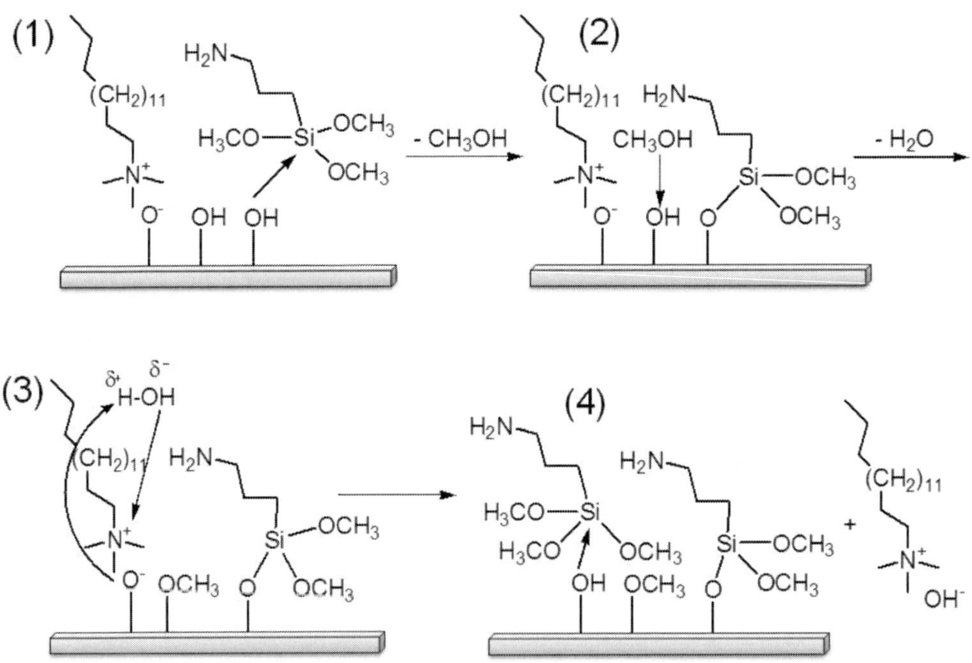

Figure 8. Grafting of organoalcoxysilane (APTS) to the interlamellar surfaces of magadiite. Adapted with permission from Ogawa, M. et al., Chem. Mater. 2000, 12, 1702-1707. Copyright 2000, American Chemical Society.

Extensive and controlled bonding of one lamella to the adjacent one can be made by intelligent design of sylilating/functional groups. These aspects are developed in the next topic. The properties associated with inorganic layered solids having organic grafted molecules lead to new organofunctionalized compounds with interesting catalyst properties. The most employed organosilanes shown in this section are summarized in Table 5.

Table 5. Main organosilanes found in grafted layered silicates

Organosilane	Silicate	References	
$ClSi(CH_3)_3$	Magadiite	[83] [85]	Ruiz-Hitzky, E. and Rojo, J. (1980) Okutomo, S. et al. (1999)
	Octosilicate	[86]	Endo, K. et al. (1994)
$ClSi(CH_2CH_3)_3$	Magadiite	[86]	Okutomo, S. et al. (1999)
$(CH_3)_3Si\text{-}NH\text{-}Si(CH_3)_3$	Magadiite	[83]	Ruiz-Hitzky, E. and Rojo, J. (1980)
$(CH_3CH_2)_3Si\text{-}NH\text{-}Si(CH_2CH_3)_3$	Magadiite	[84]	Ruiz-Hitzky, E. et al. (1985)
$(CH_3)_2ClSiCH_2Cl$	Magadiite	[84]	Ruiz-Hitzky, E. et al. (1985)
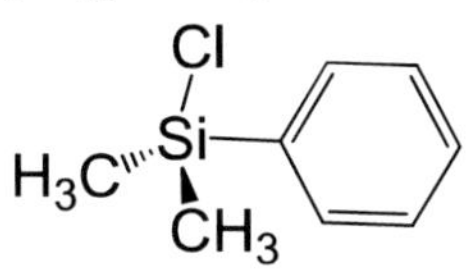	Magadiite	[84]	Ruiz-Hitzky, E. et al. (1985)
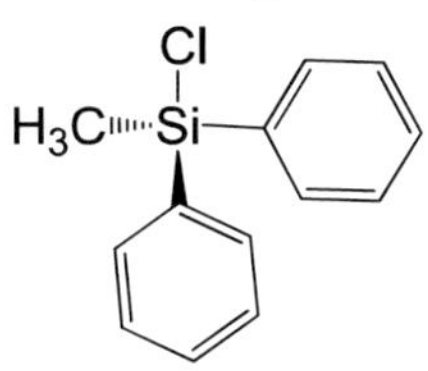	Magadiite	[86]	Okutomo, S. et al. (1999)
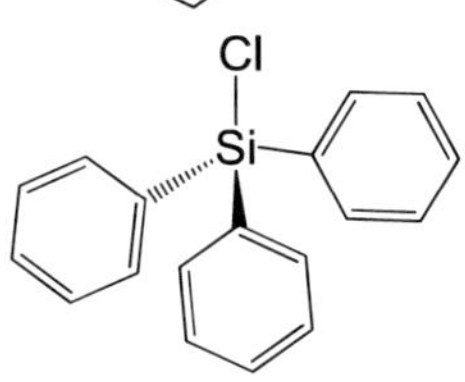	Magadiite	[84]	Ruiz-Hitzky, E. et al. (1985)
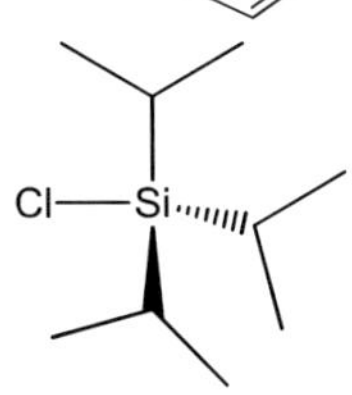	Magadiite	[86]	Okutomo, S. et al. (1999)
$(CH_3)_2ClSi(CH_2)_nCH_3$ (n=3, 7, 17)	Magadiite Kanemite	[86] [95]	Okutomo, S. et al. (1999) Shimojima, A. et al. (1996)
$(CH_3CH_2O)_3Si(CH_2)_7CH_3$	Magadiite	[87]	Park, K.-W. et al. (2009)
$Cl_3Si(CH_2)_7CH_3$	Magadiite	[88]	Ogawa, M. et al. (1998)
$(CH_3)_2ClSi(CH_2)_7CH_3$	Magadiite	[88]	Ogawa, M. et al. (1998)
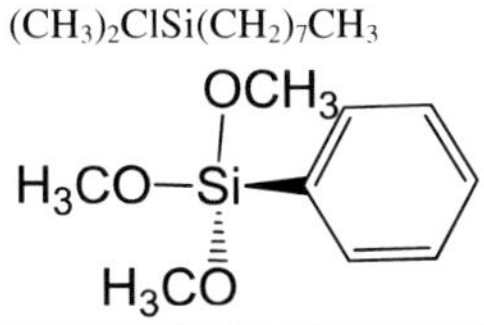	Titanate	[91]	Ide, Y. and Ogawa, M. (2007)

Table 5. (Continued)

Organosilane	Silicate	References	
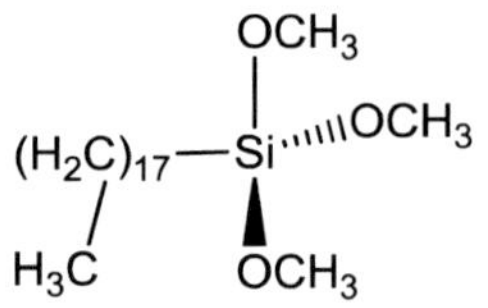	Titanate	[91]	Ide, Y. and Ogawa, M. (2007)
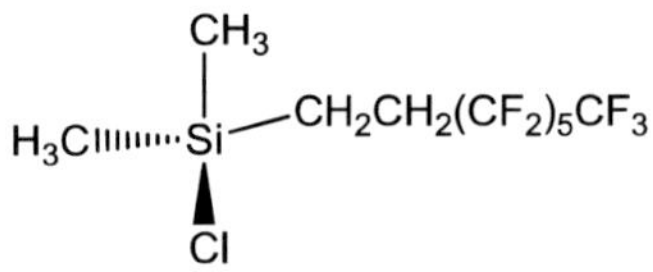	Kanemite	[92]	Toriya, S. et al. (2003)
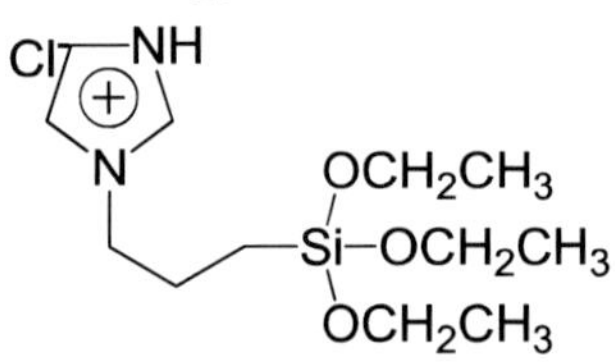	Octosilicate	[96]	Takahashi, N. et al. (2010)
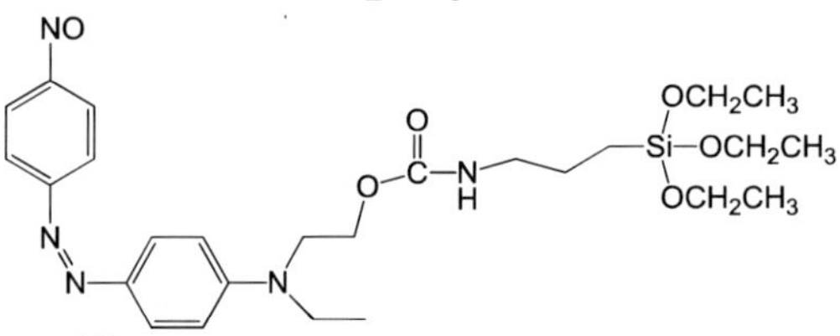	Magadiite	[97]	Guo, Y. et al. (2004)
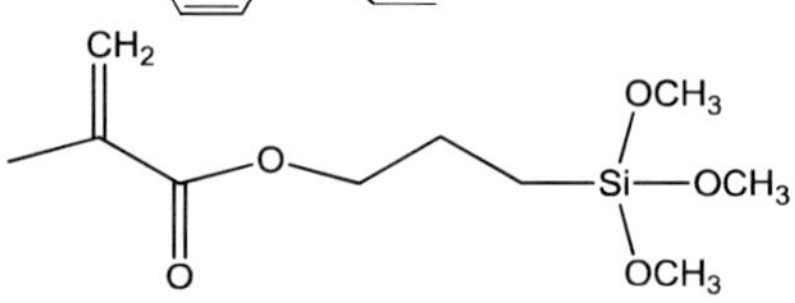	Magadiite	[99]	Isoda, K. et al. (2000)
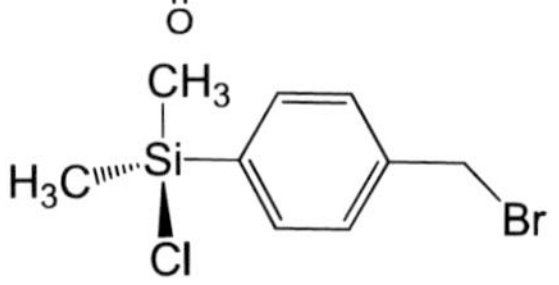	Magadiite	[100] [101]	Shindachi, I. et al. (2004 and 2007)
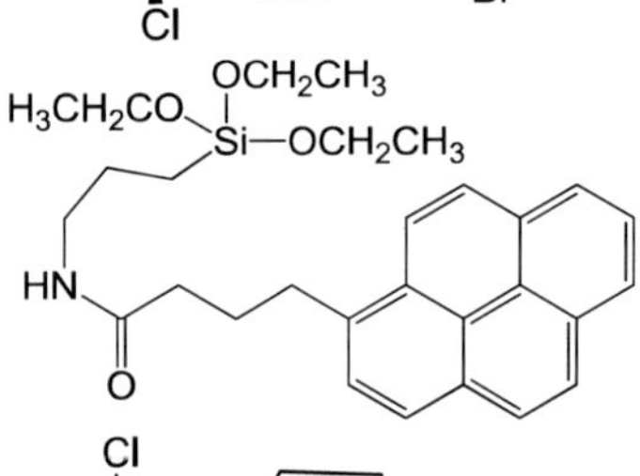	Magadiite	[102]	Matsuo, Y. et al. (2008)
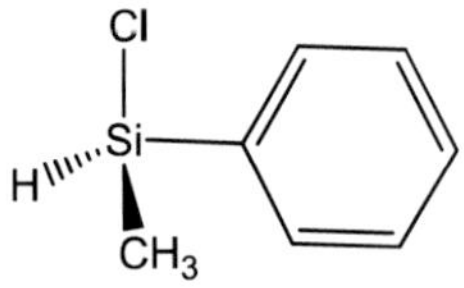	Apophyllite	[103]	Specht, K. M. et al. (2010)
$(CH_3CH_2O)Si(CH_2)_3NH_2$	Kenyaite	[104] [106]	Park, K.-W. et al. (2004) Leu, C.-M. et al. (2002)
	Magadiite	[105] [107] [109]	Wang, S.-F. et al. (2007) Zhang, Z. et al. (2003) Moura, H. M. and Pastore, H. O. (2014)

Organosilane	Silicate	References	
$(CH_3CH_2O)Si(CH_2)_3NHCH_2NH_2$	Magadiite	[107]	Zhang, Z. et al. (2003)
$(CH_3CH_2O)Si(CH_2)_3NHCH_2NHCH_2NH_2$	Magadiite	[107]	Zhang, Z. et al. (2003)
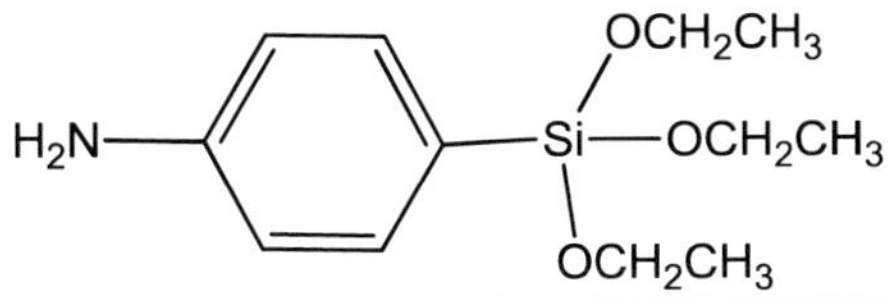	Octosilicate	[108]	Ishii, R. et al. (2009)

CROSS-LINKING LAMELLA: THE PILLARING PROCESS

According to the International Union of Pure and Applied Chemistry (IUPAC), pillaring is a process by which a layered compound is transformed to a thermally stable micro-and/or mesoporous material with retention of the layer structure. [112] The report adds that "a pillaring agent is any compound which maintains the distance between adjacent layers upon removal of the solvent and which induces an experimentally observable pore structure between the layers called 'interlayer region' accessible to molecules at least as large as N_2". Thus, a "pillared material" must fulfill at least three criteria: i) chemical and thermal stability (without further specifying the chemical or temperature conditions); ii) a certain layer ordering that enables at least a d_{001} spacing to be determined (although a rational series of basal reflections is not necessary); and iii) accessibility of the interlayer space to molecules at least as large as N_2 (thus, N_2 adsorption-desorption isotherms must be determined, and analysed via t- and α-plots; however, no order in the pillars and consequently in the pores is required). [77] From literature, a large number of works have been focused in this class of solids and a pillared material is not always a single-phase material since they may be amorphous while layers display crystallinity.

Landis et al. [113] and Dailey and Pinnavaia [114] reported the pillaring of magadiite and kenyaite with polymeric silica species by reaction of alkylamine or alkylammonium-intercalated layered silicates with tetraethoxysilane (TEOS). After calcination, silica-pillared layered silicates with micropores are obtained. Expansion with alkylamine or alkylammonium ions is necessary to obtain a large interlayer space. Increasing the TEOS content, microporous surface area reaches a maximum and then decreases with the increasing amount of TEOS, suggesting the decrease in the lateral separation (d_3 in Figure 2) of the silica pillars, probably by enlargement of pillars.

Subsequently, the effect of catalysts for polymerization of silica species was investigated by Jeong et al. [115] Pillaring with polymeric silica was reported for other layered silicates, such as octosilicate [116], MCM-22(P) [117], kanemite [118] , kenyaite [119] and potassium layered silicate [120] with kenyaite-like structure. Pillaring of various metal oxides, such as alumina, titania, and zirconia, is achieved by using MCM-22(P) and octosilicate.[121, 122]

In general, pillaring of layered silicates with silica decreased the crystallinity because of fragmentation and dissolution of silicate layers. Tsapatsis *et al.* [123] developed the pillaring process at low-temperature for the retention of the crystallinity of the layers. $[C_{16}H_{33}N(CH_3)_3]^+$-intercalated layered MCM-22(P) is repeatedly washed with water and then treated by ultrasonication. The obtained material has a new ordered layered structure. Interestingly, the swelling procedure is reversible because MCM-22(P) can be restored by acid treatment. The swollen product is pillared and a porous material with the MCM-36 structure is obtained (Figure 9). The swollen sample can be partially exfoliated *via* compounding with polymers.

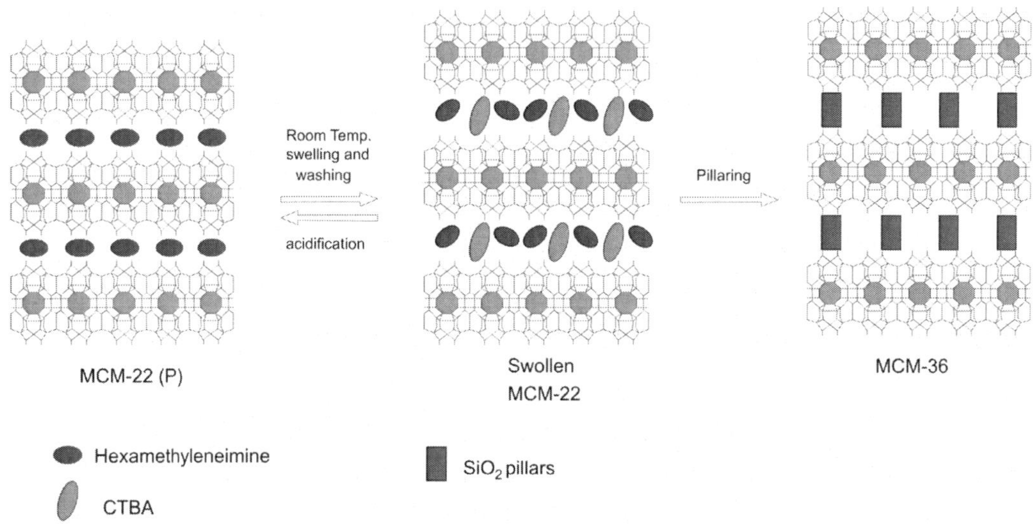

Figure 9. Reversible swelling of MCM-22(P) and pillaring of the swollen material. The swollen material can be deswollen back to MCM-22(P) by acidification, suggesting exchange of CTAB surfactant with protons. Reprinted with permission from Tsapatsis, M. et al., J. Am. Chem. Soc. 2008, 130, 1507–1516. Copyright 2008, American Chemical Society.

As mentioned before, organic clays are thermally unstable at temperatures >300°C when interlayer collapse the same way as in the unmodified clay. However, it is a good candidate to be a template for the synthesis of micro and/or mesoporous structures. The synthesis of these materials involve the use of ionic surfactants, which interact with the inorganic ions in a cooperative process to form a range of ordered mesostructures; many of these surfactants can be intercalated into the layered inorganic materials. Galarneau et al. [62] showed the possibility of using intercalated surfactants for the template synthesis of structures within the interlayer spaces of clays. In these process, the surfactants act in concert with aqueous silicate species to form a silica framework between the layers. After, removal of the surfactant by calcination leaves a mesoporous solid thermally stable, which was called 'porous clay heterostructure'. The designing of porous clay heterostructures is based on the use of intercalated quaternary ammonium cations and neutral amines as co-surfactants to direct the interlamellar hydrolysis and condensation polymerization of neutral inorganic precursor (for example, TEOS) within the interlayer space of an ionic lamellar solid.

In their work, Galarneau et al. [62] prepared $[C_{16}H_{33}N(CH_3)_3]^+$-fluorohectorite and reated it with the mixture of neutral amine and TEOS at molar ratios in the range 1:2:15 to 1:20:200, for 4h at room temperature. Under these conditions, the galleries are swollen by the co-templating amine and readily accessible to TEOS. The hydrolysis of TEOS is much faster in the amino modified clay interlayer space than in solution, and this minimizes the formation of extra gallery silica. The resulting solids were calcined at 600°C to remove the templating surfactants. Two or more orders of (001) X-ray reflections, indicating well-formed layered structures, were observed for all products. Depending on the chain length of the neutral amine, the gallery height could be varied from 14.9 Å (hexylamine) to 23.4 Å (dodecylamine). N_2 BET surface areas were in the range of 470-750 $m^2.g^{-1}$. Table 6 gives the pore sizes of these solids using alkylammonium cations and neutral amines as co-templates.

Table 6. Properties of porous clay heterostructures prepared by gallery-templated synthesis. Adapted by permission from Macmillan Publishers Ltd: Nature, Galarneau, A. et al. 1995, 374, 529–531. Copyright 1995

			Gallery height (Å)*	
Alkylammonium exchange ion (Q^+)	Amine co-template	Pore size (Å)#	Calcined heterostructure	Air dried heterostructure
$C_{16}H_{33}N(CH_3)_3^+$	$C_6H_{13}NH_2$	15	14.9	22.2
	$C_8H_{17}NH_2$	17	18.4	22.4
	$C_{10}H_{21}NH_2$	21	22.4	28.4
	$C_{12}H_{25}NH_2$	22	23.4	34.4
$C_{10}H_{21}N(CH_3)_3^+$	$C_{10}H_{21}NH_2$	14	14.3	23.9

* Gallery height is defined as the observed X-ray basal spacing minus the 9.6 Å thickness of the clay layer.

Pore size obtained by Horvath-Kawazoe analysis of N_2 adsorption data.

The chain length of the alkylammonium cation also is important in the templating process evidenced by the use of $C_{10}H_{21}N(CH_3)_3^+$ and decylamine as co-template. This surfactant system affords an average pore size of 14 Å, smaller than the 21 Å pore size provided by the mixture $C_{10}H_{21}NH_2$/ $C_{16}H_{33}N(CH_3)_3^+$.

The relationship between the pore size distribution of these solids, chain lengths of the templating surfactants and reaction stoichiometry supports a templating mechanism analogous to the M41S structure synthesis. [124, 125] The initial syntheses of M41S materials involved the so-called 'liquid-crystal templating' mechanism, in which organized micelles of cationic alkyltrimethylammonium surfactants acted as templates to direct the mesoporous structure formation by electrostatic interactions with anionic silicate precursors at 150 °C and high pH. [126, 127]

As outlined in Figure 10 by Galarneau et al., the pore structures are determined by the intrinsic structure formed from the interactions between the surfactants into the interlayer space and the silicate. The initial organic modified silicate (Figure 10a) shows gallery heights that are very near to the values expected based on the chain length of the alkylammonium cation and neutral amine. With the introduction of TEOS, some of the amines are displaced from the interlayer space, and a reactive intermediate is formed (Figure 10b). Subsequently, hydrolysis of the TEOS affords a silica template around a monolayer of micellar organization of alkylammonium cation.

The reaction with TEOS alone fails to produce a heterostructures. Spherical micelles in van der Walls interaction or, more probably (based on templating synthesis of MCM-41, for example) rod-like micelles are formed, as shown in Figure 10c. The final calcination step (Figure 10d) removes the template and complete de dehydroxylation and cross-linking of the silica structure.

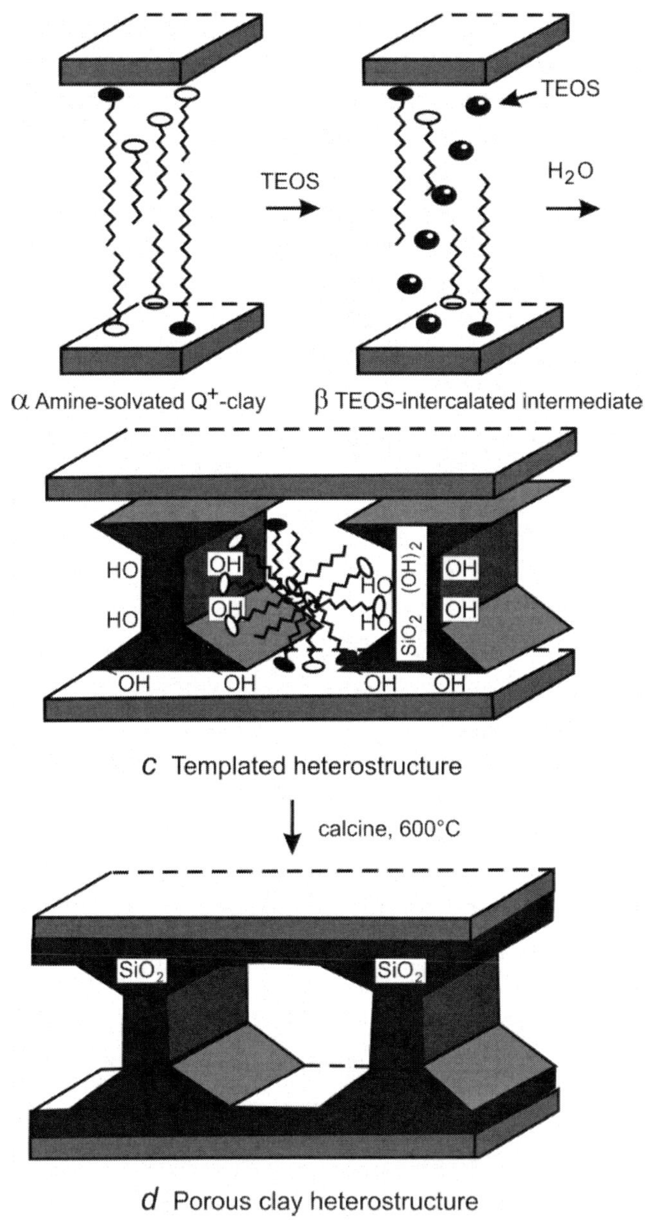

Figure 10. The synthesis of porous clay heterostructures. Reprinted by permission from Macmillan Publishers Ltd: Nature, Galarneau, A. et al. 1995, 374, 529–531. Copyright 1995.

Porous clay heterostructures are not limited to fluorohectorite. Derivatives of vermiculite, rectorite montmorillonite and magadiite by this same method, using TEOS hydrolysis in the presence of $C_{16}H_{33}N(CH_3)_3^+$ and decylamine as co-surfactants can be prepared. [62, 128, 129] Each of these derivatives had an average pore size to the pore size observed for the corresponding fluorohectorite system (~21 Å).

Pinnavaia and Dailey [114] investigated the reaction of TEOS and other metal alkoxides with layered silicates in an effort to form pillared structures with gallery height substantially larger than the thickness of the host layer, which they called the supergallery pillared clays.

Metal oxide pillared clays typically exhibit gallery heights equal to, or less than, the thickness of the host layer. Hydrolysis and condensation polymerization of metal alkoxides in a layered silicic acid, such as magadiite, is a promising route to supergallery derivatives, because the gallery height generated by the siloxane polymer can be substantially larger than the thickness of the host layers. The reaction of TEOS with octylammonium generates the silica-pillared magadiite derived from the intracrystal hydrolysis and condensation polymerization of TEOS.

The reactions of octylamine/octylammonium-magadiite with TEOS, followed by drying of the solid products in air, afford siloxane-intercalated derivatives with well-ordered basal spacings. Calcination of the siloxane intercalates at 360°C to remove organic components yields silica intercalated products with basal spacings only 2-3 Å smaller than the siloxane precursors. Table 7 summarizes the d spacing and gallery heights for the starting materials and calcined TEOS/magadiite reaction products.

Table 7. Basal spacing and gallery heights of magadiite reaction precursors and TEOS reaction products. Reprinted with permission from Dailey, J. S. and Pinnavaia, T. J., Chem. Mater. 1992, 4, 855–863. Copyright 1992, American Chemical Society

Sample	Basal spacing (Å)	Gallery height (Å)
Na^+-magadiite	15.6	4.4
H^+-magadiite	12.4	1.2
$C_8H_{17}NH_3^+$-magadiite (air dried)	14.1	2.9
$C_8H_{17}NH_3^+$-magadiite (amine solvated)	34.0	22.8
TEOS/magadiite (54:1)	23.3	12.1
TEOS/magadiite (90:1)	25.2	14.0
TEOS/magadiite (153:1)	28.1	16.9
TEOS/magadiite (54:1, calcined 360°C)	20.7	9.5
TEOS/magadiite (90:1, calcined 360°C)	22.4	11.2
TEOS/magadiite (153:1, calcined 360°C)	25.9	14.7

*Gallery height = basal spacing minus layer thickness of magadiite (11.2 Å).

The products of reaction mixtures containing 54, 90 and 153 mol of TEOS/mol of magadiite exhibit reflections corresponding to basal spacings of 23.3, 25.2, and 28.1 Å, respectively. Since the layer thickness of magadiite is 11.2 Å, the corresponding gallery heights are 12.1, 14.0 and 16.9 Å, respectively. These gallery heights for the siloxane intercalates are 5-10 Å smaller than the initial gallery height of 22.8 Å observed for the octylamine/octylammonium-magadiite precursor gel.

H^+-magadiite has a total surface area of 45 $m^2.g^{-1}$ due exclusively to adsorption at nonporous external surfaces (Table 8). The calcined products obtained from the reaction of TEOS with magadiite exhibited dramatically larger surface areas between 520-680 $m^2.g^{-1}$, depending on the amount of TEOS used in the reaction. Most of the total surface area is due to the presence of micropores <20 Å in diameter. The relationship between the microporous surface area and the gallery height indicates a dependence of microporosity on TEOS reaction stoichiometry. This can be explained by a symmetrical increase in pillar size as the gallery height increases. Thus, as the pillar size increases the lateral spacing between pillars will decrease. An intermediate value of the gallery height would yield an optimum porosity when the best compromise exists between pillar height and lateral pillar spacing (see TEOS/magadiite 90:1 in Table 7).

Table 8. Surface area analyses (m2/g) for calcined TEOS-Octylamine/Octylammonium-magadiite reaction products. Reprinted with permission from Dailey, J. S. and Pinnavaia, T. J., Chem. Mater. 1992, 4, 855–863. Copyright 1992, American Chemical Society

Sample	S_{BET}	S_{total}	S_{mic}	S
H^+-magadiite	45	45	0	45
TEOS/magadiite (54:1)	620	635	590	45
TEOS/magadiite (90:1)	680	705	670	35
TEOS/magadiite (153:1)	520	525	480	45

S_{BET} is the N_2 BET surface area; S_{total}, S_{mic} and S are the total, microporous and nonmicroporous surface areas, respectively, obtained from t-plots.

Recently, pillaring of magadiite with TEOS was studied using different CTA^+/Na^+ molar ratios in the pre-swelling step. [130] The layered materials have been synthesized by traditional methods [131] and the ion exchange procedure was performed under controlled introduction of different amounts of surfactants as spacers in the lamellar structure. After the introduction of the surfactant molecules, two more steps were necessary to obtain the pillared magadiite: i) the introduction of the pillaring agent (TEOS) which is responsible for the growth of interlamellar pillars and ii) the calcination process.

Considering the general formula of magadiite as $Na_2Si_{14}O_{29}.(5\text{-}10)H_2O$, CTA^+ was exchange with CTA/Na molar percentages equal to 25, 50, 75 and 100%. X-ray diffraction showed that small amounts of surfactant are enough to increase the interlayer space to the maximum value attainable by this process. The TEOS/CTA molar ratio is an important factor for successful pillaring. When the pillaring agent is introduced, TEOS produces firm enough silica-pillars to prop the expanded gallery after the removal of surfactants. In this step, the X-ray diffractograms show that the signals in the region of 22-30° 2θ are substituted by an amorphous halo, typical of disorganized material. For pillared magadiites in which precursor contained 25 and 50% of CTA^+ the materials present a small signal around 6.3 2θ (1.4 nm), indicating that they still present some degree of ordering that is lost after calcination. As indicated by FTIR, the characteristic vibrations of the magadiite structure, the stretching of Si-O-Si single and double rings, are observed at 620 and 575 cm^{-1}. [132]

Table 9. Si_{Q3}/Si_{Q4} ratio of CTA/Na-magadiite and pillared magadiites obtained from ^{29}Si NMR. Adapted from Pastore, H. O. et al., Eur. J. Mineral. 2012, 24, 903-912. Copyright 2012, with permission from Schweizerbart Science Publishers (www.schweizerbart.de.)

	Q^3/Q^4 ratio	
	CTA-exchanged magadiite	Pillared materials
PILC-maga25	0.78	0.43
PILC-maga50	0.92	0.55
PILC-maga75	0.80	0.57
PILC-maga100	0.87	0.54

Another aspect to be observed in the pillaring process is the local environmental of the atoms. After the pillaring step, the ^{29}Si SS-NMR spectrum shows the Q^4 sites at -110.0 ppm and the Q^3 resonances at -100.9 ppm but the lines are enlarged, indicating a high number of non-equivalent sites due the growth of the interlayer pillars: a reflection of the general disorganization observed by X-ray diffractometry. Table 9 shows that after the pillaring process, the Q^3/Q^4 ratio decreases when compared to the CTA,Na-magadiites, indicating the condensation of the Q^3 sites to Q^4 sites, regardless of the amount of cetyltrimethylammonium cation used.

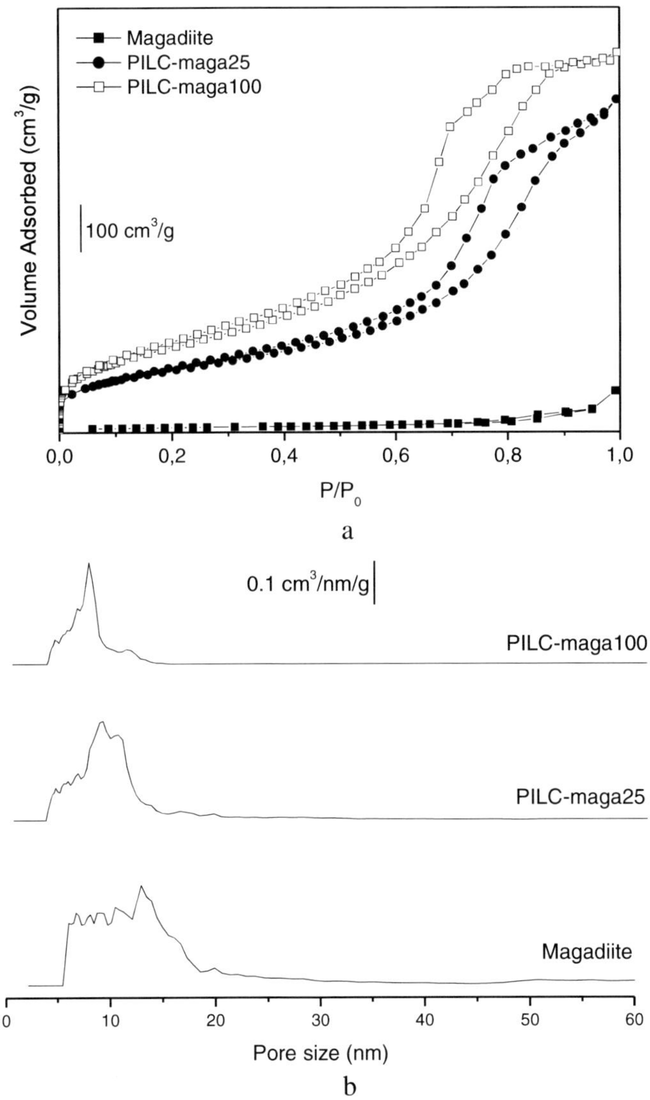

Figure 11. N_2 adsorption-desorption isotherms (A) and pore size distribution (B) for magadiite and its pillared derivatives containing 25 and 100% of CTA^+/Na^+. Adapted from Pastore, H. O. et al., Eur. J. Mineral. 2012, 24, 903-912. Copyright 2012, with permission from Schweizerbart Science Publishers (www.schweizerbart.de.).

The success of the pillaring process is confirmed by significant variations in the surface area and pore sizes (Figure 11). The N_2 adsorption data yielded pore sizes varying between 3 and 10 nm, decreasing as CTA/Na molar ratio increases, *i.e.* more surfactant in the magadiite leads to a more restricted gallery space formed (smaller d_3, see Figure 2). This indicates that surfactant concentration plays an important structure-directing role during silica pillaring. Also, as the CTA/Na molar ratios increase, a narrowing in pore size distributions can be noted and thus, more homogeneous is the porosity obtained (Table 10).

Table 10. Surface area, pore volume and pore diameter of magadiite and PILC-magadiites with different CTA/Na ratios. Adapted from Pastore, H. O. et al., Eur. J. Mineral. 2012, 24, 903-912. Copyright 2012, with permission from Schweizerbart Science Publishers (www.schweizerbart.de.)

	Surface area[a] ($m^2.g^{-1}$)	Pore volume[b] ($cm^3.g^{-1}$)	Pore diameter[b] (nm)
Magadiite	25	0.1	13.0
PILC-maga25	512	1.1	9.4
PILC-maga50	764	2.3	9.0
PILC-maga75	599	1.3	6.9
PILC-maga100	663	0.6	7.3

[a] MultiPoint BET; [b] BJH method.

Although frequently tetraethoxysilane is used as the silica source for pillars, other silanes such as phenyl or methyltriethoxysilane, may be used. [133] The alkyl/alkoxysilane can also be used as grafting reactants to modify the properties of TEOS generated pillars by post-synthesis methods using the reactive silanol groups on the surface. The development of inorganic hybrid nanocomposites by alkoxysilylation with 4,4′-biphenyl-bridged alkoxysilane compounds in ilerite was already described. [134] A porous benzene−silica hybrid clay heterostructure, prepared by condensation of bis(triethoxysilyl)benzene and tetraethylorthosilicate (TEOS), has been used for preparing solids for adsorbing volatile organic compounds. [135, 136]

It was shown that materials can be pillared and then grafted but those reactions can be performed simultaneously. This will be dealt with in the next topic.

Silylated and/or Pillared Interlayered Clays: Designing 3D Structures

Having in mind the properties of grafted structures and focusing on the reactive groups in the pillars and layered silicates (Al-OH, Si-OH, Si-O-, etc), pillared solids reactions can go beyond in modification. Actually, few works have been performed on silylation of PILCs. In 2007, Zhu et al. [137] synthesized the first silylated Al-PILC with two alkylsilane agents: trimethylchlorosilane (TMCS) and octadecyltrichlorosilane (OTS). The result was a new type of organo–inorgano composite sorbent, silylated PILC (SPILC) that possess the advantages of both organo-bentonites and PILCs.

The results of organic carbon content analysis, FTIR, XRD, and TG indicated that the silyl group can be successfully grafted to the inner surface of pillared interlayered clays (PILCs) through reaction with the –OH groups of the pillars and the *d*-spacing of synthesized PILCs and SPILCs were almost the same. SPILCs have both the higher organic carbon content relative to original bentonite and PILCs surface area and pore properties. Compared to the Al-PILC, the surface area and total pore volume for OTS-Al-PILC decreased from 263 m^2/g and 0.16 cm^3/g to 107 m^2/g and 0.065 cm^3/g, respectively, with a large decrease in the inner surface area. These findings suggest that alkylsilane groups occupied part of the interlayer space. The heat-resistance of SPILCs, 508 ºC for OTS-Al-PILC and 214 ºC for TMCS-Al-PILC, is higher than the CTA^+-bentonites. These solids were applied in the adsorption of volatile organic compounds; the sorption capacity of SPILCs is approximately the same of CTA^+-bentonites while the hydrophobicity of SPILCs is superior to that of PILCs.

The knowledge of the effects of SiO_2-pillaring and of grafting on layered silicates allow envisaging the effect of the simultaneous changes in the solids. Using magadiite and [Al]-magadiite, aminopropyl-functionalized mesoporous heterostructures with designed pore structure, accessibility and activity, can be obtained in a single step synthesis. [109] The presence of aluminum in the layered framework of these pillared heterostructures can also afford interesting catalysts.

These hybrids materials consisted of magadiite and [Al]-magadiite pillared with TEOS/APTES 1:1 (HYB-magadiites and HYB-[Al]-magadiites) and display an increase in porosity and in the surface area (values between 49-61 and 38-60 m^2/g, respectively) when compared to the starting material (25 m^2/g for magadiite and 30 m^2/g for [Al]-magadiite). This is due to the fact that TEOS, as the pillaring precursor, is converted to silica pillars and forms the rigid intercalated porous structure as in the case of pillared magadiite and [Al]-magadiite.[130]

The surface area is not as large as in the case of pillared solids (512 m^2/g for pillared magadiite and 641 m^2/g for pillared-[Al]-magadiite), probably due to the fact the organic groups populate the pores, but it is similar to the amino-grafted materials (74 m^2/g for aminopropyl grafted magadiite and 67 m^2/g for NH_2-grafted-[Al]-magadiite), indicating that the hybrids proposed are pillared and grafted solids.

The improved properties of the hybrid materials, in relation to the pillared and to the aminopropyl-grafted magadiites, were demonstrated by the exposure of the solids to an CO_2 atmosphere at 50ºC. The solids HYB-magadiite and HYB-[Al]-magadiite displayed significant improvement in carbon dioxide adsorption when compared to the aminopropyl grafted magadiite and [Al]-magadiite: *ca.* 60% (0.37 to 0.60 CO_2/N for HYB-magadiite and from 0.46 to 0.65 CO_2/N for HYB-[Al]-magadiite). The low CO_2 adsorption on aminopropyl grafted magadiite and [Al]-magadiite is the result of the approximation of the lamella caused by the dehydration process, hindering the CO_2 diffusion to the pending basic sites. In the case of hybrids, lamella are firmly separated from each other due to the presence of pillars between them and, thus, cannot collapse. As a result, the CO_2 access to the -NH_2 groups in these materials is more effective, i.e., the hybrids obtained can make use of the amino groups functionality to adsorb CO_2 in a pillared structure.

Controlled grafting of organic molecules can give, for example, hydrofobicity and gas adsorption and separation properties to the silicate. However, these grafted solids can exhibit low thermal stability and mass transfer limitation within the interlayer space. A chemical strategy to circumvent this problem is to construct pillars between the sheets, capable of keeping the lamella apart: Pillared Interlayered Clays (PILCs) with various types of metal oxide pillars have been synthesized. The ones discussed in this section are shown in Table 11.

Table 11. Pillared layered silicates and their most employed pillaring agents

Pillaring agent	Silicate	References	
SiO_2	Magadiite	[113]	Landis, M.E. et al. (1991)
		[114]	Dailey, J. S. and Pinnavaia, T. J. (1992)
		[115]	Jeong, S.-Y. et al. (1995)
		[130]	Moura, H. M. et al. (2012)
	Ilerite	[116]	Kosuge, K. et al. (1995)
		[135]	Ferreira, P. et al. (2006)
		[136]	Nunes, C. D. et al. (2008)
	Kenyaite	[113]	Landis, M.E. et al. (1991)
		[119]	Park, K.-W. et al. (2009)
		[120]	Thiesen, P. et al. (2002)
	MCM-22(P)	[117]	Maheshwari, S. et al. (2010)
		[121]	Barth, J.-O. et al. (2002)
		[123]	Maheshwari, S. et al. (2008)
	Kanemite	[118]	Toriya, S. et al. (2002)
	Fluorohectorite	[62]	Galarneau, A. et al. (1995)
	Titanates	[113]	Landis, M.E. et al. (1991)
	Vermiculite	[133]	Ishii, R. et al. (2005)
Al_2O_3	MCM-22(P)	[121]	Barth, J.-O. et al. (2002)
	Magadiite	[122]	Kosuge, K. and Singh, P. S. (2000)
$MgO+Al_2O_3$	MCM-22(P)	[121]	Barth, J.-O. et al. (2002)
$GaCl_3 + AlCl_3$	Montmorillonite	[48]	González, F. et al. (1999)
ZrO_2	Montmorillonite	[128]	Cecilia, J. A. et al. (2013)
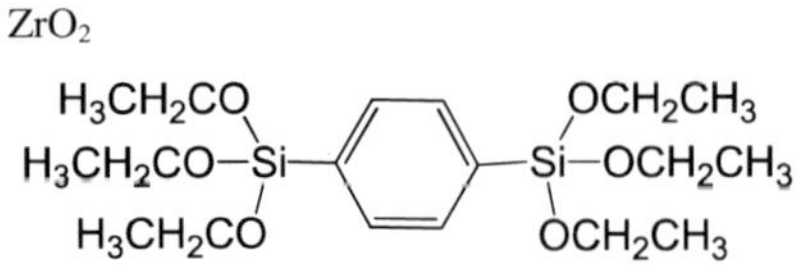 + $Si(OCH_2CH_3)_4$	Ilerite	[136]	Nunes, C. et al. (2008)
$(CH_3)_3SiCl + Cl_3Si(CH_2)_{17}CH_3$	Montmorillonite	[137]	Zhu, L. et al. (2007)
$(CH_3CH_2O)_3Si(CH_2)_3NH_2 + Si(OCH_2CH_3)_4$	Magadiite	[109]	Moura, H. M. and Pastore, H. O. (2014)

PERSPECTIVES

The concept of using intercalation chemistry to transform a lamellar solid into a porous analogous originated more than 55 years ago. Barrer described how the replacement of inorganic cations in montmorillonite by $N(CH_3)_4^+$ and $N(C_2H_5)_4^+$ cations open up the lamella and causes profound changes in the sorption and intercalation properties. Despite the promising molecular sieving properties of Barrer's alkylammonium ion-exchanged forms of smectite clays, its innovation remained essentially non-developed for two decades due primarily to the concomitant discovery of synthetic zeolites LTA and FAU. In 1977, G. W. Brindley and R. E. Sempels showed that the partial replacement of Na^+-exchange cations by hydroxyaluminum cations, in the galleries of the smectic clay beidellite, afforded intercalates with surface areas of up to 500 m^2/g after calcination at 325 ºC. Since then, different metal ions have been used as pillaring agents and much of the present interest in pillared clays and related intercalation compounds is stimulated by their properties as heterogeneous catalysts for the oil industry, production of fine chemicals and as selective adsorbents for molecular preparations, environmental remediation, and pollution control, in one word, their application in Green Chemistry.

In the 90's, M. E. Landis, J. S. Dailey and T. Pinnavaia reported the pillaring of magadiite and kenyaite with polymeric silica species by reaction of alkylamine or alkylammonium-intercalated layered silicates with tetraethoxysilane (TEOS) yelding micropores. This templating mechanism is similar to that reported for the M41S structure synthesis. Modulation of interpillar/interlayer spaces, by the optimum ratio of expanding guest species and pillaring agent, can be used as a tool for intelligent micro/mesoporous materials design from layered solids.

Due to the organized, ordered and variable voids in the interlayer space, caused by the organic cation intercalation, the interlayer region can be covalently modified, not necessarily cross-linking the lamella, but inserting functional groups on their surface and designing the physico-chemical properties desired for the layered solids.

Ryoo et al. [138] showed that appropriately designed bifunctional surfactants can direct the formation of zeolite structures on the mesoporous and microporous length scales simultaneously and thus yield MWW and MFI nanosheets from their corresponding layered precursors ITQ-1 and silicalite-1, respectively. The house of cards arrangement of the nanosheets creates a permanent network of 2-7 nm pores, which, along with the high external surface area, reduced micropore diffusion. Tsapatsis et al. [139, 140] showed how to use the MFI/MEL intergrowth to create mesoporous solids constructed by crystalline MFI layers.

The works of Gies' group [141] on the topotactic condensation of hydrous layer silicates shows promising results in generating novel zeolite materials with distinct framework types. A topotactic condensation of the layers performed at temperatures ~500 °C, with simultaneous expulsion of the interlayer constituents, is able to form fully condensed framework silicates.

Having in mind the works of these three groups, the zeotypes formation from layered n-alkylammonium aluminophosphate [142, 143, 144, 145] and the inverse σ tranformation [146] it seems that the future of layered solids in the direction of 3D micro/mesoporous materials preparation involves the topotactic transformation as a method of insightful manipulation of the lamella, as such, guided by multifunctional organic guests. In special cases, the use of crystallization defects can lead to surprising 3D structures. The field of 2D↔3D transformation is open and blooming.

REFERENCES

[1] Brenn, U.; Ernst, H.; Freude, D.; Herrmann, R.; Jähnig, R.; Karge, H. G.; Kärger, J.; König, T.; Mädler, B.; Pingel, U. T.; Prochnow, D.; Schwieger, W. *Microporous Mesoporous Mater*. 2000, 40, 43-52.

[2] Takahashi, N.; Kuroda, K. *J. Mater. Chem*. 2011, 21, 14336-14353.

[3] Pinnavaia, T. J. *ACS Adv. Chem. Ser*. 1995, 245, 283-298.

[4] Delgas, W. N.; Garten, R. L.; Boudart, M. *J. Phys. Chem*. 1969, 73, 2970-2979.

[5] Datka, J.; Abramowicz, T. *J. Chem. Soc., Faraday Trans*. 1994, 90, 2417-2421.

[6] Lewis, D. W.; Catlow, R. A.; Sankar, G.; Carr, S. W. *J. Phys. Chem*. 1995, 99, 2377-2383.

[7] Yelon, W. B.; Xie, D.; Newsam, J. M.; Dunn, J. *Zeolites* 1990, 10, 553-558.

[8] Gianetto, G.; Monque, R.; Perez, J.; Papa, J.; Garcia, L. *Zeolites* 1993, 13, 557-560.

[9] Chandwadkar, A. J.; Abdulla, R. A.; Hedge, S. G.; Nagy, J. B. *Zeolites* 1993, 13, 470-477.

[10] Bordiga, S.; Coluccia, S.; Lamberti, C.; Marchese, L.; Zecchina, A.; Boscherini, F.; Buffa, F.; Genoni, F.; Leofanti, G.; Petrini, G.; Vlaic, G. *J. Phys. Chem*. 1994, 98, 4125-4132.

[11] Corma, A.; Camblor, M. A.; Esteve, P.; Martinez, A.; Pariente, J. P. *J. Catal*. 1994, 145, 151-158.

[12] Lopez, A.; Soulard, M.; Guth, J. L. *Zeolites* 1990, 10, 134-136.

[13] Kosslick, H.; Tuan, V. A.; Fricke, R.; Peuker, C.; Pilz, W.; Storek, W. *J. Phys. Chem*. 1993, 97, 5678-5684.

[14] Vorbeck, G.; Janchen, J.; Parlitz, B.; Scheider, M.; Fricke, R. *Chem. Commun*. 1994, 1, 123-124.

[15] Mal, N. K.; Ramaswamy, V.; Ganapathy, S.; Ramaswamy, A. V. *Chem. Commun*. 1994, 17, 1933-1934.

[16] Kornatowski, J.; Sychev, M.; Goncharuk, V.; van Baur, W. H. *Stud. Surf. Sci. Catal*. 1991, 65, 581-590.

[17] Pastore, H. O.; Stein, E.; Davanzo, C. U.; Vichi, E. J. S.; Nakamura, O.; Baesso, M.; Silva, E. C.; Vargas, H. *Chem. Commun*. 1990, 10, 772-773.

[18] Spoto, G.; Bordiga, S.; Garrone, E.; Ghiotti, G.; Zecchina, A. *J. Mol. Catal*. 1992, 74, 175-184.

[19] Cichocki, A.; Datka, J.; Olech, A.; Piwowarska, Z.; Michalik, M. *J. Chem. Soc., Faraday Trans*. 1990, 86, 753-756.

[20] Lobo, R. F.; Davis, M. E. *J. Am. Chem. Soc*. 1995, 117, 3766-3779.

[21] Axon, S. A.; Klinowski, J. *J. Phys. Chem*. 1994, 98, 1929-1032.

[22] Pauling, L. *Z. Kristallogr*. 1930, 74, 213-225.

[23] Romannikov, V. N.; Chumachenko, L. S.; Mastikhin, V. M.; Ione, K. G. *J. Catal*. 1985, 94, 508-513.

[24] Bergaya, F.; Theng, B. K. G.; Lagaly, G. *Handbook of Clay Science. Development in Clay Science.* Bergaya, F.; Theng, B. K. G.; Lagaly, G. (Eds), Vol 1, Elservier, UK, 2006.
[25] Ogawa, M. and Kuroda, K. *Chem. Rev.* 1995, 95, 399–438.
[26] Parida, S. K.; Dash, S.; Patel, S.; Mishra, B. K. *Adv. Colloid Interface Sci.* 2006, 121, 77–110.
[27] Eugster, H. P. *Science* 1967, 157, 1177-1180.
[28] Brindley, G. W. *Am. Mineral.* 1969, 54, 1583-1591.
[29] Rojo, J. M.; Ruiz-Hitzky, E.; Sanz, J. *Inorg. Chem.*1988, 27, 2785–2790.
[30] Beneke, K. and Lagaly, G. *Am. Mineral.* 1983, 68, 818–826.
[31] Sheppard, R. A.; Gude, A. J.; Hay, R. L. *Am. Mineral.* 1970, 55, 358–366.
[32] Beneke, K. and Lagaly, G. *Am. Mineral.* 1977, 62, 763–771.
[33] Gies, H.; Marler, B.; Vortmann, S.; Oberhagemann, U.; Bayat, P.; Krink, K.; Rius, J.; Wolf, I.; Fyfe, C. *Microporous Mesoporous Mater.* 1998, 21, 183–197.
[34] Soler-Illia, G. J. A. A.; Sanchez, C.; Lebeu, B.; Patarin, J. *Chem. Rev.* 2002, 102, 4093-4138.
[35] Brandt, A.; Schwieger, W.; Bergk, K. H. *Cryst. Res. Tech.* 1988, 23, 1201-1203.
[36] Mochizuki, D.; Shimojima, A.; Kuroda, K. *J. Am. Chem. Soc.* 2002, 124, 12082-12083.
[37] Kim, S. J.; Lee, G.; Ryu, Y. K.; Yu, Y. B. *Res. Chem. Intermed.* 2012, 38, 1191-1202.
[38] Kawai, A.; Urabe, Y.; Itoh, T.; Mizukami, F. *Mater. Chem. Physics* 2010, 122, 269-272.
[39] Arroyabe, E.; Kaindl, R.; Többens, D. M.; Kahlenberg, V. *Inorg. Chem.* 2009, 48, 11929-11934.
[40] Schwieger, W.; Lagaly, G. Alkali Silicates and Crystaline Silicic Acids. In *Handbook of Layered Materials*, Marcel Dekker, Inc., New York, 2004, pp. 541–629.
[41] Gurvich, L.; Attapulgite, also natural and activated montmorillonites, polymeric pinene. Russian Chemical Socistey 1915, 47, 827-830. Appud: *Handbook of Clay Science. Development in Clay Science.* Bergaya, F.; Theng, B. K. G.; Lagaly, G. (Eds), Vol 1, Elservier, UK, 2006, 393-421.
[42] Houdry, E.; Burt, W. F.; Pew, A. E.; Peters, W. A. Jr. Catalytic processing by the Houdry process. *National Petroleum News*, 1938, 570-580.
[43] Brerck, D. W. Potential Uses of Natural and Synthetic Zeolites in Industry. In *The Properties and Applications of Zeolites.* Towsend, R. P. (Ed), The Chemical Society, London, pp. 391-422.
[44] Clearfield, A.; Kuchenmeister, M. E.; Wade, K.; Cahill, R.; Sylvester, P. Pillaring of layered inorganic compounds: Fundamentals. In *Expanded Clays and Other Microporous Solids.* Occeli, M. L. and Robson, H. (Eds), Van Nostrand Reinhold, New York, 1992, pp. 245.
[45] Davis, M. E.; Saldarriaga, C. M.; Montes, C.; Garces, J.; Crowder, C. Zeolites 1988, 8, 362-.
[46] Vaughan, D. E. W.; Lussier, R. J.; Magee, J. S. Pillared Interlayered Clay Materials Useful as Catalysts and Sorbents. U.S. Pat 4,176,090, Nov 27, 1979.
[47] Adams, J. M.; *Applied Clay Sci.* 1987, 2, 309-342.
[48] González, F.; Pesquera, C.; Benito, I.; Herrero, E.; Poncio, C.; Casuscelli, S.; *Applied Catal.* 1999, 181, 71–76.
[49] Bradley, S. M.; Kydd, R. A. *J. Catal.* 1993, 141, 239-249.

[50] Bradley, S. M.; Kydd, R. A. *J. Catal.* 1993, 142, 448-454.
[51] Kloprogge, J. T.; Duong, L. V.; Frost, R. L. *Environ. Geol.* 2005, 47, 967-981.
[52] Barrer, R. M.; MacLeod, D. M. *Trans. Faraday Soc.* 1955, 51, 1290-1300.
[53] Moini, A. and Pinnavaia, T. J. *Solid State Ionics* 1988, 26, 119-123.
[54] Mortland, M. M. and Berkheiser, V. *Clays and Clays Minerals* 1976, 24, 60-63.
[55] Sawhney, B. L. *Clays and Clay Minerals* 1968, 16, 157-163.
[56] Brindley, G. W.; Sempels, R. E. *Clay Minerals* 1977, 12, 229-236.
[57] Kosuge, K.; Singh, P. S. *Chem. Mater.* 2000, 12, 421-427.
[58] Kim, M. H.; Ko, Y.; Kim, S. J.; Uh, Y. J. *Appl. Catal.* 2001, 210, 345-353.
[59] Mochizuki, D.; Kowata, S.; Kuroda, K. *Chem. Mater.* 2006,18, 5223-5229.
[60] Bergaya, F. 1990. Argiles à pilers. In *Matériaux Argileux. Structures, Propriétes et Applications.* Decarreau, A. (Ed), Societé Française de Minéralogie et de Cristallographie et Groupe Français des Argiles, Paris, 513-537. Appud: Bergaya, F.; Aouad, A.; Mandalia, T. In *Handbook of Clay Science. Development in Clay Science.* Bergaya, F.; Theng, B. K. G.; Lagaly, G. (Eds), Vol 1, Elservier, UK, 393-421.
[61] Gil, A.; Gandia, L.M; Vicente, M. A. *Catal. Rev. Sci. Eng.* 2000, 42, 145-212.
[62] Galarneau, A.; Barodawalla, A.; Pinnavaia, T. J. *Nature* 1995, 374, 529–531.
[63] Kresge, C. T.; Leonowicz, M. E.; Roth, W. J.; Vartuli, J. C.; Beck, J. S. *Nature* 1992, 359, 710–712.
[64] Beck, J. S.; Vartuli, C.; Roth, W. J.; Leonowicz, M. E.; Kresge, C. T.; Schmitt, K. D.; Chu, C. T. W.; Olson, D. H.; Sheppard, E. W.; McCullen, S. B.; Higgins, J. B.; Schlenker , J. L. *J. Am. Chem. Soc.* 1992, 114, 10834–10843.
[65] Zhu, H. Y.; Ding, Z.; Lu, C. Q.; Lu, G. Q. *Appl. Clay Sci.* 2002, 20, 165-175.
[66] Inagaki, S.; Fukushima, Y.; Kuroda, K. *J. Chem. Soc., Chem. Commun.* 1993, 8, 680-682.
[67] Li, B.; Mao, H.; Li, X.; Ma, W.; Liu, Z. J. Coll. Int. Sci. 2009, 336, 244–249.
[68] Pires, J.; Araujo, A. C.; Carvalho, A. P.; Pinto, M. L.; Gonzalez-Calbet, J. M.; Castellanos, R. *J. Microporous Mesoporous Mater.* 2004, 73, 175-180.
[69] Polverejan, M.; Pauly, T. R.; Pinnavaia, T. J. *Chem. Mater.* 2000, 12, 2698-2704.
[70] Nakatsuji, M.; Ishii, R.; Wang, Z.M.; Ooi, K. *J. Colloid Interface Sci.* 2004, 272, 158-166.
[71] Pichowicz, M.; Mokaya, R.; *Chem. Commun.* 2001, 20, 2100-2101.
[72] Haushalter, R. C.; Mundi, L. A. *Chem. Mater.* 1992, 4, 31-48.
[73] Ishii, R.; Ikeda, T.; Itoh, T.; Ebina, T.; Yokoyama, T.; Hanaoka, T.; Mizukami, F.; *J. Mater. Chem.* 2006, 16, 4035-4043.
[74] Ferreira, P.; Nunes, C. D.; Pires, J.; Carvalho, A. P.; Brandao, P.; Rocha, J. *Mater. Sci. Forum* 2006, 516, 470-474.
[75] Nunes, C. D.; Pires, J.; Carvalho, A. P.; Calhorda, M. J.; Ferreira, P.; *Microporous Mesoporous Mater.* 2007, 111, 612-619.
[76] Yariv, S. and Heller, L. *Isr. J. Chem.* 1970, 8, 935-945.
[77] Yariv, S. Introduction to Organo-Clay complexes and Interactions. In *Organo-Clay Complexes and Interactions.* Yariv, S. and Cross, H. (Eds), Marcel Dekker Inc, New York, 2002, pp 39-112.
[78] Barrer, R. M. *J. Inclusion Phenomena* 1986, 4, 109-119.
[79] Barrer, R. M. *Clays Clay Miner.* 1989, 37, 385-395.
[80] Brenn, U.; Schwieger, W.; Wuttig, K. *Colloid Polym Sci.* 1999, 277, 394-399.

[81] Lagaly, G.; Beneke, K.; Weiss, A. In *Proceedings of the International Clay Conferencee*. Serratosa, J. M. (Ed), Div. Cienc. CSIC, Madrid, 1972, pp. 663. Appud: Brenn, U.; Schwieger, W.; Wuttig, K. *Colloid Polym Sci.* 1999, 277, 394-399.
[82] Boehm, H. P. *Angew. Chem., Int. Ed. Engl.* 1966, 5, 533–544.
[83] Ruiz-Hitzky, E. and Rojo, J. M. *Nature* 1980, 287, 28–30.
[84] Ruiz-Hitzky, E.; Rojo, J. M.; Lagaly, G. *Colloid Polym. Sci.* 1985, 263, 1025–1030.
[85] Endo, K.; Sugahara, Y.; Kuroda, K. *Bull. Chem. Soc. Jpn.* 1994, 67, 3352–3355.
[86] Okutomo, S.; Kuroda, K.; Ogawa, M. *Appl. Clay Sci.* 1999, 15, 253–264.
[87] Park, K.-W.; Jung, J. H.; Kim, S.-K.; Kwon, O.-Y. *Appl. Clay Sci.* 2009, 46, 251–254.
[88] Ogawa, M.; Okutomo,S.; Kuroda, K. *J. Am. Chem. Soc.* 1998, 120, 7361–7362.
[89] Fujita, I.; Kuroda, K.; Ogawa, M. *Chem. Mater.* 2003, 15, 3134–3141.
[90] Fujita, I.; Kuroda, K.; Ogawa, M. *Chem. Mater.* 2005, 17, 3717–3722.
[91] Ide, Y. and Ogawa, M. *Angew. Chem., Int. Ed.* 2007, 46, 8449–8451.
[92] Toriya, S.; Kobayashi, S.; Takei, T.; Fuji, M.; Watanabe, T.; Chikazawa, M. *Colloid Polym. Sci.* 2003, 281, 1121–1126.
[93] Ogawa, M.; Miyoshi, M.; Kuroda, K. *Chem. Mater.* 1998, 10, 3787–3789.
[94] Vortmann, S.; Rius, J.; Marler, B.; Gies, H. *Eur. J. Mineral.* 1999, 11, 125–134.
[95] Shimojima, A.; Mochizuki, D.; Kuroda, K. *Chem. Mater.* 2001, 13, 3603–3609.
[96] Takahashi, N.; Hata, H.; Kuroda, K. *Chem. Mater.* 2010, 22, 3340–3348.
[97] Guo, Y.; Wang, Y.; Yang, Q.-X.; Li, G.-D.; Wang, C.-S.; Cui, Z.-C.; Chen, J.-S. *Solid State Sci.* 2004, 6, 1001–1006.
[98] Yanagisawa, T.; Harayama, M.; Kuroda, K.; Kato, C. *Solid State Ionics* 1990, 42, 15–19.
[99] Isoda, K.; Kuroda, K.; Ogawa, M. *Chem. Mater.* 2000, 12, 1702–1707.
[100] Shindachi, I.; Hanaki, H.; Sasai, R.; Shichi, T.; Yui, T.; Takagi, K. *Chem. Lett.* 2004, 33, 1116–1117.
[101] Shindachi, I.; Hanaki, H.; Sasai, R.; Shichi, T.; Yui, T.; Takagi, K. *Res. Chem. Intermed.* 2007, 33, 143–153.
[102] Matsuo, Y.; Yamada, Y.; Nishikawa, M.; Fukutsuka, T.; Sugie, Y. *J. Fluorine Chem.* 2008, 129, 1150–1155.
[103] Specht, K. M.; Jackson, M.; Sunkel, B.; Boucher, M. A. *Appl. Clay Sci.* 2010, 47, 212–216.
[104] Park, K.-W.; Jeong, S.-Y.; Kwon, O.-Y. *Appl. Clay Sci.* 2004, 27, 21–27.
[105] Wang, S.-F.; Lin, M.-L.; Shieh, Y.-N.; Wang, Y.-R.; Wang, S.-J. *Ceram. Int.* 2007, 33, 681–685.
[106] Leu, C.-M.; Wu, Z.-W.; Wei, K.-H. *Chem. Mater.* 2002, 14, 3016–3021.
[107] Zhang, Z.; Saengkerdsub, S.; Dai, S. *Chem. Mater.* 2003, 15, 2921–2925.
[108] Ishii, R.; Ikeda, T.; Mizukami, F. *J. Colloid Interface Sci.* 2009, 331, 417–424.
[109] Moura, H. M. and Pastore, H. O. *Dalton Trans.* 2014, DOI:10.1039/C3DT53571A.
[110] ieira, R. Silicatos lamelares modificados e funcionalizados com PEI. Master thesis, University of Campinas, July 2013.
[111] Isoda, K.; Kuroda, K.; Ogawa, M. *Chem. Mater.* 2000, 12, 1702-1707.
[112] Schoonheydt, R. A.; Pinnavaia, T.; Lagaly, G.; Gangas, N. *Pure and Applied Chemistry* 1999, *71*, 2367-2371.
[113] Landis, M.E.; Aufdembrink, B.A.; Chu, P.; Johnson, I.D.; Kirker, G.W.; Rubin, M.K. *J. Am. Chem. Soc.* 1991, 113, 3189–3190.

[114] Dailey, J. S. and Pinnavaia, T. J. *Chem. Mater.* 1992, 4, 855–863.
[115] Jeong, S.-Y.; Kwon, O.-Y.; Suh, J.-K.; Jin, H.; Lee, J. M. *J. Colloid Interface Sci.* 1995, 175, 253–255.
[116] Kosuge, K. and Tsunashima, A. *Chem. Commun.* 1995, 2427–2428.
[117] Maheshwari, S.; Martínez, C.; Portilla, M. T.; Llopis, F. J.; Corma, A.; Tsapatsis, M. *J. Catal.* 2010, 272, 298–308.
[118] Toriya, S.; Tamura, Y.; Takei, T.; Fuji, M.; Watanabe, T.; Chikazawa, M. *J. Colloid Interface Sci.* 2002, 255, 171–176.
[119] Park, K.-W.; Jung, J.H.; Jeong, S.-Y.; Kwon, O.-Y. *J. Nanosci. Nanotechnol.* 2009, 9, 3160–3165.
[120] Thiesen, P.; Beneke, K.; Lagaly, G. *J. Mater. Chem.* 2002, 12, 3010–3015.
[121] Barth, J.-O.; Kornatowski, J.; Lercher, J.A. *J. Mater. Chem.* 2002, 12, 369–373.
[122] Kosuge, K. and Singh, P. S. *Chem. Mater.* 2000, 12, 421–427.
[123] Maheshwari, S.; Jordan, E.; Kumar, S.; Bates, F. S.; Penn, R. L.; Shantz, D. F.; Tsapatsis, M. *J. Am. Chem. Soc.* 2008, 130, 1507–1516.
[124] Monnier, A. *Science* 1993, 261, 1299-1303.
[125] Chen, C.; Li, H. X.; Davis, M. E. *Micropor. Mater.* 1994, 2, 27-34.
[126] Kresg, C. T.; Leonowicz, M. E.; Roth, W. J.; Vartuli, J. C.; Beck, J. S. *Nature* 1992, 359, 710-712.
[127] Beck, J. S. *J. Am. Chem. Soc.* 1992, 114, 10834-10843.
[128] Cecilia, J. A.; García-Sancho, C.; Franco, F. *Microporous and Mesoporous Mater.* 2013, 176, 95–102.
[129] *Pillared Clays and Related Catalysts*. Gil, A.; Korili, S. A.; Trujillano, R.; Vicente, M. A. (Eds), Springer, New York, 2010.
[130] Moura, H. M.; Bonk, F. A.; Pastore, H. O. *Eur. J. Mineral.* 2012, 24, 903-912.
[131] Superti, G. B.; Oliveira, E. C.; Pastore, H. O.; Gatti, G.; Marchese, L. *Chem. Mater.* 2007, 19, 4300–4315.
[132] Huang, Y.; Jiang, Z.; Schwieger, W. *Chem. Mater.* 1999, 5, 1210-1217.
[133] Ishii, R.; Nakatsuji, M.; Ooi, K. *Micropor. Mesopor. Mater.* 2005, *79*, 111-119.
[134] Ishii, R.; Ikeda, T.; Itoh, T.; Ebina, T.; Yokoyama, T.; Hanaoka, T.; Mizukami, F. *J. Mater. Chem.* 2006, 16, 4035-4043.
[135] Ferreira, P.; Nunes, C. D.; Pires, J.; Carvalho, A. P.; Brandao, P.; Rocha, J. *Mater. Sci. Forum* 2006, 516, 470-474.
[136] Nunes, C. D.; Pires, J.; Carvalho, A. P.; Calhorda, M. J.; Ferreira, P. *Micropor. Mesopor. Mater.* 2008, 111, 612-619.
[137] Zhu, L.; Tian, S.; Zhu, J.; Shi, Y. *J. Colloid and Interface Sci.* 2007, 315, 191–199.
[138] Choi, M.; Na, K.; Kim, J.; Sakamoto, Y.; Terasaki, O.; Ryoo, R. *Nature* 2009, 461, 246-249.
[139] Varoon, K.; Zhang, X.; Elvassi, B.; Brewer, D. D.; Gettel, M.; Kumar, S.; Lee, J. A.; Maheshwari, S.; Mittal, A.; Sung, C. Y.; Cococcioni, M.; Francis, L. F.; McCormick, A. V.; Mkhoyan, K. A.; Tsapatsis, M. *Science* 2011, 334, 72-75.
[140] Zhang, X.; Liu, D.; Xu, D.; Asahina, S.; Cychosz, K. A.; Agrawal, K. V.; Wahedi, Y.; Bhan, A.; Hashimi, S. A.; Terasaki, O.; Thommes, M.; Tsapatsis, M. *Science* 2012, 336, 1684-1687.
[141] Marler, B.; Gies, H. *Eur. J. Mineral.* 2012, 24, 405-428.

[142] Felix, D. L.; Strauss, M.; Ducati, L. C.; Pastore, H. O. *Micropor. Mesopor. Mater.* 2009, 120, 187-194.

[143] Martins, G. A. V.; Pastore, H. O. *Micropor. Mesopor. Mater.* 2008, 116, 131-140.

[144] Martins, G. A. V.; Strauss, M.; Pedroni, L. G.; Superti, G. B.; Oliveira, E. C.; Gatti, G.; Marchese, L.; Pastore, H. O. *Micropor. Mesopor. Mater.* 2008, 107, 81-89.

[145] Gieck, C.; Bisio, C.; Marchese, L.; Filinchuk, y.; Silva, C. E.; Pastore, H. O. *Angew. Chem., Int. Ed.* 2007, 46, 8895–8897.

[146] Chlubna, P.; Roth, W. J.; Greer, H. F.; Zhou, W. Z.; Shvets, O.; Zukal, A.; Cejka, J.; Morris, R. E. *Chem. Mater.* 2013, 25, 542-547.

In: Comprehensive Guide for Mesoporous Materials, Volume 2 ISBN: 978-1-63463-982-8
Editor: Mahmood Aliofkhazraei

Chapter 14

SCATTERING AND REFLECTION OF LIGHT BY MESOPOROUS MATERIALS

V. A. Sterligov
Institute of Semiconductor Physics, Kiev, Ukraine

ABSTRACT

Ordered mesoporous and non-porous silica films were studied by analyzing angular and spectral dependencies of reflectivity and hemispherical elastic light scattering (HELS). The real and imaginary parts of the film index were estimated from the reflection spectra. The HELS angular distribution of the mesoporous film shows a minimum, which has been interpreted as an interference pattern coming from light scattered by continuous set of the mesopores. Another mesoporous Al_2O_2 membrane (a 2D periodic array of empty and metal-filled linear nanopores) has been studied. These mesoporous membranes are characterized by noticeable differences in reflectance. A Maxwell-Garnett's (MG) model of composite material has been used to recover the dielectric function of tin nanorods. A consistent explanation of the optical properties has been achieved using multiple scattering and metallic behaviors of the nanorods.

INTRODUCTION

Table 1. Studies of mesoporous materials

Media	Porous size	Methods of producing, possible applications	Ref.
Silicat/Alumosilicat mesoporous molecular sieves	1.5-10 nm	Molecular sieves	1
2D Mesoporous silica films	-	Sol–gel dip coating with 2D hexagonal templating mesophases	4

Table 1. (Continued)

Media	Porous size	Methods of producing, possible applications	Ref.
Thin mesoporous films with covalently attached fluorescein entities	-	Optical pH sensors	7
Porous and non porous silica films	3.6 nm	Sol-gel templating	12
Empty and Sn filled porous Al_2O_3 films	15.5 nm	Anodic oxidation of Al film	14

Ordered mesoporous materials differ from bulk matter by exhibiting a highly ordered, monodispersed porosity in the 2-50 nm range giving rise to a new class of materials. Due to their unique properties, they have received increasing interest from the scientific community since the pioneering work by Beck et al. [1] in 1992. Initially obtained as powders, the thin mesoporous films are now synthesized [2-4]. These ordered films can be represented as a periodic lattice of holes in an amorphous dielectric matrix, creating an artificial material with a low dielectric constant [5,6] that may be tuned through the amount of pores and their filling. Such films are of interest for many optical applications, including the creation of devices ranging from optical sensors [7] to laser sources with promising characteristics [8,9]. Their high specific area also opens the door to catalytic or filtering processes in industrial world or environmental problems [10,11]. A good insight into the potential abilities of these new materials to the suggested applications is gained when characterizing their structure at various scales, from grazing X-rays to visible wavelengths. Visible light properties, in which possible applications are expected, are usually determined through reflectance measurements or spectroscopic ellipsometry. The latter procedure is time consuming and expensive. Reflectivity leads to the refractive index and the filling factor or pores' density q. Light scattering of a mesoporous film and new features, especially an interference pattern in the angular distribution of scattered light, have been reported first time, according to our knowledge, in [12].

In the first part of this chapter, the reflectivity spectra and HELS properties of non-porous and ordered mesoporous silica films will be presented. Reflectivity has been used to determine the film thickness as well as their refraction and extinction indices; a tentative interpretation of their optical properties is proposed [12,13]. The coherent scattering created by the ensemble of mesoporous has been linked to the structure of the film, interpreted as taking place inside the mesoporous layer, and used to interpret some optical properties of the sample.

Another nanosized structure very promising for different applications is periodical self-organized array of empty or metal-filled nanopores. Studies of reflective and scattering properties of porous alumina membranes whose pores are either empty or filled with tin, then realization of nanorods for which quantum confinement may occur in their shortest dimension are presented in [14]. From a fundamental point of view, the optical properties of such structures are attainable through methods that remain applicable although the wavelength of light is large as compared to their characteristic lengths. In particular, light scattering occurs whatever the size of the object under study is [15], whereas reflectance average the properties on a wavelength scale. In this respect, scattering should prove to be more sensitive to differences in the composition [16] of a nanostructure than reflectance.

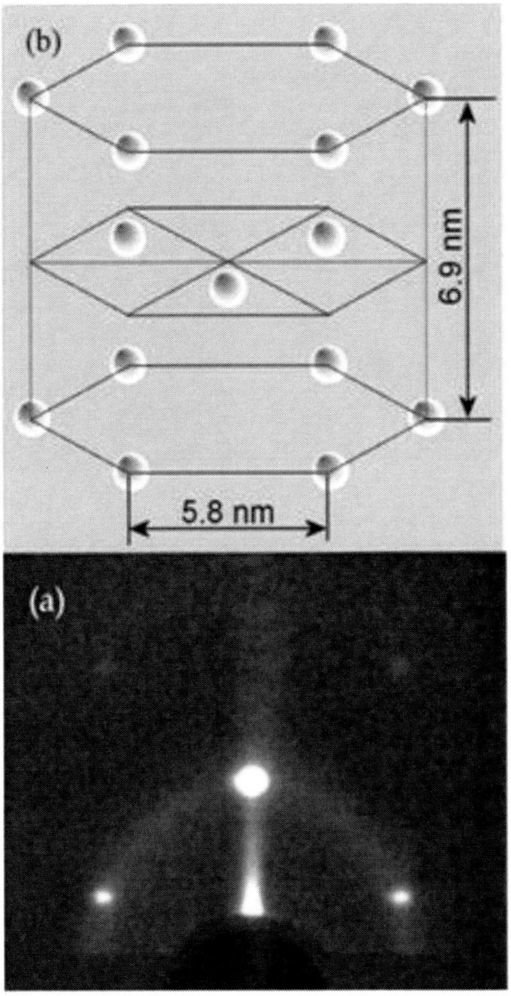

Figure 1(a). Grazing incidence small angle X-ray diffraction pattern of a mesoporous film. The film is ordered in a $P6_3/mmc$ structure with the c-axis perpendicular to the substrate. (b). Schematic presentation of the porous film structure. Spheres represent mesopores in SiO_2 matrix.

PREPARATION AND STRUCTURE OF THE SAMPLES

Table 2. Structure of the samples

Media	Mesoporous structure	Ref.
Al_2O_3, bulk material	-	14
Al_2O_3, empty pores, back Al	Pores 15.5 nm	14
Al_2O_3, empty pores	Pores 15.5 nm	14
Al_2O_3, Sn filled pores	Pores 15.5 nm	14
Non porous SiO_2 on Si	-	12
Porous SiO_2 on Si	Pores of 3.6 nm	12

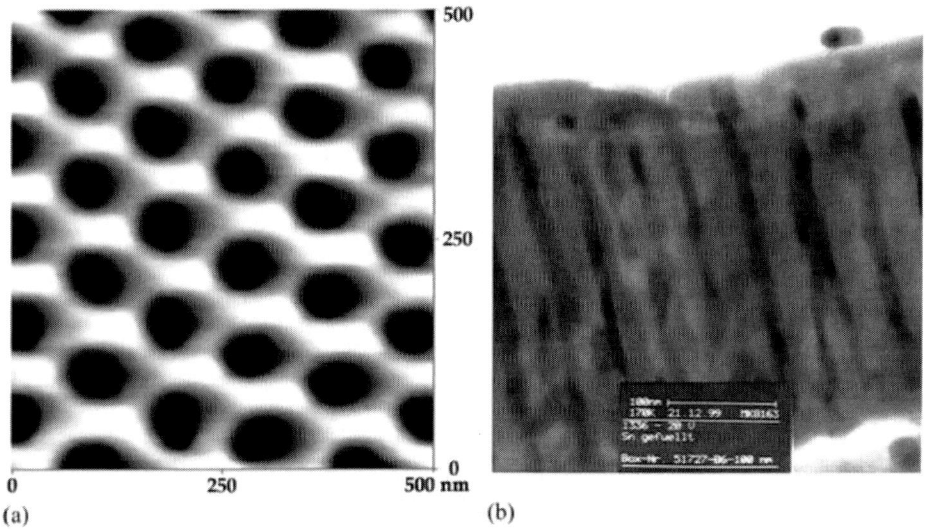

Figure 2. (a). AFM image of a typical Al_2O_3 membrane. (b) TEM cross-section of a membrane, showing tin inside the pores.

Each sample of ordered mesopores silica film consists of a silicon wafer with a thermally oxidized layer, covered by either a non-porous or mesoporous film. Three different samples were realized: the first one is Si + thermal SiO_2 (labeled later as *Th*), the second is Si + SiO_2 + a non-porous layer (*NPor*), and the third one is Si + SiO_2 + a mesoporous layer (*Por*).

The mechanisms behind the formation of mesostructured materials are still not completely understood. Therefore, the development of such materials for applications implies the iteration of cycles formulation-optimization-material characterization. Unfortunately, the process is often slowed down by the use of very heavy characterization techniques like grazing incidence x-ray scattering and electron microscopy. In this chapter, we deal with some of the optical characterization of mesoporous materials in order to characterize their light scattering properties and refractive index, as well as to determine to what extent optical techniques can bring insight into the materials structure.

The mesoporous films were obtained by hydrolysis and condensation of silica alkoxydes in presence of a templating mesophase. An extended description of the synthesis and ordered structure of these films have been reported elsewhere [12,17,18].

The ordered structure of the mesoporous film has been characterized by grazing incidence small angle X-ray diffraction. The analysis was performed with a diffractometer [4] and recorded on image plates.

The diffraction pattern of the mesoporous thin film presented in Figure 1(a) is characteristic of a $P6_3/mmc$ structure, Figure 1(b), with its *c* axis perpendicular to the substrate and its *a* axis randomly oriented in the substrate plane [18]. The cell parameters are: a = 5.8 nm and c = 6.9 nm. The ratio c/a = 1.2 differs from the theoretical value of 1.63 obtained for a close packed hexagonal structure. This is consistent with a 27% unidirectional shrinkage of the film perpendicular to the substrate [4,18]. The thin film can best be imagined as composed of grains, having all their *c* axes perpendicular to the substrate but randomly oriented in the substrate plane, see Figure 1. The mesopores form a unidirectional compressed close packed hexagonal structure. The size of the mesopores was estimated to be 3.6 nm [19,20].

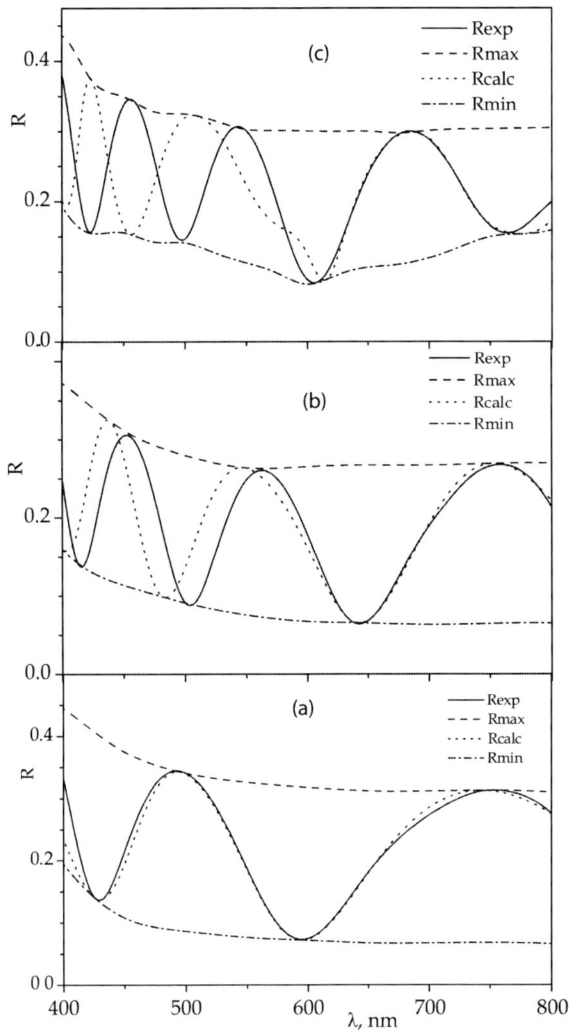

Figure 3. (a) Experimental and calculated reflection spectra of Si covered by thermal SiO_2 (*Th* film); constructive and destructive envelopes. (b) Experimental and calculated (second approximation) reflection spectra for a *NPor* film; constructive and destructive envelopes. (c) Same as for (a,b), but for a *Por* film.

The thicknesses of different samples were characterized by SEM. They are $d = 525 \pm 10$ nm for the thermal SiO_2 film, $d = 240 \pm 10$ nm for the non-porous part of the *NPor* film, and $d = 495 \pm 10$ nm for its porous part. Through AFM, a typical 3.8 nm r.m.s. roughness has been obtained for a mesoporous surface.

The samples of mesoporous Al_2O_3 membranes were prepared by anodic oxidation of high purity aluminum foils in polyprotic acids. The resulting nanoporous structure is characterized by transmission electron microscopy (TEM). The remaining aluminum can be dissolved subsequently using a solution of 0.2 m $CuCl_2$ in 7 m hydrochloric acid.

Four types of samples were realized from the same foil. They are: an Al foil, an unsupported Al_2O_3 membrane with empty pores, a membrane with Sn-filled or empty pores supported by the remaining Al foil. Detailed description of these samples preparation can be found in [14]. The alumina membranes provide pores that are nearly parallel to each other and almost perpendicular to the membranes surface [21-23]. AFM and scanning electron microscopy clearly show in Figure 2 the ordered structure of a membrane; the pore density can be as high as 10^{12} pores/cm^2. The pore diameter 2r depends on the anodising voltage applied, approx. 1.2 nm/V, and can be varied between 5 and 200 nm. In our case, 2r was estimated to be in the range of 20–25 nm.

The interpore distance d is proportional to 2r, $d \approx \alpha*2r$; according to [21,24], $\alpha \approx 1.75$. However, as we will show later, α depends on the conditions of preparation. The membrane thickness finally depends on the anodisation time and is above 10 μm for the samples studied here. The aluminum foils have been used as received after anodisation. Tin is plated into the pores using an AC plating procedure [25], with a graphite rod as counter electrode.

Figure 2(a) is an AFM image of a sample obtained at an anodisation voltage of 40 V. It shows a good reproducibility of the shapes and diameters of the pores; their arrangement in a lattice is quite good. The transmission electron micrograph in Figure 2(b) presents a cross-section of our membrane; the diameter of the pores and also the interpores' distance d can be deduced, it turns out that the aspect ratio length / diameter of Sn nanorods (NR) is important. From the same figure, the experimental parameters of the sample have been determined: 2r = (15.5 ± 2.6) nm; as well $\alpha \approx 2.9 \pm 0.37$. Our samples studied correspond to an anodisation voltage of 20 V.

The filling factor q, i.e. the ratio of tin volume to the embedding volume, has been calculated assuming an hexagonal lattice formed by cylindrical pores of radius r, separated by the distance d; then

$$q = \pi / (4(\alpha + 1)^2 \cos \pi/6);$$

with our experimental value of α, one obtains $q = 0.06$.

The difference between the diameters of rods and pores may occur from the influence of the cross-sectional process, or the presence of some oxide layer at the surface of the NRs that would be not so distinguishable from alumina. It can be added that the physical process leading to filling of the pores may also show some crystallographic or growth limitations regarding a complete filling of the pores. Later on, the pore diameter will be assumed to be the same as the rod one's.

ANALYSIS OF REFLECTION SPECTRA

Table 3. Reflectivity Data

Media	Parameters	Ref.
Al_2O_3, bulk material	R(λ)	14
Al_2O_3, empty pores, back Al	R(λ)	14
Al_2O_3, empty pores	R(λ)	14
Al_2O_3, Sn filled pores	Rp(θ), Rs(θ)	14
Non porous SiO_2 on Si	n(λ),k(λ)	12
Porous SiO_2 on Si	n(λ),k(λ)	12

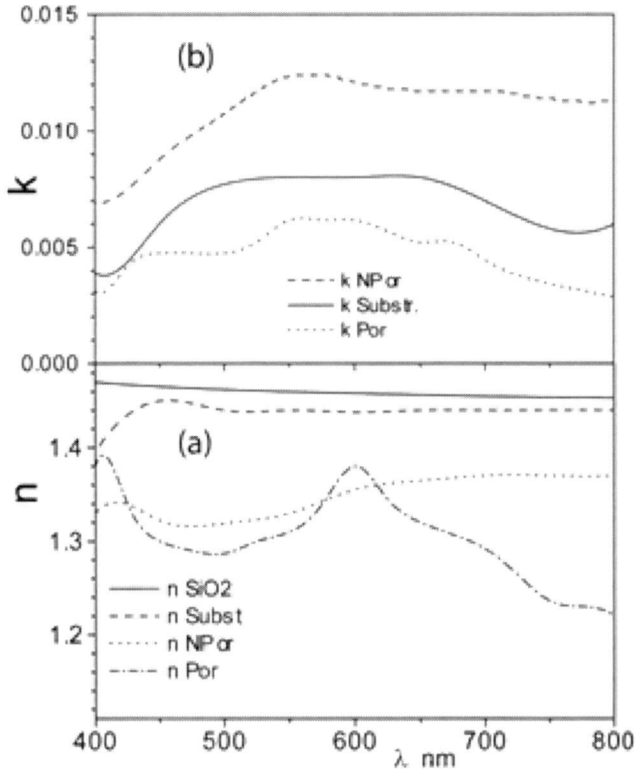

Figure 4. (a) Calculated real part of refraction indices of *Th*, *NPor* and *Por* films. SiO_2 index *n Lit* is also given for comparison. (b) Calculated (second approximation) imaginary part of indices of *Th*, *NPor* and *Por* films.

Specular reflection spectra have been measured close to normal incidence with a dedicated apparatus [26]. The spectral range extends from $\lambda = 400$ nm up to 800 nm, the spectral width of the beam is 1 nm and its angular divergence is ± 3.9°.

Numerical computations of optical parameters from reflection spectra have been and still are the subject of many publications [27-32]. For the sake of simplicity, authors have chosen to describe the real properties of the samples by a simple model of the film structure. They suppose that a homogeneous thin film, with unknown thickness d and complex index $\tilde{n}_1 = n_1 + ik_1$, lays on a semi-infinite Si substrate with a known index $\tilde{n}_2 = n_2 + ik_2$; n_i is the real part of $\tilde{n}_i$ or refractive index and k_i its imaginary part. The purpose of the calculations is the determination of $\tilde{n}_1$ and d from reflectance measurements. Solutions are not unique; an intrinsic difficulty of this kind of calculation, then solving equations goes through steps.

The procedure applied to the experimental reflection spectra of all samples, in order to determine the indices and thicknesses of their components, is given in more detail with the associated relations and acronyms in Appendix 1.

Th film, first step: the experimental reflectance $R\exp(\lambda)$ exhibits an interference modulation; then, envelopes $Rmax(\lambda)$ and $Rmin(\lambda)$ are deduced. According to relations (1)-(2) of Appendix 1, a calculated $R\text{calc}(\lambda)$ reflectance is obtained. All of them are shown in Figure 3(a). Although the difference $R\text{calc}(\lambda)$–$R\exp(\lambda)$ increased at short wavelengths, the agreement between $R\text{calc}(\lambda)$ and $R\exp(\lambda)$ is quite good from green to red spectral range.

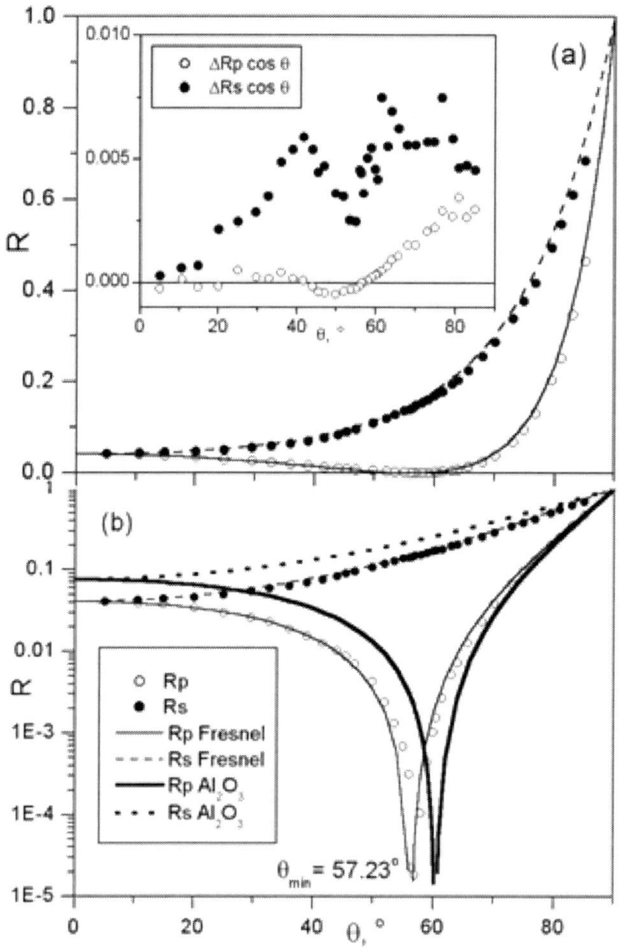

Figure 5. Angular dependence of the reflectance of an Al_2O_3 Sn-filled pores zone with back Al layer at $\lambda = 632.8$ nm. Both polarizations are shown. Dots are experimental results, lines are calculated reflectance. (a): linear scale. The inset shows the differences $\Delta Rp \cos \theta$ and $\Delta Rs \cos \theta$ between the measured values of the reflectance and those calculated using the effective index. (b): semi-logarithmic scale. Comparison between Fresnel reflectance of Sn-filled pores zone calculated with its effective index and that of Al_2O_3.

Th film, second step: *Rmax*(λ) and *Rmin*(λ) are used to extract d as follows. For each set *R*exp(λ), *Rmax*(λ) and *Rmin*(λ) of experimental reflectance, using SiO_2 published refractive index *n Lit* [33], a slight variation ± 0.1 around it, and $k_1 = 0$, a set of d solutions has been obtained. Among this set, one chooses as the physically meaningful solution the mean value $d = d_0 = 518 \pm 5$ nm that does not depend on λ although it deviates slightly in the blue part of the spectral range, see Figure 3.

Th film, third step: d_0 being known, solving system (5) leads to the unknown indices $n_1(\lambda)$ = *n Th* in Figure 4(a) and $k_1(\lambda)$ = *k Th* in Figure 4(b). The calculated values $n_1(\lambda)$ are down shifted regularly by ~0.02 as compared to *n Lit* [33]. Absorption of SiO_2 is usually supposed negligible, an assumption that would remove one unknown in system (5), in our case, however, fitting experiment and calculation does not appear possible if one neglects k, although it remains small and on the order of 0.01.

The same steps have been applied to *NPor* and *Por* films. When trying to fit *R*calc(λ) and *R*exp (λ), a very good adjustment of one interference modulation period leads to a significant disagreement for the others. According to (6), effective values of *n* and *k* for both films were estimated in order to reproduce the experimental results. The best fit over the whole spectra is obtained when the variation $\Delta R^{int}(d)$ goes through a well-determined minimum; it occurs for d_0; this value has been used to compute a second approximation of the solutions of system (5).

When applying this procedure to *NPor* film, and comparing to the case of *Th* film, it appears more difficult to realize the fit - see Figure 3(b). For *Por* films, the same procedure needs a numerical value for d_0 that should be independent of λ. A solution for *d* has been calculated, but it does not strictly converge [12,13]; however, one can assume a reasonable d_0 value located in the vicinity of 1000 nm where statistical fluctuations are smaller than those around 800 nm. With such a value, one obtains *R*calc(λ) for the *Por* film, see Figure 3(c). For this kind of film, the disagreement is stronger than for the previous one.

Measurements and associated calculations of reflectance of a Sn-filled zone of mesoporous Al_2O_3 membrane have been performed as a function of the angle of incidence. These results are presented in Figure 5.

The dielectric function of Al_2O_3 [33] is known over a wide energy range, and a careful examination of the literature showed that only ε_{Sn} (λ = 532 nm) is known [34]. Although our measurements were performed at λ = 632,8 nm, we used a MG's model of the composite at λ = 532 nm, in which $\varepsilon_{Sn} = -55 + 16.5i$ and $\varepsilon_{Al2O3} = 3.104$, and extracted a filling factor q_{calc}. Not surprisingly, the calculated value q_{calc} strongly differs from that found previously, while being negative. Such a result was to be expected, due to a possible strong spectral dependence of the optical constants of Sn and to highly probable different crystallographic and/or chemical composition of nanorods.

From another point of view, *q* has previously been determined using scanning microscopy or AFM images. This filling factor can be used in MG's model, in order to determine $\varepsilon^{MG}{}_{Sn}$. Every parameter involved in this model being real, the calculation gives a real value for $\varepsilon^{MG}{}_{Sn}$ (632.8 nm) = -2.24 ± 0.4. In one way, we gain an interesting value of the dielectric function of Sn at this new wavelength, which well describes the optical properties of a composite sample with a specific shape of NRs. Strictly speaking, Sn is a metal, and a negative ε_{Sn} is consistent with it. However, bulk Sn is also very close to a low gap semiconductor. There are two different Sn modifications in the bulk, a metallic one (above 13.2 °C) and a semi-metallic one (below 13.2°C); we know from EXAFS measurements that even at low temperatures we only have the metallic one in our pores; this point may be useful to keep in mind, as we do not know what can be the electrical properties of NRs inside the nanopores.

Fresnel reflectances $Rp^F(\theta)$ and $Rs^F(\theta)$ were calculated using the effective index n_{eff} of a Sn-filled pores zone. Experimental deviations from Fresnel reflectance have been defined by $\Delta Rp(\theta) = Rp^F(\theta) - Rp^{Exp}(\theta)$, $\Delta Rs(\theta) = Rs^F(\theta) - Rs^{Exp}(\theta)$ where $Rp^{Exp}(\theta)$ and $Rs^{Exp}(\theta)$ are the experimental values. Both $\Delta Rp(\theta)$ and $\Delta Rs(\theta)$ show a significant increase as θ increases. When θ increases, so does the illuminated area of the sample, giving rise to a proportional increase of the scattered energy that will shape both $\Delta Rp(\theta)$ and $\Delta Rs(\theta)$. One compensates this effect by taking into account ΔRp(θ)cos θ and ΔRs(θ)cos θ, as shown in the inset of Figure 2(a), instead of ΔRp(θ) and ΔRs(θ).

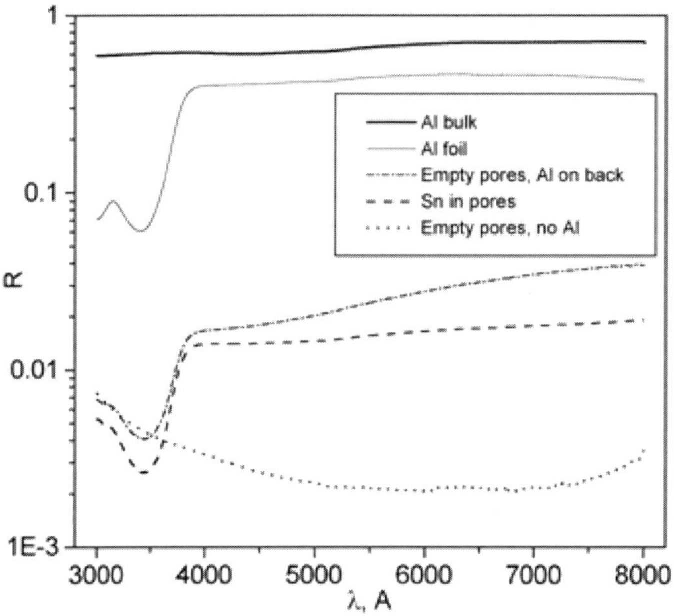

Figure 6. Spectra of reflectance R of the Al_2O_3 mesoporous membranes and other zones.

Differences for both polarizations are significant. ΔRp(θ)cosθ is smaller than 0.3%, and its sign fluctuates; errors in the measurements increase when θ > 60°. At the same time, the angular dependence of ΔRs(θ)cosθ reveals that a systematic difference, significantly higher than the absolute level of ΔRp(θ)cosθ, is involved. We attribute the origin of this difference to the strong anisotropy of our scattering object, a system of periodically arranged nanorods. We have shown that reflectance at oblique incidence was dependent on the polarization of the incident light; moreover the structure of the composite influences the variations of ΔRs and ΔRp.

Scattering is already present at normal incidence; it may depend on the relative orientations of light polarization and possible characteristic directions of the surface. In the next section, an analysis of the reflectance data for various zones of the sample will show that scattering is present and depends on the zone under study.

All the reflectance R and transmittance T measurements of various zones of the sample have been performed at normal incidence, see Figure 6. Although the surface is structured, any influence of the light polarization has been detected.

The small value of the filling factor q ensures that the samples are far from the percolation threshold, and as a consequence they behave like a dielectric with a flat response in the visible range of the light, apart a marked absorption band located at λ = 340 nm.

This absorption band appears in both cases for which Al is present on the back side of the samples: empty pores with back Al and Sn-filled pores. It disappears in the case of empty pores zone. Although its physical interpretation will not be discussed here, it is related to the presence of Al and its impurity content already mentioned since it is also found in the reflectance of an original Al foil (curve not shown). When Al is oxidized and forms Al_2O_3, the absorption band disappears and the chemical environment of the impurities is modified.

The reflectance of an empty pores zone is less than 0.01. Independently, its transmittance (spectrum not shown here) is relatively flat and varies monotonically between 0.36 and 0.52. Hence, strong scattering and absorption share the rest of the energy balance of the incident light. The empty pores' zone and the Sn-filled one's supported by a back Al layer exhibit quite comparable reflectances that differ because empty or filled nanopores in an alumina matrix realize slightly different effective media; this is consistent with the previous results shown in Figure 5.

Apart the absorption band due to Al, the reflectances spectrum are rather flat and do not show marked structures that can be attributed to a plasmon resonance.

Comparing the experimental reflectance with the calculated specular reflectivity at normal incidence shows the influence of scattering. The reflectance of Sn-filled pores has been calculated using n_{eff} (632.8 nm). The reflectance of an empty pores zone has been calculated with an effective index determined from the known filling factor q, the dielectric function of non-porous Al_2O_3 and the dielectric function of air (vacuum) as filling material. In each case, and especially for Sn-filled and empty pores zones with back Al, the measured reflectances are significantly smaller than the calculated ones. This observation demonstrates the effect of strong scattering.

ANALYSIS OF SCATTERING PROCESSES

Table 4. Scattering data

Media	Parameters	Ref.
Al_2O_3, bulk material	BSDF(θ), BSDF(θ, φ) for 632.8 nm	14
Al_2O_3, empty pores, back Al	BSDF(θ), BSDF(θ, φ) for 632.8 nm	14
Al_2O_3, empty pores	BSDF(θ), BSDF(θ, φ) for 632.8 nm	14
Al_2O_3, Sn filled pores	BSDF(θ), BSDF(θ, φ) for 632.8 nm	14
Non porous SiO_2 on Si	BSDF(f_x, f_y) for 594 and 632.8 nm	12
Porous SiO_2 on Si	BSDF(f_x, f_y) for 594 and 632.8 nm	12

In what follows, we characterize and compare the scattering properties of mesoporous and non-porous silica films deposited onto Si.

An original patented installation [35] has been used for these measurements. It allows also obtaining an optical image of the scattering area under study [12], a valuable possibility to estimate the presence of defects of all kinds on the surface and simplify the interpretation of scattering data. The light source is a He-Ne laser tunable at $\lambda = 594$ and 632.8 nm, with an optical power of ~3 mW. The laser beam impinges at normal incidence onto the sample and is focused down to ~100 μm diameter while remaining quasi parallel. Focalization determines the angular resolution, the order of 1° in our case, and enables choosing the area that exhibits the lowest possible contamination.

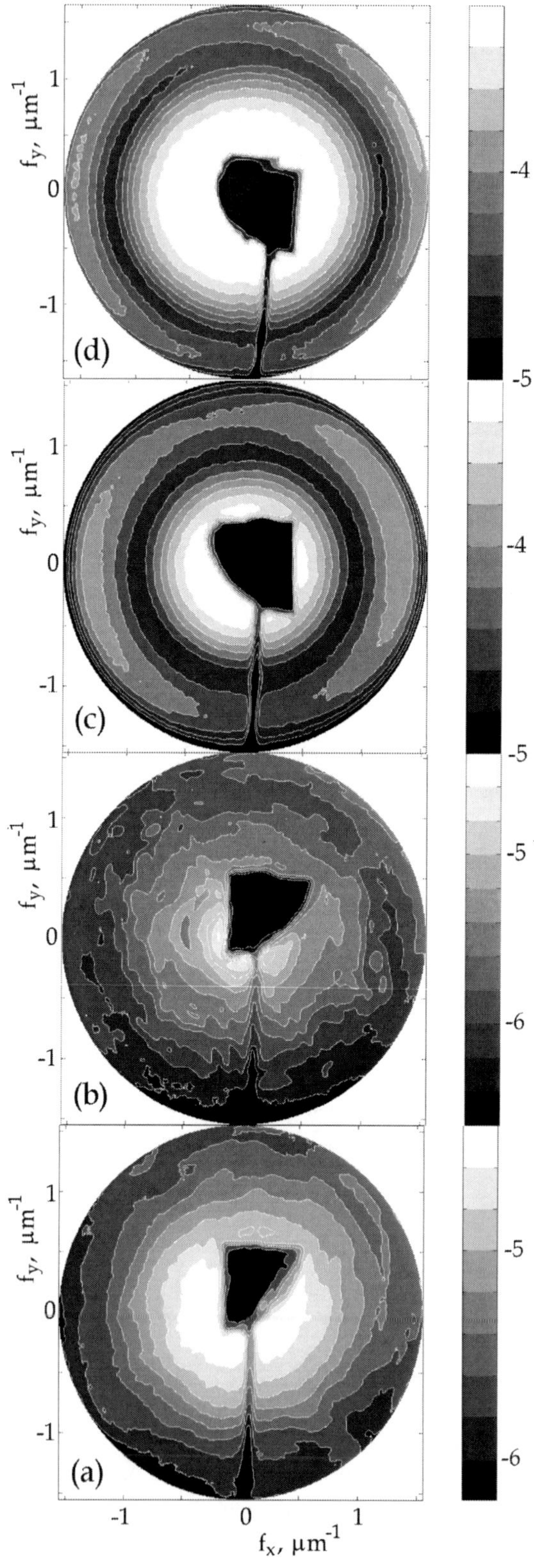

Figure 7. log BSDF as a function of the spatial frequencies f_x, f_y: (a) *Th* film; (b) *NPor* film; *Por* film: (c) $\lambda = 638.8$ nm, (d) $\lambda = 594$ nm.

Comparing the microscopy images of *NPor* and *Por* scattering areas [12,13] shows that non-porous film scatter is produced by a weak background and few point-like contaminated centers. The level of background scattering that comes from clear places is much greater for a mesoporous film: it reflects the homogeneous and very dense space distribution of mesopores.

HELS measurements on the samples have been performed. The bidirectional scatter distribution function (BSDF), function of the polar and azimuthal angles θ and φ, is defined as the scattered intensity dI normalized to the incident intensity I_0, the solid angle of the receptor $d\Omega$ and cos θ: $BSDF = \frac{1}{I_0 \cos\theta}\frac{dI}{d\Omega}$. In a 2D diagram, the BSDF coordinates are chosen as the sample's spatial frequencies f_x, f_y, for the sake of clarity. By doing so, we exclude the dependency of HELS data upon the wavelength of the incident beam. At normal incidence, $f = \sin\theta / \lambda$ for any scattering plane.

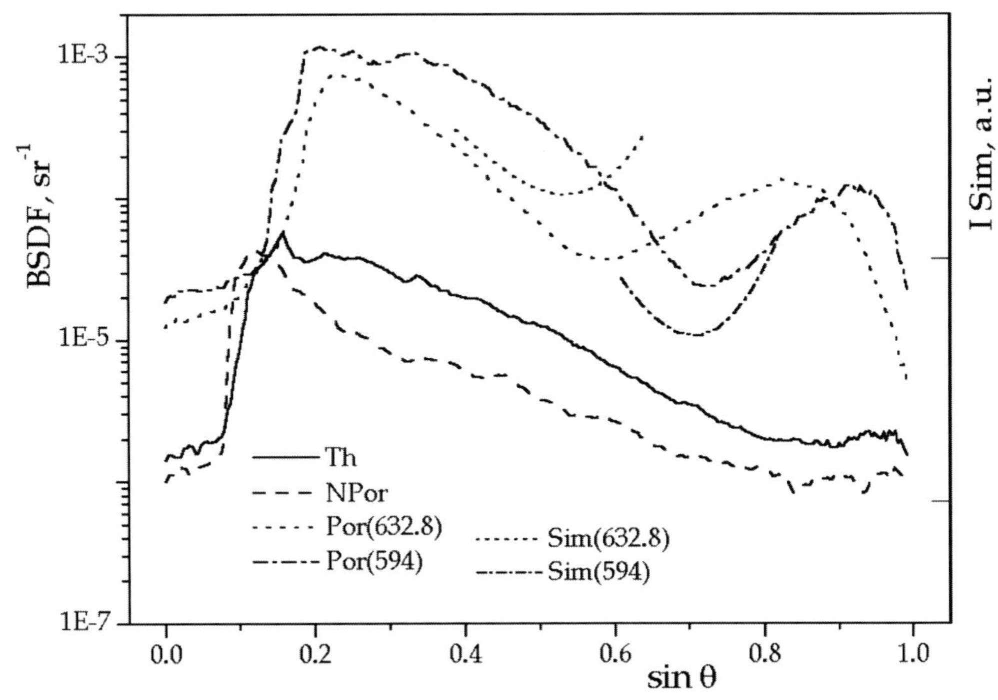

Figure 8. Azimuthally averaged cross-section of BSDF(f_x, f_y) as a function of sin θ, for a *Por* film at 594 and 632.8 nm, and for *Th* and *NPor* films at 632.8 nm. Simulated minima (*Sim*) are also drawn.

The data: log BSDF(f_x, f_y) presented in Figure 7 are mapped in grey levels on a disc. Its center represents the light intensity scattered normally to the sample's surface; its borderline represents scattering intensity in grazing directions.

Figures 7(a, b) show that a non-porous film decreases the average scattering. Moreover, the increase in the speckle structure of a non-porous film produces an important modulation of the azimuthal dependence of scattered light; this modulation will be discussed later.

Figures 7(c, d) show, for a porous film, an important increase of the scattered intensity and the presence of well-pronounced minima in its angular distribution; they are given by sin

θ_{min}(632.8 nm) = 0.582 ± 0.02 and sin θ_{min}(594nm) = 0.721 ± 0.02. The Total Integrated Scatter (TIS) enables comparison of scattering properties; measurements show that TIS_{Por} /TIS_{NPor} = 54 ± 20, while TIS_{Por} / TIS_{Th} = 21 ± 8; errors margins quantify the surface homogeneity but not the measurements' errors by themselves. At high scattering angles, the azimuthal structure of the maximum probably reflects some azimuthal anisotropy of the mesoporous structure.

Surface relief and total scattering do not depend on λ once they are presented in their Fourier decomposition.

Figures 7(c, d) also show that the spatial frequency modulus associated with BSDF's minimum is significantly shifted when λ varies from 632.8 nm to 594 nm. An azimuthal averaging of BSDF, Figure 8, performed for a φ-range in which scattering data represent all samples, facilitates intensity levels comparison. Of course, the angular minimum is still present.

We have to question the physical process able to shift significantly this angular minimum for so small a change of λ. Interference is a good candidate for such a process, and its origin may lay in surface relief and bulk scattering centers. The pores are not only homogeneously distributed inside the film but exhibit order; they can act as coherent scatterers, giving rise to interference.

Scattering properties of mesoporous Al_2O_3 membranes and other zones of the sample are presented in Figure 9; they are mapped on a plane as a function of θ and φ. The gray levels represent log BSDF(θ, φ) intensity.

BSDF's data reveal an azimuthal anisotropy that can be divided in two parts: a strong almond shaped one seen in the central part of Figure 9(a) and another part at high polar angles of scattering. Figure 9(a) shows the strongest anisotropy which results from the influence of foil's rolling and polishing treatments on scattering. Azimuthal anisotropy at high polar angles of scattering takes its origin in a residual curvature of the sample, which remains even under vacuum sticking. At mean and small polar angles of scattering, anisotropy remains but becomes very light; the influence of curvature is less important on one hand. On the other hand, the creation of the ordered nanopores array tends to render scattering more isotropic.

Sectors where the influence of anisotropy is small have been selected from the data given in Figure 9. An averaging procedure has been applied along the azimuthal angle φ to these selected data. Such a procedure minimizes the influence of the remaining parasitic curvature of the sample and the speckle structure. The averaged data are given in Figure 10(a).

Whatever the sample's zone is, the scattered intensity decreases at high polar angles. In Figure 10(a), the scattered intensities of the porous membranes differ from the bare Al foil, labeled as "Al foil". Scattering by an empty pores zone is significantly stronger than for an Al foil; this difference increases at high scattering angles. As compared to the scattering of empty pores zones, filling the pores by Sn (data labeled by "Sn fill. rear Al") decreases the scattering level by more than one order of magnitude.

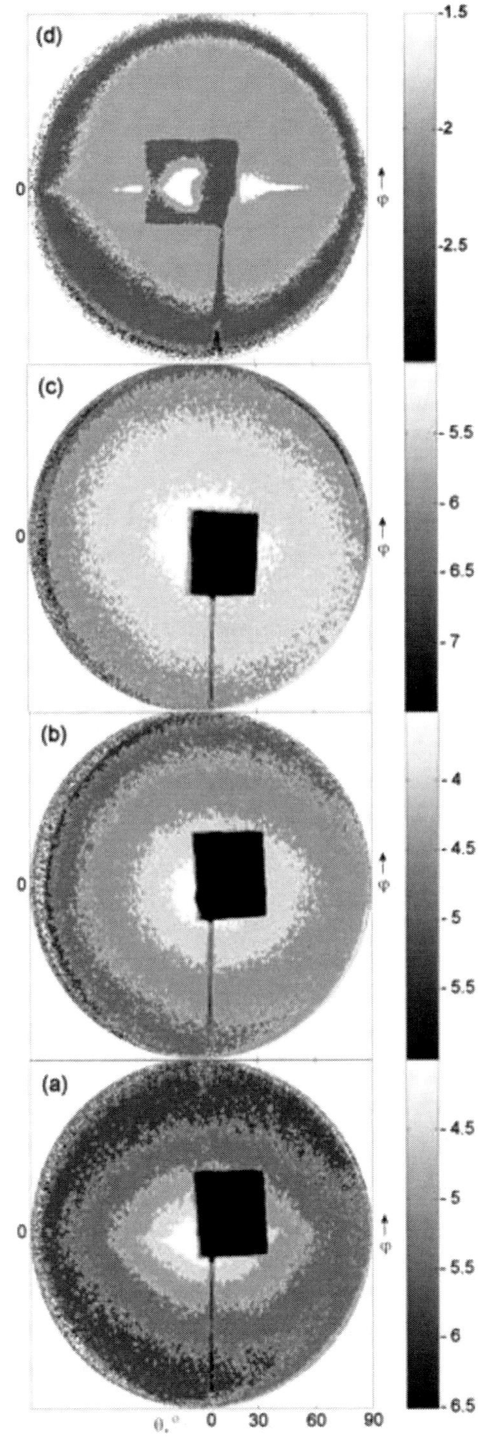

Figure 9. Hemi-spherical distribution of log BSDF(θ, φ) for different zones of the sample. Values between isophotes separated by a 0.5 step have been set to constant gray colors. Note the different intensity scales. The black rectangular zone is the shadow of the sample. (a): Al foil. (b): empty pores with back Al. (c): tin-filled pores with back Al.(d): empty pores.

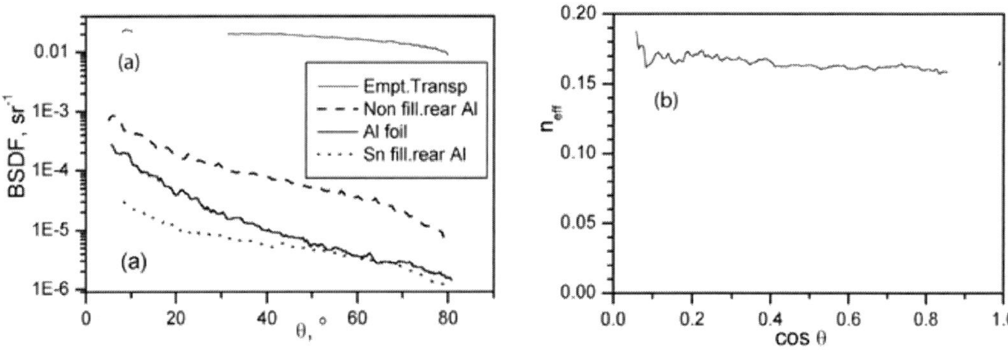

Figure 10. (a): BSDF(θ) defined as the average of BSDF(θ,φ) along φ for Al foil, empty and Sn-filled zones. (b): Calculation of an effective index of an empty pores zone according to the theoretical model of Vera *et al.* [37] See text below.

Few articles are devoted to scattering of nanorods [36], and calculations for such structures are relatively new; to our knowledge, nobody allowed a direct comparison between calculations and our experimental data.

Multiple scattering by empty pores is important due to the small distance between them. An attempt to estimate it has been realized, using a diffusion theory from [37]. Assuming multiple scattering and a theoretical model, [37] predicts the angular distribution of the transmitted scattered light by inhomogeneous media. The physical parameter returned by this theory is the effective index $n_{eff}^{(22)}$ of the scattering medium:

$$n_{eff}^{(22)} = n \cdot \sqrt{\frac{2 \cdot BSDF(\mu)}{3 \cdot (z_e + \mu)(1 - R(\mu))}}.$$

It comes from the experimental BSDF data of the transmitted scattered energy, $\mu = \cos(\theta)$, n is the index of bulk Al_2O_3 matrix, z_e is an extrapolation length ratio and R the average reflectivity of the medium [37]. Because of multiple scattering and neglecting coherent backscattering, we have assumed that the angular distribution of scattering should be similar for reflection and transmission hemispheres. With our BSDF data, the calculated $n_{eff}^{(22)} = 0.165 \pm 0.004$ is presented in Figure 10(b). It does not depend practically on μ . This important fact corroborates that multiple scattering takes a significant place in shaping the BSDF(μ) dependence and that the diffusion theory we applied describes correctly the scattering process.

Thicknesses of our samples are on the order of 10 μm, and they exhibit a relatively high value of transmittance: T ~ 0.5 (see the text above). Small thickness and high transmittance are not sufficient to force all incident light that goes through to participate in scattering. The main assumption of [37] being that all incident light is scattered, it explains why $n_{eff}^{(22)}$ calculated value is much smaller than could be expected.

INTERPRETATION AND DISCUSSION

For a homogeneous film, a unique solution for both indices and thickness that would give the experimental reflectance is rather seldom, multiple solutions being generally the rule. However, reflectance measurements at normal incidence have led to the indices $n_1(\lambda) = n\ Th$ and $k_1(\lambda) = k\ Th$ (see Figure 4(a, b)) and thickness d_0 of the thermal SiO_2 film (*Th* film). The indices are in good agreement with the published values [27] while $d_0 = 518 \pm 5$ nm corresponds to $d = 525 \pm 10$ nm measured by SEM.

For *NPor* and *Por* films, we have assumed that a unique and homogeneous layer would conveniently describe their properties. This reduces the number of unknowns and equations and gives a tractable and useful, although not perfect, analytical model. Of course, such layer could no longer describe the real structure of the film; some mismatch between calculated and measured reflectance values had to be expected and have been found. However, the analytic treatment of reflectance data has given the following thicknesses: *d Por* = 1019 - 518 = 501 nm and *d NPor* = 826 - 518 = 308 nm, values, in reasonable agreement with those given above by SEM.

This is not only interesting because it indicates that the used model sounds physically but also because the determination of film thickness is crucial when optimizing the synthesis and deposition of mesoporous films. In general, this quantity is determined by SEM, a heavy technique compared to reflectance, and it drastically limits the number of characterized samples. Therefore, it is of interest that a fairly simple and inexpensive technique, contrary to SEM, leads to similar results as it opens way for a simple table top apparatus for direct use in the laboratory.

As a surprising result, the porous film refractive index $n_1(\lambda)$ exhibits two maxima (see Figure 4) located near 400 and 600 nm. There is no *a priori* information about their origin. One can suggest two possibilities. Ellipsometric data [38-40] show that the pores are empty and filled by air immediately after preparation. After a few hours in the laboratory environment, water vapor remaining in air is adsorbed and represents a non negligible quantity. This condensed water may contain some dissolved contaminations or impurities, giving rise to these maxima. Another possible origin may be some remaining impurities of the template that would not be removed entirely after the thermal treatment. This explanation can be additionally proved by presence of some maximum near 400 nm in NPor spectra as well. However, we were not able to put this in evidence by other characterization techniques like infrared spectroscopy or x-ray photoemission spectroscopy. This suggests that optical properties are very sensitive to adsorbates or doping impurities.

It is well known that refractive indices of various SiO_2 films in the wavelength range from 400 nm to 800 nm are described by the Cauchy dispersion formula with a high accuracy. The spectral dependencies of indices of the films under investigation were approximated by the Cauchy model:

$$n(\lambda) = A0 + A1/\lambda + \ldots + A7/\lambda^7 + A8/\lambda^8.$$

The obtained Ai values are presented in Table 5. To estimate precision of such approximation, the rms differences between the values calculated and approximated by this model are shown in the last column of Table 5. The complicated shape of n Por dependence increases the corresponding rms error value relative to the simpler shapes of n Lit and n Th spectra.

Table 5.

	A0	A1	A2	A3	A4	A5	A6	A7	A8	rms error
n Lit	3.26E00	-6.57E03	1.02E07	-8.92E09	4.77E12	-1.61E15	3.34E17	-3.91E19	1.98E21	4.97E-07
n Th	1.09E02	-4.21E05	7.04E08	-6.57E11	3.74E14	-1.33E17	2.87E19	-3.46E21	1.78E23	7.80E-05
n Npor	6.91E01	-2.21E05	2.85E08	-1.79E11	4.57E13	6.19E15	-6.81E18	1.57E21	-1.25E23	1.67E-04
n Por	5.86E02	-2.16E06	3.40E09	-2.97E12	1.57E15	-5.18E17	1.03E20	-1.14E22	5.27E23	7.84E-04

Ordered mesoporous films are composite materials that could satisfy all the basic assumptions of the effective medium theories. Such physical processes as important scattering for mesoporous film as well as modification of film optical properties by some dopants after chemical treatment and adsorbates disturb correct application of the Cauchy model.

The resulting optical properties of mesoporous film calculated in the framework of Rayleigh-Gans approximation were presented in [41]. Scattering and reflection by the films with spherical, ellipsoidal and cylindrical pores of different orientations have been calculated. The authors have found pores parameters that reproduce the observed experimental dependences. A disagreement between the experimentally measured and calculated dependencies is due to influence of scattering by mesoporous structure.

As is shown in Figure 5, *k Th* is bigger than *k Por*, due to higher density of the thermal oxide. The imaginary parts of *k Por* and *k NPor* demonstrate similar behavior, suggesting a common origin; both films come from the same material chemically treated in a different way. However, *k Por* is smaller than *k NPor* because of their different densities. For the same densities, *Th* and *NPor* films exhibit different *k*, due to their different chemical contents or absorbing centers: the *Th* film is purer than the *NPor* film. One can estimate the filling factor q of the mesoporous film; assuming the same origin of absorption, q should be proportional to the ratio of absorptions. A reasonable value of $q \sim 0.44 \pm 0.09$ is obtained by averaging of *k Por* / *k NPor* over the spectral range. This value is in very good agreement with previous determination on different samples by another method [40] ($q \sim 0.43$). This result is of interest because q value is a key parameter when optimizing the synthesis of mesoporous materials. We have hereby showed that it is possible to characterize it using simple visible light optics.

Figure 4 shows that the index of *Th* film is bigger than the *NPor*'s one, which in turn is also bigger than the *Por*'s one, the result consistent with the decrease of the silica volume at mesopores production inside the film.

The results presented in Figure 7 show that a porous film scatters stronger than a non-porous one. For the same material, the filling factor q decreases from a porous film to a non-porous one, so does the amount of scattered energy as well as the absorption presented in Figure 4. As the chemical composition of the samples is almost the same, this difference is attributed to their structure, and to significant increase of the index contrast for the *Por*'s film.

Several reasons may explain the decrease of scattering level when going from a *Th* to a *NPor* film (see Figure 7). For the *Th*-mesoporous film, coverage by a layer does not reproduce the roughness of the interface, and generally the roughness of the outer surface is less than the original one's. Moreover, the index contrast interface decreases. Both reasons lead to scattering reduction.

Another reason can be the so-called anti-scattering effect [43-45] that, for light scattered by interfaces, has the same physical origin as antireflection coating.

The speckle structure exhibited by the *NPor*'s BRDF distribution (see Figure 7(b)) is originated from interferences of point-like contamination centers, presented in Figure 9(a) of [12].

Figures 7 and 8 show, to our knowledge, the first report on light scattering by a mesoporous film. A plane wave incident normally to the surface of a mesoporous film is scattered and exhibits a deep minimum in the θ_{min} direction. The value of θ_{min} is very sensitive to λ and is not represented by the same point in the spatial frequencies space. Then, the position of θ_{min} cannot originate in a Fourier decomposition of the surface relief and should be related to interference. We assume that coherent scattering takes place inside the mesoporous layer and results in a fixed direction θ_{min} of the interference's minimum The intensity *I Sim* (λ, sin θ) of all beams scattered in θ direction has been modeled and is presented in Figure 8 according to the geometrical layout in Figure 11 (see Appendix 2).

It goes through a minimum that depends on indices and thicknesses of the films. The simulated positions and relative values of the minima are close to the experimental ones.

We are led to conclude that the above model of scattering by ordered mesoporous silica film adequately describes the BSDF(θ) dependence, although a perfect agreement is not obtained.

First result dealing with reflective properties of mesoporous Al_20_3 films is that Sn-filled membranes show a typical dielectric reflectance for both polarizations of light at variable incidence. This behavior comes from the very small filling factor q compared to the percolation threshold. Tin, if still metallic, represents a relatively small amount of dispersed metal inside alumina, unable to strongly modify its optical properties as a dielectric.

The pores array produces a modulation of the surface structure in a wide range of spatial frequencies but our measurements attain only the lower part of this range; this creates a stochastic surface relief which is the origin of the observed scattered light and its speckle structure.

As compared with the wavelength of light, the small period of mesoporous structures does not allow a direct observation of diffraction. However, sub-wavelength sized periodical array made of silver nanoparticles has a significant influence on reflection properties [46]. The main physical mechanism responsible for such an influence is the excitation of collective oscillations due to mutual interactions between particles provided the correct choice of the excitation wavelength.

Such an interaction could be realized in the Sn-filled zone of our membrane and would influence its scattering properties. Our observations do not show a direct evidence of diffraction by a sub-wavelength sized array of tin nanorods in the distribution of scattered light. One explanation of such an absence lays in the weak electrical conductivity of tin that is sometimes considered as a zero gap semiconductor. It implies that collective oscillations of electrons may be not well defined as in the case of well conducting metals.

By analogy with [46], future use of Ag or Au- filled pores should show a significant appearance of the influence of the sub-wavelength array on scattering.

Scattering properties are influenced by interface properties described by surface roughness, local variations of indices, and volume properties related also to local variations of indices.

Looking at interface scattering, filling of the pores modifies the surface roughness and creates a spatial modulation of the index. On one hand, it is reasonable to think that filling would decrease the surface roughness; hence part of scattering originating from the interface should be less important for filled pores than for empty ones.

On the other hand, the index contrast for Sn-filled zones $|1.76 - 1.5\ i| = 2.31$ is higher than that $|1.76 - 1| = 0.76$ of empty pores zones [14]; hence filled pores zone should scatter more than empty one's. Then, the first argument is consistent with the experimental results while the second is not.

The second result about Al_2O_3 scattering properties, on the contrary to what could be expected, is that a filled pores zone scatters less than empty one's. Let us look now at the influence of the volume and its composite structure on scattering.

As deduced from Figure 5, the reflective properties are clearly those of a dielectric, and the sample should be more or less transparent. Moreover, the Al substrate does not influence much the reflection process; hence the conclusion would be that light does not penetrate the layer up to the substrate. Then, propagation of light through the membrane up to its back Al is prevented because of absorption and scattering.

The interpores distance is small compared to the wavelength, and the pores present some deviations from perfect cylinders and cannot be strictly considered as straight and perpendicular to the surface of the sample. Then, the incident plane wave is multiply scattered in the volume. As we have seen earlier, length of NRs is at least on the order of one wavelength, which is larger than the mean free path of electrons. Hence, the component of scattered light parallel to the length of the NRs should be absorbed. The Al substrate reflects the rest of energy that impinges it, and the same process of absorption occurs again. With empty pores, scattering also occurs but no metal is present and able to absorb the created parallel component; light propagates through the layer. As a consequence, more energy remains available for scattering by empty pores zone than from Sn-filled zone, in agreement with experiment.

CONCLUSION

In this chapter, the reflection and light scattering properties of a Si substrate covered by different SiO_2 layers with a non-porous or mesoporous structure have been studied. We showed that it was possible from reflectivity measurements through a simple model, to obtain both the index of the film and its thickness found in good agreement with that given by SEM. These results are interesting as they demonstrate that a part of the complicated tools (SEM, TEM, GISAXS) used for characterization during the optimization of mesoporous film can be replaced by a simple optical equipment in the laboratory. The index of a mesoporous medium is lower than that of the corresponding bulk material, a result that was expected. Furthermore, from the determined index it is possible to estimate a very important parameter for mesoporous systems, namely, the filling factor q. The light scattering properties of the mesoporous film proved to be one order of magnitude higher than those of homogeneous films. Most surprisingly, the light scattered by the mesostructured film exhibits a well marked minimum in the angular distribution of intensity. This property is linked with the film structure and has been explained by coherent scattering taking place in the thickness of the film and leading to interference. Future work will settle if the mesopores or the texture of the films is the origin of the phenomenon. In all cases these studies will enhance the insight in the structure and formation of mesoporous films.

It was shown that a nanoporous Al_2O_3 membrane with a sub-wavelength periodic array of Sn-filled or empty pores can be described to a very good approximation as a dielectric with a very low metal content.

Its reflectance shows a Brewster angle; a reasonable estimate of the real part of the dielectric function ε^{MG}_{Sn} of tin at $\lambda = 632.8$ nm has been deduced using a Maxwell-Garnett's effective medium theory and the filling factor. Although our procedure does not enable finding the imaginary part of ε^{MG}_{Sn}, $Rp = 10^{-5}$ ensures that it should remain extremely small. Despite the rather restricted range of diameter of the nanorods, no plasmon resonance has been found in the wavelength range under study.

The scattering properties of filled and empty pores zones have been described. Scattering of empty pores zone has been found much greater than that of filled pores. This unexpected result has been interpreted using multiple scattering and absorption of the component of the electric field parallel to the nanorods. This strong absorption comes to be possible because tin nanorods are supposed to be metallic, at least along their major dimension.

APPENDIX 1.

One should find here in more detail the determination, from the reflectance spectra, of the index $\tilde{n}_1$ and thickness d of a transparent film deposited onto an absorbing substrate. Most of the times, the existing methods have to be adapted, making each resolution a problem by itself. Non-porous and mesoporous films are deposited on an underlying SiO_2 layer that covers a Si substrate. Our main assumption, which appeared valid, is that index difference between either the non-porous or mesoporous films and the underlying SiO_2 is small. Then, the optimized envelope method is appropriate for determining the optical parameters of the upper film. This method enables one to calculate optical parameters of thin film not only for the points of reflectivity extremes, but for all measured between them points. Many publications [27,29,30,31,32] present the reflection coefficient R of a film covering an absorbing substrate with index $\tilde{n}_2 = n_2 + ik_2$ as follows (we used expressions close to those given by [32]):

$$R = \frac{A + Bx + Cx^2}{D + Ex + Fx^2}; \quad x = \exp(-\alpha d); \quad \alpha = (4\pi k_1)/\lambda \tag{1}$$

These non-linear equations are solved through a method, proposed in [27-32], that consists in drawing envelopes: smooth curves passing through experimental extrema of $R(\lambda)$; those passing through $R(\lambda)$'s maxima (respectively minima) are called *Rmax* and *Rmin*.

$$
\begin{gathered}
A = [(1 - n_1)^2 + k_1^2][(n_1 + n_2)^2 + (k_1 + k_2)^2] \\
B = 2[A' \cos\phi + B' \sin\phi]; \ \phi = 4\pi n_1 d/\lambda \\
C = [(1 + n_1)^2 + k_1^2][(n_1 - n_2)^2 + (k_1 - k_2)^2] \\
D = [(1 + n_1)^2 + k_1^2][(n_1 - n_2)^2 + (k_1 - k_2)^2] \\
E = 2[C' \cos\phi + D' \sin\phi]; \\
F = [(1 - n_1)^2 + k_1^2][(n_1 + n_2)^2 + (k_1 + k_2)^2] \\
A' = (1 - n_1^2 - k_1^2)(n_1^2 - n_2^2 + k_1^2 - k_2^2) + 4k_1(n_1k_2 - n_2k_1) \\
B' = 2(1 - n_1^2 - k_1^2)(n_1k_2 - n_2k_1) - 2k_1(n_1^2 - n_2^2 + k_1^2 - k_2^2) \\
C' = (1 - n_1^2 - k_1^2)(n_1^2 - n_2^2 + k_1^2 - k_2^2) - 4k_1(n_1k_2 - n_2k_1) \\
D' = 2(1 - n_1^2 - k_1^2)(n_1k_2 - n_2k_1) + 2k_1(n_1^2 - n_2^2 + k_1^2 - k_2^2)
\end{gathered} \tag{2}
$$

If $n_1 < n_2$, then analytic expressions of *Rmax*(λ) are obtained replacing *B* and *E* by $\tilde{B}$ and $\tilde{E}$.

$$
\begin{aligned}
\tilde{B} &= 2[(1 - n_1^2 - k_1^2)(n_1^2 - n_2^2 + k_1^2 - k_2^2) + 4k_1(n_1k_2 - n_2k_1)] \\
\tilde{E} &= 2[(1 - n_1^2 - k_1^2)(n_1^2 - n_2^2 + k_1^2 - k_2^2) - 4k_1(n_1k_2 - n_2k_1)]
\end{aligned} \tag{3}
$$

Rmin (λ) is obtained replacing $\tilde{B}$ by $-\tilde{B}$ and $\tilde{E}$ by $-\tilde{E}$

Si indices n_2(λ) and k_2(λ) are represented by analytic polynomial interpolations of published data [33].

The first step: spectra of n_1 (λ) and k_1 (λ) are determined by iteration. In the 400 - 800 nm range, the references points n_1 (λ_j) and k_1 (λ_j) are taken with a 50 nm step and a spline approximation is established. *Rmax*(λ), *Rmin*(λ) and the calculated reflectance *R*calc(λ) of the film are calculated. New values of $n_1(\lambda_j)$ and $k_1(\lambda_j)$ that realize the coincidence between the experimental extrema of *R*exp(λ) and *Rmax*(λ) and *Rmin*(λ) data are determined. Then *R*calc(λ) is deduced.

The second step: the film thickness *d* is calculated using the obtained envelopes *Rmax*(λ) and *Rmin*(λ). For any wavelength, three non-linear equations link experimental and calculated values:

$$
\begin{aligned}
&R\max(n_1, k_1, d, \lambda) - R\max^{exp} = 0; \\
&Rcalc(n_1, k_1, d, \lambda) - Rexp = 0; \\
&R\min(n_1, k_1, d, \lambda) - R\min^{exp} = 0;
\end{aligned} \tag{4}
$$

System (4) is solved numerically, using Levenberg-Marquardt algorithm, in order to obtain d_0: thickness of the upper film. This value is used to determine the solutions n_1(λ), "*n* film" and k_1(λ) "*k* film" of system (5):

$$R\max(n_1,k_1,d_0,\lambda) - R\max^{exp} = 0;$$
$$R\min(n_1,k_1,d_0,\lambda) - R\min^{exp} = 0; \quad (5)$$

The deviation ΔR^{int} between Rcalc and Rexp is given by (6) and it enables finding effective values of the real and imaginary parts of the index that reproduce the experimental reflectance of the film under study:

$$\Delta R^{int}(d) = \sum_{\lambda} \left| Rcalc(n_1,k_1,d,\lambda) - Rexp(\lambda) \right| \quad (6)$$

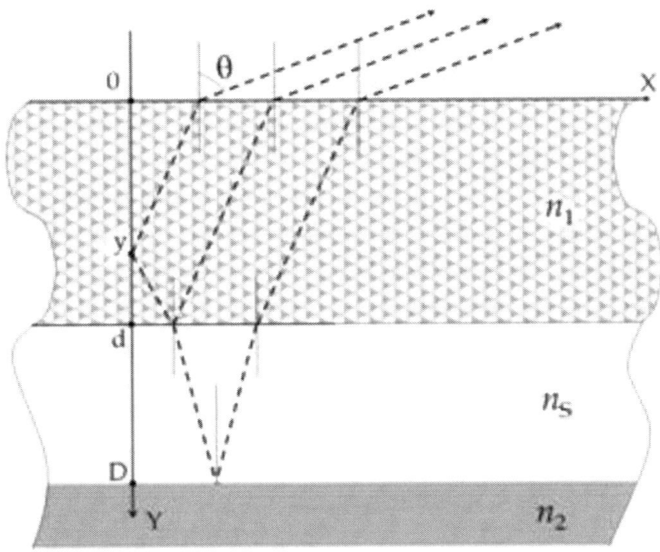

Figure 11. Scattered beams involved in the interference coming from a film with homogeneously distributed scattering centers.

APPENDIX 2. INTERFERENCE OF SCATTERED LIGHT

This appendix is intended to give in more detail the origin of the observed BSDF's minimum, see Figure 7(b, c). We assume that it is due to coherent interferences of light scattered by point sources, homogenously distributed inside the porous film, see Figure 11. Assumed also is an isotropic scattering by all point sources. Taking into account paths differences, reflection and transmission coefficients by the various interfaces, the analytical expression of the amplitude of the scattered beams has been obtained. This amplitude is integrated along (over?) the mesoporous film thickness. The scattered intensity is deduced but not given here because of its length. Unfortunately, the analytic complexity of this expression and its minima does not enable resolution in terms of optical parameters of the film. A simulation $I\ Sim(\lambda, \sin\theta)$ of the dependence of this scattered intensity is presented in Figure 8.

REFERENCES

[1] Beck, J. S.; Vartuli J. C., Roth W. J., Leonowicz M. E., Kresge C. T., Schmitt K. D., Chu C. T. W., Olson D. H., Sheppard E. W., McGullen S. B., Higgins J. B., Schlenk J. L. *J. Am. Chem. Soc.* 1992, *114*, 10834-10843.

[2] Lu, Y.; Ganguli R., Drewien C.A., Anderson M.T., Brinker C.J., Gong W., Guo Y., Soyez H., Dunn B., Huang M. H., Zink J. I. *Nature* 1997, *389*, 364-368.

[3] Klotz, M.; Ayral A., Guizard C., Cot L. *J. Mater. Chem.* 2000, *10*, 663-669.

[4] Klotz, M.; Albouy P-A., Ayral A., Ménager C., Grosso D., Van der Lee A., Cabuil V., Babonneau F., Guizard C. *Chem. Mater.* 2000, *12*, 1721-1728.

[5] Fan, H.; Bentley H.R., Kathan K.R., Clem P., Lu Y., Brinker C.J. *J. of Non-Crystalline Solids* 2001, *285*, 79-83.

[6] Yang, C-M.; Cho A-T., Pan F-M., Tsai T-G., Chao K-J. *Adv. Mater.* 2001, *13*, 1099-1102.

[7] Wirnsberger, G.; Scott B.J., Stucky G.D. *Chem. Commun.*, 2001, *119-120.*

[8] Wirnsberger, G.; Yang P., Scott B. J., Chmelka B. F., Stucky G. D. *Spectrochimica Acta A* 2001, *57,* 2049-2060.

[9] Scott, B.J.; Wirnsberger G., Stucky G.D. *Chem. Mater.* 2001, *13,* 3140 -3150.

[10] Dufaud, V.; Davis M. E. *J. of Am. Chem. Soc.* 2003, *125,* 9403-9413

[11] Rodriguez-Castellon, E.; Jimenez-Lopez A., Maireles-Torres P., Jones D.J., Roziere J., Trombetta M., Busca G., Lenarda M., Storaro L. *J. of Sol. St. Chem.* 2003, *175,* 159-169

[12] Sterligov, V. A.; *Appl. Opt.* 2005, *44*, 4538-4546.

[13] Cheyssac, P.; M.Klotz M., E.Sondergard E., Sterligov V.A. *Opt. Comm.* 2005 *252*, 344-354.

[14] Sterligov, V. A.; Cheyssac P., Blau W., Kröll M. *Opt. Comm.* 2003, *226,* 125-134.

[15] Sterligov, V. A.; Cheyssac P., Lysenko S. I., Kofman R. *Opt. Comm.* 2000, *186*, 27-33.

[16] Sterligov, V. A.; Cheyssac P., Kofman R., Lysenko S. I., Lytvyn P. M., Vohnsen B., Bozhevolnyi S. I., Maradudin A. A. *Phys.Stat.Sol.(b)* 2002, *229*, 1283-1294.

[17] Besson, S.; Gacoin T., Jacquiod C., Ricolleau C., Babonneau D., Boilot J.-P. *J. Mater. Chem.* 2000, *10,* 1331-1336.

[18] Besson, S.; Ricolleau C., Gacoin T., Jacquiod C., Boilot J.-P. *J. Phys. Chem.* 2000, *B 104*, 12095-12097.

[19] Klotz, M.; Besson S., Ricolleau C., Bosc F., Ayral A. *Mat. Res. Soc. Symp. Proc.* 2003, *752*, AA8.6.1.

[20] Besson, S.; Gacoin T., Ricolleau C., Jacquiod C., Boilot J-P. *Nanoletters* 2002, *2,* 409-414.

[21] Furneaux, R. C.; Rigby W. R., Davidson A. P., *Nature* 1989, *337,* 147-149.

[22] Sawitowski, T.; Ph. D. Thesis, University of Essen (Germany), 1999.

[23] Heilmann, A.; Altmann F., Katzer D., Müller F., Sawitowski T., Schmid G., *Appl. Surf. Sci.* 1999, *144-145*, 682-685.

[24] Hanaoka, T. A.; Heilmann A., Kröll M., Kormann H. P., Sawitowski T., Schmid G., Jutzi P., Klipp A., Kreibig U., Neuendorf R. *Appl. Organomet. Chem.* 1998, *12,* 367-373.

[25] Grandjean, D.; R. Benfield E., Nayral C., Maisonnat A., Chaudret B. *J. Phys. Chem. B* 2004, *108*, 8876-8887.
[26] Kofman, R.; Cheyssac P., Richard J. *Phys. Rev. B* 1977, *16*, 5216-5224.
[27] Minkov, D. A.; *J. Phys. D: Appl. Phys.* 1989, *22*, 1157 - 1161.
[28] Minkov, D. A.; *J. Phys. D: Appl. Phys.* 1989, *22*, 199-205.
[29] Epstein, K. A.; Misemer D. K., Vernstrom G. D. *Appl. Opt.* 1987, *26*, 294 - 299.
[30] Rusli, R.; Amaratunga G. A. J. *Appl. Opt.* 1995, *34*, 7914 - 7924.
[31] González-Leal, J. M.; Márquez E., Bernal-Oliva A. M., Ruiz-Pérez J. J., Jiménez-Garay R. *Thin Sol. Films* 1998, *317*, 223 - 227.
[32] Müllerová, J.; Mudro J. *Acta physica slovaca* 2000, *50,* 477 - 488.
[33] Palik, D. Ed.; "*Handbook of Optical Constants of Solids*", Academic Press, New York, 1985.
[34] Boyd, G. T.; Rasing Th., Leite J. R. R., Shen Y. R. *Phys. Rev. B* 1984, *30*, 519–526.
[35] Sterligov, V. A.; Cheyssac P. *Patent CNRS* 23.11.2001, n° 0115232.
[36] Boustimi, M. ; Baudon J., Féron P., Robert J., *Opt. Comm.* 2003, *220,* 377-381.
[37] Vera, M.U.; Durian D.J., *Phys. Rev.E* 1996, *53*, 3215-3224.
[38] Brunet-Bruneau, A.; Fisson S., Gallas B., Vuye G., Rivory J. *Thin. Sol. Films* 2000, *377 - 378*, 57 - 61.
[39] Grosso, D.; Balkenende A. R., Albouy P. A., Lavergne M., Mazerolles L., Babonneau F. *J. Mater. Chem.* 2000, *109*, 2085 - 2089.
[40] Besson, S.; PhD thesis, 2002, Ecole Polytechnique, Paris, in French.
[41] Cheyssac, P.; ten Bosch A. *Thin. Sol. Films* 2008, *516*, 1503–1507.
[42] Brunet-Bruneau, A.; Besson S., Gacoin T., Boilot J.P., Rivory J. *Thin Sol. Films* 2004, 447-448, 51-55.
[43] Roche, P.; Pelletier E., Albr G. *J. Opt. Soc. Am. A*, 1984, *1*, 1032-1033.
[44] Amra, C.; Albr G., Roche P. *Appl. Opt.* 1986, *16*, 2695–2702.
[45] Giovannini, H.; Amra C. *Appl. Opt.* 1997, *36*, 5574-5579.
[46] Taleb, A.; Russier V., Courty A., Pileni M.P. *Appl. Surf. Sci.* 2000, *162-163,* 655-661.

In: Comprehensive Guide for Mesoporous Materials, Volume 2 ISBN: 978-1-63463-982-8
Editor: Mahmood Aliofkhazraei

Chapter 15

Novel Functionalities of Mesoporous Molecular Sieves to Extend Their Applications in Emerging Fields

***Aurelia Vasile*[1]*, Alina Maria Tomoiaga*[1*] *and Maria Ignat*[1,2]**
[1]Department of Chemistry, "Alexandru Ioan Cuza" University of Iasi, Iasi, Romania
[2]Petru Poni Institute of Macromolecular Chemistry, Iasi, Romania

Abstract

The interest of worldwide scientists on developing functional materials with special properties that allow their use in emerging fields as biomedicine and environmental remediation has continuously grown in the last several years. Due to their structure and surface properties mesoporous molecular sieves offer possibilities to be modified in a controlled manner to obtain important materials with specific applications in emerging fields. Silica- and carbon-based mesoporous molecular sieves modified with iron oxides (magnetite and hematite), titanium oxide (anatase), bismuth oxide (alpha-Bi_2O_3) by controlled encapsulation were developed. Their functionalities and specific applications in biomedicine and environmental remediation are discussed in separate sections.

Introduction

Mesotextured materials constitute a challenging domain in materials chemistry, which is experiencing explosive growth. The potential of these structures has been recognized in the various fields as adsorption, catalysis, separation, controlled delivery, optical devices and sensors. They are characterized by a regular distribution of pores within the mesoporous range, large surface areas and pore volumes, as well as a chemically-reactive surface that can be easily functionalized to extend their application in various fields.

* Author of correspondence: Alina Tomoiagă, Department of Chemistry, "Alexandru Ioan Cuza" University of Iasi, Carol I Bvd., No. 11, 700506, Iasi, Romania, T:+40232201314, F:+40232201313, email: alinamaria.tomoiaga@gmail.com.

Over the years, pure siliceous and pure carbon mesoporous molecular sieves have been extensively studied and successfully used in a wide range of applications. Nowadays, worldwide scientists are intensely working on advancing the nano-manipulation of pure mesoporous materials in order to imprint novel functionalities and widen their use in emerging fields as catalysis, biomedicine and environment protection. Therefore, this chapter refers to the development of silica and carbon-based mesoporous molecular sieves modified with metal-oxide nanoparticles and their specific applications in drug delivery and wastewaters treatment. Some reports on development of metal oxide-mesoporous materials and their applications in emerging fields are summarized in table 1.

Table 1. Some recent advances on the development of metal oxide-modified mesoporous molecular sieves for specific applications in emerging fields

Modified Molecular sieve	Mesoporous support	Application	Ref
$Fe_3O_4@SiO_2$	MCM-41 mesoporous silica	Targeted delivery and sustained release	[1-4]
	Hollow mesoporous silica	Drug delivery	[5]
γ- Fe_2O_3/SiO_2	Mesoporous silica	Hyperthermia therapy	[6, 7]
MgO/SiO_2	Hollow mesoporous silica	Cancer diagnosis and therapy	[8]
MgO/SiO_2	SBA-15 mesoporous silica	Drug storage and sustained release	[9]
Fe_2O_3/SiO_2	MCM-41 mesoporous silica	Ultrasound-assisted degradation of organic pollutants in wastewaters	[10, 11]
TiO_2/SiO_2	SBA-15 mesoporous silica	Photocatalytic degradation of organic pollutants in wastewaters	[12-16]
TiO_2/SiO_2	SBA-15 mesoporous silica	catalytic degradation of aqueous pollutants and VOC's	[17]
ZnO/SiO_2	SBA-15 mesoporous silica	Photocatalytic degradation of organic pollutants in wastewaters	[18]
TiO_2/MC	Mesoporous carbon	Photocatalytic degradation of phenols	[19]
TiO_2/MC	Mesoporous carbon	Photocatalytic degradation of dyes in wastewaters	[20, 21]
Bi_2O_3/MC	Mesoporous carbon	Photocatalytic degradation of organic pollutants in wastewaters	[22]
VOx/SiO_2	SBA-15 mesoporous silica	Catalysis	[23, 24]
CuO/SiO_2;	MCM-41 mesoporous silica	Catalysis	[25]
Fe_2O_3/SiO_2	MCM-48-, SBA-15-, MCF-, MSU mesoporous silicas	Catalysis	[26]
TiO_2/CNT	Multi wall carbon nanotubes	Photoelectrochemical cell	[27]

In section 1 we present recent findings of our research in development of magnetic silica nanoparticles with potential applications in targeted delivery and sustained release of low-molecular anticancer drugs. Section 2 is dedicated to recent results obtained in our research group in the field of nanodesign of mesoporous silica modified with hematite dedicated to wastewater treatment by ultrasound-assisted Fenton-like process. In section 3 are presented our results in mesoporous silica- based titanium dioxide photocatalysts for wastewater remediation.

Finally, chapter 4 focuses metal oxide- modified mesoporus carbon materials for water treatment.

MAGNETIC FUNCTIONALITIES ON MESOPOROUS SILICA NANOPARTICLES FOR BIOMEDICAL APPLICATIONS

Mesoporous silica nanoparticles can be specifically designed to effectively deliver drugs or other biological materials, with the following features: (i) biocompatibility, (ii) ordered pore network, with uniform tunable mesopores that allows fine control of drug load and release kinetics; (iii) high pore volume to host required amount of pharmaceuticals, (iv) high specific surface area with good potential for drug adsorption (v) a rich silanol groups-containing surface that can be functionalized to allow better control over drug loading and release [28-33]. However, introduction of magnetic properties to these marvelous materials opens new possibilities of their use especially in cancer early diagnosis and treatment.

Herein, we focus on development of magnetic silica nanoparticles in which hydrophilic superparamagnetic nanoparticles are coated with mesoporous silica layer to be used for the targeted delivery and sustained release of chemotherapeutic agents. Within such a therapeutic nanosystem, the superparamagnetic nanoparticles act as delivery tool of the drug from the administration point to the target site under the influence of an external magnetic field, while the mesoporous silica layer plays the role of host for large amounts of antineoplastic substance, allowing at the same time control of the drug release rate. In order to be applicable in the biomedical field it is highly important that the iron oxide nanoparticles are coated with mesoporous silica, thus protecting the human body against the toxicity of high local concentrations of metal cations resulting from dissolution of surface Fe cations.

The magnetic silica nanoparticles (MSN) developed herein were prepared under ultrasonic irradiation using various amounts of citrate-capped magnetite nanoparticles and well defined reaction conditions for growth of mesoporous silica layer. The use of ultrasonic irradiation made possible the formation of core-shell nanoparticles with well-defined particle shape in only 1.5 h due to the remarkable reaction conditions. Applicability of prepared magnetic mesoporous silica nanoparticles in sustained release of chemotherapeutics, was tested using 5-Fluorouracil (5-FU) (5-fluoro-1H,3H-pyrimidine-2,4-dione), a small molecule chemotherapeutic agent appertaining to the family of antimetabolites as probe molecule.

The synthesis of magnetic mesoporous silica nanoparticles (MSNs) is presented schematically in Scheme 1. The procedure is described in details elsewhere [1, 2].

The initial concentration of Fe_3O_4 nanoparticles was varied from 0.25 mg/ml to 2 mg/ml in the reaction sol containing constant dosages of TEOS, CTAB, EtOH, NH_3 and H_2O. Throughout this section, samples are encoded as $xFe_3O_4@mSiO_2$ NPs, where x represents the initial amount of Fe_3O_4 nanoparticles, in mg/ml. All samples have been analyzed using FIB-SEM, N2-sorption and VSM techniques in order to assess their morphological, textural and magnetic properties. The results are summarized in Table 2.

FIB-SEM microcrographs are displayed in figure 1. In all images Fe_3O_4 nanoparticles are clearly visible with lighter color. When high initial concentrations of magnetite NPs are used magnetite NPs agglomerates are observed on the outer surface of the silica nanoparticles. This is due to high tendency of aggregation and low dispersion of magnetite nanoparticles, making the coating process difficult. Thus spherical mesoporous silica nanoparticles are formed separately with no magnetic cores, as observed for samples $1.0Fe_3O_4@mSiO_2$ and $2.0Fe_3O_4@mSiO_2$.

Table 2. Physico-chemical features of developed $Fe_3O_4@mSiO_2$ nanoparticles

MSN	Morphologic features	Porous Features			Magnetic features*
		S_{BET} (m^2/g)	V_p (cm^3/g)	D_p (nm)	
0.25 $Fe_3O_4@mSiO_2$	Multicore-shell spherical NPs	1546	1.21	2.36	10
0.5 $Fe_3O_4@mSiO_2$	Multicore-shell spherical agglomerated NPs	1451	1.21	2.36	7.5
1.0 $Fe_3O_4@mSiO_2$	Agglomerated nanoparticles with Fe_3O_4 on the outer surface	1376	1.29	2.36	5
2.0 $Fe_3O_4@mSiO_2$	Agglomerated large nanoparticles with Fe_3O_4 only on the outer surface	1245	0.98	2.53	1.75

*expressed as saturation magnetization (Ms) in emu/g.

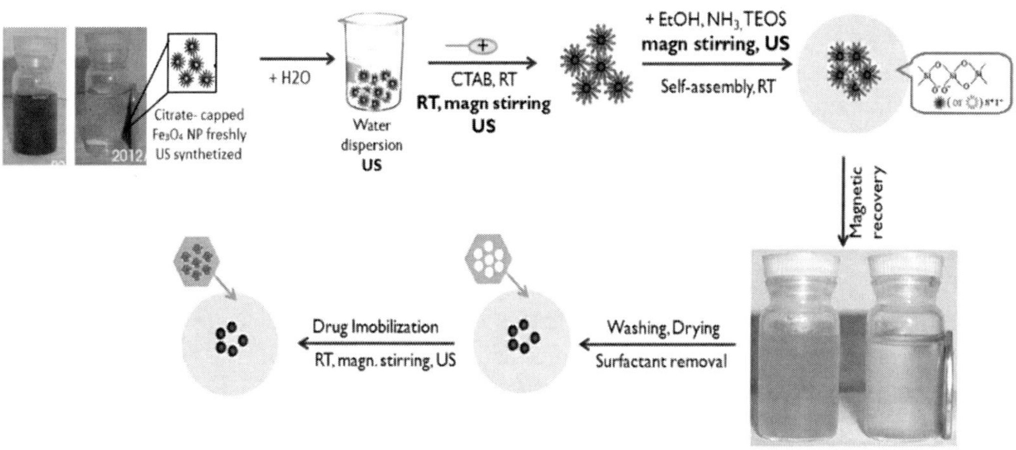

Scheme 1. Preparation of 5FU-containing therapeutic nanosystems based on magnetic silica nanoparticles.

This is highly undesirable since all the magnetite nanoparticles are exposed to the body environment once administered, and high local concentrations of metal ions resulting from dissolution of surface Fe cations can occur causing toxicity. As the concentration of Fe_3O_4 NPs decreases from 2 to 0.25 mg/ml, completely coverage by the mesoporous silica layer is obtained resulting in spherical nanoparticles with sizes in the range of 500 nm (sample $0.25Fe_3O_4@mSiO_2$). However, it looks like some of these nanoparticles are glued together, forming particle agglomerates. The FIB-SEM micrograph recorded on particle agglomerate, prove core-shell nature these nanoparticles. The core consists of multiple small magnetite nanoparticles covered by a thick silica shell.

Magnetic behavior of prepared samples was investigated using VSM (Vibrating sample Magnetometer) technique in applied magnetic field at room temperature. The saturation magnetization (M_S) decreases from 10 to 1.75 emu/g as the initial concentration of Fe_3O_4 NPs decreases from 2.0 to 0.25 mg/ml. This is due, partly to the surface contributions: stoichiometry deviation, spin canting, surface disorder and cation distribution [34], partly to the presence of mesoporous silica matrix. To prove the capacity to respond to magnetic field of $0.25Fe_3O_4@mSiO_2$ sample, having the lowest saturation magnetization value, two vials were loaded with 50 mg in 3 ml of Phosphate Buffer Solution (PBS, 100.0 mM, pH 7.4).

Without the presence of magnetic field, the drug-loaded nanoparticles are homogeneously dispersed in PBS solution. After applying an extern magnetic field, most MSNs were attracted to one side of the vial where the magnet is fixed, in 30 minutes. By removing the magnet, the nanoparticles revert to dispersed state (figure 1).

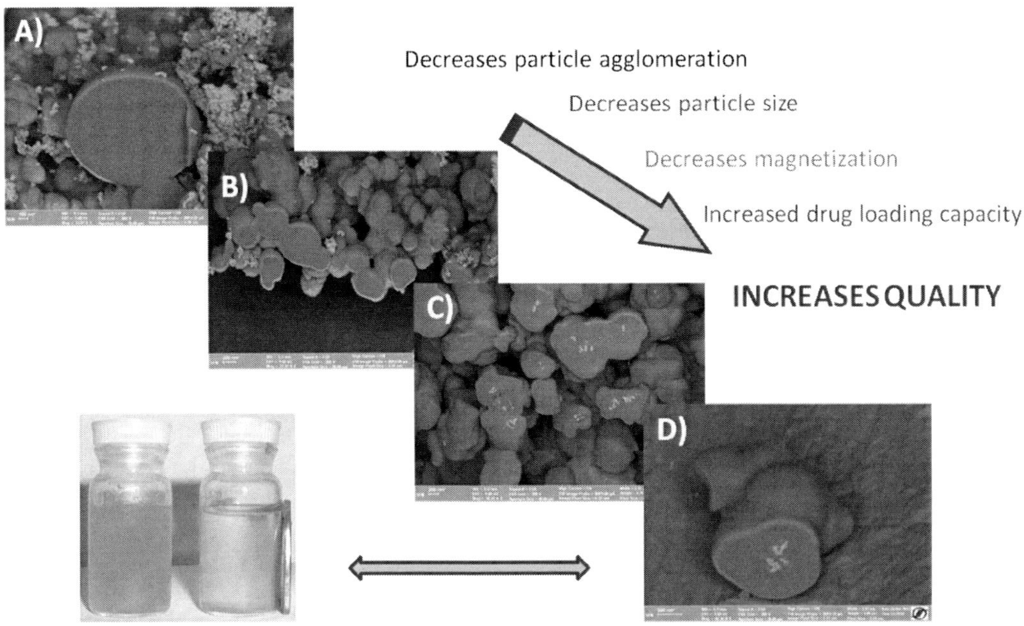

Figure 1. FIB-SEM images of MSNs prepared with different concentrations of Fe_3O_4: A) $0.25Fe_3O_4@mSiO_2$; B) $0.5Fe_3O_4@mSiO_2$; C) $1.0Fe_3O_4@mSiO_2$; D) $2.0Fe_3O_4@mSiO_2$;

N_2-sorption measurements were performed to investigate the porous features of prepared MSNs. The calculated porous features are summarized in table 2. Pore sizes are in in the nanometric range characteristic to hexagonally ordered mesoporous silica MCM-41 [35]. More, the calculations show that surface area decreases significantly as the initial concentration of Fe_3O_4 NPs increases, while this decrease is less important for pore volume and the pore diameter is mainly the same. Modification is noticeable for pore volume of sample $2.0Fe_3O_4@mSiO_2$, probably because exist magnetite NPs blocking the access of N_2 into the pores. Also, due to its higher surface area, the $0.25Fe_3O_4@mSiO_2$ shows best loading capacity towards low-molecular 5FU.

Further, 5-FU was immobilized on MSNs prepared with the lowest Fe_3O_4 NPs initial concentration. Thus we synthesized larger amount of $0.25Fe_3O_4@mSiO_2$ sample and subjected to drug immobilization tests, investigating many parameters that can influence the effectiveness of the uptake process. Nanoconfinement of 5-FU was performed also on $Fe_3O_4@mSiO_2$ samples in which the mesopores of the silica layer were functionalized with amino-groups in order to assess the influence of the surface chemistry and porous features on the drug uptake and release. The results of full characterization realized using modern techniques are described in details in previous publication [1].

Figure 2 shows the release of 5-FU from the unmodified and amino-modified $Fe_3O_4@mSiO_2$ nanoparticles in close comparison with its release from commercially available product Adrucil (50 mg/ml).

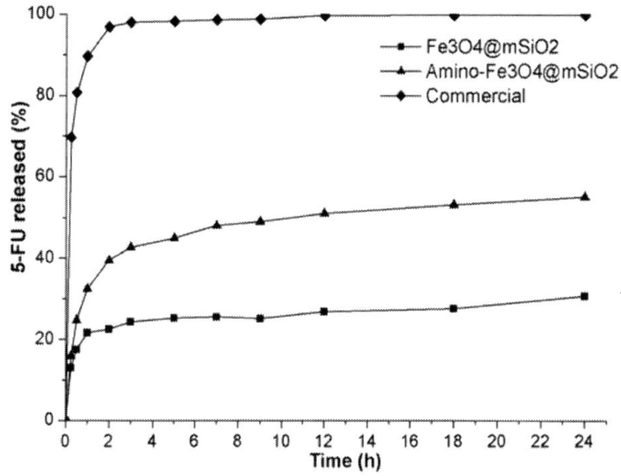

Sample	Released 5-FU (%)
Commercial	97
Amino-Fe_3O_4@ $mSiO_2$	54
Fe_3O_4@$mSiO_2$	30

Figure 2. Release of 5-FU from commercially available product and unmodified and amino modified MSNs in PBS solution.

The release experiments were performed at 37 °C in PBS solution with a pH of 7.4. All samples show a quicker release in the beginning and a slower release in the late releasing process. However, as clearly observed, 97 % of 5-FU is released from the commercially available product in only 3h, while delayed release profiles are registered for newly developed therapeutic nanosystems. Particularly, only 42 % of 5-FU was released by amino-modified MSN in the first 3h. Then, the release of 5-FU is significantly slowed down and the Amino-Fe_3O_4@$mSiO_2$ sample released 54 % of its cargo in 24 h. On the other hand, the release of 5-FU is even more delayed from unmodified MSN, reaching a total release of 30% after 24 h. The slower release property of these new therapeutic nanosystems is not only related to the difference in concentration in each sample, but rather to the nature of interactions between the drug and its carrier.

In the amino-modified therapeutic nanosystems, the interaction drug - carrier is quite strong hydrogen bonding between the atoms (O, N and F) with strong negativity in 5-FU and the hydrogen of – NH_2 groups present on the mesopore surface [1]. Thus delayed release profile is obtained. In unmodified magnetic therapeutic nanosystems (Fe_3O_4@$mSiO_2$) apparently, the much stronger electrostatic attraction between the silanolate groups ($\equiv$Si-O^-) present on the silica surface at this pH values and positively charged 5-FU molecules, prevents the easy-release 5-FU, dictating its significantly delayed release.

In conclusion, mesoporous silica nanoparticles with magnetic properties were prepared successfully by a facile ultrasound-assisted method implying coating of 20 nm magnetite nanoparticles with mesoporous silica layer. The formation of the silica layer is controlled by careful tune of experimental parameters to control the competition between nucleation and growth of mesoporous silica.

The use of ultrasonic irradiation made possible the formation of well-defined core-shell nanoparticles particle in only 1.5 h due to the remarkable reaction conditions which lead to unique properties. The applicability of prepared magnetic mesoporous silica nanoparticles in drug storage and modified delivery systems was demonstrated.

3. NANOENCAPSULATION OF Fe_2O_3 ON MESOPOROUS SILICA FOR ULTRASOUND-ASSISTED ENVIRONMENT REMEDIATION

With the increase in the public awareness and concern about environmental pollutants, novel, stable, cheap, non-toxic and high catalytic activity treatment methods are in demand for the detoxification and degradation of harmful organic compounds in both aqueous and gaseous media. Since their discovery in 1992 [36, 37] silica-based mesoporous molecular sieves were intensely exploited for specific applications related to environment pollution control [38-43].

In this section we report recent researches on nano-manipulation of silica-based mesoporous molecular sieves towards their efficient applicability in innovative environmental remediation processes. Specifically, we studied the stabilization of magnetic nanoparticles within the channels of ordered mesoporous molecular sieve silica MCM-41 in order to obtain nanocomposites catalytically active in heterogeneous Fenton-like processes.

Thanks to the uniform and well-defined pores, MCM-41 mesoporous silica provides an ideal environment for controlled dispersion of nanoparticles, achieving size growth and morphology control. Among the iron oxides, hematite (α-Fe_2O_3) is the most stable under ambient conditions and is often the end product of the transformation of other oxides. The stabilization of hematite nanoparticles within the channels of MCM-41 ordered mesoporous silica leads to nanostructured magnetic materials with unique and reproducible properties, allowing the study of theoretical and technological issues regarding potential applications in various fields such as information storage, sensors, medical imaging and controlled drug release [44, 45].

Successful stabilization of hematite nanoparticles initially involves modifying mesoporous MCM-41 silica surface with active organic functional groups to facilitate encapsulation. Fuctionalization of mesoporous silica MCM-41 and hematite stabilization within its channels can be achieved either *in situ*, either by post-synthesis treatment. The *in situ* nanoencapsulation method (co-condensation) maybe applied for obtaining the organic-inorganic precursor of the α-Fe_2O_3/MCM-41 nanocomposites. Unfortunately, parts of the iron species may be incorporated into the MCM-41 mesoporous silica framework by isomorphous substitution [10], diminishing significantly the applicability of the materials. This incorporation is avoided by using a post-synthesis wet impregnation method for the controlled dispersion of α-Fe_2O_3 nanoparticles within the pores of MCM-4 [46].

The phenomenon of acoustic cavitation provides unique local conditions of extremely high temperature and pressure (hot-spot temperature of ~5000 K and a pressure of ~1000 bar [47]), inducing the formation of hydroxyl radical (HO•) and hydrogen atoms [48]. Hydrophilic non-volatile pollutants can be degraded by (HO•) at the bubbles surface or in the solution bulk, but the efficiency of degradation process is limited due to the high stability of these pollutants [49]. The improvement of degradation rate of hazardous organic pollutants may be achieved by proper coupling ultrasonic irradiation with other oxidation processes such as photocatalysis, Fenton-type processes or ozonization. Promising results were obtained recently for degradation of phenol [50] and methylene blue [48].

We targeted the preparation and testing of α-Fe_2O_3/MCM-41 nanocomposite catalysts in ultrasound-assisted Fenton-like heterogeneous processes for remediation of wastewaters containing non-volatile resistant organic compounds, avoiding homogeneous Fenton, photo-

Fenton and photocatalytic processes. Methylene blue (MB) was selected as probe molecule in this study, as it is a highly stable, frequent, nonvolatile organic pollutant in textile wastewaters.

Pure ordered MCM-41 mesoporous silica supports were prepared by ultrasound-assisted synthesis and by conventional method. Both methods are described in details in our previous publications [10, 11]. The samples are encoded as C for MCM-41 prepared by conventional method and U for MCM-41 prepared under ultrasonic irradiation. Modification of pure silica samples with amino groups was carried out post-synthesis as described earlier [10, 11]. Samples are encoded as CF and UF for amino-modified MCM-41 prepared by conventional method, respectively by sonochemical method. To disperse iron oxide inside the pores, the amino-modified MCM-41 supports were impregnated with an aqueous solution containing $Fe(NO_3)_3 \cdot 6H_2O$ at room temperature for 4 h, under vigorous stirring, followed by solvent evaporation to dryness. The dried powder were calcined at 550ºC for 6 h. Samples are encoded as H-CF, respectively H-UF. The evaluation of catalytic activity of α-Fe_2O_3/MCM-41 nanocomposites towards degradation of methylene blue (MB) by an ultrasound-assisted heterogeneous Fenton-like process was carried out under controlled temperature of $20 \pm 2°C$, using a 0,03 mM aqueous solution of MB, at pH = 2-5 (adjusted using NaOH), under pulsed ultrasound cycle (4s on/1s off). Monitorization of discoloration process was followed by means of UV-vis spectroscopy, measuring the absorbance of dye solution at 665 nm.

Understanding the structural, textural, morphological and magnetic properties of a catalyst is a point of major importance for the evaluation of its performances in specific applications.

Therefore, complete characterization of unmodified MCM-41, amino-modified MCM-41 and α-Fe_2O_3/MCM-41 samples was realized and using modern techniques in order to assess their features. Results on unmodified and amino-modified MCM-41, before the dispersion of iron oxide nanoparticles are presented in details in our previous works.

Table 3 summarizes the structural and textural features as well as magnetic behavior showing that amino-functionalization of silica MCM-41 leads to lower interplanar distances, pore narrowing down to the microporous range and to significant decrease of the surface area and pore volume. These findings are in agreement with results previously reported in the literature [51, 52].

Large surface areas and pore volumes are obtained for the final nanocomposite materials, suggesting that, after calcination, the organic moieties are removed leaving space for the iron oxide nanoparticles.

The presence of iron oxide nanoparticles inside the mesopores of MCM-41 is indicated by the low pore volumes in spite of high surface area of these samples.

Overall, decreases in surface areas, pore volumes and pore diameters are obvious on α-Fe_2O_3/ MCM-41 nanocomposites, compared with unmodified MCM-41 supports. N_2-sorption isotherms registered for MCM-41 and α-Fe_2O_3/ MCM-41 materials are of type IV in IUPAC classification, characteristic to MCM-41-like mesoporous materials. After nanoencapsulation of α-Fe_2O_3 nanoparticles, the isotherms show the same allure, indicating that the ordered mesoporous structure is maintained.

However, the inflection characteristic to capillary condensation in the mesopores is slightly visible for nanocomposite material based on MCM-41 support prepared by conventional synthesis.

Table 3. Structural and textural calculated characteristics of prepared α-Fe_2O_3/ MCM-41 nanocomposites

Sample	Porous Features (N_2-sorption)				Structural Features (XRD)			Magnetic behavior
	S_{BET}[a] (m^2/g)	V_p[b] (cm^3/g)	(D_p[c] (nm)	W_t[d] (nm)	d_{100}[e] (nm)	a_0[f] (nm)	Phase identification	
C	1601	1.129	2.24	1.77	3.48	4.01	MCM-41	Diamagnetic
U	1699	1.086	2.16	1.74	3.38	3.9	MCM-41	Diamagnetic
CF	106	0.131	0.98	2.92	3.38	3.9	MCM-41	Diamagnetic
UF	232	0.449	1.00	2.93	3.41	3.93	MCM-41	Diamagnetic
H-CF	1042	0.555	1.86	2.17	3.49	4.03	α-Fe_2O_3; MCM-41	Weak ferromagnetic
H-UF	1456	0.788	1.99	2.92	3.39	3.91	α-Fe_2O_3; MCM-41	Weak ferromagnetic

[a]BET surface area, calculated from the linear part of the BET plot; [b]total pore volume, taken from the volume of N_2 adsorbed at $p/p_0 = 0.95$; [c]average pore diameter, estimated using the desorption branch of the isotherm and the Barrett-Joyner-Halenda(BJH) method; [d]wall pore thickness estimated from difference ($a_0 - D_p$);

[e]d-spacing of (100) reflection; [f]unit cell constant, $a_0=(2d_{100})/\sqrt{3}$.

This indicates more pronounced filling of the mesopores with iron oxide nanoparticles or loss of ordering. Also the pore size distribution for all materials is narrow, suggesting that porosity was not affected by dispersion of metal oxide nanoparticles.

Structural characterization of α-Fe_2O_3/ MCM-41 nanocomposites using XRD analysis focused on identification of iron oxide crystalline phases (wide angle XRD), as well as on the effect of nanoparticles presence on the structure of MCM-41 support (low angle XRD). The registered low angle XRD patterns for all samples (Figure 3) show that diffraction peaks, characteristic to ordered hexagonal lattice like MCM-41.

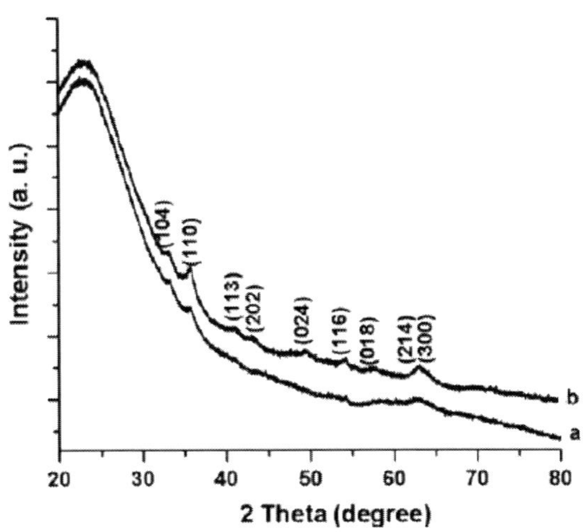

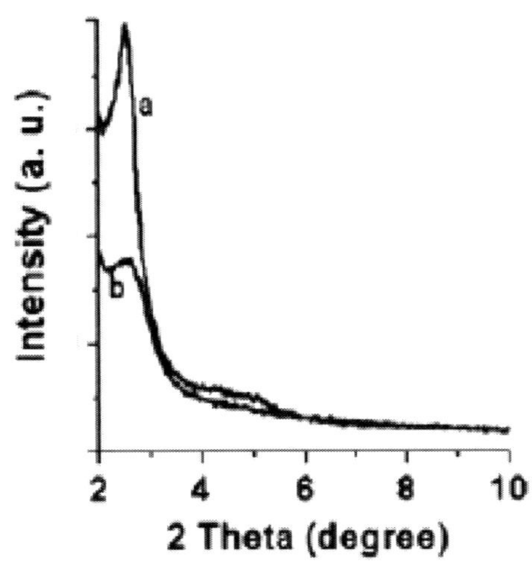

Figure 3. Wide – angle and small – angle XRD patterns of α-Fe_2O_3/ MCM-41 nanocomposites: a: H-UF sample; b: H-CF sample [11].

This diffraction peak is sharper for the nanocomposite material prepared using MCM-41 support prepared in sonochemical conditions, suggesting that the hexagonal ordering of the mesoporous channels is better maintained in this case. Also, a decrease in peak intensity attributed to diminishing of the electron density contrast between the silica wall and mesopore, due to the location of iron oxides particles inside mesopores of silica matrix [53] is observed.

The wide-angle XRD patterns of prepared nanocomposites show diffraction peaks characteristic to hematite, spatial group *R3c* ((167), Nr. Card 72-0469) coresponding to (104), (110), (113), (202), (024), (116), (018) and (214) reflections. The broad diffraction peak observed at 22° is characteristic amorphous silica.

Magnetic properties of α-Fe_2O_3/MCM-41 nanocomposites were investigated by VSM technique, revealing weak ferromagnetic behavior for α-Fe_2O_3/MCM-41 nanocomposite samples show [11] and predictable diamagnetic behavior of pure siliceous supports.

Further, the research study aimed to test the prepared nanocomposites towards the removal of methylene blue (MB) from aqueous solutions by an ultrasound-assisted heterogeneous Fenton-like process. The results of the degradation tests reveal that after 270 minutes, only 17% of the initial amount of MB was degraded in the presence of H_2O_2, under the influence of ultrasounds and in the absence of any catalyst (Figure 4) Also, in the absence of ultrasounds, MB adsorption occurs only at the surface of nanocomposite which colors in green.

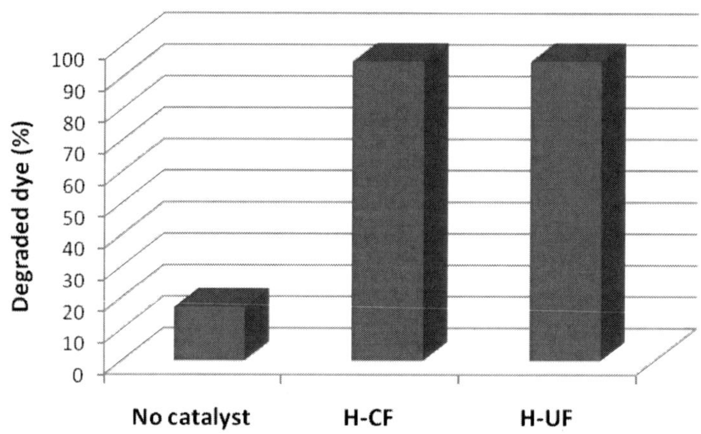

Figure 4. Degradation of MB by ultrasound-assisted heterogeneous Fenton-like process.

The catalytic behavior of nanocomposite samples, containing α-Fe_2O_3 on MCM-41 support prepared by the two methods, towards the US-assisted degradation of MB is similar with some small differences. Specifically, although after 270 min the two catalysts (H-CF and H-UF) contribute to the degradation of similar amounts of MB (95% from initial MB disappears after 90 minutes) the process is slightly faster when using H-UF catalyst. Secondly, a significant decrease of the absorbance at λ = 200 nm (attributed to H_2O_2) is observed after 30 minutes of sonication when H-UF catalyst, was used, in close comparison with H-CF catalyst. This significant decrease in absorbance at 200 nm indicates a higher consumption of H_2O_2 [11].

The obtained results indicate that the α-Fe_2O_3/MCM-41 nanocomposites exhibit high catalytic activity towards the degradation of methylene blue in aqueous solutions. Under the unique conditions created by the cavitation phenomenon, α-Fe_2O_3 nanoparticles acts as a

Fenton-like catalyst, leading to the degradation of methylene blue, with all the advantages of a heterogeneous catalytic process.

The study demonstrated that mesoporous silica molecular sieves MCM-41 are ideal materials for stabilization of hematite nanoparticles with narrow size distribution. The uniform pore system with tunable molecular sizes constitutes an appropriate confinement space for controlled dispersion of hematite nanoparticles, facilitating the development of materials with predefined properties and target applications. Nanoencapsulation of α-Fe_2O_3 within the channels of MCM-41 has been shown to imprint a new function, with highly efficient potential application in a simple ultrasound-assisted heterogeneous catalytic process for wastewater remediation.

4. Photocatalytically Active Silica-Based Mesoporous Molecular Sieves for Wastewater Treatment

In previous section was shown that with proper functionalities, mesoporous silica is highly efficient for removal of organic dyes under ultrasonic irradiation, opening new possibilities towards the applications of silica-based materials. However, photocatalysis is one of the most efficient processes used for wastewaters purification. The main advantage of photocatalysis is that leads to the complete mineralization of organic pollutants [54-56]. Due to their wonderful properties, mesoporous silica materials are very good candidates to be used in photocatalytic remediation of wastewaters when properly modified with photocatalytically active entities. Titanium dioxide is the best representative of the photocatalysts class due to its high quantum yield, chemical stability, low hazard to human body and easy processability [57, 58]. Thus, huge efforts are made by scientists worldwide to develop effective photocatalytic materials based on titania and mesoporous silica [12, 13, 15, 16, 23, 59-62]. It is the scope of this section to contribute to these efforts and bring some light on the parameters influencing the design of a titania-silica photocatalyst and the effect of its structural features towards its effectiveness for removal of organic dyes. Therefore, preparation, characterization and photocatalytic behavior of TiO_2/MCM-41 and SBA-15 materials are described.

We followed two main research lines: firstly, we focussed on understanding the influence of the structure modification on the photocatalytic behaviour of the TiO_2/SiO_2 supported oxides. The effect of the preparation conditions on the structural modification of the support structure, and implicitly on photocatalytic behaviour was investigated. Secondly, we studied the role of the silica support on titania nanostructuring for specific applications in adsorption and photocatalysis. The pore architecture of the silica support plays an essential role on the dispersing degree, final loading and growth of TiO_2 particles. Therefore, two different porous silica supports, specifically mesoporous MCM-41 and SBA-15 were used for the immobilization of titania nanoparticles. This section focuses on understanding the effect of the temperature modification in various stages of composite photocatalyst preparation on the phase formation and growth of the titania nanoparticle. Also, the efficacy of these photocatalysts in the removal of cationic organic dyes from aqueous solutions is given in details in terms of adsorption capacity and photocatalytic performance.

Titania nanoparticles were deposited onto mesoporous silica support (MS) by simply filling with colloidal dispersion prepared by acid-catalyzed sol-gel process (ACSG) [63-65] using Ti-isopropoxide ($Ti^{i}OPr$) in the presence of HNO_3 (1M). The procedure step-by-step, characterization methods and the experimental setup for photocatalytic tests are described in details elsewhere [12, 13, 14].

Herein we show interesting results obtained by changing the temperature in different stages of TiO_2/MCM-41 and TiO_2/SBA-15 preparation: during the formation of the titania sol (before the addition of the MS support) and during the incorporation of the colloidal suspension in the support, from room temperature (RT) to 70°C. The samples are encoded TiO_2/MCM (*y/z*), respectively TiO_2/SBA (*y/z*), where *y* represents the formation temperature of the titania sol and *z* represents the deposition temperature.

Table 4 summarizes the structural characteristics of the TiO_2-activated MCM-41 and SBA-15 materials prepared at different temperatures. Overall, when the temperature increases from RT to 70°C, lower TiO_2 loadings on the final catalyst are obtained. This phenomenon is more pronounced when the temperature is modified during the formation and growth of titania nanoparticles. Also, by comparison can be observed that the calculated amount of titania loaded of MCM-41 is smaller when the deposition temperature is modified from RT to 70°C, thus the lowest loading is obtained for the sample TiO_2/MCM (70/70). This can be explained by the formation of larger nanoparticles at higher temperatures. Moreover, higher loading are registered when MCM-41 is used as support, in comparison with SBA-15. Powder XRD and Raman spectra show that only anatase is present in the samples, when TiO_2 nanoparticles are formed at room temperature, while rutile and brookite are formed at more elevated temparatures.

Table 4. Structural features of the TiO_2/MCM-41 and TiO_2/SBA-15 supported oxides prepared by changing the temperature in different stages of preparation

Sample	TiO_2 loading (wt %)	S_{BET} (m^2/g)	V_{tot} (cm^3/g)	V_{μ} cm^3/g	V_{open} cm^3/g	V_{plug} cm^3/g	Crystalline form of TiO_2
TiO_2/MCM (RT/RT)	3.54	407	0.39	-	-	-	A
TiO_2/MCM (RT/70)	2.25	529	0.5	-	-	-	A
TiO_2/MCM (70/RT)	1.43	382	0.47	-	-	-	A+R+B
TiO_2/MCM (70/70)	1.14	452	0.49	-	-	-	A+R+B
TiO_2/SBA (RT/RT)	2.75	366	0.38	0	0.09	0.29	A
TiO_2/SBA (RT/70)	2.02	371	0.41	0	0.11	0.29	A
TiO_2/SBA (70/RT)	1.85	269	0.39	0	0.39	0	A+R+B
TiO_2/SBA (70/70)	1.82	460	0.56	0.04	0.51	0	A+R+B
TiO_2*		166	0.1	0.00	0.11	0.00	A

* refers to titania nanoparticles formed by ACSG method; A = anatase; R = rutile; B = brookite.

N_2-sorption analysis was used to investigate the modifications occurred in terms of porosity and textural properties of prepared supported oxides. The recorded isotherms and pore size distributions are shown in figure 5. All materials show type IV isotherms in IUPAC classification with a steep inflection characteristic to mesoporous materials. The sharpness of this inflection is related to the pore size homogeneity, expressed in terms of narrow pore size distributions and its position is determined by the characteristic pore sizes. As it is clearly observed, the inflection step is situated at higher relative pressures for SBA-15 ($P/P_0 = 0.7$)

than for MCM-41 (P/P_0 = 0.35), indicating larger pores. Moreover, for the mesoporous materials with pores larger than 4 nm, a steep hysteresis loop occurs. This is the case of SBA-15 with pores of about 7 nm (Figure 5). Specifically similar one-step capillary condensation isotherms are observed for the samples TiO_2/MCM-41 (RT/RT) and TiO_2/MCM-41 (RT/70) and a hysteresis loop on the horizontal part, indicating formation of slit-shaped mesopores due to deposition of titania within cylindrical channels (Figure 5A). The pore size distribution diagram indicates that MCM-41 support possess homogeneous pore sizes of ~ 3.4 nm.

When TiO_2 nanoparticles are immobilized onto SBA-15 mesoporous silica, the N_2-sorption analysis suggests that the growth of these nanoparticles is controlled by their confinement within the pores. N_2-sorption measurements revealed similar isotherm for the samples TiO_2/SBA-15 (RT/RT) and TiO_2/SBA-15 (RT/70) indicating important pore-blocking effects (Figure 5B curves e), f)). It can clearly be observed that the presence of TiO_2 into SBA-15 leads to a marked change in the shape of the hysterezis loop and a second step on desorption branch appears, indicating the existence of both open and plugged mesopores. This two-step desorption is similar to that of the PHTS materials (Plugged Hexagonal Templated Silica) with amorphous microporous silica nanoparticles (plugs) in the mesopores [66-68]. Therefore, one can conclude that titania nanoparticles have been dispersed inside of the channels of SBA-15 (with an average pore diameter of 7.3 nm) thereby narrowing parts of the mesopores.

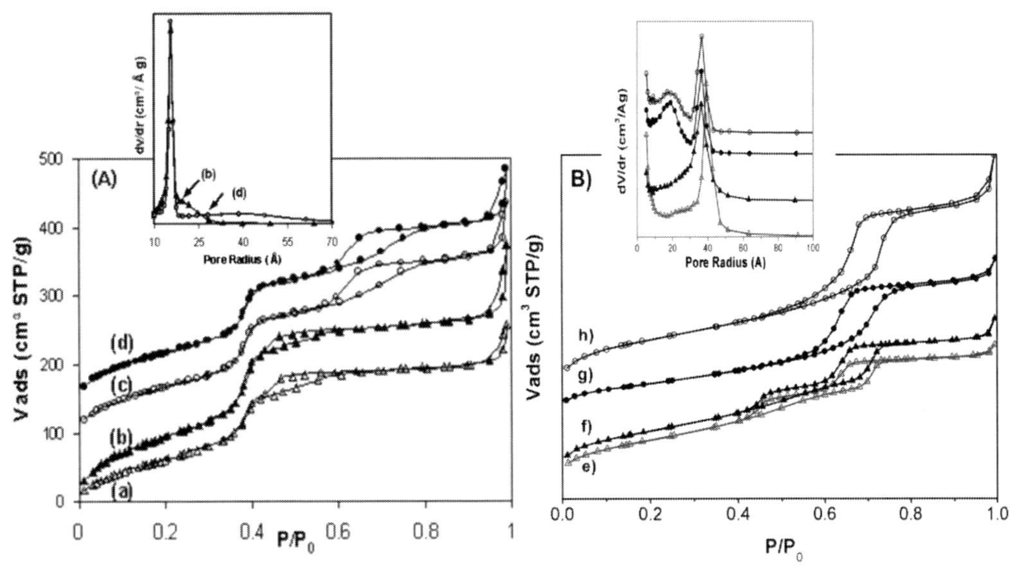

Figure 5. N_2-sorption isotherms and PSD curves for: (A) (a) TiO_2/MCM-41 (RT/RT); (b) TiO_2/MCM-41 (RT/70); (c) TiO_2/MCM-41 (70/RT); (d) TiO_2/MCM-41 (70/70); (B) (e) TiO_2/SBA (RT/RT); (f) TiO_2/SBA (RT/70); (g) TiO_2/SBA (70/RT); (h) TiO_2/SBA (70/70).

In contrast to PHTS materials, the plugs are now composed of photocatalytically active sites TiO_2. Nitrogen that is adsorbed between two of these nanoparticles can only be desorbed through the narrowed sections of the pores. Therefore, desorption is postponed to lower p/p_0 values resulting in a second desorption step. At room temperature, titania nanoparticles with an average size of 4.5 nm are formed, as was determined by XRD measurements. Part of these nanoparticles enter inside the mesopores when SBA-15 is added to the turbid solution and will narrow the pores to such an extent that nitrogen will only be desorbed from the plugged sections at p/p_0 0.42. The diffusion of these nanoparticles to the active sites where they can bond to the silica framework is improved by increasing the deposition temperature to 70°C, resulting in higher surface area and pore volume.

However, when formation temperature of titania nanocrystals was changed from RT to 70°C, major modifications were observed. On one hand, the N_2 sorption measurements show similar isotherms with two adsorption and two desorption steps for the samples TiO_2/MCM-41 (70/RT) and TiO_2/MCM (70/70), suggesting the existence of two different pore systems. The first one it is related to the small intra-particle regular mesopores organized in a hexagonal array (MCM-41-like mesopores). The second one consists of large inter-particles mesopores, which correspond to the voids between the large titania nanoparticles and MCM-41 particles (cage-like mesopores) [69]. The sizes of the titania nanoparticles formed were of about 6.5 nm, as determined by XRD. These nanoparticles are too big to enter inside the mesopores of SBA-15. Therefore, these particles will be randomly linked on the external surface resulting in low loadings of titania since large amounts on the external surface were removed due to the washing procedure. The N_2 sorption measurements confirm this hypothesis, since no plugs are formed inside the mesopores of the SBA-15, which can be explained by a faster growth of the titania crystals. Moreover these materials exhibits very low surface areas and pore volumes indicating that there are large nanoparticles blocking completely the entrance of the pore.

Powder XRD and Raman measurements revealed the coexistence of anatase, rutile and brookite crystalline phases in the cases where the temperature was set at 70 °C during the titania nanoparticles formation and growth, regarding the mesoporous support used. XRD patterns and Raman spectra may be viewed in our previous publications [12, 13]. The size of the titania nanoparticles formed was of 6.5 nm, larger than those obtained at room temperature. Due to the increased temperature, the small anatase crystallite seeds experienced a faster growth, resulting in larger nanoparticles. This transformation A to R, accompanied by rapid growth of the grains, was previously reported in the literature [70]. This has been attributed by the authors, to that the rutile grain nucleate and grow on anatase particle and the processes are accompanied by interparticle sintering.

As seen, the modification of the temperature in different stages of the sample preparation has a significant influence on the structural characteristics of the final catalyst and the TiO_2 crystalline phase. Further, it was investigated the overall capacity of the prepared TiO_2/MCM-41 and TiO_2/SBA-15 photocatalysts to remove cationic dyes from simulated contaminated water, based on the registered structural modifications. Figure 6 shows the catalytic performances of prepared composite photocatalysts towards removalof Rhodamine 6G in aqueous solutions. The adsorption capacities of all TiO_2/MCM-41 samples are significantly higher than those of TiO_2/SBA-15 and pure TiO_2 prepared by ACSG method. TiO_2/MCM (RT/70) shows the lowest adsorption capacity towards R6G from aqueous solution, in spite of its higher surface area and pore volume. This is due to the size of the adsorbed molecule: R6G is a bulk organic molecule and it is adsorbed on the surface or the open pores of the catalysts which are not blocked by TiO_2 nanoparticles. In general it is observed that the samples in which the temperature used during the formation of the titania sol was modified from RT to 70°C show slightly increased adsorption capacities towards R6G. At the same time, the adsorption capacities decrease in the case of the samples in which the formation temperature was maintained constant and the immobilisation temperature was modified to 70°C. Good activities in the photo-oxidation of R6G reaction is observed for all samples. The photocatalytic activity of prepared TiO_2/MCM-41 composites follows the same trend as their adsorption capacity.

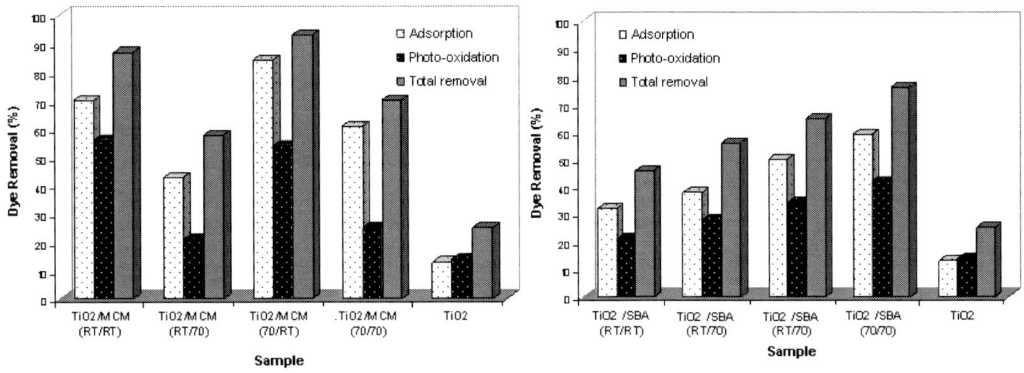

Figure 6. Catalytic performances of pure TiO_2/MCM-41 (left) and TiO_2/SBA-15 (right) supported catalysts obtained by changing the temperature in different stages of sample preparation.

For TiO_2/SBA-15 materials, the adsorption capacities of the samples TiO_2/SBA (70/RT) and TiO_2/SBA (70/70) are high since no plugs are formed in the mesopores resulting in the completely open mesopores as shown in figure 6 right. Since the titania nanoparticles in these samples are located only on the outer surface, the accessibility of the R6G molecules to the active sites is higher, resulting in high photocatalytic activity. It is possible that the co-existence of anatase, rutile and brookite phases in these samples is responsible for the increased photocatalytic activity. Higher adsorption capacities and photocatalytic activities were observed in the case of the samples in which the deposition of titania nanoparticles onto SBA-15 support took place at higher temperatures independent of the formation temperature. This higher adsorption capacity is due to the slightly higher open pores volumes, surface areas and total pore volumes.

Through this study is demonstrated that the TiO_2 loading on the final material is just one of the main factors governing the efficiency of a composite photocatalyst towards removal of organic dyes. Accessibility of R6G molecules to the photocatalytically active sites is another key factor governing the photocatalytic behaviour of TiO_2/SiO_2 supported oxides. It is influenced mainly by the location of titania nanoparticles on the inner or outer surface of the support channels. In the case of MCM-41 supported oxides, most of titania nanoparticles are located on the outer surface. Therefore these catalysts show the best catalytic activity. When SBA-15 is used as support, part of the formed nanoparticles are located inside the channels and the access of the R6G molecules to the active sites is limited, resulting in lower photocatalytic activity and total removal capacity in comparison with the MCM-41 titania supported materials. However, the accessibility of R6G molecules to the active sites is determined also by the size of the nanoparticles. This is well reflected in the case of TiO_2/SBA-15 materials. The titania nanoparticles, act as plugs inside the nanotubular channels of SBA-15 resulting in a new type of PHTS materials, where the plugs are now composed of photocatalytically active sites. On the other hand, once few nanoparticles are located at the pore mouth, the R6G molecules cannot pass through and their access to the active sites located inside the channels is forbidden. Thus, the total capacity of a TiO_2-

activated PHTS-like material to remove R6G from aqueous solution decreases with the growth of the titania nanoparticles inside the mesopores

Furthermore, by careful modification of temperature during the formation of titania nanoparticles mixed phases was observed. The coexistence of anatase, rutile and brookite crystalline phases was identified by both XRD and Raman measurements.

5. Changing the Properties of Carbon-Based Mesoporous Molecular Sieves with Metal Oxides for Photocatalytic Environment Remediation

As well known, carbon is the second most abundant element on the Earth and is one of the most important for all living organisms, as most of the organic compounds are made out of it. Due to the ability of carbon atoms to bond with each other, carbon-based materials are exhibit fascinating properties and have a variety of forms.

In the last decades, carbon materials have revolutionized the field of material science and engineering. More specific, carbon nanomaterials offer a wide range of excellent mechanical, electrical and thermal properties, making them very interesting materials to a broad range of applications.

In one way or another, carbon-based materials touch every aspect of our daily lives. For example, such materials are the most used for water and air purification, in tire reinforcing, for manufacturing ultra-light graphite sporting goods and aircraft brakes, or as fire retardant insulators. Carbon molecular sieves are made out of coal and are specialized in sorption processes. They have a predominantly amorphous phase with an extraordinarily large internal surface area and pore volume. Nowadays, it is still widely used in different fields being produced as a powder, granule, pellet, extrudate, or block for application in different fields [71].

As well known, heterogeneous photocatalysis using metals oxides is widely recognized as an effective technology for treating waters containing some refractory organic compounds. Generally, in photocatalytic reactions, the photocatalyst is suspended in a liquid phase using slurry reactors. For practical reasons, there are some usual shortcomings of the process, namely the recovery of the photocatalytic fine particles and effective absorption of UV-vis radiation. In order to facilitate the use of photocatalytic nanoparticles, immobilization on a suitable substrate is desired.

Recently, carbon-based materials have been used as ideal support materials for the deposition metal oxide nanoparticles for photocatalytic environment remediation [72, 73]. A broad range of applications for these innovative materials with different desirable characteristics have been reported, such as energy conversion/fuel storage, catalysis, biotechnology, optoelectronic nanodevices, etc [74]. Discovering the photoinduced charge transfer between the photocatalyst acting as donor and carbon-based mesoporous materials as acceptor, Peng and co-workers used successfully these type of molecular sieves as photochemical energy conversion systems with the aim of increasing photoresponse of the photocatalysts [73].

Therefore, the stability, surface chemistry, and textural properties of carbon-based materials play an important role in photocatalyst supporting and could enhance their photocatalytic activity of by increasing its effective surface area and surface properties. Actually, the photocatalytic activity of metal oxide nanoparticles is limited by the very fast electron-hole recombination that decreases the photocatalyst efficiency. Therefore, the electronic properties of the carbon-based material support will help to improve the efficiency of the photocatalytic species, by stabilizing the charge separation and will reduce the charge recombination process.

Our research focuses on understanding the unique role that plays the properties of carbon-based mesoporous molecular sieves for the photocatalytic activity improvement. On a more applied level we used mesoporous carbon molecular sieves as support for metal oxide nanoparticles exhibiting photocatalytic activity. Titanium oxide was chosen as UV-responsive photocatalyst, while bismuth oxide was chosen as Vis-responsive species in order to tune their photocatalityc properties, thus changing the properties of carbon-based mesoporous molecular sieves.

Highly dispersed titanium oxide nanoparticles have been deposited on mesoporous carbon structure by ultrasound assisted impregnation method followed by calcination in a N_2-flow at 400 °C. The detailed preparation method is presented in [19, 20].

The supported titanium oxide nanoparticles on carbon-based mesoporous molecular sieve, exhibited an enhanced photocatalytic activity for the phenol photodegradation process [19, 20]. In this case, the improved activity was attributed to the reversible adsorption and concentration of phenol molecules on the hydrophobic sites of the carbon molecular sieve support. The results showed that the mesoporous carbon molecular sieve improves the photocatalytic activity of titania nanoparticles, and its outstanding performance (figure 7).

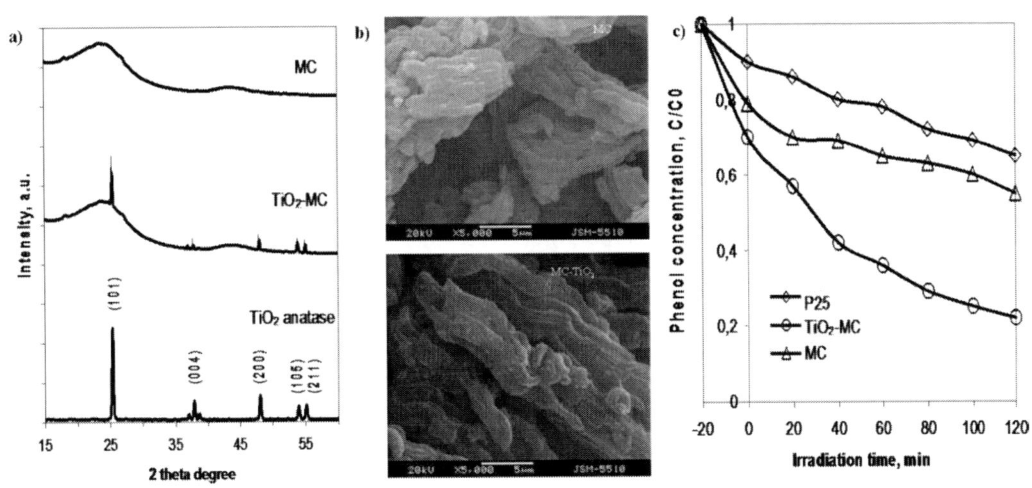

Figure 7. (a) XRD patterns, (b) SEM images and comparative photocatalytic degradation of phenol in the presence of MC, MC-TiO_2 and TiO_2 (Degussa P25) samples [20].

In another context, when the microwaves have been used for the calcination process, and was observed that under microwave treatment textural properties of the composite sample are improved (higher surface area and total pore volume) and smaller crystallites are obtained compare to that of calcined samples in a conventional oven. This way, we demonstrate that the microwave-assisted synthesis of TiO_2-supported mesoporous carbon molecular sieve lead to the improvement of its textural properties and photocatalytic activity, as could be observed from the figure 8. The enhanced photocatalytic activity could be assigned to the porosity of the mesoporous carbon materials, as well as the graphitic structures present in their structures (the presence of graphitized carbon structures have been determined by Raman spectroscopy, not shown here). The composites obtained by this way have been tested for methylene blue photodegradation, and the results showed that by microwave treatment and more efficient photocatalyst could be obtained.

a)

	P25	Furnace	MW
S_{BET}, m^2/g	109	172	357
V_t, cc/g	0.16	0.29	0.37
d_{pore}, nm	5.0	3.5 5.6	3.5

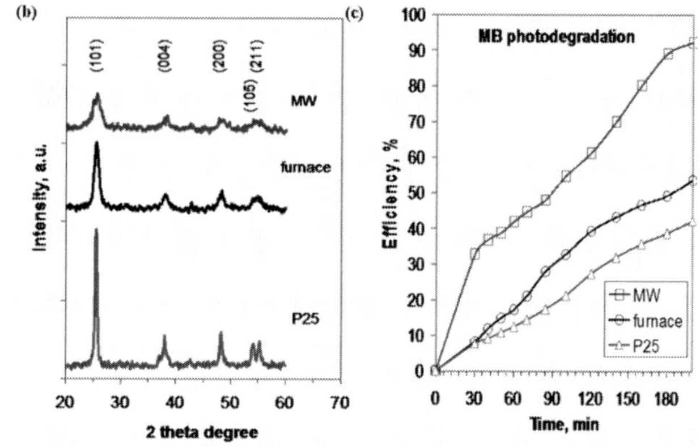

Figure 8. (a) Textural properties and crystallite sizes (b) XRD patterns, and (c) Efficiencies of TiO_2 (Degussa P25) and TiO_2-coated mesoporous carbon materials calcined in a conventional furnace and microwave treated samples.

Our studies were aimed not only improve the photocatalytic activity of titanium oxide nanoparticles, but also we tried to support on mesoporous carbon matrix bismuth oxide nanoparticles. These nanocomposites have been prepared by an ultrasound-assisted method and were characterized by various techniques that showed the composites have very good structural, textural and morphological features. The photocatalytic experiments showed that the bismuth oxide nanoparticles supported on carbon-based mesoporous molecular sieve exhibit an increased photoresponse, mainly due to the presence of graphitic carbon structures which are able to promote a rapid photoinduced charge separation and a slow charge recombination, by accepting the photogenerated electrons from photocatalytic nanoparticles. But also, again the influence of the porosity of carbon structure has been observed (figure 9).

Considering the above experiments and results, it was found that the resulting carbon-based molecular sieves exhibit newer property. This property is given by the photocatalytic nanoparticles supported on carbon structure. It was observed that carbon-based mesoporous molecular sieve increases significantly photocatalytic activity of the photocatalysts on degradation of various organic pollutants (as phenols, dyes) in water under UV light irradiation.

Due to their high surface area, porosity, structural and morphological features and relatively easy synthesis carbon-based molecular sieves have been successfully applied as catalyst support, changed their properties. It was demonstrated that these materials are useful for photocatalytic environment remediation, contributing to the sorption of the organic pollutant on the photocatalyst surface and the electron transport during the photocatalytic reaction is enhanced.

a)

	P25	Furnace	MW
S_{BET}, m^2/g	109	172	357
V_t, cc/g	0.16	0.29	0.37
d_{pore}, nm	5.0	3.5 5.6	3.5

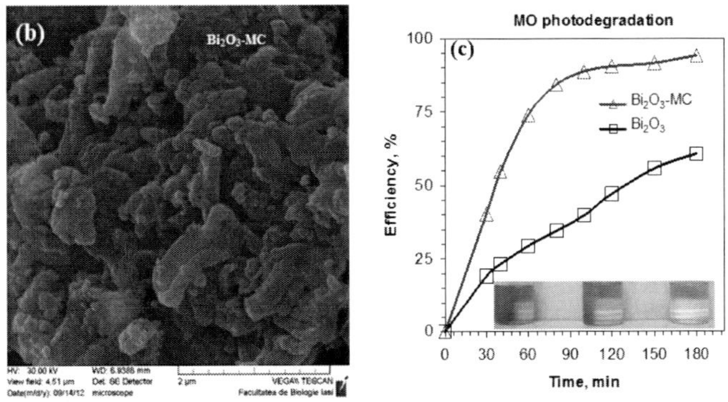

Figure 9. (a) Textural properties of the carbon-based mesoporous molecular sieve and its composite with Bi_2O_3 (b) SEM image of Bi_2O_3-MC composite, (c) Efficiency of Bi_2O_3-MC sample on methylene orange photodegradation compare to photocatalytic efficiency of Bi_2O_3 alone.

Actually, upon excitation of the photocatalytic nanoparticles, photogenerated electrons are injected into graphitic structures of carbon molecular sieves. Fine tuning the competition between electrons transferred from the excited photocatalyst to the collecting electrode surface (in this case carbon structures acts as an electrode) and the recombination probability of the electron–hole pairs, the resultant photoresponse is improved. Thus, the improvement of the interface morphology between the donor and acceptor of such nano-heterojunctions could be an important issue in developing also novel and versatile optoelectronic devices.

The photocatalytic activity of obtained carbon-based photocatalysts was evaluated by quantifying the degradation rate of organic pollutants in aqueous media under UV light. As mentioned, mesoporous carbon molecular sieves play an important role in increasing the concentration capacity of the organic molecules around the photocatalyst nanoparticles and the electronic structure of the photocatalyst is tuned. Therefore, the carbon-based composite photocatalysts appear to be interesting alternatives for dye bleaching and removal of phenolic compounds from waste waters. Because of the short and simplified synthetic procedure, the use of carbon-based photocatalysts, being an energy-saving product, can be easily scaled up for photocatalytic environment remediation at industrial level.

CONCLUSION

In this chapter were presented recent advances of our research group on nano-manipulation of mesoporous silica and carbon molecular sieves using metal oxides for specific applications in emerging fields. In this line of research it is highlighted the usefulness of mesoporous silica in drug storage, targeted delivery and sustained release by proper combination with superparamagnetic iron oxide nanoparticles. Also, it is presented the importance of using the right amount of magnetite nanoparticles in order to obtain magnetic mesoporous silica matrices proper for biomedical applications. Secondly, it is shown that mesoporous silica can find interesting application in environment remediation by ultrasounds, if it is properly modified with weak ferromagnetic moieties and specific organic functionalities. Further, interesting results on the role played by the pore architecture and synthesis temperature on nanostructuring of titania nanoparticles are presented in section 4 of this chapter. By combining marvelous mesoporous features of ordered silica materials with photocatalytically active nanoparticles advantages lead to great materials that can be successfully used for photodegradation of bulk organic pollutants in wastewaters. Fifth section relates the use of mesoporous carbon materials as support for TiO_2 and Bi_2O_3 nanoparticles, demonstrating that the carbon enhances the activity of metal oxide photocatalysts. The role of the carbon support is to suppress charge recombination and to facilitate a hydrophobic environment for the reactant concentration around the photoactive sites. Thus, the obtained experimental results support the theory of improvement the photocatalytic activity of photocatalysts, changing the properties of carbon-based mesoporous molecular sieves. In conclusion, the carbon-based photocatalyst is a new promising composite material for photocatalytic environment remediation.

Acknowledgments

Part of this work was financed by European Special Found POSDRU 2007-2013, COMMSCIE – POSDRU /89/1.5/S/63663 project. This work was also part of the NoE project "Inside Pores" and the GOA project funded by the Special Fund for Research of the University of Antwerp.

References

[1] Tomoiaga, A. M., Cioroiu, B. I., Nica, V., Vasile, A., *Coll. Surf. B: Biointerf.* 2013, 111, 52 – 59.

[2] Tomoiaga, A. M., Ochiuz L, Vasile A, *J. Nanotechnol. Diagn. Treat.* 2013, 1, 21 – 26.

[3] S. Huang, C. Li, Z. Cheng, Y. Fan, P.Yang, C. Zhang, K. Yang, J. Lin, *J. Coll. Interf. Sci.*, 2012, 376, 312 – 321.

[4] Kim J., Kim H. S., Lee N., Kim T., Kim H. et al. *Angew Chem Int Ed.*, 2008, 47, 8438-8441.

[5] Zhu Y. F., Ikoma T., Hanagata N., Kaskel S., *Small*, 2010, 6, 471 – 478.

[6] Martin-Saavedra F.M., Ruiz-Hernandez E., Bore A., Arcos D., Vallet-Regi M., Vilabora N., *Acta Biomater.* 2010, 6, 4522 – 4531.

[7] Vallet-Regí, M., *ISRN Materials Science*, vol. 2012, Article ID 608548, 20 pages, 2012. doi:10.5402/2012/608548.

[8] Y. Chen, Q. Zin, X. Ji, S. Zhang, H. Chen. Y. Zheng, Y. Sun, H. Qu, Z. Wang, Y. Li, X. Wang, K. Zhang, L. Zhang, J. Shi, *Biomaterials,* 2012, 33, 7126 – 7137.

[9] Alexa I.F., Păstrăvanu C.G., Ignat M., Popovici E., *Colloids and Surf B: Biointerf* 2013, 106, 135 – 139.

[10] Ursachi, I., Vasile, A., Ianculescu, A., Vasile, E., Stancu, A., *Mater. Chem. Phys.* 2011, 130, 1251–1259.

[11] Ursachi, I., Vasile, A., Stancu, A., *J. Coll. Interf. Sci.* 2012, 377, 184 – 190.

[12] Busuioc, A. M., Meynen, V., Beyers, E., Mertens, M., Cool, P., Bilba, N., Vansant, E.F. *Appl. Cat A: Gen.* 2006, 312, 153 – 164.

[13] Busuioc, A.M. *Micro- and Mesoporous oxide materials for removal of organic pollutants from wastewaters*, PhD thesis, University of Antwerpen, Belgium, 2007.

[14] Meynen, V., Busuioc, A. M., Beyers, E., Cool, P., Vansant, E. F., Bilba, N., Mertens, M., Lebedev, O. I., Van Tendeloo, G. in Buckely R. W. (ed.) Solid State Chemistry Research Trends, Nova Science Publishers Inc, New York, USA, 2007.
[15] de Witte K., Busuioc A. M., Meynen V., Mertens M., Bilba N., van Tendeloo G., Cool P., Vansant E.F., *Microp. Mesop. Mater.* 2008, 110, 100-110
[16] Beyers E., Biermans E., Ribbens S., de Witte K., Mertens M., Meynen V., Bals S., van Tendeloo G., Vansant E., Cool P., *Appl. Cat. B : Env.*, 2009, 88, 515 – 524.
[17] deWitte K., Meynen V., Mertens M., Lebedev O., van Tendeloo G., Sepúlveda-Escribano A., Rodríguez-Reinoso F., Vansant E. F., Cool P., *Appl. Cat. B : Env.* 2008, 125 – 132.
[18] Mihai G.D., Meynen V., Mertens M., Bilba N., Cool P., Vansant E.F., *J. Mater. Sci.* 2010, 45, 5786 – 5794.
[19] Ignat, M., Coromelci, C., Popovici, E., Rev. Chim. (Bucharest) 63, 2012.
[20] Ignat, M., Păstrăvanu, C., Popovici, E., *IEEE CAS 2010 Proceedings*, 2010, 2, 379.
[21] Coromelci C., Ignat M., Popovici E., Harabagiu V., *J. Haz. Mater.,* under review.
[22] Ignat M., Sacarescu L., Vasile A., Oxide semiconductors loaded mesoporous carbon with improved, *J. Chem. Sci.*, under review.
[23] Segura, Y., Cool, P., Kutrowski, P., Chmierlaz, L., Dziembaj. R., Vansant, E.F., *J. Phys. Chem. B*, 2005, 109, 12071 – 12079.
[24] Kuśtrowski P., Segura Y., Chmielarz L., Surman J., Dziembaj R., Cool P., Vansant E., *Cat. Today,* 2006, 114, 307 – 313.
[25] Hanu A.M., Liu S., Meynen V., Cool P., Popovici E.,Vansant E. Microp. Mesop. Mater., 2006, 95, 31 – 38.
[26] Chmielarz L., Kustrowski P., Dziembaj R., Cool P., Vansant E.. *Appl. Cat. B: Env.* 2006, 62, 369 – 380.
[27] Kongkanand, A.; Dominguez, R.M.; Kamat, P. V., *Nano Lett.,* 2007, *7*, 676 – 680.
[28] Lu, J., Liong, M., Li, Z.X., Zink, J. J., Tamanoi, F., *Small*, 2010, 6, 1794 – 1805.
[29] Vallet-Regi, M., Balas, F., Arcos, D., *Angew Chem. Int. Ed*, 2007, 46, 7548 – 7558.
[30] Slowing, I. I., Trewyn, B.G., Lin, V. S. Y., *J. Am. Chem. Soc.* 2007, 129, 8845 – 8849.
[31] Rosenholm, J. M., Linden, M., *J Control Rel.*, 2008, 128, 157 – 164.
[32] Rosenholm, J. M., Sahlgren C., Linden, M., *Nanoscale,* 2010, 10, 1870 – 1883.
[33] Chen Z., Li X., He H., Ren Z., Liu Y., Wang J., Li Z., Shen G., Han G., *Coll. Surf. B: Biointerf.*, 2012, 95, 274 – 278.
[34] Lv, B., Xu, Y., Wu, D., Sun, Y., *Mat Res Bull* 2009, 44, 961 – 965.
[35] Rouquerol, F., Rouquerol, J., Sing, K. *Adsorption by powders and porous solids*, Acad. Press, 1999.
[36] Kresge, C. T., Leonowicz, M. E., Roth, W. J., Vartuli, J.C., Becks, J. S., *Nature*, 1992, 359, 710.
[37] Beck, J. S., Vartuli, J. C., Roth, W. J., Leonowicz, M. E., Kresge, C. T., Schmidt, K. D., Chu, C.T.W., Olson, D.H., Sheppard, E.W., McCullen, S.B., Higgins, J.B., Schlenker, J. L., *J Am Chem Soc* 1992, 114, 10834.
[38] Nooney, R. I., Kalyanaraman, M., Kennedy, G., Maginn, E. J., *Langmuir*, 2001,17,528–533.
[39] Ho, K. Y., McKay, G., Yeung, K. L., *Langmuir,* 2003, 19, 3019 - 3024.
[40] Busuioc, A. M., de Mondt, R., Moisio, H., Cool, P., Vasile, A., Bilba, N., Vansant, E. F., Van Vaeck L., *Rapid Commun Mass Spectro.* 2005, 19, 2809 – 2818.

[41] Juang, L. C., Wang C. C., Lee, C. K. *Chemosphere*, 2006, 64, 1920 – 1928.

[42] Cestari, A. R., Vieira, E. F. S., Vieira, G. S., da Costa, L. P., Tavares, A. M. G., Loh, W., Airoldi, C., *J Haz Mater,* 2009, 16, 307 – 316.

[43] Zhang, L., Lv, J., Xu, T., Yang, L., Jiang, X., Li, Q. *Sep. Purif.Technol* 2013, 116,145–153.

[44] Son, S. J., Reichel, J., He, B., Schuchman, M., Lee, S.B. *J. Am. Chem. Soc* .127, 2005, 7316-7317.

[45] Lee, J.-H., Huh, Y.-M., Jun, Y. W., Seo, J. W., Jang, J.-tak, Song, H.-T., Kim, S., Cho, E.-J., Yoon, H.-G., Suh, J.-S., Cheon J. *Nat. Med,* 2007, 13, 95-99.

[46] Cassiers, K., Linssen, T., Mathieu, M., Benjelloun, M., Schrijnemakers, K., Van Der Voort, P., Cool, P., Vansant, E. F., *Chem Mater* 2002, 14, 2317 - 2324.

[47] Suslick, K. S., *Science* 1990, 247, 1439 – 1445.

[48] Weissler, A., *Nature* 1962, 193, 1070.

[49] Tang, X., Liu, S., Wang, Y., Huang, W., Sominski, E., Palchik, O., Koltypin Y., Gedanken, A., *Chem. Commun.,* 2000, 2119 - 120.

[50] Run, M., Wu, S., Wu, G. *Microp. Mesop. Mater.* 2004, 74, 37 – 47.

[51] Melero, J. A., van Grieken, R., Morales, G., *Chem. Rev.*, 2006, 106, 3790-3812.

[52] Meynen, V., Cool, P., Vansant, E.F., *Microp. Mesop. Mater.* 2009, 125, 170 – 223.

[53] Tsoncheva, T., Rosenholm, J., Linden, M., Tiemann, M., Ivanova, L., Dimitriv, M., Paneva, D., Mitov, I., Minchev, C., *Microp. Mesop. Mater.* 2008, 112, 327.

[54] Fox, M. A., Dulay, M. T., *Chem. Rev.* 1993, 93, 341 - 357.

[55] N. Serpone, *Res. Chem. Interm.*, 20, 1994, 953 - 992.

[56] Herrmann, J. M. *Environmental Catalysis,* in: Jansen, F., Van Santen R.A. (Eds), *Catalysis Science Series*, Vol. 1, Imperial College Press, London, 1999, 171.

[57] Chen, X., Mao, S.S., *Chem. Rev.* 2007, 107, 2891 – 2959.

[58] Ohno, T., Tagawa, S., Itoh, H., Suzuki, H., Matsuda, T., *Mater Chem Phys*, 2009, 113, 119–123.

[59] Srinivasan, S., Datye, A.K., Smith, M. H., Peden, C.H.F., *J Catal,* 1994, 145, 565-573.

[60] Hsien, Y. H., Chang, C.F., Chen Y. H., Cheng, S., *Appl. Cat B: Environ*, 2001, 31, 241– 249.

[61] Hilonga, A., Kim, J. K., Sarawade, P. B., Kim, H. T., *Powder Technol.* 2009, 196, 286 – 291.

[62] Tasbihi, M., Lavrenčič Štangar, U., Sever Škapin, A., Ristić, A., Kaučič, V., Novak N., Tuša, *J. Photochem Photobiol A: Chem.*, 2010, 216, 167 – 178.

[63] Bhattacharyya, A., Kawi S., Ray, M. B. *Cat Today*, 2004, 98, 431 - 439.

[64] Lakshmi, B. B., Dorhout P. K, Martin, C.R. *Chem. Mater.,* 1997, 9, 857.

[65] Lakshmi, B. B., Patrissi C. J., C. R. Martin, *Chem. Mater.*, 1997, 9, 2544.

[66] Van Der Voort, P., Ravikovitch, P. I., De Jong, K. P., Neimark, A. V., Janssen, A. H., Benjelloun, M., Van Bavel, E., Cool, P., Weckhuysen, B.M., Vansant, E.F. *Chem Commun*, 2002, 1010.

[67] Van Bavel, E., Cool, P., Aerts, K., Vansant, E.F., *J. Phys. Chem. B*, 2004, 108, 5263.

[68] Kruk, M., Jaroniec, M., Joo S.H., Ryoo, R. *J. Phys. Chem. B*, 2003, 107, 2205.

[69] Busuioc-Tomoiaga, A. M., Mertens, M., Cool, P., Bilba, N., Vansant, E.F. *Stud. Surf. Sci. Catal.,* 2008, 174, 377 – 380.

[70] Xia, B., Huang H., Xe, Y. *Mat Sci Eng,* 1999, B57, 150.

[71] Derbyshire, F., Jagtoyen, M., Thwaites, M., *Activated Carbon-Production and Applications, Porosity in Carbon* (Patrick, J.W., ed.) Edward Arnold 1995, Chapter 9, 227-252.

[72] Pulickel, M. A., Tour, J. M., *Nature,* 2007, 447, 1066.

[73] Peng, X., Chen, J., Misewichb, J. A., Wong, S. S., *Chem. Soc. Rev.,* 2009, 38, 1076.

[74] Kongkanand, A., Dominguez, R. M., Kamat, P.V., *Nano Lett,* 2007, 7, 676.

In: Comprehensive Guide for Mesoporous Materials, Volume 2 ISBN: 978-1-63463-982-8
Editor: Mahmood Aliofkhazraei

Chapter 16

A COMPARISON STUDY ON THE CATALYTIC ACTIVITY OF POLYMER-PROTECTED PALLADIUM NANOPARTICLES SUPPORTED ON CMK-8 AND KIT-6 ORDERED MESOPOROUS STRUCTURES

Roozbeh Javad Kalbasi[1*] and Neda Mosaddegh [2]
[1]Faculty of Chemistry, Kharazmi University, Tehran, Iran
[2]Department of Chemistry, Science and Research Branch, Islamic Azad University, Tehran, Iran

ABSTRACT

The synthesis and characterization of Pd-poly(N-vinyl-2-pyrrolidone)/CMK-8 (Pd-PVP/CMK-8) composite and its application in Suzuki-Miyaura cross-coupling reaction as a novel heterogeneous catalyst is reported. The catalyst was characterized by FT-IR, XRD, XPS, BET, TGA, DRS UV-Vis, SEM and TEM techniques. The catalytic activity of this purely organic hybrid catalyst was compared with Pd-PVP/KIT-6 to clarify the advantages of mesoporous carbon on mesoporous silica as support. Surprisingly, the composite prepared by mesoporous carbon showed much higher activity than that of Pd-PVP/KIT-6. In addition, the results showed that the stability of Pd-PVP/CMK-8 was excellent and that it could be reused 13 times without much loss of activity in Suzuki reaction. This unique result opens new perspectives for application of mesoporous carbons as structurally defined hydrophobic catalyst support in catalytic reactions.

INTRODUCTION

Palladium based catalysts, particularly nanoscale palladium particles, have recently attracted much attention due to their versatile role in organic synthesis. One of the well-

* Corresponding author. Tel.: +98 21 88848949; Fax: +98 21 88820992; E-mail: rkalbasi@gmail.com, rkalbasi@khu.ac.ir.

known applications of palladium nanoparticles as catalysts for the coupling of carbon-carbon molecules is the Suzuki-Miyaura reaction (Table 1).

Table 1. Suzuki–Miyaura reaction catalyzed by palladium based catalysts

Catalyst	Yield (%)	Ref.
Palladium-polyaniline hybrid	93 [a]	[1]
Pd nanoparticles supported on grapheme	100 [b]	[2]
Chitosan-g-mPEG-supported palladium (0)	96 [c]	[3]
copper oxide-supported Pd(0) nanoparticles	90 [d]	[4]
Phosphine free diamino-diol based palladium	100 [e]	[5]
Iminophosphine-Pd	99 [f]	[6]

[a]Reaction conditions: Aryl bromide (1 mmol), phenylboronic acid (1 mmol), K_2CO_3 (1.5 mmol), toluene (5 ml), catalyst (polymer supported Pd 0.02 mol%), 80°C, 8 h.

[b]Reaction conditions: Aryl bromide (0.32 mmol), boronic acid (0.382 mmol), K_2CO_3 (0.96 mmol), Pd/G cataltst (1.3 mg), 4 mL (H_2O:EtOH) (1:1), 80°C (MWI), 5 min.

[c]Reaction conditions: Catalyst (0.5 mmol), 4-bromoacetophenone (0.5 mmol), phenylboronic acid (0.75mmol), NaOH (1.5mmol), H_2O (4mL), 100°C, 1 h.

[d]Reaction conditions: Catalyst (0.5 mmol), Aryl iodid (0.5 mmol), phenylboronic acid (0.75mmol), DMF (1.5mmol), DMF (4mL), 110°C, 8 h.

[e]Reaction conditions: Aryl bromide (0.5 mmol), phenylboronic acid (0.75 mmol), K_2CO_3 (1 mmol), MeOH (5 mL), r.t.

[f]Reaction conditions: 4-bromoanisole (0.2 mmol), phenylboronic acid (0.24 mmol), 0.5 mol% of Fe_3O_4@SiO_2–iminophosphine–Pd catalyst (8 mg), KOH (0.4 mmol), toluene (3 mL), 100 °C, 2 h in inert atmosphere.

Heterogeneous catalysts can be easily separated from the reaction mixture by simple filtration and they can be reused in successive reactions provided that the active sites have not become deactivated. Heterogeneous catalysis also helps to minimize wastes derived from reaction workup, contributing to the development of green chemical processes (Table 2).

In catalytic applications, a uniform dispersion of nanoparticles and an effective control of particle size are usually expected. However, nanoparticles frequently aggregate to yield bulk-like materials, which greatly reduce the catalytic activity and selectivity. Therefore, they must be embedded in a matrix such as polymer or macromolecular organic ligands [13, 14] or immobilized in the pores of heterogeneous supports, like ordered mesoporous silicas [15-17]. Recently, remarkable progress has been made in the synthesis of carbons with periodic ordered structures in the range of micropores (2 nm), mesopores (2-50 nm), and macropores (>50 nm) using inorganic templates such as zeolites, mesoporous molecular sieves, and silica gels (Table 3). In particular, a new class of mesoporous carbon materials that has attracted great interest has been reported using both cubic and hexagonal silica mesostructures as templates. This new group of materials designated as CMK-n, possess several advantages compared to mesoporous silica molecular sieves (SBA-15, MCM-48, etc), i.e. high thermal stability up to 1600 °C in N_2 atmosphere, high stability in strong acids and bases, high mechanical stability, especially electric conductivity and many more due to their interesting properties, such as high surface areas, narrow pore size distributions and regular frameworks [18].

Table 2. Suzuki–Miyaura reaction catalyzed by heterogeneous palladium based catalysts

Entry	Catalyst	Yield (%)	Ref.
1	Pd-poly(N-vinyl-2-pyrrolidone)/KIT-5	97 [a]	[7]
2	Solid-supported nano and microparticles of Pd(0), SS-Pd	85 [b]	[8]
3	Pd-poly(N-vinyl-2-pyrrolidone)/SBA-15	98 [c]	[9]
4	palladium(0)-MCM-41	93 [d]	[10]
5	Pd/SiO_2–TiO_2	88.6 [e]	[11]
6	Palladium nanoparticles-polyacrylamide containing phosphinite ligand	96 [f]	[12]

[a] Reaction conditions: Catalyst (0.12 g), iodobenzene (1 mmol), phenylboronic acid (1.5 mmol), K_2CO_3 (5 mmol), H_2O (5 mL), r.t, 5 h.

[b] Reaction conditions: SS-Pd Catalyst, arylebenzene (1 mmol), phenylboronic acid (1.5 mmol), K_2CO_3, DMF, 110 °C, 6-12 h.

[c] Reaction conditions: Catalyst (0.12 g), iodobenzene (1 mmol), phenylboronic acid (1.5 mmol), K_2CO_3 (5 mmol), H_2O (5 mL), r.t, 8 h.

[d] Reaction conditions: Pd(0)-MCM-41 (0.05 g), aryl halide (3 mmol), phenylboronic acid (3.6 mmol), K_2CO_3 (5 mmol), 20% H_2O/EtOH, 80 °C, 20 h.

[e] Reaction conditions: Catalyst (176.7 mg), bromobenzene (5 mmol), phenylboronic acid (7.5 mmol), K_2CO_3 (10 mmol), ethanol (10 mL), 78 °C, 6 h.

[f] Reaction conditions: Catalyst (7 mg, 0.1 mol%), aryl halide (1 mmol), phenylboronic acid (1.2 mmol), K_2CO_3 (2 mmol), DMF (3 mL), 80 °C, 40 min.

Table 3. Various types of mesoporous carbon materials based on the structure and form of inorganic template [18-20]

Mesoporous carbon	Inorganic template	Carbon precursor	Pore size (nm)
CMK-1	MCM-48	Sucrose/Furfuryl alcohol	3.5
CMK-2	SBA-1	Sucrose/Furfuryl alcohol	4
CMK-3	SBA-15	Sucrose/Furfuryl alcohol	4.5
CMK-5	SBA-15	Furfuryl alcohol	5
CMK-8	KIT-6	Sucrose	4
Mesocarbon microwires	Fe-silica	Sucrose/Furfuryl alcohol	3
SNU-1	MCM-48	Sucrose/Furfuryl alcohol	4
CKT-3	KIT-5	Sucrose	4

Therefore, the mesoporous carbons are of great interest for, e.g. adsorption, catalyst, catalyst supports, host–guest chemistry, and offer potential applications such as size and shape-selective adsorption media or chromatography separation systems [19,20]. Ordered mesoporous carbon materials have been obtained by the carbonization of sucrose, furfuryl alcohol, polyacrylonitrile, polystyrene, and some other carbon sources in various types of silica, particularly MCM-48 and SBA-15 [21, 22].

Very recently, large-pore cubic *Ia3d* mesoporous materials were synthesized by several detailed procedures using block copolymers. The use of KIT-6 silica template, depending on the size of complementary pores, afforded a faithful inverse replica, CMK-8. The structure of KIT-6 silica (cubic *Ia3d*) consists of an enantiomeric pair of interpenetrating 3-D networks of mesoporous channels, into which the carbon source is filled. The carbonization of the sucrose results in the formation of rod-type carbon inside silica channels [23, 24].

Nowadays, mesoporous carbon has been used as a carbon support for catalysts [25]. Carbon-based catalysts have several advantages as solid catalysts because of the high surface area and large porosity as well as hydrophobic surface nature. Because of these advantages, research has increased using carbon as a catalyst [26-30].

In our previous works, the effect of polymer-inorganic hybrid materials as heterogeneous catalyst for Suzuki-Miyaura reaction, using mesoporous materials such as KIT-6, was studied [31]. Herein, we will introduce a novel heterogeneous organic hybrid catalyst based on a carbon mesoporous material. In this condition, mesoporous carbon CMK-8, which can be replicated from mesoporous silica KIT-6 and whose surface is hydrophobic, was prepared and used as an ideal support for Pd nanoparticle-PVP. The catalytic activity of this novel hybrid composite was tested for Suzuki-Miyaura cross-coupling reaction in water medium under aerobic conditions. In addition, the catalytic activity of this hybrid catalyst was compared with Pd-PVP/KIT-6 to investigate the effect of carbon mesoporous material on the activity and stability of the catalyst. Green media for this reaction, catalyst reusability and simple work-up of the product were the main advantages for these nanoparticles supported on the polymers/mesoporous carbon. Therefore, the main goal of this catalytic synthesis is to introduce a novel and efficient organic composite to expand the use of these types of composites for organic reactions, and to compare it with our previous works.

Polymer-Mesoporous Silica/Carbon Nanocomposites

The synthesis of CMK-8 was performed according to recently reported procedure using KIT-6 as the hard template and sucrose as the carbon source [32]. Mesoporous silica KIT-6 materials with large-pore cubic *Ia3d* structure were prepared following our recently reported procedure [31]. The preparation of CMK-8 was carried out as follows:

0.5 g KIT-6 was added to 4 mL aqueous solution containing 0.6 g (3.65 mmol) sucrose and 0.1 g (1.42 mmol) of H_2SO_4 (98%). The resulting mixture was heated in an oven at 100°C for 6 h and then 160 °C for another 6 h. In order to obtain fully polymerized sucrose inside the pores of the KIT-6 template, 2 mL aqueous solution containing 0.4 g (2.33 mmol) sucrose and 0.05 g (0.917 mmol) of H_2SO_4 were added again and the mixture was subjected to the thermal treatment described above one more time. Then, it was carbonized under N_2 gas flow at 900 °C for 6 h with a heating rate of 5°C min^{-1}. Finally, the resulting solid was washed with 1 M NaOH solution (50 vol. % ethanol-50 vol. % H_2O) twice to remove the silica template, filtered, washed with ethanol until pH = 7, and dried at 100°C for 4 h. Thus, mesoporous carbon CMK-8 was obtained (Scheme 1).

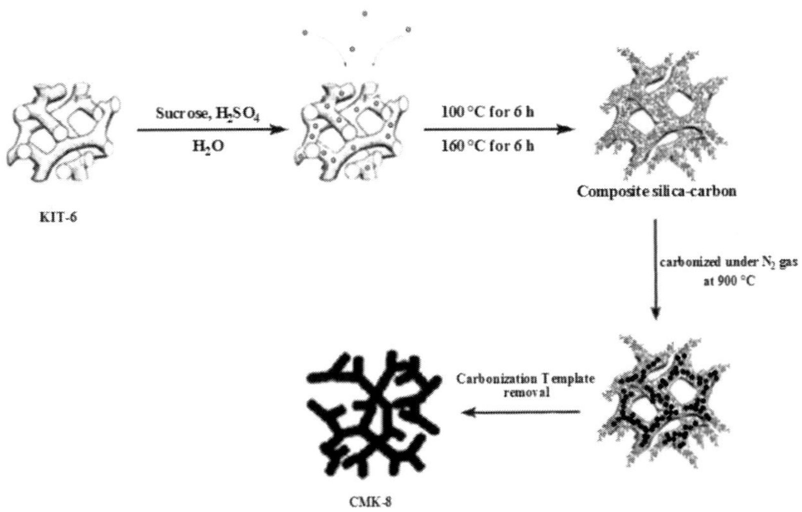

Scheme 1. Preparation of CMK-8.

CMK-8 (0.5 g) and N-vinyl-2-pyrrolidone (NVP) (0.5 mL, 4.6 mmol) in 7 mL tetrahydrofuran (THF) were placed in a round bottom flask. Benzoyl peroxide (3 % mol, 0.034 g) was added and the mixture was heated to 65-70 ˚C for 5 h while being stirred under N_2 gas. The resulting white fine powder composite (PVP/CMK-8) was collected by filtration, washed several times with THF, and finally dried at 60 ˚C under reduced pressure (Scheme 2). Pd-PVP/CMK-8 composite was synthesized as follows (Scheme 2):

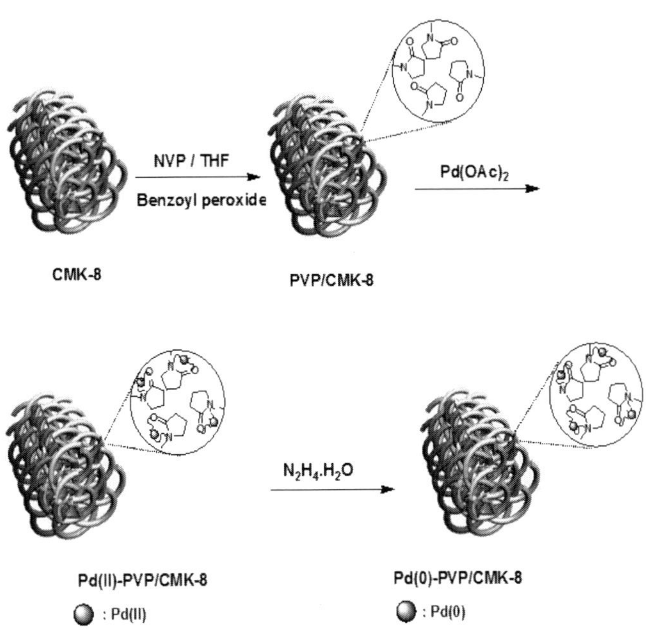

Scheme 2. Preparation of Pd-PVP/CMK-8.

PVP/CMK-8 (0.1 g) and 10 mL of an aqueous acidic solution (C_{HCl} = 0.09 M) of $Pd(OAc)_2$ (0.025 g, 0.236 mmol) were placed in a round bottom flask. The mixture was heated to 80 °C for 5 h while being stirred under N_2 gas. Then, 0.6 mL, (9.89 mmol) aqueous solution of hydrazine hydrates ($N_2H_4.H_2O$) (80 vol. %) was added to the mixture drop by drop in 15-20 minutes. After that, the solution was stirred at 60 °C for 1 h. Afterwards, it was filtered and washed sequentially with chloroform and methanol to remove excess $N_2H_4.H_2O$ and was dried in room temperature to yield Pd-PVP/CMK-8 composite (Scheme 2). The Pd content of the catalyst was estimated by decomposing the known amount of the catalyst by perchloric acid, nitric acid, fluoric acid, hydrochloric acid, and the resulting mixture was analyzed by inductively coupled plasma atomic emission spectrometry (ICP-AES). The Pd content of the catalyst estimated by ICP-AES was 1.02 mmol g^{-1}.

Pd-PVP/KIT-6 was prepared with the same procedure in order to compare the effect of mesoporous materials on the catalytic activity of the catalysts. The Pd content of Pd-PVP/KIT-6 catalyst estimated by ICP-AES was 0.98 mmol g^{-1}.

The prepared nanocomposites was tested in the Suzuki-Miyaura reaction between iodobenzene and phenylboronic acid at room temperature as the standard reaction (Scheme 3).

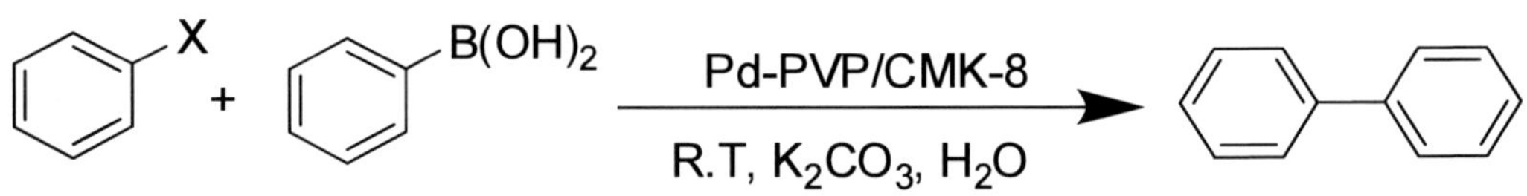

Scheme 3. Suzuki-Miyaura reaction of aromatic arylhalides and phenylboronic acid catalyzed by Pd-PVP/CMK-8.

The general procedure was as follows: a mixture of iodobenzene (1 mmol), phenylboronic acid (1.5 mmol), K_2CO_3 (5 mmol), and catalyst (0.12 g, Pd-PVP/CMK-8) in H_2O (5 mL) was placed in a round bottom flask. The suspension was stirred at room temperature for 5 h. The progress of the reaction was monitored by Thin Layer Chromatography (TLC) using n-hexane as eluent. After completion of the reaction (monitored by TLC), for the reaction work-up, the catalyst was removed from the reaction mixture by filtration, and then the reaction product was extracted with CH_2Cl_2 (3×5 mL). The solvent was removed under reduced pressure. The crude product was purified by flash column chromatography (hexane or hexane/ethyl acetate) to afford the desired coupling product (97% isolated yield). The product was identified with ^1H NMR, ^{13}C NMR and FT-IR spectroscopy techniques.

RESULTS AND DISCUSSION

Catalyst Characterization

The XRD patterns of ordered mesoporous material, CMK-8, the functionalized sample, PVP/CMK-8 and Pd-PVP/CMK-8 in the range of $2\Theta = 0.7–8$, are shown in Figure 1.

The XRD pattern of CMK-8 showed three diffraction peaks that can be indexed to (110), (210), and (220) in the 2Θ range from 0.7° to 8°, representing well-ordered cubic pores [32].

The PVP/CMK-8 and Pd-PVP/CMK-8(2Θ = 0.7-8) samples show the same pattern, indicating that the structure of the CMK-8 (110) was retained after the support of the surface of the CMK-8 with PVP and Pd (Figure 1). However, the intensity of the characteristic reflection peaks of the PVP/CMK-8 and Pd-PVP/CMK-8 (2Θ = 0.7-8) samples are found to be reduced (Figure 1). This may be attributed to the symmetry destroyed by the hybridization of CMK-8, which is also found in the ordered mesoporous carbon loading with guest matter. In addition, composites contain much less CMK-8 due to the dilution of the carbon material by PVP and Pd, therefore, this dilution can also account for a decrease in the peak intensity.

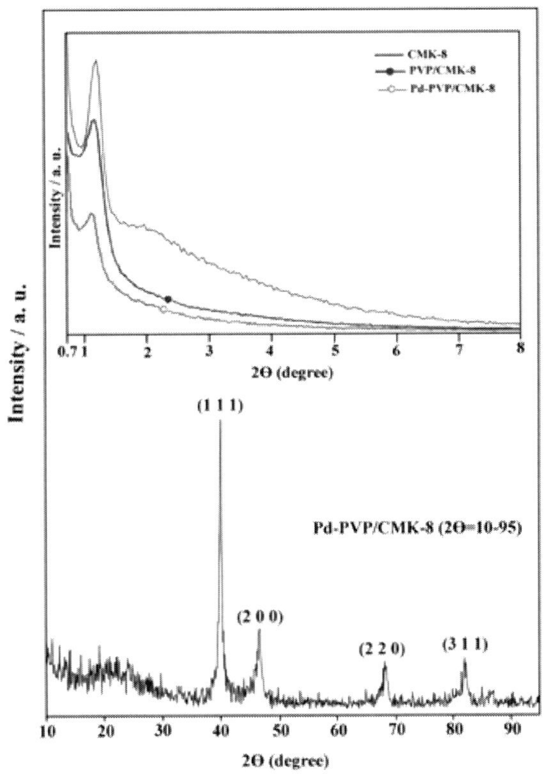

Figure 1. The powder XRD pattern of mesoporous silica CMK-8, PVP/CMK-8 and Pd-PVP/CMK-8.

As can be seen from Figure 2, the XRD patterns of CMK-8, PVP/CMK-8 and Pd-PVP/CMK-8 samples are similar to those of KIT-6 (Figure 2) [31], which indicates that CMK-8 is a true replica of the mesoporous silica KIT-6, and that supporting the surface of CMK-8 with PVP and Pd nanoparticles almost does not damage the structure of CMK-8.

Compared to d-spacing in CMK-8, the incorporation of the PVP chains into the CMK-8 pores leads to the increase of CMK-8 pore array (Table 4). Furthermore, with incorporation of Pd nanoparticles into the composite structure, the d-spacing is increased more. These results are in good agreement with those obtained from the BET observations and show that some polymer chains and Pd nanoparticles are incorporated in the pores of the mesoporous carbon which can cause some pressure on the pore structure of mesoporous carbon leading to an increase in the d-spacing of structures. Similar results can be seen for KIT-6 and related composites (Table 4).

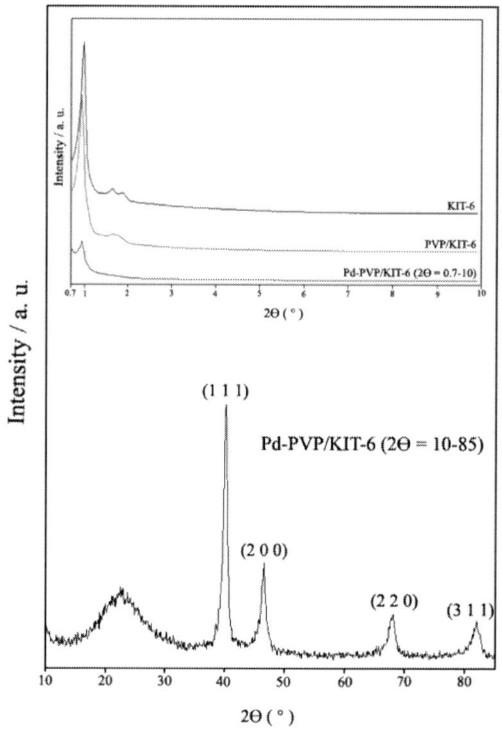

Figure 2. The powder XRD pattern of mesoporous silica KIT-6, PVP/KIT-6 and Pd-PVP/KIT-6.

The wide-angle XRD pattern of the Pd-PVP/CMK-8 composite (2Θ = 10-95) (Figure 1) shows the reflections at 39.92°, 46.54°, 67.90° and 82.0°. These peaks correspond to (111), (200), (220) and (311) lattice planes of Pd nanoparticles [33]. Planes were assigned by comparing them with the standard Pd, and these planes correspond to the *fcc* crystal lattice structure of Pd (JCPDS, Card No. 05-0681). It should be mentioned that the wide-angle XRD pattern of Pd-PVP/KIT-6 composite is similar to Pd-PVP/CMK-8 composite (Figure 2).

Figure 3 presents the FT-IR spectra of CMK-8 (a), PVP/CMK-8 (b), and Pd-PVP/CMK-8 (c). A broad band at around 3408-3423 cm^{-1} was observed in all samples. It was mainly caused by the O–H stretching vibration of the adsorbed water molecules. In the FT-IR spectrum of CMK-8 (Figure 3a), there aren't any signals of the organic bonds, resulting from the complete carbonization of sucrose (Figure 3a). The existence of PVP in the PVP/CMK-8 composite is evidenced by the appearance of typical PVP vibration on the FTIR spectrum (Figure 3b). In the FT-IR spectrum of PVP/CMK-8 (Figure 3b), the new band at 1663 cm^{-1} is corresponding to the carbonyl bond of PVP [34]. Moreover, the presence of peaks at around 2800-3000 cm^{-1} and 1433 cm^{-1} correspond to the aliphatic C–H stretching and bending in PVP/CMK-8, respectively (Figure 3b). The appearance of the above bands shows that PVP has been attached to the surface of CMK-8 and the PVP/CMK-8 has been obtained.

As shown in Pd-PVP/CMK-8 spectrum (Figure 3c), the band around 1663 cm^{-1} which corresponds to carbonyl bond of PVP, is shifted to lower wave numbers (1636 cm^{-1}) (red shift). Moreover, the peak intensity of the carbonyl bond in the spectrum of Pd-PVP/CMK-8 is lower than that of PVP/CMK-8. This may be due to the interaction between the Pd nanoparticles and C=O groups. This means that the double bond CO stretches become weak by coordinating to Pd nanoparticles. Thus, it is confirmed that PVP molecules exist on the surface of Pd nanoparticles and coordinate to Pd nanoparticles [34-36].

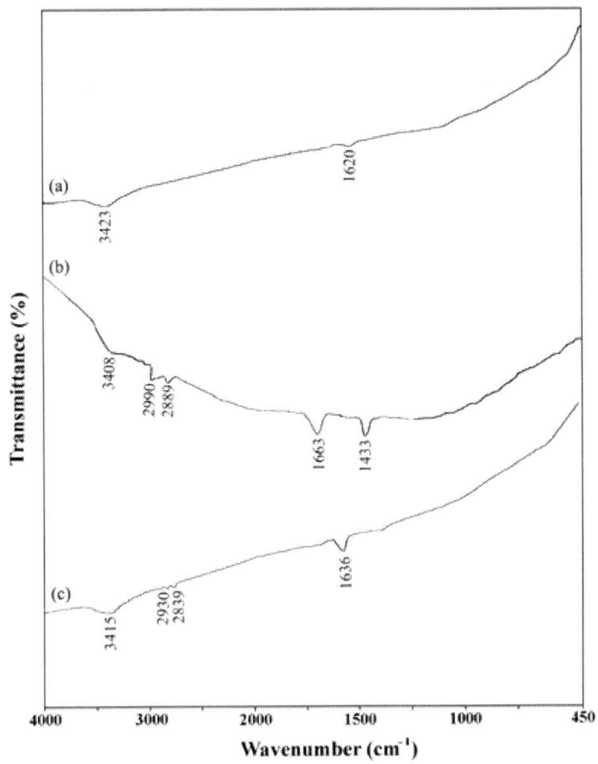

Figure 3. FT-IR spectra of (a) mesoporous carbon CMK-8, (b) PVP/CMK-8 and (c) Pd-PVP/CMK-8.

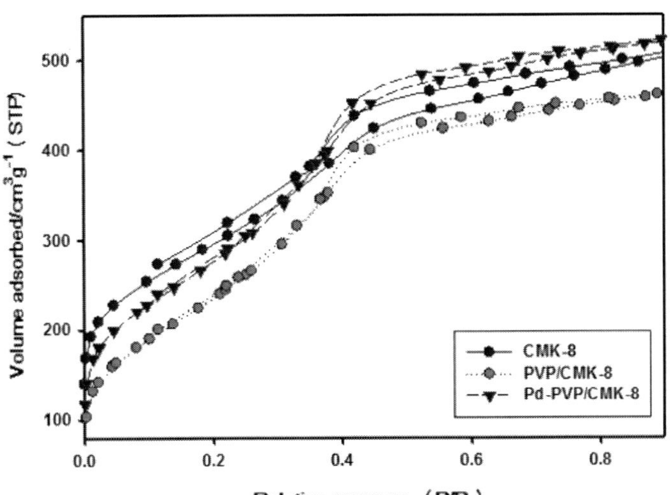

Figure 4. N_2 adsorption-desorption isotherms of mesoporous carbon CMK-8, PVP/CMK-8 and Pd-PVP/CMK-8.

Table 4. Physicochemical properties of mesoporous carbon samples as well as mesoporous silica samples obtained from XRD and N_2 adsorption

Sample	BET surface Area (m^2 g^{-1})	V_P (cm^3 g^{-1})	BJH pore diameter (nm)	D_{211} (nm)
Mesoporous carbon CMK-8	1043	0.80	3.06	11.26
PVP/CMK-8	889	0.72	3.26	11.48
Pd-PVP/CMK-8	103	0.82	3.16	11.83
Mesoporous silica KIT-6	989	1.35	4.03	9.23
PVP/KIT-6	794	1.17	3.53	9.11
Pd-PVP/KIT-6	500	1.32	4.61	9.5

[a] Total pore volume.

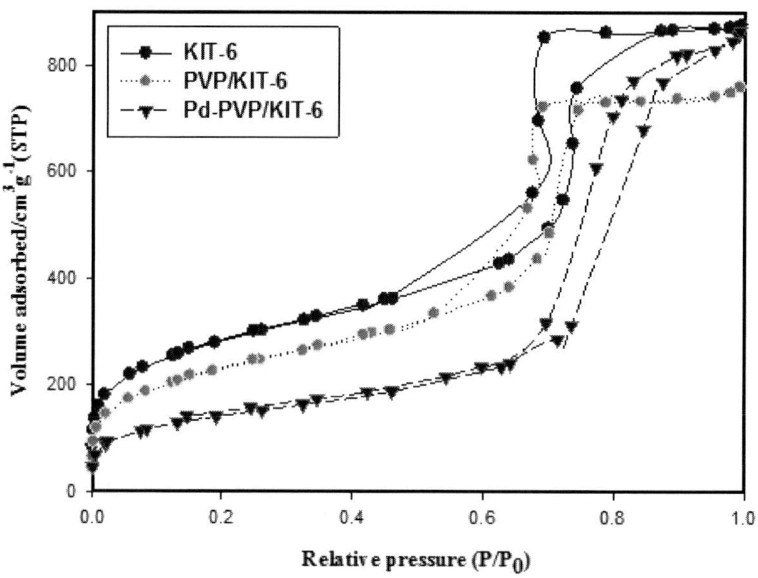

Figure 5. N_2 adsorption-desorption isotherms of mesoporous silica KIT-6, PVP/KIT-6 and Pd-PVP/KIT-6.

The BET specific surface areas and the pore sizes of CMK-8, PVP/CMK-8, Pd-PVP/CMK 8, KIT 6, PVP/KIT-6 and Pd-PVP/KIT-6 were calculated using BET and BJH methods (Table 4). The nitrogen sorption isotherms for the KIT-6 materials are of type IV according to the IUPAC classification, and exhibit an H1 hysteresis loop [31] (Figure 4) and those for the CMK-8 materials are of H2 hysteresis loop (Figure 5). For KIT-6, capillary condensation occurs at around $P/P_0 = 0.6–0.8$, whereas condensation occurs at about $P/P_0 = 0.4–0.7$ in both cases of the CMK-8, PVP/CMK-8 and Pd-PVP/CMK-8. The CMK-8, PVP/CMK-8 and Pd-PVP/CMK-8 obtained from KIT-6 possess pores with diameter of about 3.06, 3.26 and 3.16 nm respectively, which are smaller than that of KIT-6 (4.03 nm). However, the BJH surface areas of CMK-8 (1043 m^2 g^{-1}), is higher than that of KIT-6 (988 m^2 g^{-1}).

The morphologies of the KIT-6 and Pd-PVP/KIT-6 are shown in Figure 6. KIT-6 (Figure 6a,b), and Pd-PVP/KIT-6 (Figure 6c,d) are agglomerations of small irregular particles. As we can see, there is negligible difference in particle surface morphology between the KIT-6 and Pd-PVP/KIT-6 composite, which indicates that the supporting of Pd nanoparticles takes place more in the channels and less on the outer surfaces of the catalyst.

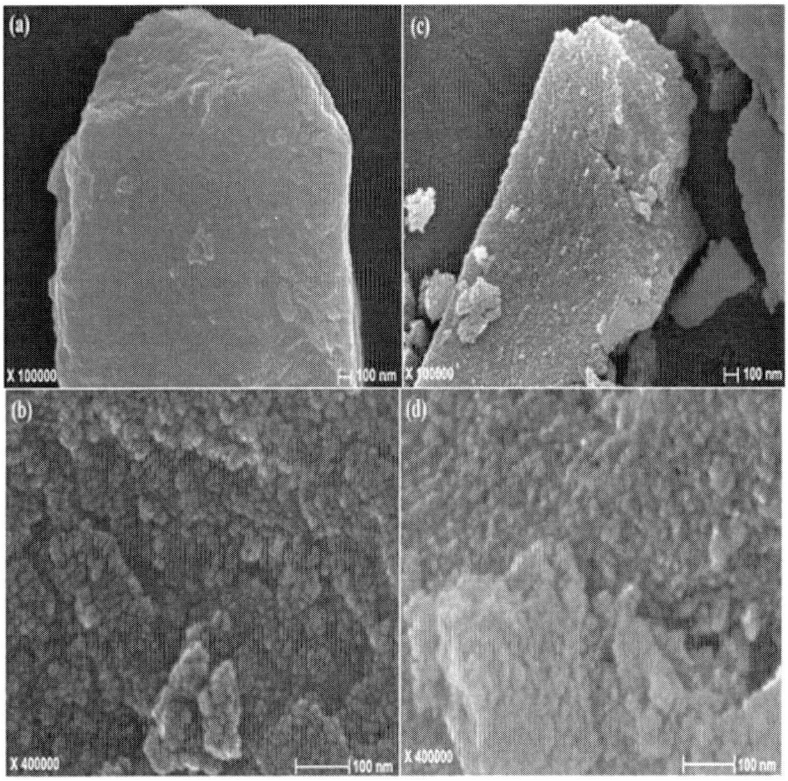

Figure 6. SEM photographs of (a,b) mesoporous silica KIT-6 and (c,d) Pd-PVP/KIT-6.

The surface morphologies of the obtained CMK-8, PVP/CMK-8 and Pd-PVP/CMK-8 were observed by SEM, as shown in Figure 7. As it can be seen from Figure 7a-c, which are with magnification of 10000, there is no significant difference between their morphologies. However, by comparing images with higher magnification (Figure 7d-f), we can see that there are many agglomerate particles on the outer surface of Pd-PVP/CMK-8 which can be related to the Pd nanoparticles that are distributed on the outer surface of the support (Figure 7f). Furthermore, these observations can be seen much better by comparing SEM image of CMK-8 and Pd-PVP/CMK-8 with magnification of 400000 (Figure 7g,h).

These observations display that in Pd-PVP/KIT-6 the Pd nanoparticles are mostly distributed in the channels of KIT-6, while they are distributed on the outer surface as well as inner surface of CMK-8. It can be related to the different structure between KIT-6 and CMK-8. KIT-6 has a cubic structure with interconnected tube-type channels, but CMK-8 has a cubic structure with a rod-type carbon structure. Therefore, the Pd nanoparticles can be distributed to channels of KIT-6, while they have to distribute on the surface of rod-type of CMK-8.The TEM micrographs of CMK-8 (a), Pd-PVP/CMK-8 (b) and Pd-PVP/KIT-6 (c) were depicted in Figure 8. The ordered cubic *Ia3d* mesostructure of CMK-8 can be seen from Figure 8a.

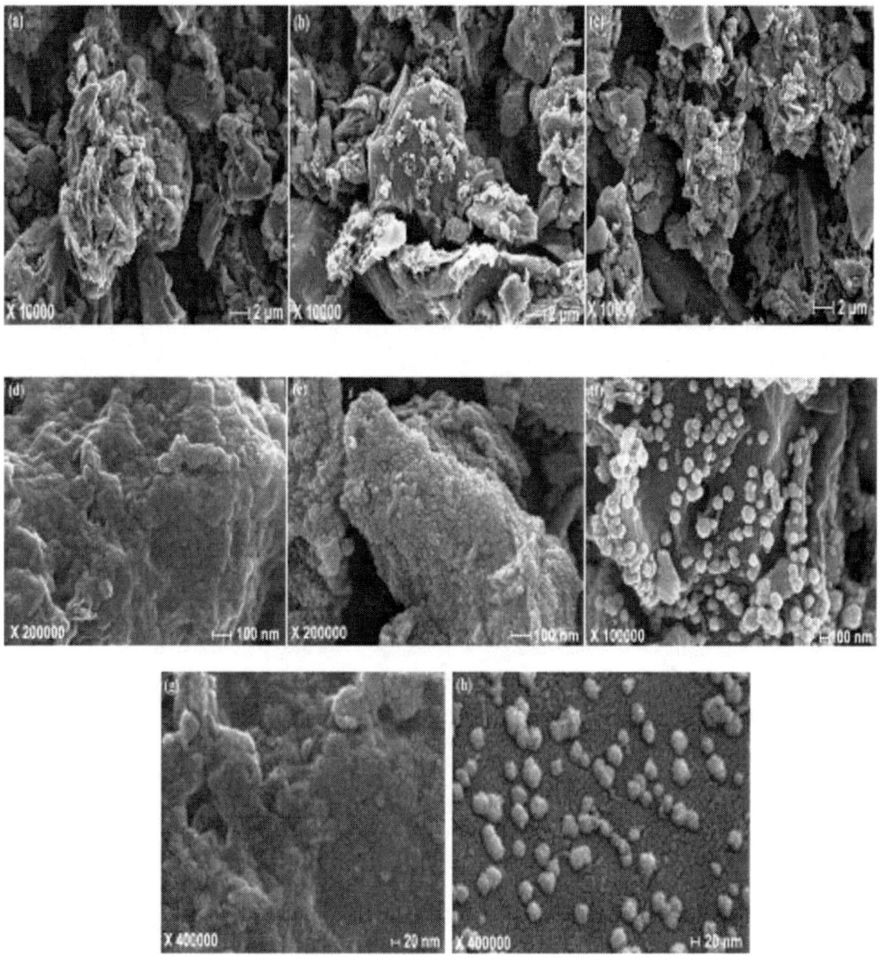

Figure 7. SEM photographs of (a,d,g) CMK-8, (b,e) PVP/CMK-8 and (c,f,h) Pd-PVP/CMK-8.

The TEM micrograph of Pd-PVP/CMK-8 can be seen from Figure 8b. The places with darker contrast could be assigned to the presence of Pd particles with different dispersion. The small dark spots in the images could be ascribed to Pd nanoparticles with an average diameter of ~ 3 nm, probably located into the support channels. The larger dark areas over the channels most likely correspond to Pd nanoparticles agglomerates on the external surface with an average diameter of ~ 5-10 nm (Figure 8b). These are in accordance with SEM results.

Figure 8c shows the TEM images of Pd-PVP/KIT-6 composite and the distribution of the Pd nanoparticles in the composite. The image shows that the ordered cubic *Ia3d*mesostructure of KIT-6 is retained, and no damage of the periodic structure of the silicate framework is observed. The small dark spots in the images could be ascribed to Pd nanoparticles, located into the support channels (average diameter of ~ 4 nm) (Figure 8c).

The DRS UV–Vis spectrum of Pd(0)-PVP/CMK-8 was studied after the reduction of Pd(II)- PVP/CMK-8 by hydrazine hydrate. According to the literature results, the UV-Vis spectrum of $Pd(OAc)_2$ revealing a peak at 400 nm refers to the existence of Pd(II) [37]. However, there wasn't any peak at 400 nm in the DRS UV-Vis spectrum of Pd-PVP/CMK-8, which indicated complete reduction of Pd(II) to Pd(0) nanoparticles.

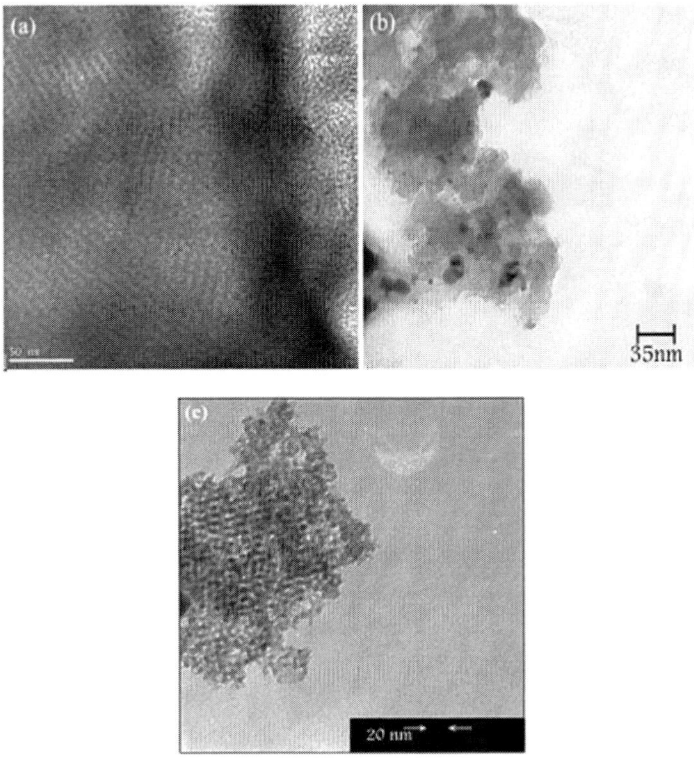

Figure 8. TEM of (a) CMK-8, (b) Pd-PVP/CMK-8 and (c) Pd-PVP/KIT-6.

The TGA results of the CMK-8 (a), PVP (b), PVP/CMK-8 (c) and Pd-PVP/CMK-8 (d) under N_2 atmosphere are presented in Figure 9. The mass loss at temperature <100 °C (around 2%, w/w) is attributed to desorption of water present on the surfaces of the CMK-8 (Figure 9a). The TGA curves of PVP shows a small mass loss (around 7.5%, w/w) in the temperature range 50-150 °C, which is apparently associated with adsorbed water (Figure 9b). At temperatures above 200 ˚C, PVP shows one main stage of degradation. The mass loss for PVP in the second step is equal to 77% (w/w) which corresponded to the effective degradation of the polymer (Figure 9b). Thermo analysis of PVP/CMK-8 shows two steps of mass loss (Figure 9c). The first step (around 3%, w/w) that occurs at 95 °C is related to desorption of water. The second step (around 10%, w/w) which appeared at 200 °C is attributed to degradation of the polymer and the degradation ended at 560°C (Figure 9c). By comparing the PVP and PVP/CMK-8 curves, one can find that the weight loss of PVP/CMK-8 occurs at a higher temperature, it illustrates that PVP/CMK-8 has higher thermal stability and slower degradation rate than PVP (Figure 9b,c). Therefore, after hybridization, the thermal stability is enhanced greatly. However, for Pd-PVP/CMK-8 sample two separate weight loss steps are seen (Figure 9d). The first step (around 2.4%, w/w) appearing at temperature <100°C corresponded to the loss of water. The second weight loss (about 300–480 °C) amounts around 7% (w/w) is related to the degradation of the polymer. Obviously, the hybrid Pd-PVP/CMK-8 had higher thermal stability than PVP/CMK-8 which may be attributed to the presence of Pd nanoparticles in the composite structure. Therefore, it is very important for the catalyst application that the thermal stability was enhanced greatly after hybridization.

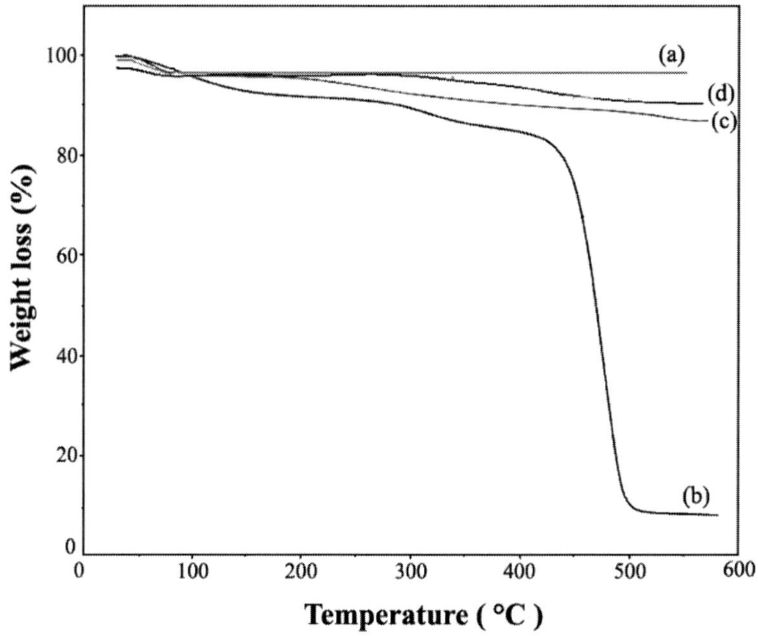

Figure 9. TGA curves of (a) mesoporous carbon CMK-8, (b) PVP, (c) PVP/CMK-8 and (d) Pd-PVP/CMK-8.

To obtain information on the structural features of the Pd-PVP/CMK-8 catalyst, XPS study was carried out on the Pd-PVP/CMK-8 catalyst. The XPS result of Pd nanoparticles was dispersed in PVP/CMK-8 media for Pd 3d spectrum with the binding energies of Pd 3d5/2 and Pd 3d3/2 lying at about 334.5 and 339.6 eV, respectively (see Figure 10).

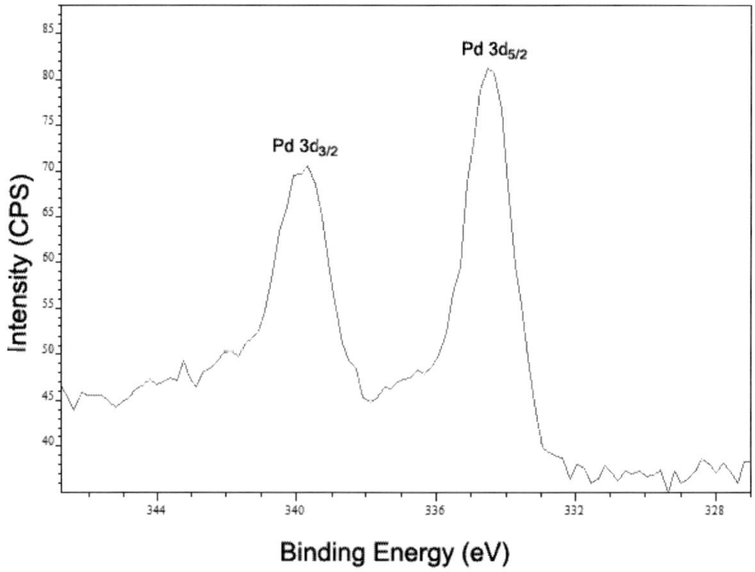

Figure 10. XPS spectrum of Pd 3d of Pd-PVP/CMK-8.

It means that Pd nanoparticles are stable as metallic state in PVP/CMK-8 media (the XPS spectrum of Pd(II) ions reveal two peaks at about 337 and 342 eV refer to Pd 3d5/2 and 3d3/2 [38]). In comparison to the standard binding energy (Pd0 with Pd 3d5/2 of about 335 eV and Pd3/2 of about 340 eV [39]), it can be concluded that the Pd peaks in Pd-PVP/CMK-8 were shifted to lower binding energy than Pd0 standard binding energy. As we know, the position of Pd 3d peak is usually influenced by the local chemical/physical environment around Pd species besides the formal oxidation state, and shifts to lower binding energy when the charge density around it increases [39]. So, the peaks at 334.5 and 339.6 eV should be due to Pd0 species bound directly to O of carbonyl group or N of PVP. The contribution of the charge transfer of N and O of PVP to Pd0 results in the increase in the charge density around Pd0 and the decrease in binding energy.

The full XPS spectrum of Pd-PVP/CMK-8 showed peaks of carbon, oxygen and palladium. Carbon peaks correspond to mesoporous carbon CMK-8 structure and PVP structure (Figure 11).

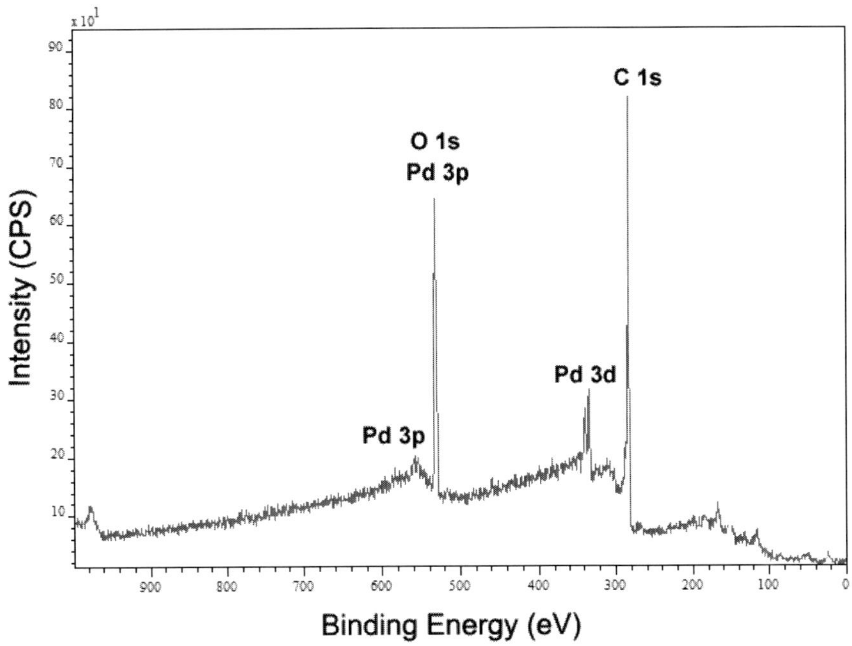

Figure 11. Full XPS spectrum of Pd-PVP/CMK-8.

The XPS spectrum of Pd-PVP/KIT-6 in the range of 334 to 343 eV showed two weak peaks. In addition, the full XPS spectrum of Pd-PVP/KIT-6 showed strong peaks for Si and oxygen and weak peaks for palladium (Figure 12). As we know, the XPS analysis is a surface analysis technique and if the elements were not on the surface, they wouldn't be shown in the XPS spectrum. As found from the TEM and SEM results, the Pd nanoparticles are mostly distributed in the channels of PVP/KIT-6, while they are distributed on the outer surface of PVP/CMK-8. Therefore, it is obvious that Pd nanoparticles should not show strong peaks in the XPS spectrum. These results are in good agreement with those obtained from TEM and SEM analysis.

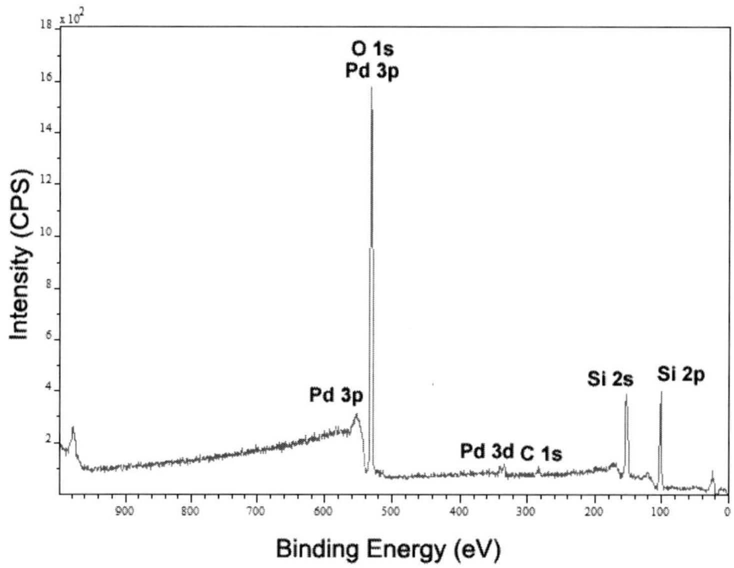

Figure 12. Full XPS spectrum of Pd-PVP/KIT-6.

3.2. Catalytic Activity

This novel composite was synthesized and fully characterized with different methods. The main goal of this catalytic synthesis was to compare this polymer-organic composite (Pd-PVP/CMK-8) with a polymer-inorganic composite (Pd-PVP/KIT-6) and to expand the use of these types of composites for organic reactions [31]. This composite was prepared by a very simple method without using any pre-modification on CMK-8 (pre-modification of CMK-8 involves complicated synthesis and purification method). We have chosen CMK-8 to support the PVP and Pd nanoparticles because the structure of open-framework CMK-8 has some specific characteristics compared to mesoporous silica KIT-6:

- High thermal stability up to 1600 °C in N_2 atmosphere, high stability in strong acids and bases, high mechanical stability, especially electric conductivity and many more due to their interesting properties such as high surface areas, narrow pore size distributions and regular frameworks [18].
- Hydrophobic surface nature of CMK-8 compared to KIT-6 will allow the organic reactants to have a higher tendency to be adsorbed on the surface of Pd-PVP/CMK 8 and consequently the yield will be increased.

Our initial investigation started with the cross-coupling of phenylboronic acid and iodobenzene as a model system. The reaction was carried out in the presence of Pd-PVP/CMK-8 composite in water at room temperature.

Since the proper combination of base and solvent is extremely important, we have examined several different bases and solvents for the Suzuki–Miyaura cross-coupling reaction at room temperature (Table 5). Solvent plays a crucial role in the rate and the product distribution of Suzuki-Miyaura coupling reactions. Since water is known to increase the activity of the Suzuki–Miyaura catalyst [40,41], two kinds of solvents were used: H_2O and $MeOH/H_2O$ (3:1 v/v) and the reaction was catalyzed with 0.12 g of Pd-PVP/CMK-8 catalyst in the presence of 5 equivalents of base and with an iodobenzene/phenylboronic acid ratio of 1 to 1.5 (Table 5).

Table 5. Effect of different bases and solvents on Suzuki-Miyaura reaction [a]

Base	Yield (%) [b]	Time (h)
K_2CO_3 [c]	97	4
K_2CO_3 [d]	98	3
Na_3PO_4 [c]	70	6
Et_3N [c]	43	6

[a]Reaction conditions: Pd-PVP/CMK-8 (0.12 g), iodobenzene (1 mmol), phenylboronic acid (1.5 mmol), base (5 mmol), H_2O (5 mL), room temperature.
[b]Isolated yield.
[c]Solvent: H_2O.
[d]Solvent: $MeOH/H_2O$ (3:1 v/v).

As it can be seen from Table 5, the time of the completion of the reaction is shorter in the case of using $MeOH/H_2O$ (3:1 v/v) as solvent, but neat H_2O was chosen because of the advantages of using water that was mentioned above. We then turned our attention to investigate the effect of base on the Suzuki–Miyaura cross-coupling reaction. Among the bases used [K_2CO_3, Na_3PO_4, and Et_3N], K_2CO_3 was the best choice. The miserable yields obtained with Et_3N indicated the necessity of addition of stronger bases to the reaction media, and hence the economically cheaper K_2CO_3 was chosen as base for the coupling reactions.

In further studies, Suzuki reactions of a wide range of other aryl halides with phenylboronic acid with 0.12 g of Pd-PVP/CMK-8 as catalyst were conducted. The results are shown in Table 6 [42, 43]. The Suzuki reaction of iodobenzene with phenylboronic acid could reach the complete conversion within 4 h. Various aryl bromides (entry 5, 7) and aryl chlorid (entry 8) were used in the reactions under optimized conditions, and relatively high yields were obtained, though longer reaction periods were needed. The more easily accessible and cheaper aryl chlorides have not been employed much, in palladium-catalyzed coupling reactions, primarily because the oxidative addition of C–Cl bond to Pd(0) species is usually difficult. Few heterogeneous Pd catalysts were found to convert activated aryl chlorides at high temperature [44, 45]. However, chlorobenzene afforded excellent coupling products (entry 8) but needed more time than that of aryliodides and arylbromides.

Aryliodides with electron donating substituents (entries 2, 3) as well as the iodobenzene (entry 1) afforded excellent coupling products at room temperature.

The coupling of 4-iodoanisol with phenylboronic acid resulted in moderate yields (70%) at room temperature and the yield did not improve at elevated temperature. Thus the catalyst afforded average to excellent yields of the biaryl products even at room temperature. In the literature, only a few catalysts are known for affecting the Suzuki-Miyaura cross-coupling reactions under mild conditions [46-58]. Another important issue concerning the application of a heterogeneous catalyst is its reusability and stability under reactionconditions. To gain insight into this issue, catalyst recyclingexperiments were carried out using a Suzuki reaction of iodobenzene and phenylboronic acid.

Table 6. Suzuki-Miyaura reaction of aromatic arylhalides and phenylboronic acid catalyzed by Pd-PVP/CMK-8 [a]

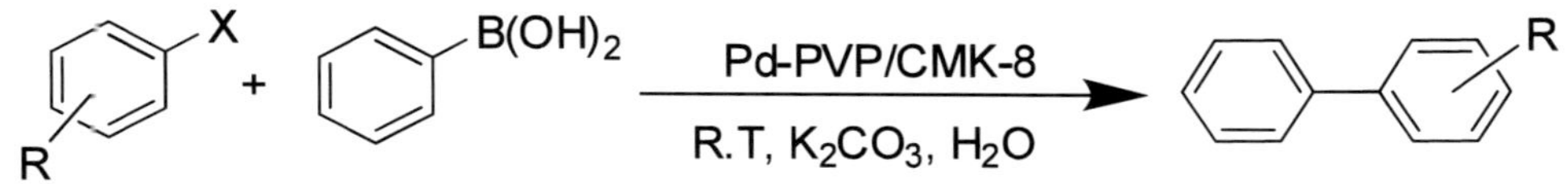

Entry	Substrate	Product	Yield (%) [b]	Time (h)	Mp (˚C) Found	Mp (˚C) Reported
1	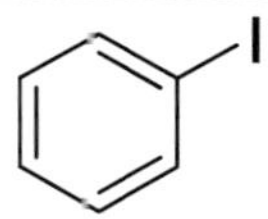	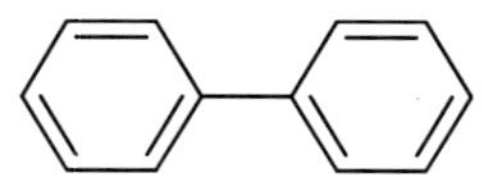	97	4	71-73	70-72 [42]
2	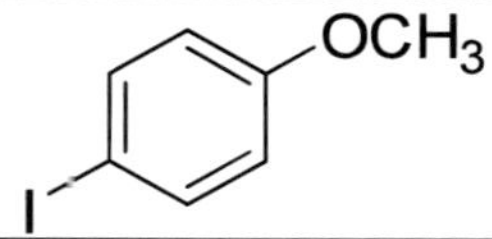	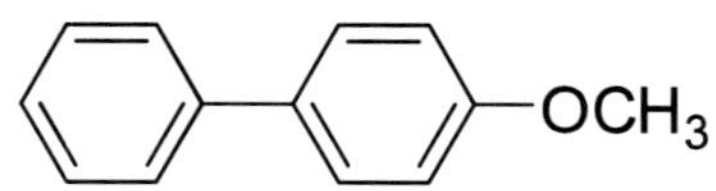	70	12	92-94	91-92 [42]
3	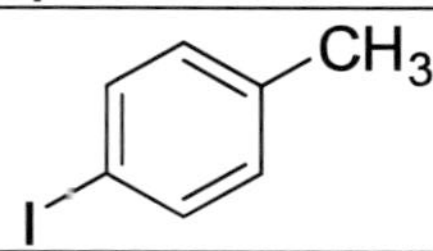	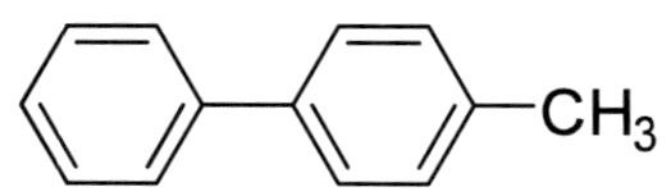	97	5	47-49	45-50 [42]
4	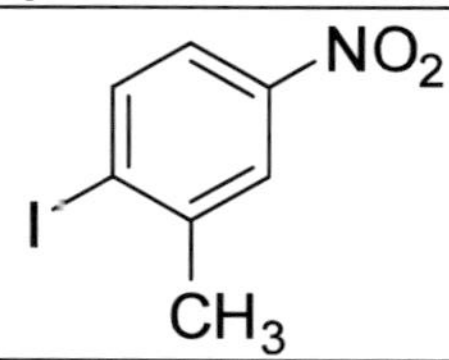	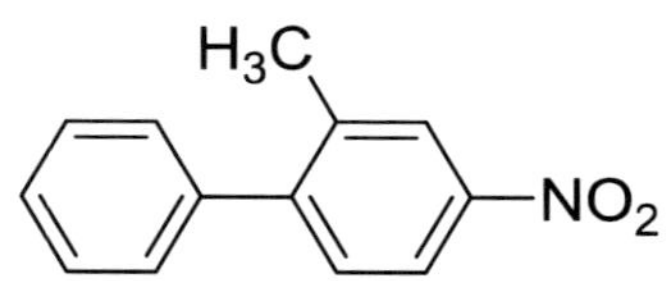	80	7	53-55	55-56 [43]
5	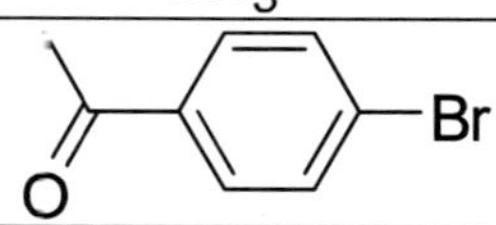	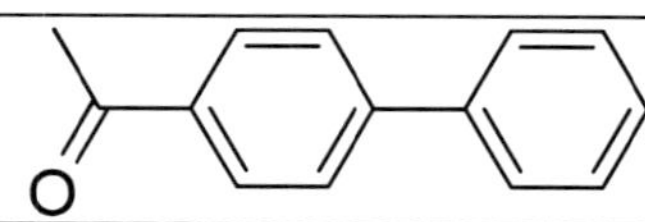	95	12	121-123	120-122 [42]

Entry	Substrate	Product	Yield (%) [b]	Time (h)	Mp (˚C) Found	Mp (˚C) Reported
6	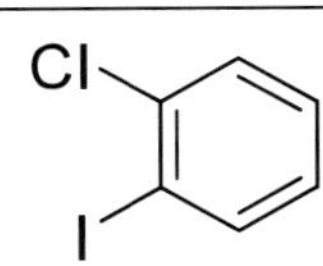	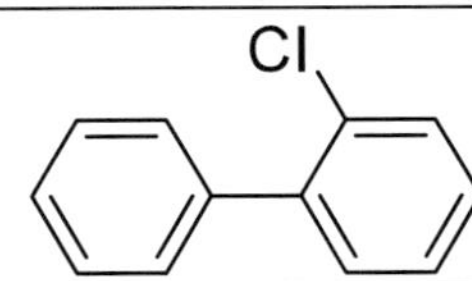	98	2	31-35	32.5-33.5 [42]
7	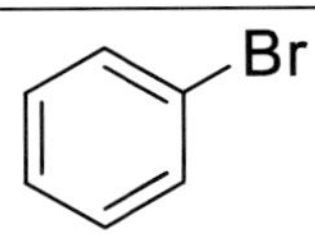	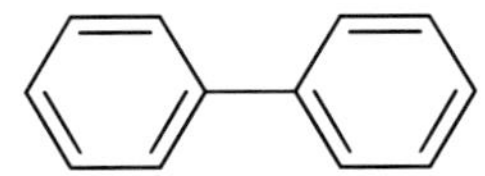	95	9	71-73	70-72 [42]
8	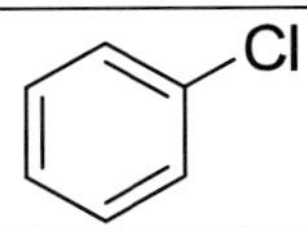	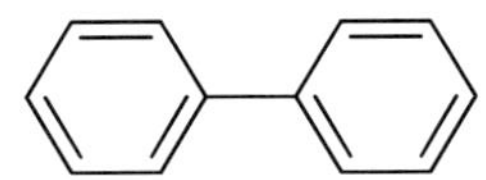	90	14	71-73	70-72 [42]

[a] Reaction conditions: Pd-PVP/CMK-8 (0.12 g), iodobenzene (1 mmol), phenylboronic acid (1.5 mmol), K_2CO_3 (5 mmol), H_2O (5 mL), room temperature.
[b] Isolated yield.

The results are shown in Table 8. After each cycle, the catalyst was filtered off, washed with water (10 mL), diethylether and acetone (3×5 mL). Then it was dried in oven at 60°C and reused in the Suzuki-Miyaura reaction. The results show that this catalyst can be reused without any modification, 13 times and no significant loss of activity/selectivity performance is observed. It should be mentioned that there are very low Pd leaching during the reaction and the catalyst exhibits high stability even after 13 recycles (Table 7).

Table 7. The catalyst reusability for the Suzuki-Miyaura reaction [a]

Cycle	Yield(%) [b]	Pd content of catalyst (mmol/0.12 g) [c]
Fresh	97	0.1231
1	97	0.1231
2	97	0.1231
3	97	0.1231
4	96	0.1231
5	96	0.1230
6	~96	0.1229
7	~96	0.1229
8	95	0.1228
9	~95	0.1228
10	~93	0.1228
11	~93	0.1226
12	~93	0.1226
13	~90	0.1224

[a]Reaction conditions: Pd-PVP/CMK-8 (0.12 g), iodobenzene (1 mmol), phenylboronic acid (1.5 mmol), K_2CO_3(5 mmol), H_2O (5 mL).
[b]Isolated yield.
[c]Estimated by ICP-AES technique.

The catalytic activity of Pd-PVP/CMK-8 in Suzuki-Miyaura cross-coupling reaction was compared with Pd-PVP/KIT-6 to investigate the effect of carbon mesoporous material on the activity and stability of the catalyst. The results are presented in Table 8. As can be seen, Pd-PVP/CMK-8 shows higher activity in Suzuki-Miyaura cross-coupling reaction at shorter time.

Table 8. The catalyst activity for the Suzuki-Miyaura reaction [a]

Catalyst	Time (h)	Yield(%) [b]	TOF (h^{-1})
Pd-PVP/CMK-8	4	97	1.97
Pd-PVP/KIT-6	4 7	60 98	- 1.02

[a] Reaction conditions: catalyst (0.12 g), iodobenzene (1 mmol), phenylboronic acid (1.5 mmol), K_2CO_3 (5 mmol), H_2O (5 mL), room temperature.
[b] Isolated yield.

It can be related to the hydrophobic nature of Pd-PVP/CMK-8 compared to Pd-PVP/KIT-6. It should be mentioned that CMK-8 is completely hydrophobe and KIT-6 is completely hydrophile. Therefore, the organic reactants have a higher tendency to be adsorbed on the surface of Pd-PVP/CMK-8, and consequently the yield will be higher in the case of using Pd-PVP/CMK-8 as catalyst.

A comparison of the catalytic activity values of various catalysts for the Suzuki-Miyaura cross-coupling reaction is given in Table 9. As can be seen, the Pd-PVP/CMK-8 catalyst shows good yield for coupling reaction of iodobenzene with phenylboronic acid at reasonable time and room temperature.

Table 9. Comparison of the efficiency of Pd-PVP/CMK-8 with other catalysts in the Suzuki-Miyaura reaction

Entry	Catalyst	Arylhalide	Solvent	T (°C)	Time	Base	Yield (%)	Ref.
1	Pd-polyacrylamide phosphinite ligand	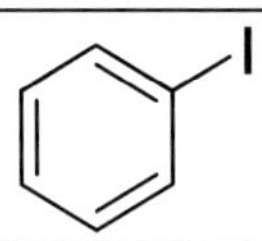	DMF	80	40 min	K_2CO_3	96	[12]
2	Pd-polyacrylamide phosphinite ligand	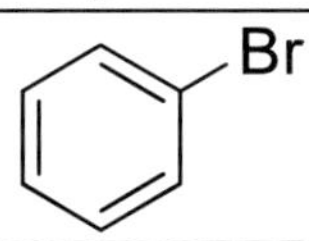	DMF	80	1 h	K_2CO_3	85	[12]
3	$Pd/SiO_2–TiO_2$	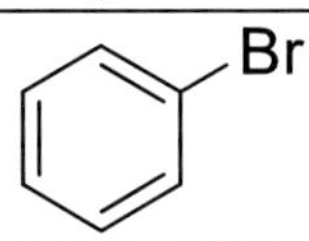	Ethanol	78	6 h	K_2CO_3	88.6	[11]
4	$Pd/SiO_2–TiO_2$	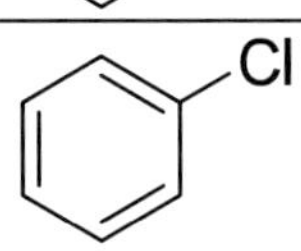	Ethanol	78	6 h	K_2CO_3	13.2	[11]
5	$Pd/SiO_2–TiO_2$	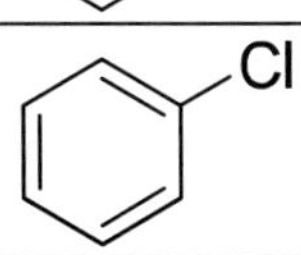	1,4-Dioxane	13.2	24 h	Cs_2CO_3	41.9	[11]
6	Chitosan-g-mPEG-Pd	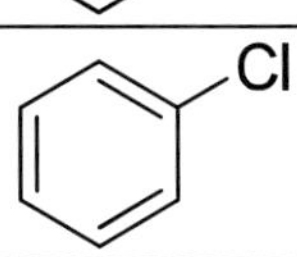	H_2O	150	3 h	NaOH	13	[3]
7	Chitosan-g-mPEG-Pd	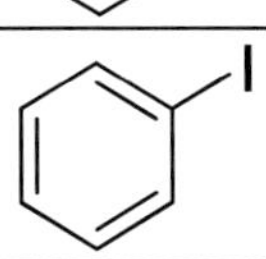	H_2O	150	1 h	NaOH	81	[3]
8	Chitosan-g-mPEG-Pd	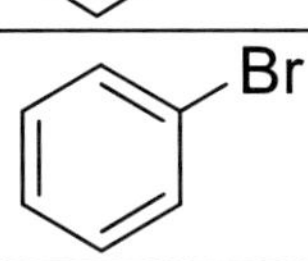	H_2O	150	2 h	NaOH	79	[3]
9	Chitosan-g-mPEG-Pd	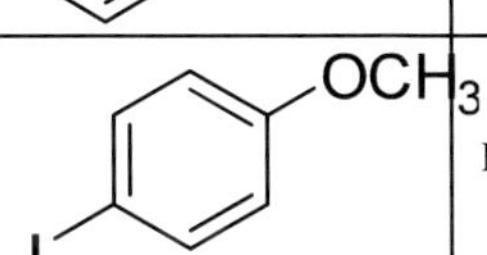	H_2O	150	3 h	NaOH	95	[3]

Table 9. (Continued)

Entry	Catalyst	Arylhalide	Solvent	T (°C)	Time	Base	Yield (%)	Ref.
10	copper oxide-supported Pd(0)	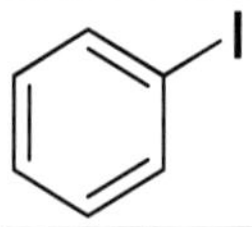	DMF	110	8 h	K_2CO_3	90	[4]
11	Phosphine free diamino-diol based palladium	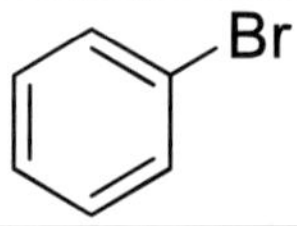	MeOH	150	2 h	K_2CO_3	100	[5]
12	Palladium-polyaniline hybrid	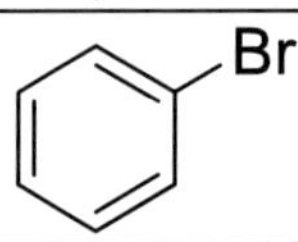	Toluene	80	8 h	K_2CO_3	93	[1]
13	Pd-PVP/CMK-8	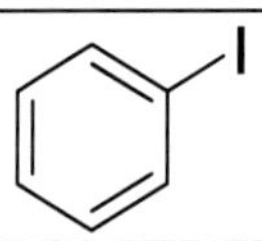	H_2O	RT	4	K_2CO_3	97	This work
14	Pd-PVP/CMK-8	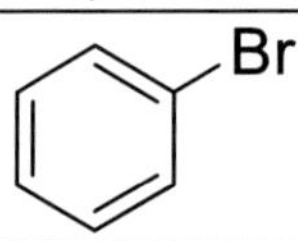	H_2O	RT	9	K_2CO_3	95	This work
15	Pd-PVP/CMK-8	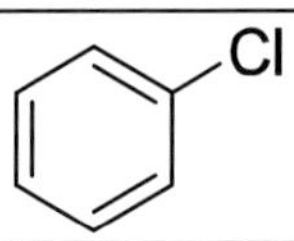	H_2O	RT	14	K_2CO_3	90	This work

CONCLUSION

This chapter described the preparation of CMK-8 with hydrophobic surface and its using as an ideal support for PVP and Pd nanoparticles. The catalytic activity of this novel catalyst was tested for Suzuki-Miyaura reaction of aryl chloride, bromide and iodides at room temperature under aerobic conditions. Most importantly, water has been chosen as a green solvent for this reaction. Further, easy catalyst recovery and excellent recycling (at least 13 times) efficiency of the catalyst made it an ideal system for coupling reactions in aqueous phase.

Comparison of catalytic activity of the composite was done with KIT-6 supported catalyst. A comparison between KIT-6 and CMK-8 supported catalysts showed superior activities and reusability of CMK-8 supported catalyst. This higher activity and reusability is attributable to high surface area of CMK-8.

In addition, hydrophobic surface nature will allow the organic reactants to have a higher tendency to be adsorbed on the surface of Pd-PVP/CMK-8, and consequently the yield will be increased. These unique results open new perspectives for application of these types of hybrid catalysts.

REFERENCES

[1] Islam, R. U.; Witcomb , M.; Lingen , E.; Scurrell, M.; Van Otterlo,W.; Mallick, K. *J. Organomet. Chem.* 2011, 696, 2206-2210.
[2] Siamaki, A. R.; Khder, A. E. R. S.; Abdelsayed, V.; El-Shall, M. S.; Gupton, B. F. *J. Catal.* 279 (2011) 1–11.
[3] Sin, E.; Yi, S. S.; Lee, Y. S. *J. Mol. Catal. A: Chem.* 315 (2010) 99–104.
[4] Chattopadhyay, K.; Dey, R.; Ranu, B. C. Tetrahedron Lett. 50 (2009) 3164–3167.
[5] Mohanty, S.; Suresh, D.; Balakrishna, M. S.; Mague, J. T. *J. Organomet. Chem.* 694 (2009) 2114–2121.
[6] Costa, N. J. S.; Kiyohara, P. K.; Monteiro, A.; Coppel, Y.; Philippot, K.; Rossi, L. M. *J. Catal.* 276 (2010) 382–389.
[7] Kalbasi, R. J.; Mosaddegh, N. *J. Solid State Chem.* 2011, 184, 3095–3103.
[8] Das, P.; Sharma, D.; Shil, A. K.; Kumar, A. Tetrahedron Lett. 52 (2011) 1176–1178.
[9] Kalbasi, R. J.; Mosaddegh, N. in: M.A. Farrukh (Ed.), Atomic Absorption Spectroscopy, E-Publishing, In Tech–Open Access Publisher, 2012, pp. 89–104.
[10] Jana, S.; Haldar, S.; Koner, S. *Tetrahedron Lett.* 50 (2009) 4820–4823.
[11] Fan, G.; Zou, B.; Cheng, S.; Zheng, L. *J. Ind. Eng. Chem.* 16 (2010) 220–223.
[12] Tamami, B.; Ghasemi, S. *J. Mol. Catal. A: Chem.* 322 (2010) 98–105.
[13] Sanchez-Delgado, R. A.; Machalaba, N.; Ng-a-Qui, N. *Catal. Commun.* 2007, 8, 2115–2118.
[14] Luo, Y.; Sun, X. *Mater. Lett.* 2007, 61, 2015–2107.
[15] Thomas, J. M.; Johnson, B. F. G.; Raja, R.; Sankar, G.; Midgley, P. A. *Acc. Chem. Res.* 2003, 36, 20–30.
[16] Jacquin, M.; Jones, D. J.; Roziere*, J. Appl. Catal. A: Gen.* 2003, 251, 131–141.
[17] Macquarrie, S.; Nohair, B.; Horton, J. H.; Kaliaguine, S.; Crudden, C. M. *J. Phys. Chem.* 2010, 114, 57–64.
[18] Darmstadt, H.; Roy, C.; Kaliaguine, S.; Choi, S. J.; Ryoo, R. *Carbon* 2002, 40, 2673-2683.
[19] Ryoo, R.; Joo, S. H.; Kruk, M.; Jaroniec, M. *Adv. Mater.* 2001, 13, 677-681.
[20] Joo, S. H.; Choi, S. J.; Oh, I.; Kwak, J.; Liu, Z.; Terasaki, O. *Nature* 2001, 412, 169-172.
[21] Joo, S. H.; Jun, S.; Ryoo, R. Micropor. *Mesopor. Mater.* 2001, 44-45, 153-158.
[22] Li, Z.; Zhang, J.; Li, Y.; Guan, Y.; Feng, Z.; Li, C. *J. Mater. Chem.* 2006, 16, 1350–1354.
[23] Kim, T. W.; Solovyov.; L. A. *J. Mater. Chem.* 2006, 16, 1445–1455.
[24] Dai, W.; Zheng, M.; Zhao, Y.; Liao, S.; Ji, G.; Cao, *J. Nanoscale Res. Lett.* 2010, 5, 103–107.
[25] Antolini, E. *Appl. Catal. B: Environ.* 2009, 88, 1-24.
[26] Wang, X.; Liu, R.; Waje, M. M.; Chen, Z.; Yan, Y.; Bozhilov, K. N. *Chem. Mater.* 2007, 19, 2395-2397.
[27] Liu, R.; Wang, X.; Zhao, X.; Feng, P. *Carbon* 2008, 46, 1664-1669.
[28] Tanaka, S.; Nishiyama, N.; Egashira, Y.; Ueyama, K. *Chem. Commun.* 2005, 16, 2125-2127.

[29] Xing, R.; Liu, Y.; Wang, Y.; Chen, L.; Wu, H.; Jiang, Y. *Micropor. Mesopor. Mater.* 2007, 105, 41-48.
[30] Peng, L.; Philippaerts, A.; Ke, X.; Van Noyen, J.; De Clippel, F.; Van Tendeloo, *G. Catal. Today* 2010, 150, 140-146.
[31] Kalbasi, R. J.; Mosaddegh, N. *J. Inorg. Organomet. Polym.* 2012, 22, 404–414.
[32] Dubey, A.; Mishra, B. G.; Sachdev, D. *Appl. Catal. A: Gen.* 2008, 338, 20–26.
[33] Wang, P.; Wang, Z.; Li, J.; Bai, Y. *Micropor. Mesopor. Mater.* 2008, 116, 400–405.
[34] Iwamoto, T.; Matsumoto, K.; Matsushita, T.; Inokuchi, M.; Toshima, *N. J. Colloid Interface Sci.* 2009, 336, 879–888.
[35] Metin, O.; Ozkar, S. *J. Mol. Catal. A: Chem.* 2008, 295, 39–46.
[36] Hirai, H.; Chawanya, H.; Toshima, N. *React. Polym.* 1985, 3, 127–141.
[37] Ahmadian Namini, P.; Babaluo, A. A.; Bayati, B. *Int. J. Nanosci. Nanotechnol.* 2007, 3, 37-43.
[38] Gniewek, A.; Trzeciak, A. M.; Ziolkowski, J. J.; Kepinski, L.; Wrzyszcz, J.; Tylus, W. *J. Catal.* 2005, 229, 332–343.
[39] Teranishi, T.; Miyake, M. *Chem. Mater.* 1998, 10, 594–600.
[40] Bedford, R. B.; Cazin, C. S. J.; Coles, S. J.; Gelbrich, T.; Horton, P. N.; Hursthouse, M. E. *Organometallics* 2003, 22, 987–999.
[41] Lamblin, M.; Nassar-Hardy, L.; Hierso, J. C.; Fouquet, E.; Felpin, F. X. *Synth. Catal.* 2010, 352, 33–79.
[42] Xu, Q.; Duan, W. L.; Lei, Z. Y.; Zhu, Z. B.; Shi, M. *Tetrahedron* 2005, 61, 11225-11229.
[43] Yu, Y.; Hu, T.; Chen, X.; Xu, K.; Zhang, J.; Huang, *J. Chem. Commun.* 2011, 47, 3592-3594.
[44] Prockl, S.; Kleist, W.; Gruber, M. A.; Kohler, K. *Angew. Chem. Int. Ed.* 2004, 43, 1881–1882.
[45] Kohler, K.; Kleist, W.; Prockl, S. S. *Inorg. Chem.* 2007, 46, 1876–1883.
[46] Littke, A. F.; Dai, C.; Fu, G. C. *J. Am. Chem. Soc.* 2000, 122, 4020-4028.
[47] Weissman, H.; Milstein, D. *Chem. Commun.* 1999, 1901-1902.
[48] Wolfe, J. P.; Buchwald, S. L. *Angew. Chem. Int. Ed.* 1999, 38, 2413-2416.
[49] Cuia, X.; Zhoua, Y.; Wanga, N.; Liu, L.; Guo, Q. *Tetrahedron Lett.* 2007, 48, 163-164.
[50] Marion, N.; Navarro, O.; Mei, J.; Stevens, E. D.; Scott, N. M.; Nolan, S. P. *J. Am. Chem. Soc.* 2006, 128, 4101-4111.
[51] Navarro, O.; Marion, N.; Mei, J.; Nolan, S. P. *Chem. Eur. J.* 2006, 12, 5142-5148.
[52] Fairlamb, I. J. S.; Kapdi, A. R.; Lee, A. F. *Org. Lett.* 2004, 6, 4435-4438.
[53] Beeby, A.; Bettington, S.; Fairlamb, I. J. S.; Goeta, A. E.; Kapdi, A. R.; Niemela, E. H. *New J. Chem.* 2004, 28, 600-605.
[54] Guo, M.; Jian, F.; He, R. *Tetrahedron Lett.* 2006, 47, 2033-2036.
[55] Li, J. H.; Hu, X. C.; Xie, Y. X. *Tetrahedron Lett.* 2006, 47, 9239-9243.
[56] Ryu, J. H.; Jang, C. J.; Yoo, Y. S.; Lim, S. G.; Lee, M. *J. Org. Chem.* 2005, 70, 8956-8962.
[57] Savarin, C.; Liebeskind, L. S. *Org. Lett.* 2001, 3, 2149-2152.
[58] Zim, D.; Gruber, A. S.; Ebeling, G.; Dupont, J.; Monteiro, A. L. *Org. Lett.* 2000, 2, 2881-2884.

INDEX

A

B

C

D

E

F

G

H

I

J

K

L

M

O

P

Q

R

S

T

U

V

W

X

Y

Z